LONDON MATHEMATICAL SOCIETY
MONOGRAPHS NEW SERIES

Series Editors
H. G. Dales   Peter M. Neumann

LONDON MATHEMATICAL SOCIETY MONOGRAPHS
NEW SERIES

Previous volumes of the LMS monographs were published by Academic Press, to whom all enquiries should be addressed. Volumes in the New Series will be published by Oxford University Press throughout the world.

NEW SERIES

Spectral Decompositions and Analytic Sheaves

Jörg Eschmeier

Department of Pure Mathematics
University of Leeds

Mihai Putinar

Department of Mathematics
University of California at Riverside

CLARENDON PRESS · OXFORD
1996

Oxford University Press, Walton Street, Oxford OX2 6DP

Oxford   New York
Athens   Auckland   Bangkok   Bombay
Calcutta   Cape Town   Dar es Salaam   Delhi
Florence   Hong Kong   Istanbul   Karachi
Kuala Lumpur   Madras   Madrid   Melbourne
Mexico City   Nairobi   Paris   Singapore
Taipei   Tokyo   Toronto
and associated companies in
Berlin Ibadan

Oxford is a trade mark of Oxford University Press

Published in the United States
by Oxford University Press Inc., New York

A catalogue record for this book is available from the British Library

Library of Congress Cataloging in Publication Data
Eschmeier, Jörg.
Spectral decompositions and analytic sheaves / Jörg Eschmeier, Mihai Putinar.
(London Mathematical Society monographs; new ser., 10)
Includes bibliographical references and index.
1. Operator theory. 2. Spectral theory (Mathematics)
3. Decomposition (Mathematics) 4. Analytic sheaves. I. Putinar,
Mihai, 1955–   . II. Title. III. Series: London Mathematical
Society monographs; new ser., no. 10.
QA329.E83   1996   515'.7246--dc20   95-37748

ISBN 0 19 853667 4

Typeset by Technical Typesetting Ireland
Printed in Great Britain by
Biddles Ltd., Guildford & King's Lynn

To our parents

Preface

The spectral data of a linear operator are traditionally arranged into a discrete array of numbers such as the Jordan form of a finite matrix, a vector-valued measure in the case of self-adjoint operators, or a vector-valued distribution in the case of generalized scalar operators. In Fredholm theory the spectral picture of an operator is given by a family of open sets and associated multiplicities, or, more generally, by an element of a topological or analytic K-group. A common principle in these examples is that the spectral data are translated into properties of a suitable mathematical structure supported by the spectrum. Thus it seems quite natural to use a vector bundle, or a sheaf, as the carrier for the same type of information.

It is the aim of the present monograph to develop this latter idea and to describe abstract classes of operators to which sheaf-theoretical methods apply, and which correspond via localization to specific classes of analytic Fréchet sheaves on Stein spaces. This approach relies on a few standard homological methods from the theory of analytic sheaves, which in turn has its origin in the theory of functions of several complex variables. Sheaf models have been used to clarify the local spectral properties of certain commuting families of Fréchet-space operators. Results originally proved by using integral representations, or *ad hoc* arguments, find a natural explanation in the context of analytic sheaves. Secondly, the sheaf-theoretic approach to multivariable spectral theory allows a variety of applications to problems in function theory and complex analytic geometry.

We introduce the analytic functional calculus for several commuting operators on Fréchet spaces in the sense of J. L. Taylor. Both the approach via Cauchy–Weil type integrals and the homological approach are described. We develop basic concepts of a topological homology theory in the category of Fréchet modules over Fréchet algebras. The idea central to the whole book is the notion of quasi-coherent sheaves on Stein spaces, and an associated transversality concept based on topological module tensor products. We present a coordinate-free solution of the analytic functional calculus problem for analytic modules over Stein spaces.

We develop the theory of spectral decompositions and duality for systems of commuting operators on Banach spaces. Using the Scott Brown technique, we obtain general invariant-subspace results for single operators and for systems of commuting operators.

We give applications of the general theory to division problems for distributions (including a new operator-theoretic proof of a classical result of L.

Schwartz), the flatness theorem of Malgrange, the Gleason problem, and the corona problem. We study the spectral properties of Toeplitz systems on multivariable Bergman spaces. We give a complete proof of Grauert's theorem on the coherence of direct image sheaves. We describe the Fredholm and index theory for commuting systems of Banach-space operators, giving topological and analytic index formulas, and leading to a proof of a general Riemann–Roch theorem on complex spaces with singularities.

The selection of the material in this book was mainly dictated by our aim to present a complete account of multivariable spectral theory from the perspective of homological algebra and sheaf theory. The reader is assumed to have minimal experience in functional analysis, elementary homological algebra, and complex analytic geometry. Some necessary background material from these areas is collected in an appendix.

Chapter 1 contains a general survey on the topics of the monograph as well as a short history, and some motivation for the field now known as multivariable spectral theory. Each succeeding chapter has a short introduction which complements the general presentation given in Chapter 1. The notes at the end of each chapter contain brief historical comments, references, and a few remarks indicating results not covered in the text. Far from being exhaustive, the bibliography at the end of the book contains only articles cited or very closely related to the topics of the book.

We wish to express our gratitude to our teachers, Heinz-Günter Tillmann and Constantin Bănică, who introduced us to most of the subjects treated in this book, and who stimulated our interest to study them further. We are grateful to Garth Dales for his interest and support, and for numerous valuable remarks. We want to thank George Maltese for reading drafts of the whole book and for many helpful suggestions. We are thankful to the Mathematics Departments of the University of Münster and the University of California, Berkeley and Riverside, for creating the environments in which this book could be completed.

We are indebted to Mrs Hildegard Eissing, Mrs Danielle McQueen, and Mrs Jan Patterson for their expert typing of large portions of the manuscript.

The first author wishes to acknowledge the support that he received as a Heisenberg Fellow from the German Science Foundation, and the second author wishes to express his gratitude to the Alexander von Humboldt Foundation of Germany and the National Science Foundation of the USA for their generous support.

Leeds   J. E.
Riverside   M. P.
June 1995

Contents

1
Review of spectral theory

This first chapter has an introductory nature; it contains a review of a few classical spectral theories, and then informally explains the main themes of the book. Among these classical concepts are E. Bishop's ideas on local spectral theory and spectral duality, the theory of decomposable operators of C. Foiaş, the multidimensional spectral theory of J. L. Taylor, and its homological ramifications. Each of these subjects will be treated in detail in different chapters of the book, and sometimes a topic will appear in various places, treated from different points of view. The bibliographical sources listed in the notes at the end of the present chapter complement our review and, at the same time, they contain some indispensable first references.

1.1 CLASSICAL SPECTRAL THEORIES

The notion of *spectrum* has so many widely-used meanings that some explanation of the terminology and of the contents of this book is necessary from the very beginning. Our attention will be focused on bounded linear operators acting on Banach spaces, commutative n-tuples of such operators, and certain generalizations of these objects to the setting of Fréchet spaces. At a certain point the basic object to be studied will be a Fréchet module over a Stein algebra, which will eventually turn out to be a natural extension of the notion of a commuting n-tuple of operators on Fréchet spaces. All spectra appearing in the text will refer to these objects, and not for instance to unbounded operators or differential operators. Physical interpretations or other concrete applications are not mentioned. However, the spectral theory of Fréchet modules developed in Chapter 5 contains almost every abstract spectral theory known so far, and it is aimed to be a general model of this kind.

The first and best understood example of a theory of spectral decompositions is the *Jordan form* of a finite matrix. Let V be a finite-dimensional complex vector space, and let T be a linear transformation of V. The algebra of all linear transformations of V will be denoted by $L(V)$. Then it is well known that there exists a basis of V with respect to which T is represented by a matrix of the form

$$T \sim \begin{pmatrix} \begin{matrix} \lambda & 1 & & \\ & \lambda & 1 & \\ & & \ddots & \ddots \\ & & & \end{matrix} & \mathbf{0} & & \mathbf{0} \\ \mathbf{0} & \begin{matrix} & & 1 \\ & & \lambda \end{matrix} & & \\ \hline & \begin{matrix} \mu & 1 & \\ & \mu & 1 \\ & & \ddots & \ddots \end{matrix} & & \\ \mathbf{0} & & \begin{matrix} 1 \\ \mu \end{matrix} & \\ & & & \ddots \end{pmatrix}. \tag{1.1.1}$$

The complex numbers $\lambda, \mu, \ldots$ (not necessarily distinct) form the *spectrum* of T, denoted by $\sigma(T)$. Another definition of the spectrum is

$$\sigma(T) = \{\lambda \in \mathbb{C}; \lambda - T \text{ is not invertible}\}.$$

A basis in which the linear transformation T has the representation (1.1.1) consists of *generalized eigenvectors*. More precisely, each block of the matrix (1.1.1) corresponds to a linear subspace of V generated by vectors of the form $(\lambda - T)^{p-1}\xi, \ldots, (\lambda - T)\xi, \xi$, where $\xi \in V$ is a non-zero solution of the equation

$$(\lambda - T)^p \xi = 0 \tag{1.1.2}$$

for a certain positive integer p. The block structure of the matrix (1.1.1) is unique up to the order, and it is known as the *Jordan canonical form* of T. An operator of the form STS^{-1}, with S an invertible linear transformation in $L(V)$, is said to be *similar* to T, and it has the same Jordan canonical form as T. Conversely, any two linear operators on V with the same canonical form are similar.

On infinite-dimensional (normed) spaces, the classification of bounded linear operators is very subtle (see Dunford and Schwartz (1958, 1963, 1971) for a comprehensive treatment of the main achievements, and Apostol *et al.* (1984) for some recent advances in the same field). There are only a few classes of operators whose spectral theories and classification is well understood. Our next aim is to recall some of these classes.

Let H be a separable, infinite-dimensional complex Hilbert space, and let $T \in L(H)$ be a compact operator acting on H. Then the spectrum of T is countable, with zero as the only possible accumulation point. The corresponding matrix structure of T is similar to (1.1.1). There is a closed invariant subspace M of T such that the restriction $T|M$ has a matrix representation of the form (1.1.1), but this time with possibly infinitely many blocks of Jordan form, and such that the quotient $T_0 = T/M$ is a compact *quasi-nilpotent* operator (an operator T is quasi-nilpotent if $\sigma(T) = \{0\}$). The 'nice' part of this model is again spanned by the generalized eigenvectors, that is, by the vectors satisfying an equation of the form (1.1.2). On the other hand, the quasi-nilpotent part T_0 is very inaccessible to any form of classification.

However, there are quasi-nilpotent operators acting on Hilbert spaces whose structure is completely understood, such as the Volterra operators (see for example Gohberg and Krein (1970)). A description of their refined structure, for instance that of their invariant subspace lattices, would go beyond the scope of this book.

The richest spectral theory developed so far, and the one most used in applications, is the spectral theory of *self-adjoint operators* on Hilbert spaces. The basic model was discovered independently at the beginning of this century by Hahn and Hellinger, and it brings measure-theoretic concepts into this field for the first time. Let A be a bounded self-adjoint operator acting on a Hilbert space H. In the model of Hahn and Hellinger the space H is isometrically identified with an orthogonal direct sum

$$H \cong \bigoplus_{k=1}^{N} L^2(\mathbb{R}, \mu_k) \quad (N \in \mathbb{N} \cup \{\infty\}).$$

Via this identification, the operator A is diagonal, that is, it acts as the multiplication by the variable $x \in \mathbb{R}$ on each of the spaces $L^2(\mathbb{R}, \mu_k)$. Here each μ_k is a compactly-supported positive Borel measure. In this case the spectrum of A is the closure of the union of the supports of the measures μ_k $(1 \le k \le N)$. Moreover, in a measure-theoretic sense, the spectral multiplicity of each point of the real axis classifies the original operator up to unitary equivalence.

As a typical example, let us consider the operator M of multiplication by the variable x on the space $L^2[0,1]$ equipped with Lebesgue measure. Then, for each positive integer p and real number λ, the equation $(M - \lambda)^p f = 0$ has only the trivial solution $f = 0 \in L^2[0,1]$. Consequently the operator M has no generalized eigenvectors, and there is no hope of obtaining a model for it as in the case of finite matrices. It was P. A. M. Dirac who resolved this anomaly by introducing certain generalized eigenvectors for M not belonging to the space $L^2[0,1]$. In modern terms Dirac observed that distributions of the form

$$u(x, y) = f(y)\delta(x - y) \in \mathscr{D}'(\mathbb{R}) \mathbin{\hat{\otimes}} L^2[0,1]$$

satisfy the equation

$$(M - x)u(x, y) = 0 \tag{1.1.3}$$

and that they span the underlying space, in the sense that any element f of $L^2[0,1]$ is of the form

$$\int u(x, y)\,\mathrm{d}y = f(x).$$

Similarly, any bounded self-adjoint operator has sufficiently many solutions of equation (1.1.3) in the sense of distributions.

These 'eigendistributions' diagonalize the original operator in a continuous way; moreover they can be arranged into a continuous matrix (more precisely, a kernel) which classifies the operator.

Quite unexpectedly, any continuous linear operator acting on a function or distribution space admits such a generalized diagonalization, and it has a resulting model. This is the core of the *kernel theorem* of L. Schwartz, and it has proved to be the essential tool in many applications involving integro-differential operators. In general the computation of the kernel of a given operator is very difficult, but the effort is rewarding because this kernel encodes the complete refined (spectral) structure of the original operator. The notes at the end of this chapter contain some bibliographical references in this direction.

Returning to the theory of self-adjoint operators, let us recall the notion of spectral measure. Let $A \in L(H)$ be a self-adjoint operator. The correct substitute for the space of generalized eigenvectors attached to a fixed eigenvector (as in the case of finite matrices) is a space of vectors supported by an arbitrary measurable set rather than a single point. For example, if F is a closed subset of $\mathbb{R}$, then the corresponding space of vectors in H supported by F is

$$H_F(A) = \overline{\left\{ u(\varphi);\ (x - A)u = 0, \qquad u \in \mathscr{D}'(\mathbb{R}) \,\hat{\otimes}\, H, \qquad \varphi \in \mathscr{D}_F(\mathbb{R}) \right\}}.$$

Here $\mathscr{D}_F(\mathbb{R})$ denotes the space of test functions supported on F. Let $E(F)$ be the orthogonal projection of H onto the closed subspace $H_F(A)$. It turns out that $E(\cdot)$ is a strongly σ-additive, projection-valued measure with the property that $E(\delta)^* = E(\delta)$ and $E(\delta \cap \tau) = E(\delta)E(\tau)$ for any pair of Borel sets δ, τ in $\mathbb{R}$. This *spectral measure* of the operator A diagonalizes A in a continuous way;

$$A = \int_{\mathbb{R}} t E(\mathrm{d}t), \tag{1.1.4}$$

and it carries all the spectral data of A.

The same construction can be carried out for more general classes of operators. First, any *normal operator* $N \in L(H)$ (that is, $N^*N = NN^*$) can be diagonalized by a spectral measure as above. A natural explanation for the last assertion is given by the fact that any normal operator N is a sum $N = A + iB$, where A and B are commuting self-adjoint operators. More generally, any finite system of commuting self-adjoint or normal operators can be simultaneously diagonalized by a unique spectral measure.

The theory of spectral measures has been generalized to Banach-space operators, relaxing the orthogonality condition for the projections occurring in the range. The Banach-space operators which admit a representation of the form (1.1.4) are called *scalar*, and they generalize the diagonalizable matrices. The Jordan form has its analogue in the class of *spectral operators* (in the sense of Dunford). A spectral operator T acting on a Banach space has a unique Jordan decomposition

$$T = S + Q,$$

where S is scalar, Q is quasi-nilpotent, and both operators commute.

Most of the spectral theory of self-adjoint operators extends without essential modifications to the class of spectral operators.

1.2 LOCAL SPECTRAL THEORY

At the end of the fifties, probably under the influence of the Bourbaki movement, there was a general desire to enlarge as much as possible the class of Banach-space operators whose spectral properties resemble those of self-adjoint operators. Later it became clear that this abstract and mostly axiomatic approach led to some pathologies which are artificial and difficult to handle. On the other hand, the same abstract theories have isolated a few classes of concrete operators with a rich spectral theory that are not necessarily very close to the self-adjoint operators. The aim of the present section (and actually of the whole book) is to extract from this so-called abstract local spectral theory the essential facts and to discuss applications.

The pioneering work of Errett Bishop in this field of axiomatic spectral theories remains as important today as it was at the time of its publication (Bishop 1957, 1959). The second of these papers contains both a synthesis of the previous results in abstract spectral theories as well as the basic principles needed for further developments. Bishop asserts (1959, p. 379):

'We seek a spectral theory which will be valid independently of any of the usual restrictions (such as normality or complete continuity). It is, of course, not to be expected, in view of many known counterexamples, that such a theory will even approach in power the spectral theory of a Hermitian or normal operator on a Hilbert space. In fact, it is surprising that a spectral theory for an arbitrary operator exists at all. The results obtained here are incomplete, but it seems likely that any spectral theory which is valid for an arbitrary operator will be closely related to the theory developed here'.

The achievements of (local) spectral theory, from this date until today, strongly support Bishop's intuition. Although not stated in this form, his main idea was to use analytic functionals instead of distributions in the search for generalized eigenvectors of a given Banach space operator. This apparently minor change considerably extends the class of operators that can be treated, and it explains practically all the new results in local spectral theory obtained during the last three decades. One of the basic constructions of Bishop is reproduced below.

Let $T \in L(X)$ denote a bounded linear operator acting on a complex Banach space X. The Fréchet space of X-valued analytic functions defined on an open subset U of $\mathbb{C}$ is denoted by $\mathcal{O}(U, X)$. The corresponding sheaf of Fréchet spaces is denoted by $\mathcal{O}^X$.

In the presence of a spectral measure E, relative to the operator T, the space of those vectors in X which are supported by a closed set F is $E(F)X$. One may immediately check that each vector $x \in E(F)X$ has a representation of the form

$$x = (T - z)f(z) \qquad (z \in \mathbb{C} \backslash F),$$

where f is a suitable X-valued analytical function on $\mathbb{C} \backslash F$.

Thus we are naturally led to the following construction. For $T \in L(X)$ and F a closed subset of $\mathbb{C}$, one defines a map

$$J_F \colon X \to \mathcal{O}(\mathbb{C} \backslash F, X) / (T - z)\mathcal{O}(\mathbb{C} \backslash F, X)$$

by $J_F(x) = [1 \otimes x]$ (that is, $J_F(x)$ is the equivalence class of the function identically equal to x). A candidate for the spectral space $E(F)X$ is then

$$M(F, T) = \mathrm{Ker}(J_F).$$

Actually this is only one of several types of spectral subspaces originally considered by Bishop.

The operator T would have satisfactory spectral-decomposition behaviour if there existed sufficiently many non-trivial subspaces $M(F, T)$ such that, for each fixed partition of the complex plane into small pieces F, these subspaces spanned the whole space. Bishop observed that this condition is related to certain natural properties of the dual operator $T' \in L(X')$ acting on the dual space X' of X. To be more precise, we state below the following fundamental theorem of Bishop.

Theorem 1.2.1 (Bishop) *Let T be a bounded linear operator on a reflexive Banach space X. Suppose that the dual T' has the property*:

> *for each open set U in $\mathbb{C}$, the map*
> $T' - z \colon \mathcal{O}(U, X') \to \mathcal{O}(U, X')$ *is injective*   (β)
> *with closed range.*

Then for each finite open cover $(U_i)_{i \in I}$ of $\mathbb{C}$, the subspace $\sum_{i \in I} M(\overline{U}_i, T)$ is dense in X.

Proof A sketch of a proof of this theorem, which in fact yields more than is stated here, runs as follows. Since the sheaf $\mathcal{O}^{X'}$ has vanishing cohomology on each open subset of $\mathbb{C}$ there is, for each finite open cover $(U_i)_{i \in I}$ of the complex plane $\mathbb{C}$, an exact augmented Čech complex of the following form:

$$0 \to \mathcal{O}(\mathbb{C}, X') \to \prod_{i \in I} \mathcal{O}(U_i, X') \to \prod_{i, j \in I} \mathcal{O}(U_i \cap U_j, X') \to \cdots . \quad (1.2.1)$$

Let us suppose that the operator T' satisfies property (β). Then the complex obtained from (1.2.1) by taking quotients modulo the range of the operator of multiplication by $T' - z$,

$$0 \to \mathscr{O}(\mathbb{C}, X')/(T' - z)\mathscr{O}(\mathbb{C}, X') \to \prod_{i \in I} \mathscr{O}(U_i, X')/(T' - z)\mathscr{O}(U_i, X') \to \cdots,$$

is still exact (use diagram chasing or Lemma A2.1). A simple computation with power series can be used to identify the space X' with the quotient $\mathscr{O}(\mathbb{C}, X')/(T' - z)\mathscr{O}(\mathbb{C}, X')$ via the map J_ϕ defined above.

Because of the reflexivity of X, the spaces $\mathscr{O}(U_i)' \,\hat{\otimes}\, X$ ($i \in I$) can be regarded as the strong dual spaces of the Fréchet spaces $\mathscr{O}(U_i) \,\hat{\otimes}\, X'$ (see Theorem A1.6). Therefore the kernels

$$K_i = \mathrm{Ker}\!\left(T - z: \mathscr{O}(U_i)' \,\hat{\otimes}\, X \to \mathscr{O}(U_i)' \,\hat{\otimes}\, X\right)$$

can be identified algebraically with the dual spaces of the spaces

$$C_i = \mathrm{Coker}\!\left(T' - z: \mathscr{O}(U_i) \,\hat{\otimes}\, X' \to \mathscr{O}(U_i) \,\hat{\otimes}\, X'\right).$$

Let us denote by $\sigma: \oplus_{i \in I} K_i \to X$ the adjoint of the operator

$$j = (J_{F_i})_{i \in I}: X' \to \prod_{i \in I} C_i,$$

where $F_i = \mathbb{C} \setminus U_i$ for each index i. By general duality theory (Köthe 1979, §33.4.(2)) the operator σ is surjective if and only if j is injective with closed range. But each element $u_i \in K_i \subset \mathscr{O}(U_i)' \,\hat{\otimes}\, X$ is an X-valued analytic functional supported by $\overline{U}_i$ which, in addition, is an eigendistribution of the operator T, that is,

$$(T - z)u_i = 0.$$

Moreover, since $u_i(1) = (T - \lambda)u_i(1/(z - \lambda))$ for any $\lambda \in \mathbb{C} \setminus \overline{U}_i$, it follows that $u_i(1) \in M(\overline{U}_i, T)$. Hence the observation that

$$\sigma((u_i)_{i \in I}) = \sum_{i \in I} u_i(1)$$

implies that $X = \sum_{i \in I} M(\overline{U}_i, T)$ under the hypotheses of Theorem 1.2.1. $\qquad\square$

Although the above proof does not give the best possible result in this particular case, the general scheme of proof used here will appear several times in the book. For the definitive form of the result stated as Theorem 1.2.1 the reader is referred to Albrecht and Eschmeier (1987).

Bishop's property (β) replaces several well-known pointwise defined properties of the resolvent of Banach-space operators. On the other hand, the above proof indicates the importance of duality arguments in the theory of spectral decompositions.

An operator $T \in L(X)$ with the property that the spaces $M(F, T)$ are norm-closed in X for each closed set $F \subset \mathbb{C}$ and satisfy the equation

$$X = \sum_{i \in I} M(\overline{U}_i, T)$$

for each finite open cover $(U_i)_{i \in I}$ of $\mathbb{C}$ is called *decomposable* by Foias (1963). This spectral decomposition property is satisfied by most of the concrete examples of multipliers on Lebesgue, Sobolev, or similar function spaces. However, as shown by Albrecht (1978), there are decomposable operators which do not possess a functional calculus with smooth functions, or any other algebra of functions admitting partitions of unity.

A useful restatement of the above definition of decomposability asserts that the operator $T \in L(X)$ is decomposable if and only if, for each finite open cover $(U_i)_{i \in I}$ of $\mathbb{C}$, there are closed invariant subspaces X_i for T such that

$$\sigma(T \mid X_i) \subset U_i \quad (i \in I) \qquad \text{and} \qquad X = \sum_{i \in I} X_i$$

(Albrecht 1979). Moreover, it suffices to demand this condition for open covers consisting of two open sets.

Two recent results complement Theorem 1.2.1 by revealing natural connections between decomposability and Bishop's property (β).

Theorem 1.2.2 *Let T be a bounded linear operator on a Banach space.*

(a) *The operator T is decomposable if and only if both T and T' possess Bishop's property (β).*

(b) *The operator T is similar to the restriction of a decomposable operator to a closed invariant subspace if and only if T satisfies property (β).*   □

The proof of part (a) again exploits standard duality theory for spaces of analytic functions, and it relies on the existence of a sheaf model for operators which satisfy Bishop's condition (β). A proof of the above results will be given in Chapter 6.

A Banach-space operator that is similar to the restriction of a decomposable operator is called *subdecomposable*. Thus according to part (b) of Theorem 1.2.2, Bishop's property (β) characterizes precisely the class of subdecomposable operators. Bishop's property (β) occurs naturally in connection with spectral decompositions and duality. Another interesting application is the following general invariant-subspace result.

Theorem 1.2.3 *Let T be a bounded linear operator on a complex Banach space. Suppose that T satisfies Bishop's property (β) and that the spectrum of T has non-empty interior. Then T possesses non-trivial invariant subspaces.*   □

This theorem is proved by combining methods from local spectral theory with ideas originally used by Scott W. Brown (1978) to prove the existence of invariant subspaces for subnormal operators.

Bishop's property (β) also offers a natural framework for localizing the spectral behaviour of an operator T in $L(X)$. Although all of the spaces

$M(F, T)$ may be trivial, a certain localization of T is still possible. To understand this, let us consider the sheaf associated with the presheaf

$$\mathscr{F}_T(U) = \mathscr{O}^X(U)/(T - z)\mathscr{O}^X(U) \qquad (U \subset \mathbb{C} \text{ open}).$$

If the operator T has property (β), then $\mathscr{F}_T$ is a sheaf of Fréchet spaces. The natural multiplication by analytic functions turns $\mathscr{F}_T$ into an analytic Fréchet sheaf. Obviously,

$$\mathscr{F}_T(\mathbb{C}) \cong X,$$

and via this identification T becomes the multiplication by the independent variable $z \in \mathbb{C}$. Moreover, the sheaf $\mathscr{F}_T$ is supported by the spectrum $\sigma(T)$ of T, and the spectral subspaces of T can be described as

$$M(F, T) = \{u \in \mathscr{F}_T(\mathbb{C}); \text{supp}(u) \subset F\}.$$

In this way any spectral decomposition property for T becomes an intrinsic property of the associated sheaf $\mathscr{F}_T$. In particular, one can prove that the operator T is decomposable if and only if it can be modelled as the multiplication by the complex variable on the global section space of a soft analytic Fréchet sheaf.

The above sheaf model carries in a condensed form all the spectral data of the original operator, in exactly the same way as the Jordan canonical form or the spectral measure. This fact will be constantly exploited throughout this book.

A couple of elementary, but typical, examples of explicit sheaf models will clarify the above assertions.

Let V be a finite-dimensional complex vector space, and let $T \in L(V)$ be a linear map with spectrum $\sigma(T) = \{\lambda_1, \ldots, \lambda_p\}$. Then the Jordan form of T can be read from the following algebraic structures. Let $\mathscr{O}^V$ denote the sheaf of V-valued analytic functions on the complex plane, and for $\lambda \in \mathbb{C}$, let $\mathscr{O}_\lambda$ be the ring of germs of analytic functions defined near λ.

For each fixed point λ_j $(1 \leq j \leq p)$ in the spectrum of T, the $\mathscr{O}_{\lambda_j}$-module $\mathscr{O}_{\lambda_j}^V/(T - z)\mathscr{O}_{\lambda_j}^V$ is non-trivial, and it has finite dimension as a vector space. In virtue of the structure of Artinian modules over a local Noetherian ring of dimension one, we derive the following direct sum decomposition of the preceding $\mathscr{O}_{\lambda_j}$-module:

$$\mathscr{O}_{\lambda_j}^V/(T - z)\mathscr{O}_{\lambda_j}^V \cong \bigoplus_{k=1}^{l_j} \mathscr{O}_{\lambda_j}/(m_{\lambda_j})^{\nu_{jk}},$$

where m_{λ_j} denotes the maximal ideal of $\mathscr{O}_{\lambda_j}$. The integers l_j represent the number of Jordan blocks corresponding to the fixed eigenvalue λ_j, while ν_{jk} is the size of each block. In fact the entire space V can be decomposed as

$$V \cong \bigoplus_{j=1}^{p} \bigoplus_{k=1}^{l_j} \mathscr{O}_{\lambda_j}/(m_{\lambda_j})^{\nu_{jk}}.$$

Thus the sheaf model of T is correspondingly the $\mathcal{O}$-module

$$\mathscr{F}_T = \bigoplus_{j=1}^{p} \bigoplus_{k=1}^{l_j} \mathcal{O}/\left(\overline{m}_{\lambda_j}\right)^{\nu_{jk}},$$

where $\overline{m}_\lambda$ denotes the ideal sheaf in $\mathcal{O}$ of functions that vanish at λ. In particular we remark that $\mathscr{F}_T$ is a coherent analytic sheaf supported by $\sigma(T)$.

As a second example we consider the localization of a space of analytic functions defined by a typical global growth condition. Let Ω be a bounded open set in $\mathbb{C}$, and let $L_a^2(\Omega)$ denote the space of all analytic functions in $L^2(\Omega)$. Here $L^2(\Omega)$ is formed with respect to planar Lebesgue measure. The operator M given by the multiplication by the complex coordinate z is obviously decomposable on $L^2(\Omega)$. Therefore its restriction T to $L_a^2(\Omega)$ satisfies Bishop's condition (β) by Theorem 1.2.2(b). An elementary computation shows that the sheaf $\mathscr{F}_M$ is exactly $L_{\mathrm{loc}}^2 \mid \Omega$, while

$$\mathscr{F}_T = \mathcal{O}_\Omega \cap (L_{\mathrm{loc}}^2 \mid \Omega).$$

In other terms, the section space of the latter sheaf on an arbitrary open set U in $\mathbb{C}$ is given by

$$\mathscr{F}_T(U) = \{f \in \mathcal{O}(U \cap \Omega);\ \|f\|_{2, K \cap \Omega} < \infty,\ K \Subset U\}.$$

Therefore, $\mathscr{F}_T$ coincides on Ω with the ordinary sheaf of analytic functions, near the boundary of Ω it reflects the L^2-boundedness condition defining the space $L_a^2(\Omega)$, and it vanishes outside $\overline{\Omega}$. One can read from this picture the essential spectrum of T, the values of the Fredholm index on the Fredholm domain, and so on. Details will appear in Chapter 6.

In summary, the class of Banach-space operators with Bishop's property (β) provides a natural, and very general, setting for studying abstract spectral theories. For operators belonging to this class, one can define an analytic sheaf containing all information on the spectral properties of the operators.

1.3   MULTIVARIABLE SPECTRAL THEORY

Originally, a joint spectrum and a multivariable analytic functional calculus were introduced for n-tuples of elements in a commutative Banach algebra. It is interesting to remark that some theorems involving single elements of a commutative Banach algebra, as for instance the implicit function theorem, needed multivariable techniques for their proofs. A breakthrough in the field occurred in the early seventies with the ideas of J. L. Taylor. He defined a new, spatial, joint spectrum for commuting n-tuples of operators, and he derived the basic properties of this spectrum, including the existence of an analytic functional calculus for functions defined in neighbourhoods of the spectrum (Taylor 1970a, b). Soon afterwards he recognized the homological nature of the constructions, and subsequently he developed a general homological framework for an arbitrary functional calculus, not necessarily commutative (see Taylor 1972a, b). This new and more refined spectral theory has naturally led to new developments both in single and multivariable operator theory. In some respects, it is the aim of the present book to develop and to describe the foundations of this algebraic method of Taylor, and to give some of its applications. For some questions of multivariable spectral theory the homological approach is essential, just as it was for the development of the function theory of several complex variables in the fifties.

First let us recall Taylor's definition of a joint spectrum. Let $T = (T_1, \ldots, T_n)$ be a commuting system of bounded linear operators on a Banach space X. A point $\lambda = (\lambda_1, \ldots, \lambda_n) \in \mathbb{C}^n$ does not belong to the *joint spectrum* $\sigma(T, X)$ if and only if the cochain Koszul complex $K^\bullet(T - \lambda, X)$ is exact. This finite complex consists of the spaces

$$K^p(T - \lambda, X) = X \otimes_\mathbb{C} \Lambda^p(\mathbb{C}^n) \qquad (0 \leq p \leq n)$$

and the coboundary operators

$$\delta^p: K^p(T - \lambda, X) \to K^{p+1}(T - \lambda, X), \qquad \delta^P(\xi) = \tau \wedge \xi,$$

where $\tau = (T_1 - \lambda_1) \otimes e_1 + \cdots + (T_n - \lambda_n) \otimes e_n$ and $\{e_1, \ldots, e_n\}$ is the canonical basis of $\mathbb{C}^n$. It should be pointed out that the Koszul complex appeared in a similar context even before Taylor's definition of the joint spectrum in a paper by Hörmander (1967) on the Corona problem.

With this definition in hand, one proves the basic properties of this notion of joint spectrum by standard manipulations with exact sequences. The new spectrum is contained in any joint spectrum relative to a commutative Banach algebra of operators on X containing the n-tuple T. The analytic functional calculus problem, that is, the problem of defining a unital algebra homomorphism

$$\Phi: \mathcal{O}(\sigma(T, X)) \to L(X), \qquad \Phi(f) = f(T) \quad (f \in \mathcal{O}(\sigma(T, X)))$$

which extends the natural polynomial functional calculus, was solved in two different ways by Taylor.

One possibility is to use a refinement of the Cauchy–Weil integral formula in several complex variables (by constructing an operator-valued kernel which gives the map Φ), or one can solve the same existence problem by abstract homological methods. Each of these possibilities has its own advantages. They will be discussed in detail in Chapters 2 and 5, respectively.

Next we focus on the second, homological, method of treating the functional calculus problem. The basic idea is to regard a commutative n-tuple of bounded operators $T = (T_1, \ldots, T_n)$ acting on a Banach space X as a topological $\mathcal{O}(\mathbb{C}^n)$-module structure on X such that the multiplication by the coordinate function z_i corresponds to the action of the operator T_i ($1 \leq i \leq n$). This point of view is not new; it can and has been used, for instance, in the study of the Jordan form of a matrix via the structure of finitely-generated modules over principal ideal rings. In this setting the joint spectrum $\sigma(T, X)$ is nothing other than the 'homological support' of the module X. Quite specifically, one proves that an open polydisc Δ in $\mathbb{C}^n$ is disjoint from $\sigma(T, X)$ if and only if

$$\text{Tôr}_q^{\mathcal{O}(\mathbb{C}^m)}(\mathcal{O}(\Delta), X) = 0 \qquad (q \geq 0). \qquad (1.3.1)$$

Here Tôr denotes the derived functors of a topological tensor product over the algebra $\mathcal{O}(\mathbb{C}^n)$ (see Chapter 3).

The relative homology theory for topological modules over topological algebras which is needed at this stage was constructed by Taylor (1972a). At this point we have to notice the first in a series of fortunate coincidences which facilitated the proper understanding and continuation of Taylor's ideas. Namely the same topological homology theory was developed independently, and for different reasons, by A. Ya. Helemskii in Moscow and by the French school of complex analytic geometry (B. Malgrange, A. Douady, J.-L. Verdier, J.-P. Ramis). Chapters 3 and 4 contain full details on this subject.

With the characterization (1.3.1) of the joint spectrum properly understood, the existence proof for the analytic functional calculus is immediate: one has to prove that X is actually a Banach $\mathcal{O}(\sigma(T, X))$-module. We explain this idea in the one-variable case.

Let $T \in L(X)$ be a bounded linear operator acting on the complex Banach space X. The spectrum $\sigma(T)$ of T (in the usual sense or in Taylor's theory) consists of all points $z \in \mathbb{C}$ such that the operator $z - T \colon X \to X$ is not invertible. The set $\sigma(T)$ is compact, and the operator function

$$\mathbb{C} \setminus \sigma(T) \to L(X), \qquad z \mapsto (T - z)^{-1}$$

is analytic. In particular it follows that the linear map

$$T - z \colon \mathcal{O}(U, X) \to \mathcal{O}(U, X) \qquad (1.3.2)$$

is a bijection for each open set U in $\mathbb{C}$ which is disjoint from $\sigma(T)$.

Conversely, let U be an open set in $\mathbb{C}$ such that the operator (1.3.2) is bijective. Then for each $x \in X$, there is a unique function $\tilde{x} \in \mathcal{O}(U, X)$ with $(z - T)\tilde{x}(z) = x$ on U; thus for each fixed point $\lambda \in U$ the operator $\lambda - T$ has the inverse $x \mapsto \tilde{x}(\lambda)$. Therefore a given open set U in $\mathbb{C}$ is disjoint from $\sigma(T)$ if and only if the operator (1.3.2) is bijective. We shall see that, on the other hand, the kernel and the cokernel of the multiplication operator $z - T \colon \mathcal{O}(U, X) \to \mathcal{O}(U, X)$ represent the only possibly non-zero Tôr spaces that can occur in (1.3.1).

To construct an analytic functional calculus for T, we first remark that the quotient space $\mathcal{O}(\mathbb{C}, X)/(T - z)\mathcal{O}(\mathbb{C}, X)$ is canonically isomorphic to X via the map

$$\rho_T \colon \mathcal{O}(\mathbb{C}, X) \to X, \ \sum_{n=0}^{\infty} x_n z^n \mapsto \sum_{n=0}^{\infty} T^n x_n.$$

In particular, if $f(x) = \sum_{n=0}^{\infty} x_n z^n$ with $\rho_T(f) = 0$, then f can be divided by $T - z$, since

$$f(z) = f(z) - \rho_T(f) = \sum_{n=1}^{\infty} (z^n - T^n)x_n = (T - z)g(z),$$

where $g \in \mathcal{O}(\mathbb{C}, X)$ is defined in an obvious way. Thus we obtain an exact sequence

$$0 \to \mathcal{O}(\mathbb{C}, X) \xrightarrow{T-z} \mathcal{O}(\mathbb{C}, X) \to X \to 0.$$

One way to construct the analytic functional calculus is to use the observation that this sequence can be localized to arbitrary open neighbourhoods of the spectrum of T.

Lemma 1.3.1 *One can associate with each open neighbourhood V of the spectrum of T an exact sequence*

$$0 \to \mathcal{O}(V, X) \xrightarrow{T-z} \mathcal{O}(V, X) \xrightarrow{\rho_T^V} X \to 0$$

in such a way that the resulting family $(\rho_T^V)_V$ is compatible with restrictions.

Proof Let U be an open subset of $\mathbb{C}$ with $U \cap \sigma(T) = \varnothing$ and $U \cup V = \mathbb{C}$. Then we have the exact Čech complex

$$0 \to \mathcal{O}(\mathbb{C}) \xrightarrow{\alpha} \mathcal{O}(U) \oplus \mathcal{O}(V) \xrightarrow{\beta} \mathcal{O}(U \cap V) \to 0 \qquad (1.3.3)$$

given by the maps $\alpha(f) = (f|U, f|V)$ and $\beta(g, h) = g|U \cap V - h|U \cap V$ (Hörmander 1966, Theorem 1.4.5). Since all occurring spaces are nuclear, the sequence (1.3.3) remains exact when forming the topological tensor product with the identity operator on X (see Appendix 1). Hence the rows in the commutative diagram

$$
\begin{array}{ccccccccc}
0 \to & \mathcal{O}(\mathbb{C}, X) & \to & \mathcal{O}(U, X) \oplus \mathcal{O}(V, X) & \to & \mathcal{O}(U \cap V, X) & \to 0 \\
& \downarrow{\scriptstyle T-z} & & \downarrow{\scriptstyle (T-z) \oplus (T-z)} & & \downarrow{\scriptstyle T-z} & \\
0 \to & \mathcal{O}(\mathbb{C}, X) & \to & \mathcal{O}(U, X) \oplus \mathcal{O}(V, X) & \to & \mathcal{O}(U \cap V, X) & \to 0
\end{array}
$$

are exact.

By using the serpent's lemma (Appendix 2) one can show that the induced restriction operator

$$\mathcal{O}(\mathbb{C}, X)/(T-z)\mathcal{O}(\mathbb{C}, X) \xrightarrow[\sim]{r_V} \mathcal{O}(V, X)/(T-z)\mathcal{O}(V, X)$$

is an isomorphism.

Suppose that W is another open set such that $\sigma(T) \subset W \subset V$. Then the diagram given by the natural restriction maps

$$
\begin{array}{ccc}
& \mathcal{O}(\mathbb{C}, X)/(T-z)\mathcal{O}(\mathbb{C}, X) & \\
{\scriptstyle r_V}\swarrow & & \searrow{\scriptstyle r_W} \\
\mathcal{O}(V, X)/(T-z)\mathcal{O}(V, X) \longrightarrow & & \mathcal{O}(W, X)/(T-z)\mathcal{O}(W, X)
\end{array}
$$

commutes. Therefore, using the identification $X \cong \mathcal{O}(\mathbb{C}, X)/(T-z)\mathcal{O}(\mathbb{C}, X)$, one can define the map ρ_T^V by

$$\rho_T^V(f) = r_V^{-1}([f]) \qquad (f \in \mathcal{O}(V, X)).$$

Clearly, the resulting family $(\rho_T^V)_V$ satisfies all assertions. $\qquad\square$

The existence of an analytic functional calculus for the operator T is now established. Indeed, for each open neighbourhood V of $\sigma(T)$, one can use the canonical identifications

$$X \cong \mathscr{O}(\mathbb{C}, X)/(T-z)\mathscr{O}(\mathbb{C}, X) \xrightarrow[\sim]{r_V} \mathscr{O}(V, X)/(T-z)\mathscr{O}(V, X)$$

and the natural topological $\mathscr{O}(V)$-module structure of the space on the right to turn X into a topological $\mathscr{O}(V)$-module. This $\mathscr{O}(V)$-module structure is compatible with restrictions, and it yields an analytic functional calculus for the operator T.

In a similar manner one can deduce other properties of the analytic functional calculus, as for instance its uniqueness, and the superposition principle (see Chapter 5).

With the correct homological interpretation one can extend the above construction to the case of several commuting operators. The key lemma is a transversality result of Douady and Verdier. For technical reasons, originating in the necessity to pass from an open subset of $\mathbb{C}^n$ to its envelope of holomorphy (which is not necessarily contained in $\mathbb{C}^n$), we shall replace $\mathbb{C}^n$ by an arbitrary Stein space in Chapter 5.

The multivariable counterpart of the local spectral theory can be developed in both ways, either by using the Cauchy–Weil integral (Chapter 2), or in a topological–homological framework (Chapter 6). Retaining the above notations we say that a commuting n-tuple T satisfies Bishop's property (β) if, for every open polydisc Δ in $\mathbb{C}^n$, the Koszul complex $K_\bullet(T-z, \mathscr{O}(\Delta, X))$ is exact in all positive degrees and has separated homology in degree zero. Using homological language, the last condition means that

$$\mathrm{T\hat{o}r}_q^{\mathscr{O}(\mathbb{C}^n)}(\mathscr{O}(\Delta), X) = 0$$

for $q > 0$ and that the space

$$\mathscr{F}_T(\Delta) = \mathrm{T\hat{o}r}_0^{\mathscr{O}(\mathbb{C}^n)}(\mathscr{O}(\Delta), X) = \mathscr{O}(\Delta, X)/\sum_{i=1}^{n}(T_i - z_i)\mathscr{O}(\Delta, X)$$

is Hausdorff in its canonical quotient topology.

The resulting analytic Fréchet sheaf $\mathscr{F}_T$ is a model for the n-tuple T in the sense of Section 1.1. Furthermore, it turns out that the sheaves arising in this way are precisely the quasi-coherent analytic Fréchet sheaves. Sheaves of this type have been studied by the French school of analytic geometry, since they form a natural generalization of the class of coherent sheaves. Thus a commuting n-tuple of operators with Bishop's property (β) is cohomologically equivalent to a quasi-coherent sheaf. This observation is essential for establishing many of the results involving systems with property (β), as for instance in the multidimensional variant of Theorem 1.2.2.

In this framework the analogue of the class of decomposable operators consists of the commuting tuples of operators possessing a Fréchet soft sheaf

model. These systems will be shown to satisfy condition (β). Interpreted at
the level of sheaves, the last remark can be transformed into a useful
quasi-coherence criterion. Details in this direction will be presented in
Chapters 4 and 6.

To complete the sheaf-theoretical interpretation of the Jordan form of a
finite matrix presented at the end of the preceding section, we discuss below
a similar construction for systems of commuting matrices. Let V be a
finite-dimensional vector space, and let $T = (T_1, \ldots, T_n)$ be an n-tuple of
commuting linear maps on V. Then Taylor's joint spectrum $\sigma(T, V)$ is a finite
subset of $\mathbb{C}^n$, and one can show that T possesses property (β). Exactly as in
the single operator case we have a direct sum decomposition

$$V = \bigoplus_{\lambda \in \sigma(T)} \mathcal{O}_\lambda^V \Big/ \left(\sum_{j=1}^n (T_j - z_j)\mathcal{O}_\lambda^V \right). \tag{1.3.4}$$

Each of the direct summands is a finite-dimensional Artinian module over
the Noetherian local ring $\mathcal{O}_\lambda$, but in this case the structure of these modules
cannot be further simplified.

Thus (1.3.4) is the analogue of the Jordan form for several commuting
matrices.

The corresponding sheaf model is

$$\mathcal{F}_T(U) = \bigoplus \left\{ \mathcal{O}_\lambda^V \Big/ \left(\sum_{j=1}^n (T_j - z_j)\mathcal{O}_\lambda^V \right); \lambda \in U \cap \sigma(T) \right\}.$$

Hence $\mathcal{F}_T$ is a coherent analytic sheaf supported by finitely many points in $\mathbb{C}^n$
(a so-called skyscraper sheaf). Two systems of commuting matrices are
simultaneously similar if and only if their sheaf models (formed as above) are
isomorphic.

Typical examples of tuples with property (β) are systems of analytic
multipliers on certain Banach spaces of analytic functions. Although in
general the concrete spectral problems on these spaces may be very difficult,
such problems usually correspond to classical questions in complex analysis.
This relationship is exploited in Chapters 7 and 8 of the book.

To end this section, we should make clear the limits of our aims, and
implicitly the limits of the present monograph. The spectral theory developed
in these pages represents only a first (and most rudimentary) level. Questions
like the unitary invariants of systems of commuting operators, the K-
homology of these objects, sharp functional models, harmonic analysis, or
positivity conditions are touched only tangentially. Moreover, we rarely go
beyond the comfortable cohomological assumption of property (β), though it
would be extremely interesting for applications to relax this condition. Also,
most of the classical domains in several complex variables possess a rich
group of symmetries which should be reflected at the level of any spectral

theory on these domains. We do not pursue this direction here. However, we shall try to show that this first level of spectral theory, *à la* Bishop, has deep and quite unexpected applications. Some of these will be recorded in the next section.

1.4 SUMMARY OF THE APPLICATIONS CONTAINED IN THE BOOK

The last four chapters of the book are devoted to applications of the general theory outlined above. The applications can be grouped into two distinct classes: applications to function theory, and to complex analytic geometry. They correspond roughly to the main tools used in developing the general theory.

The localization principle in spectral theory is used, via the sheaf model approach, for the computation of the character space of certain function algebras defined on open pseudoconvex domains in $\mathbb{C}^n$. Although most of these results are not new, this point of view allows a unified treatment of the subject and various generalizations. For instance, a solution of the corona problem for finitely connected domains in $\mathbb{C}$ is explained in this context. Interpreted from a different perspective, the same computations offer various pseudoconvexity criteria for open subsets of $\mathbb{C}^n$ in terms of vanishing conditions for certain cohomologies with bounds. These results appear in Sections 7.3 and 7.4.

An analogue of Bishop's property (β) for smooth vector-valued functions, rather than analytic functions, is shown to characterize the class of *subscalar operators*. These are operators which are, up to similarity, the restriction of an operator that is generalized scalar in the sense of Colojoară and Foiaş.

By duality one finds that the map

$$T - z : \mathscr{E}'(\mathbb{C}) \,\hat{\otimes}\, X \to \mathscr{E}'(\mathbb{C}) \,\hat{\otimes}\, X$$

is surjective and allows the lifting of bounded sets if and only if T is a quotient of a generalized scalar operator (see Theorem 6.4.10). Here $\mathscr{E}'(\mathbb{C})$ stands for the space of compactly supported distributions on $\mathbb{C}$. This observation explains a series of classical results concerning the division of distributions by analytic functions, as for instance a celebrated theorem of L. Schwartz, which asserts that distributions can be divided by non-zero complex analytic functions. Moreover, the above characterization of the quotients of generalized scalar operators combined with the general theory of Fréchet quasi-coherent sheaves will be related to a deep result of Malgrange concerning the closedness of ideals generated by analytic functions in the space of smooth functions. The first part of Chapter 7 contains details about these subjects.

Chapter 8 is entirely devoted to spectral computations on the Bergman space associated with a bounded pseudoconvex domain in $\mathbb{C}^n$. Based on the sheaf model of the tuple of coordinate functions on the Bergman space, as described in the one-variable case in Section 1.1 above, one computes the joint and the essential joint spectrum for systems of analytic Toeplitz operators with bounded symbols. The main result is an L^2-corona theorem for arbitrary pseudoconvex domains (a result which seems to be optimal so far). Besides the homological and sheaf-theoretical methods, this part of Chapter 8 exploits L^2-estimates for the $\bar{\partial}$-operator due to Hörmander and Skoda. The same computations yield, without additional effort, the Fredholm index data for such systems of Toeplitz operators. It is known for instance that, modulo some standard results from K-theory, these index data form the root of any general index theorem.

The second part of Chapter 8 deals with the classification problem for the closed, analytically invariant subspaces of Bergman spaces. Having as a model the well-known theorem of Beurling, which describes all analytically invariant subspaces of the Hardy space on the unit circle, we show how to approach the problem using spectral localization techniques. Thus a generic class of invariant subspaces is isolated; roughly speaking these subspaces are characterized by the fact that they are not too seriously affected by the boundary behavior of their elements. Thus the classification of these subspaces becomes a purely algebraic problem. In this approach the work of A. Douady and his collaborators on deformation theory for coherent analytic sheaves is essential. Moreover, following R. G. Douglas and his school, we prove rigidity results for the same class of analytically invariant subspaces of the Bergman space.

One of the basic concepts in this book is the notion of an analytic quasi-coherent sheaf. The general theory of this class of sheaves is presented in Chapter 4, while various applications of quasi-coherence criteria appear throughout the second part of the book. These sheaves are more flexible under algebraic and geometric operations than the coherent sheaves. This explains their importance in the deformation theory of analytic spaces. On the other hand, the topological homology theory presented in Chapter 3 and used intensively throughout all chapters is intimately related to quasi-coherent sheaves. Based on this relationship we give proofs of some basic finiteness results in complex analytic geometry in Chapter 9. First we discuss the Cartan–Serre theorem on the finiteness of the cohomology groups of coherent sheaves on compact analytic spaces, and then we focus on a proof of a relative version of this result given by Grauert's theorem on the coherence of direct image sheaves. Our approach follows closely Kiehl and Verdier (1971). Indeed, this field formed one of the origins of topological homology theory and of the theory of quasi-coherent sheaves.

A slightly more elaborated study of complexes of quasi-coherent sheaves with coherent cohomology offers the natural framework for studying the index theory of commuting systems of operators. The basic properties of the Fredholm index for single operators are thus extended to the multivariable case. However, in this setting the index is no longer a number, but an element of the Grothendieck K-theory group of coherent sheaves. A detailed study of these concepts is contained in the first part of Chapter 10.

In the second part of Chapter 10 we present the main lines of a proof of the Riemann–Roch theorem for coherent sheaves on arbitrary complex spaces with singularities. The argument closely follows Levy (1987a). Stated as the existence of a functor

$$\alpha: K^{\mathrm{an}} \to K^{\mathrm{top}},$$

which is compatible with proper direct images, this theorem contains as particular cases all known Riemann–Roch results on complex spaces, for instance the Hirzebruch–Riemann–Roch theorem (for compact complex manifolds), and the theorem of Baum–Fulton–MacPherson (for projective varieties). Levy receives the credit for having discovered that multivariable Fredholm theory can be adapted, modulo some standard methods from K-theory and complex analytic geometry, to give a proof of a very general Riemann–Roch theorem. As a subject of independent interest, in this last part of Chapter 10, we mention a canonical construction of almost complex embeddings of complex analytic spaces into Euclidean spaces.

The introduction to each of Chapters 7–10 contains additional information about the applications mentioned above.

1.5 REFERENCES AND COMMENTS

Section 1.1

The basic reference for the classical spectral decomposition theories is the three-volume monograph of Dunford and Schwartz (1958, 1963, 1971). It contains, of course, much more material than is needed for our book. The modern classification theory of linear operators is treated in the monographs by Apostol *et al.* (1984) and Herrero (1982).

The idea of diagonalizing self-adjoint operators by using eigendistributions appears in Dirac (1930). It seems that this was the origin of the famous δ-distribution. The generalized eigenfunction expansions for various concrete operators appear in classical studies of Weyl, Titchmarsh, Gelfand, and many others. A good reference for classical aspects of this theory is the monograph of Berezanskii (1968). The kernel theorem and some of its applications appear in Treves (1967) and Hörmander (1983).

The theory of spectral measures in Banach spaces and of spectral operators in the sense of Dunford is presented in detail in the third part of Dunford and Schwartz (1971). The same volume contains some valuable applications of the theory of spectral operators to perturbation and scattering problems.

Section 1.2

The essentials of local spectral theory are contained, besides in the papers of

Bishop mentioned in the text, in Dunford and Schwartz (1971), Apostol (1968), Colojoară and Foiaş (1968), Eschmeier (1987b). For the theory of decomposable operators see Colojoară and Foiaş (1968), Albrecht (1978, 1979), Radjabalipour (1978), and Vasilescu (1982).

Theorem 1.2.2(a) was proved in Eschmeier and Putinar (1984), and in a particular case in Lange (1980); part (b) is due to Albrecht and Eschmeier (1987). For a proof of Theorem 1.2.3, see Eschmeier (1989) and Eschmeier and Prunaru (1990).

The sheaf model for an operator with Bishop's property (β) was introduced in Putinar (1983a). The subsequent papers by Putinar (1986, 1990a) and Eschmeier and Putinar (1984) develop the basic properties of this sheaf model, both for a single operator and for tuples of commuting operators.

Section 1.3

A comprehensive reference for the multivariable spectral theory in Banach algebras is Bourbaki (1967). For the spatial theory of the joint spectrum of Taylor we strongly recommend the fundamental articles by Taylor (1970a, b; 1972a, b).

The topological homology needed for the multivariable spectral theory is extensively discussed in *Séminaire de géométrie analytique* (1974) and in the monograph by Helemskii (1986). The book of Helemskii is a valuable reference for the cohomology of Banach algebras. The coordinateless spectral theory for Fréchet modules over Stein algebras is treated in Putinar (1980, 1984a).

Multivariable local spectral theory was developed immediately after the publication of Taylor's works. An early synthesis of the main contributions in this field is contained in the monograph by Vasilescu (1982). For questions related to joint spectra and multivariable analytic functional calculi, see Curto (1988).

A few recent works on related aspects of spectral theory not covered by our book are Douglas and Paulsen (1989), Salinas *et al.* (1989), and Zhu (1990).

Section 1.4

The closedness theorem for ideals generated by analytic functions in spaces of smooth functions was originally proved by Malgrange (1966). The relation between spectral theory and the division of distributions by analytic functions is made explicit in Eschmeier and Putinar (1988, 1989).

For questions of spectral theory on Bergman spaces see the survey articles by Axler (1988), Salinas *et al.* (1989), and Zhu (1990). The classification theory of Bergman submodules is treated in Douglas and Paulsen (1989), Douglas and Yan (1990, 1992), and Douglas *et al.* (1993).

The theory of quasi-coherent sheaves appeared implicitly in *Séminaire de géométrie analytique* (1974) and explicitly in a paper by Ramis and Ruget (1974). The article of Leiterer (1978) contains a similar idea on how to enlarge the class of coherent sheaves to a more flexible category of analytic modules.

The index theory for commutative tuples of operators was developed by Curto (1981), Douglas and Voiculescu (1981) Vasilescu (1982), Putinar (1985), Carey and Pincus (1985), Levy (1989), and Salinas (1989), to mention only a few of the principal papers. An early article on Fredholm complexes, which is basic for the whole of Chapter 10, is Segal (1970). The two papers by Levy (1987a, 1989) are fundamental for the multivariable index theory as well as for the Riemann–Roch theorem. The literature on the Riemann–Roch theorem and its application is vast. We mention only Borel and Serre (1958), Atiyah and Hirzebruch (1959, 1962), Hirzebruch (1965), and Fulton and Lang (1985).

2
Analytic functional calculus via integral representations

In a unital, commutative Banach algebra A there is a very natural notion of a joint spectrum for systems of elements in A. Let $a = (a_1, \ldots, a_n) \in A^n$. Then the *joint spectrum* $\sigma(a)$ of a is defined as the set of all n-tuples $z \in \mathbb{C}^n$ for which the ideal generated by $z_1 - a_1, \ldots, z_n - a_n$ in A is proper, or equivalently, as the set $\sigma(a) = \hat{a}(\Delta_A)$, where

$$\Delta_A = \{\lambda; \lambda: A \to \mathbb{C} \quad \text{non-trivial, multiplicative linear functional}\}$$

is the *character space* of A, and $\hat{a} = (\hat{a}_1, \ldots, \hat{a}_n)$ denotes the tuple of Gelfand transforms

$$\hat{a}_i: \Delta_A \to \mathbb{C}, \qquad \lambda \mapsto \lambda(a_i) \quad (i = 1, \ldots, n).$$

In the fifties, Shilov (1953), Waelbroeck (1954), and Arens and Calderon (1955) showed how to construct an analytic functional calculus for finite systems of elements in A.

For $a = (a_1, \ldots, a_n) \in A^n$, let us equip the algebra $\mathscr{O}(\sigma(a))$ of all germs of analytic functions defined in a neighbourhood of $\sigma(a)$ with its canonical inductive limit topology. The *analytic functional calculus* associates with each tuple a an extension of the polynomial functional calculus of a to a continuous algebra homomorphism

$$\Phi: \mathscr{O}(\sigma(a)) \to A, \qquad a \mapsto f(a).$$

The functional calculus of Shilov, Waelbroeck, and Arens and Calderon is uniquely determined by the additional condition that

$$f \otimes 1(a, b) = f(a)$$

for any pair of finite tuples $a \in A^n$, $b \in A^m$ and each function $f \in \mathscr{O}(\sigma(a))$ (see Bourbaki 1967). The analytic functional calculus has been used to prove, among other things, implicit function theorems for elements of commutative Banach algebras, and used to study the connections between the algebraic structure of a commutative Banach algebra and topological properties of its character space (see, for example, Shilov 1953; Arens 1963; Royden 1963; Forster 1974; Taylor 1976).

The aim of the present chapter is to define an analytic functional calculus for commuting systems of continuous linear operators. One possible method of defining a joint spectrum for a commuting system $T = (T_1, \ldots, T_n)$ of continuous linear operators on a Banach space X is to regard $T_1, \ldots, T_n$ as elements of a commutative Banach subalgebra A of $L(X)$, and to define $\sigma(T)$ to be $\sigma_A(T)$, the joint spectrum of T in the algebra A. The drawback of this approach is that, in general, the resulting joint spectrum for T will depend on the choice of the subalgebra A. In the early seventies, J. L. Taylor succeeded in defining a new, spatial, notion of joint spectrum which directly generalizes the idea of an operator being not injective or not surjective to the case of several commuting operators. This 'Taylor' spectrum is contained in all joint spectra obtained in the fashion described above, and it may be a proper subset of each such $\sigma_A(T)$. Some extreme examples demonstrating this phenomenon can be found in Albrecht (1979). Moreover, using a Cauchy–Weil-type integral, Taylor (1970b) constructed an analytic functional calculus based on the new notion of joint spectrum. Applied to the special case of multiplication tuples in commutative Banach algebras, Taylor's functional calculus gives back the analytic functional calculus of Shilov, Waelbroeck, and Arens and Calderon.

Shortly afterwards, Taylor gave a second, alternative, proof for the existence of a multidimensional analytic functional calculus using homological methods (see Taylor 1972a,b). To construct an analytic functional calculus for a commuting tuple T means to isolate a singularity set $\sigma(T)$ in $\mathbb{C}^n$, called the spectrum of T, such that for each open set U containing $\sigma(T)$ the global topological $\mathcal{O}(\mathbb{C}^n)$-module structure of X, given by

$$\mathcal{O}(\mathbb{C}^n) \times X \to X, \qquad (f, x) \mapsto f(T)x,$$

extends to a topological $\mathcal{O}(U)$-module structure satisfying certain natural conditions (for example, spectral mapping theorems, the superposition principle, uniqueness conditions). Taylor observed that many of the constructions from homological algebra concerning the extension of module structures still make sense if the objects under consideration carry an additional topological structure, and that these techniques can be used to solve the analytic functional calculus problem.

In the present chapter, we shall describe the first approach of Taylor, that is, the construction of an analytic functional calculus for several commuting operators by means of a Cauchy–Weil integral. The homological approach will be discussed in Chapter 5 after we have prepared the necessary tools from topological homology and complex analysis. We shall consider commuting systems of continuous linear operators on Fréchet spaces using a notion of joint spectrum for such tuples which also goes back to Taylor (1972b). Apart from the inclusion of some simplifications mainly due to Frunză (1975b) and Putinar (1979) we follow closely the original constructions of Taylor. In the final section of this chapter we shall prove spectral mapping theorems for subspectra of the Taylor spectrum due to Słodkowski (1977), Fainshtein (1980), and Eschmeier (1987). For an alternative description of the constructive solution of the multidimensional analytic functional calculus problem, the reader is referred to the monograph of Vasilescu (1982).

The parts of this chapter concerning the Cauchy–Weil integral are relatively independent of the rest of the book. They serve as a basis of comparison for Chapter 5, where the same results will be proved by different methods. Historically, the contents of this chapter formed the starting point for the multidimensional axiomatic theory of spectral decompositions which forms the subject of Chapter 6. Finally, we should mention that some results on functional calculi for systems of non-commuting operators have been obtained. The interested reader is referred to Taylor (1972b) and Albrecht (1982) for results and further references.

2.1 PARAMETRIZED COMPLEXES

The aim of this section is to describe some exactness results for parametrized complexes of Banach and Fréchet spaces. These results will be used in later sections to give different characterizations of the joint spectrum of commuting systems of continuous linear operators. We start with a result concerning continuously parametrized complexes of Fréchet spaces.

Let X and Y be Fréchet spaces. We denote by $L_s(X,Y)$ the space $L(X,Y)$ of all continuous linear operators from X to Y equipped with the locally convex topology of pointwise convergence. The topology of $L_s(X,Y)$ is generated by the seminorms

$$\|T\|_{p,x} = p(Tx) \qquad (x \in X, p \text{ continuous seminorm on } Y).$$

We write $L_b(X,Y)$ for the space $L(X,Y)$ equipped with the locally convex topology generated by the seminorms

$$\|T\|_{p,B} = \sup_{x \in B} p(Tx) \qquad (B \subset X \text{ bounded}, p \text{ continuous seminorm on } Y).$$

Let X, Y, and Z be Fréchet spaces, and let Ω be a Hausdorff topological space. We fix continuous operator functions $\alpha \in C(\Omega, L_s(X,Y))$ and $\beta \in C(\Omega, L_s(Y,Z))$ with $\beta(\lambda) \circ \alpha(\lambda) = 0$ for all $\lambda \in \Omega$.

Definition 2.1.1 Let A be an arbitrary subset of Ω. Then

$$X \xrightarrow{\alpha(\lambda)} Y \xrightarrow{\beta(\lambda)} Z \tag{2.1.1}$$

is *uniformly exact* on A if, for each zero neighbourhood U in X, there is a zero neighbourhood V in Y such that, for all $\lambda \in A$,

$$\operatorname{Ker} \beta(\lambda) \cap V \subset \alpha(\lambda)U.$$

We shall say that the complex (2.1.1) is *locally uniformly exact* if, for each point $\lambda \in \Omega$, it is uniformly exact on some neighbourhood of λ in Ω.

Let us recall that a Hausdorff topological space Ω is *paracompact* if each open cover of Ω possesses a locally finite refinement. Each metrizable topological space is paracompact. By Urysohn's metrization theorem, each second countable locally compact Hausdorff space is metrizable, and hence paracompact. Paracompact spaces are easily seen to be normal. Moreover, one of the most important properties of a paracompact space Ω is that, for each open cover of Ω, there is a locally finite continuous partition of unity relative to this open cover.

Theorem 2.1.2 *Let X, Y, and Z be Fréchet spaces, and let $\alpha \in C(\Omega, L_s(X, Y))$ and $\beta \in C(\Omega, L_s(Y, Z))$ be continuous operator functions on a paracompact space Ω such that $\beta(\lambda) \circ \alpha(\lambda) = 0$ for all $\lambda \in \Omega$. Suppose that*

$$X \xrightarrow{\alpha(\lambda)} Y \xrightarrow{\beta(\lambda)} Z$$

is locally uniformly exact on Ω. Then, for each function $g \in C(\Omega, Y)$ with $\beta(\lambda) g(\lambda) = 0$ for all $\lambda \in \Omega$, there is a function $f \in C(\Omega, X)$ with $g(\lambda) = \alpha(\lambda) f(\lambda)$ for all $\lambda \in \Omega$.

Proof Let $g \in C(\Omega, Y)$ be a function with $\beta(\lambda) g(\lambda) = 0$ for all λ.

(a) We first suppose that our given complex is uniformly exact on Ω. Let $(U_n)_{n \geq 1}$ and $(V_n)_{n \geq 1}$ be decreasing sequences of absolutely convex, closed sets forming a neighbourhood basis of zero in X and Y, respectively, such that $\operatorname{Ker} \beta(\lambda) \cap V_n \subset \alpha(\lambda) U_n$ for all $n \geq 1$ and all $\lambda \in \Omega$. We denote by $p_n = p_{U_n}$ and $q_n = p_{V_n}$ the corresponding Minkowski functionals.

Let $\epsilon > 0$ be arbitrary. For each $\lambda \in \Omega$ we choose a vector $x_\lambda \in X$ with $\alpha(\lambda) x_\lambda = g(\lambda)$ and $p_1(x_\lambda) \leq \max(q_1(g(\lambda)), \epsilon)$ and an open neighbourhood W_λ of λ with the property that

$$q_2(g(\mu) - \alpha(\mu) x_\lambda) \leq \epsilon \qquad (\mu \in W_\lambda).$$

Since Ω is paracompact, there is a locally finite, continuous partition of unity $(\varphi_\lambda)_{\lambda \in \Omega}$ relative to the open cover (W_λ) of Ω. The functions $f_1 = \sum_{\lambda \in \Omega} \varphi_\lambda x_\lambda$ and $g_1 = g - \alpha f_1$ are continuous, and we obtain the estimate

$$q_2(g_1(\mu)) \leq \sum_{\lambda \in \Omega} \varphi_\lambda(\mu) q_2(g(\mu) - \alpha(\mu) x_\lambda) \leq \epsilon \qquad (\mu \in \Omega).$$

Now one can repeat the same procedure with g replaced by g_1.

Inductively, one obtains a sequence $(f_n)_{n \geq 1}$ in $C(\Omega, X)$ such that, for all $n \geq 2$ and $\lambda \in \Omega$,

$$p_n(f_n(\lambda)) \leq 2^{-n} \quad \text{and} \quad q_n\left(g(\lambda) - \sum_{j=1}^{n-1} \alpha(\lambda) f_j(\lambda)\right) \leq 2^{-n}.$$

Set $f = \sum_{n=1}^{\infty} f_n$. Clearly $f \in C(\Omega, X)$ is a well-defined function with $\alpha f = g$.

(b) We turn to the general case. Since Ω is a normal topological space, we can choose an open cover $(U_i)_{i \in I}$ of Ω such that our given complex is uniformly exact on $\overline{U}_i$ for each i. Since the sets $\overline{U}_i$ are paracompact, for each i there is a function $f_i \in C(\overline{U}_i, X)$ with $\alpha f_i = g$ on $\overline{U}_i$. If $(\varphi_i)_{i \in I}$ is a continuous partition of unity relative to the open cover $(U_i)_{i \in I}$, then $f = \sum_{i \in I} \varphi_i f_i$ is a solution of the equation $\alpha f = g$ in $C(\Omega, X)$. $\qquad\square$

An inspection of the above proof shows that, in the setting of Theorem 2.1.2, for a given function $g \in C(\Omega, Y)$ with $\beta g = 0$, a given point $\lambda \in \Omega$, and an arbitrary element $x \in X$ with $\alpha(\lambda)x = g(\lambda)$, there is a solution f of the equation $\alpha f = g$ in $C(\Omega, X)$ with $f(\lambda) = x$. In particular, for each point λ in Ω and each vector $x \in \text{Ker}\ \alpha(\lambda)$, there is a function $f \in C(\Omega, X)$ with $\alpha f = 0$ and $f(\lambda) = x$.

Let X and Y be Banach spaces. We equip $L(X, Y)$ with its usual norm topology. For a continuous linear operator $T \colon X \to Y$ between Banach spaces X and Y, with closed range, we denote by $k(T)$ the norm of the inverse of the topological isomorphism $X/\text{Ker}\ T \xrightarrow{T} \text{Im}\ T$.

Lemma 2.1.3 *Let $\alpha_0 \in L(X, Y)$ and $\beta_0 \in L(Y, Z)$ be continuous linear operators between Banach spaces such that $\text{Im}\ \beta_0$ is closed and $\text{Ker}\ \beta_0 = \text{Im}\ \alpha_0$. Take $r > \max(k(\alpha_0), k(\beta_0))$, and set $\delta = (6r)^{-1}$. Suppose that $\alpha \in L(X, Y)$ and $\beta \in L(Y, Z)$ are operators with $\text{Im}\ \alpha \subset \text{Ker}\ \beta$ and*

$$\max(\|\alpha - \alpha_0\|, \|\beta - \beta_0\|) < \delta.$$

Then $\text{Im}\ \alpha = \text{Ker}\ \beta$ and $k(\alpha) \le 4r$.

Proof Suppose that α and β satisfy the above hypothesis. If $y \in \text{Ker}\ \beta$, then $\|\beta_0 y\| \le \delta \|y\|$. Hence there is a vector $y' \in Y$ with $\beta_0 y' = \beta_0 y$ and $\|y'\| \le r\delta \|y\|$. By the same argument one can choose a vector $x_1 \in X$ with

$$\alpha_0 x_1 = y - y' \quad \text{and} \quad \|x_1\| \le 2r\|y\|.$$

Then $y_1 = y - \alpha x_1 \in \text{Ker}\ \beta$ satisfies the estimate

$$\|y_1\| \le \|y - \alpha_0 x_1\| + \|(\alpha_0 - \alpha)x_1\| \le \|y\|/2.$$

Inductively, one obtains sequences $(x_n)_{n \ge 1}$ in X and $(y_n)_{n \ge 1}$ in Y with $\|x_n\| \le 2^{-n+2} r\|y\|$, $\|y_n\| \le 2^{-n}\|y\|$, and

$$y = y_n + \alpha \sum_{i=1}^{n} x_i \quad (n \ge 1).$$

Thus $x = \sum_{i=1}^{\infty} x_i$ is a well-defined vector in X with $\alpha x = y$ and $\|x\| \le 4r\|y\|$. $\qquad \square$

As an application of Theorem 2.1.2 and Lemma 2.1.3, we obtain the following exactness result for continuously parametrized complexes of Banach spaces.

Corollary 2.1.4 *Let X, Y, and Z be Banach spaces, and let Ω be a paracompact topological space. Suppose that $\alpha \in C(\Omega, L(X, Y))$ and $\beta \in C(\Omega, L(Y, Z))$ are continuous maps such that $\text{Im}\ \beta(\lambda)$ is closed and $\text{Ker}\ \beta(\lambda) = \text{Im}\ \alpha(\lambda)$ for all $\lambda \in \Omega$. Then*

$$\text{Ker}(C(\Omega, Y) \xrightarrow{\beta} C(\Omega, Z)) = \text{Im}(C(\Omega, X) \xrightarrow{\alpha} C(\Omega, Y)). \qquad \square$$

We next turn to the case of analytically parametrized complexes of Banach spaces on open sets Ω in $\mathbb{C}^n$.

Lemma 2.1.5 *Let X, Y, and Z be Banach spaces, and let Ω be an open set in $\mathbb{C}^n$. Suppose that $\alpha \in \mathcal{O}(\Omega, L(X,Y))$ and $\beta \in \mathcal{O}(\Omega, L(Y,Z))$ are analytic maps with $\beta(z) \circ \alpha(z) = 0$ for all $z \in \Omega$, and that w is a point in Ω with $\mathrm{Ker}\,\beta(w) = \mathrm{Im}\,\alpha(w)$. Then there is a real number $r_0 > 0$ such that*

$$\mathrm{Ker}(\mathcal{O}(V,Y) \xrightarrow{\ \beta\ } \mathcal{O}(V,Z)) = \mathrm{Im}(\mathcal{O}(V,X) \xrightarrow{\ \alpha\ } \mathcal{O}(V,Y))$$

for each open polydisc $V = P(w, r)$ with centre w and radius $r < r_0$.

Proof For simplicity, we shall suppose that $w = 0$. Let $\rho > 0$ be a real number such that the power series expansion of α,

$$\alpha(z) = \sum_k a_k z^k,$$

converges on the open polydisc $P(0, \rho)$. We choose real numbers $r_0, c > 0$ with $r_0 < \rho$ and

$$\sum_k{}' \|a_k\| r_0^{|k|} < c^{-1} < (k(\alpha(0)) + 1)^{-1},$$

where the notation $\sum'$ is used to indicate that the summation has to be performed over all indices different from zero. We fix an arbitrary real number r with $0 < r < r_0$, and define $V = P(0, r)$.

Suppose that $f \in \mathcal{O}(V, X)$ and $g \in \mathcal{O}(V, Y)$ are given by their power series expansions

$$f(z) = \sum_k f_k z^k \quad \text{and} \quad g = \sum_k g_k z^k \qquad (z \in V).$$

Then the relation $\alpha f = g$ holds if and only if

$$\alpha(0) f_k = g_k - \sum_{\mu \le k}{}' a_\mu f_{k-\mu} \tag{2.1.2}$$

holds for all $k \in \mathbb{N}^n$. We use this observation to construct the desired analytic solution of the equation $\alpha f = g$ for a given function $g \in \mathcal{O}(V, Y)$ with $\beta g = 0$.

For this purpose, we fix an arbitrary vector $x \in X$ with $\alpha(0) x = g_0$ and define $f_0 = x$. Let us suppose that, for a given integer $m \ge 0$, coefficients $(f_k)_{|k| \le m}$ in X have been chosen in such a way that 2.1.2 holds for all multi-indices k with $|k| \le m$. Since the function $h \in \mathcal{O}(V, Y)$ defined by

$$h(z) = g(z) - \alpha(z) \sum_{|\nu| \le m} f_\nu z^\nu$$

possesses no non-zero Taylor coefficients h_ν of order $|\nu| \le m$, and since $\beta h = 0$ on V, we obtain, for each $k \in \mathbb{N}^n$ with $|k| = m + 1$, the relation

$$0 = \beta(0)\left(g_k - \sum_{\mu \le k}{}' a_\mu f_{k-\mu} \right).$$

Hence for each such multi-index k there is a solution $f_k \in X$ of (2.1.2) with

$$\|f_k\| \le c \Big\| g_k - \sum_{\mu \le k}{}' a_\mu f_{k-\mu} \Big\|. \tag{2.1.3}$$

Therefore, by induction on $|k|$, one can define a family $(f_k)_{k \in \mathbb{N}^n}$ such that (2.1.2) holds for all $k \in \mathbb{N}^n$ and such that (2.1.3) holds for all $k \in \mathbb{N}^n \setminus \{0\}$.

To prove the convergence of the power series $f(z) = \sum_k f_k z^k$ on V, we consider the analytic function $u \in \mathscr{O}(V)$ given by

$$\left(\frac{1}{c} - \sum_k{}' \|a_k\| z^k \right) u(z) = \frac{\|f_0\|}{c} + \sum_k{}' \|g_k\| z^k.$$

The Taylor coefficients of u satisfy $u_0 = \|f_0\|$ and

$$u_k = c \left(\|g_k\| + \sum_{\mu \le k}{}' \|a_\mu\| u_{k-\mu} \right) \qquad (|k| \ge 1).$$

An elementary induction using (2.1.3) yields that $\|f_k\| \le u_k$ for all multi-indices k. Hence the power series with coefficients f_k converges on V. $\qquad \square$

It is worthwhile to remark that in the previous proof we could have chosen any vector $x \in X$ with $\alpha(0)x = g(0)$ as the value of f at 0.

Let Ω be a topological space. We consider complexes of the form

$$\cdots \xrightarrow{\alpha^{p-2}(z)} X^{p-1} \xrightarrow{\alpha^{p-1}(z)} X^p \xrightarrow{\alpha^p(z)} X^{p+1} \xrightarrow{\alpha^{p+1}(z)} \cdots,$$

where $(X^p)_{p \in \mathbb{Z}}$ is a sequence of Banach spaces (Fréchet spaces, Hilbert spaces,...) and the operators $\alpha^p(z) \in L(X^p, X^{p+1})$ depend on the parameter $z \in \Omega$. We shall abbreviate this situation by saying that $(X^\bullet, \alpha^\bullet)$ is a *parametrized complex*. Suppose that $\alpha^p \in C(\Omega, L(X^p, X^{p+1}))$ for each p. Then we call $(X^\bullet, \alpha^\bullet)$ a *continuously parametrized complex* on Ω. In the Banach-space case, $L(X^p, X^{p+1})$ will always be equipped with its norm topology. For more general locally convex spaces, we shall specify the topology used on $L(X^p, X^{p+1})$ in each particular case. If Ω is an open set in $\mathbb{C}^n$ and $\alpha^p \in \mathscr{O}(\Omega, L(X^p, X^{p+1}))$ for each p, then we call $(X^\bullet, \alpha^\bullet)$ an *analytically parametrized complex* on Ω. In the latter case, we shall denote by $(\mathscr{O}(\Omega, X^\bullet), \alpha^\bullet)$ the induced complex of spaces of analytic functions, and we denote, for a given point $z \in \Omega$, by $(\mathscr{O}_z^{X^\bullet}, \alpha^\bullet)$ the complex induced between the spaces of all germs of analytic functions at z. Similar notations will be used for complexes depending continuously or smoothly, or in any other well-defined sense, on a given parameter. A parametrized complex $(X^\bullet, \alpha^\bullet)$ is said to be *bounded* (*to the right*, respectively, *to the left*) if there is an integer $r \ge 0$ such that $X^p = 0$ for $|p| \ge r$ ($p \ge r$, respectively, $p \le -r$).

Lemma 2.1.6 *Let $(X^\bullet, \alpha^\bullet)$ be a continuously parametrized complex of Banach* spaces on a paracompact space Ω. Suppose that $(X^\bullet, \alpha^\bullet)$ is bounded to the right. Then the following conditions are equivalent:

(i) $(C(\Omega, X^\bullet), \alpha^\bullet)$ is exact;
(ii) $(\mathscr{C}^{X^\bullet}_\Omega, \alpha^\bullet)$ is exact;
(iii) $(X^\bullet, \alpha^\bullet(z))$ is exact for all $z \in \Omega$.

Proof Obviously condition (i) implies condition (ii). According to Corollary 2.1.4, condition (iii) implies condition (i).

Let us suppose that condition (ii) holds and that $X^p = 0$ for $p > n$. Then the complexes $(X^\bullet, \alpha^\bullet(z))$ $(z \in \Omega)$ are exact in degree $p = n$. By the remark following Theorem 2.1.2, for a given point $z \in \Omega$ and a given element $x \in X^{n-1}$ with $\alpha^{n-1}(z)x = 0$, there is a function $f \in C(\Omega, X^{n-1})$ with $\alpha^{n-1}f = 0$ and $f(z) = x$. The proof is completed by descending induction. $\square$

To obtain a corresponding result for continuously parametrized complexes of Fréchet spaces, pointwise exactness of the complex $(X^\bullet, \alpha^\bullet(z))$ is in general not sufficient (see Levy (1987a, Section 1)). But under suitable conditions on the space Ω, it suffices to replace pointwise exactness by local uniform exactness.

Theorem 2.1.7 *Let Ω be a second countable, locally compact Hausdorff space. Consider a parametrized complex $(X^\bullet, \alpha^\bullet)$ of Fréchet spaces on Ω such that $\alpha^p \in C(\Omega, L_b(X^p, X^{p+1}))$ ($p \in \mathbb{Z}$). Suppose that $(X^\bullet, \alpha^\bullet)$ is bounded to the right. Then the following conditions are equivalent:*

(i) $(C(\Omega, X^\bullet), \alpha^\bullet)$ *is exact;*
(ii) $(\mathscr{C}^{X^\bullet}_\Omega, \alpha^\bullet)$ *is exact;*
(iii) $(X^\bullet, \alpha^\bullet)$ *is locally uniformly exact on Ω.*

Proof The equivalence of (i) and (ii) follows from an elementary partition of unity argument. Theorem 2.1.2 shows that condition (iii) implies condition (i).

To complete the proof, let us suppose that (i) holds and that $X^p = 0$ for $p > n$. Each of the spaces $C(\Omega, X^p)$ is a Fréchet space with neighbourhood basis at zero given by the sets of the form

$$W_{K,U} = \{f \in C(\Omega, X^p); f(K) \subset U\},$$

where K runs through all compact subsets of Ω and U runs through all zero neighbourhoods in X^p. In particular, $(C(\Omega, X^\bullet), \alpha^\bullet)$ is a complex of continuous linear operators between Fréchet spaces. Hence the open mapping principle for Fréchet spaces implies that $(X^\bullet, \alpha^\bullet(z))$ is uniformly exact in degree n on each compact subset of Ω.

Let $K \subset \Omega$ be compact, and let U be a zero neighbourhood in X^{n-2}. Again by the open mapping principle, there is a compact set L in Ω and an absolutely convex, open neighbourhood V of zero in X^{n-1} with

$$\alpha^{n-1}W_{K,U} \supset \mathrm{Ker}\, \alpha^{n-1} \cap W_{L,V}.$$

By the remark following Theorem 2.1.2, for each $\lambda \in K$ and for each $x \in \operatorname{Ker} \alpha^{n-1}(\lambda)$, there is a function $f \in C(\Omega, V)$ with

$$\alpha^{n-1}f = 0 \qquad \text{and} \qquad f(\lambda) = x.$$

Hence $(X^\bullet, \alpha^\bullet)$ is uniformly exact in degree $n - 1$ on K.

The proof can be completed by an obvious descending induction. $\qquad\square$

Similar results hold for analytically parametrized complexes on suitable domains in $\mathbb{C}^n$. We begin by considering a local situation.

Theorem 2.1.8 *Let $(X^\bullet, \alpha^\bullet)$ be a bounded, analytically parametrized complex of Banach spaces on an open set Ω in $\mathbb{C}^n$. For a fixed point z in Ω, the following statements are equivalent:*

- (i) *there is an open neighbourhood U of z such that the complex $(\mathscr{O}(U, X^\bullet), \alpha^\bullet)$ is exact;*
- (ii) *there is an open neighbourhood U of z such that, for each open polydisc $P = P(z, r) \subset U$ with radius $r > 0$, the complex $(\mathscr{O}(P, X^\bullet), \alpha^\bullet)$ is exact;*
- (iii) *the complex $(\mathscr{O}_z^{X^\bullet}, \alpha^\bullet)$ is exact;*
- (iv) *the complex $(X^\bullet, \alpha^\bullet(z))$ is exact.*

Proof We first prove the equivalence of (ii), (iii), and (iv).

By Lemma 2.1.5, we know that (iv) implies (ii). Obviously, condition (ii) implies condition (iii). Let us suppose that condition (iii) is satisfied and that p is an integer with $H^{p+1}(X^\bullet, \alpha^\bullet(z)) = 0$. If $x \in X^p$ is an element with $\alpha^p(z)x = 0$, then by the remark following Lemma 2.1.5 applied to the sequence

$$X^p \xrightarrow{\alpha^p} X^{p+1} \xrightarrow{\alpha^{p+1}} X^{p+2},$$

there is an analytic function $f \in \mathscr{O}(V, X^p)$ on a suitable open neighbourhood V of z with $\alpha^p f = 0$ and $f(z) = x$. But then condition (iii) is applicable, and implies, in particular, the existence of a vector $y \in X^{p-1}$ with $\alpha^{p-1}(z)y = x$. Thus (iv) follows inductively.

Finally, we prove the equivalence of (i) and (ii). Obviously, condition (ii) implies condition (i). To prove the converse, we use a result from topological homology theory which will be proved in Chapter 3. Let $(\mathscr{O}(U, X^\bullet), \alpha^\bullet)$ be exact for some open neighbourhood U of z, and let $V \subset U$ be an arbitrary open subset. Corollary 3.1.11 shows that $\mathscr{O}(V) \perp_{\mathscr{O}(U)} \mathscr{O}(U, X^\bullet)$. The remark following Corollary 3.1.16 implies that

$$\left(\mathscr{O}(V) \hat{\otimes}_{\mathscr{O}(U)} \mathscr{O}(U, X^\bullet), 1 \otimes \alpha^\bullet \right)$$

remains exact. But this complex is isomorphic to the complex $(\mathscr{O}(V, X^\bullet), \alpha^\bullet | V)$ via the identifications explained in Section 3.1. $\qquad\square$

The reader should note that the last part of the above proof yields more than is stated in the corresponding part of Theorem 2.1.8. Namely, it shows that the exactness of the complex $(\mathscr{O}(U, X^{\bullet}), \alpha^{\bullet})$ on some open subset U of Ω implies the exactness of $(\mathscr{O}(V, X^{\bullet}), \alpha^{\bullet})$ for each open subset V of U.

Methods from complex analysis and sheaf theory can be used to obtain a global version of the result stated in Theorem 2.1.8. Recall that an open set U in $\mathbb{C}^n$ is called *pseudoconvex* if there is a continuous plurisubharmonic function $u: U \to \mathbb{R}$ such that

$$U_c = \{z \in U; u(z) < c\}$$

is a relatively compact subset of U for each real number c. For the definition and properties of plurisubharmonic functions and pseudoconvex sets we refer the reader to Hörmander (1966), Range (1986), Klimek (1991), or any other standard monograph on several-variable complex analysis. A sheaf $\mathscr{F}$ on a topological space Ω is called *acyclic* if its canonical cohomology groups $H^p(\Omega, \mathscr{F})$ ($p \geq 1$) vanish (see Bredon 1967). Let X be a complex Banach space. Since the X-valued $\bar{\partial}$-sequence

$$0 \to \mathscr{O}(U, X) \xrightarrow{i} C^{\infty}_{0,0}(U, X) \xrightarrow{\bar{\partial}} C^{\infty}_{0,1}(U, X) \xrightarrow{\bar{\partial}} \cdots \xrightarrow{\bar{\partial}} C^{\infty}_{0,n}(U, X) \to 0$$

is exact on each pseudoconvex open set U in $\mathbb{C}^n$ (see Hörmander 1966, Corollary 4.2.6, for the scalar-valued case and Appendix 1 for the vector-valued case), the sheaf $\mathscr{O}^X_{\mathbb{C}^n}$ of germs of analytic X-valued functions on $\mathbb{C}^n$ is acyclic on each pseudoconvex open set in $\mathbb{C}^n$ (see for instance Bredon 1967, Sections II.4 and II.9). Thus the missing implication in the following global version of Theorem 2.1.8 follows from standard sheaf theory (Bredon 1967, Corollary II.4.3).

Corollary 2.1.9   *For a bounded, analytically parametrized complex $(X^{\bullet}, \alpha^{\bullet})$ of Banach spaces on a pseudoconvex open set Ω in $\mathbb{C}^n$ the following statements are equivalent*:

 (i) $(\mathscr{O}(\Omega, X^{\bullet}), \alpha^{\bullet})$ *is exact*;
 (ii) $(\mathscr{O}(U, X^{\bullet}), \alpha^{\bullet})$ *is exact for each open subset U of Ω*;
 (iii) $(\mathscr{O}^{X^{\bullet}}_{\Omega}, \alpha^{\bullet})$ *is exact*;
 (iv) $(X^{\bullet}, \alpha^{\bullet}(z))$ *is exact for all $z \in \Omega$.*   □

In Corollary 2.1.9 the equivalence of conditions (i), (ii), and (iii) remains true for bounded, analytically parametrized complexes of Fréchet spaces, since all the arguments needed to prove these parts remain true in the Fréchet space case.

2.2   THE KOSZUL COMPLEX

Let K be a commutative unital ring, and let $s = (s_1, \ldots, s_n)$ be the canonical basis of K^n over K, where $n \geq 1$ is a given integer. We denote by $\Lambda(s) = \Lambda(K^n)$ the alternating algebra of the free K-module K^n (see Lang (1984, p. 589)). The space $\Lambda(s)$ is a graded K-algebra

$$\Lambda(s) = \bigoplus_{p=0}^{n} \Lambda^p(s),$$

where the submodules $\Lambda^p(s)$ consist of all homogeneous elements of degree p. Each of the spaces $\Lambda^p(s)$ is a free K-module with basis formed by the elements

$$s_{i_1} \wedge \cdots \wedge s_{i_p} \qquad (1 \leq i_1 < \cdots < i_p \leq n).$$

For an arbitrary K-module X we define

$$\Lambda(s, X) = X \otimes_K \Lambda(s), \qquad \Lambda^p(s, X) = X \otimes_K \Lambda^p(s).$$

The elements of $\Lambda^p(s, X)$ are called forms of degree p in indeterminates $s_1, \ldots, s_n$ with coefficients in X. Each such form has a unique representation

$$\sum_{|i|=p} x_i s_i = \sum_{1 \leq i_1 < \cdots < i_p \leq n} x_{i_1 \ldots i_p} s_{i_1} \wedge \cdots \wedge s_{i_p},$$

where the left-hand side should be regarded as an abbreviation of the sum on the right. To simplify the notation we always omit the tensor sign. Suppose that $X \times Y \to Z, (x, y) \mapsto xy$, is a K-bilinear map. Then the *exterior product* $\Lambda(s, X) \times \Lambda(s, Y) \to \Lambda(s, Z)$ denoted by $(\varphi, \psi) \mapsto \varphi \wedge \psi$ is by definition the unique K-bilinear map with

$$\left(x s_{i_1} \wedge \cdots \wedge s_{i_p} \right) \wedge \left(y s_{j_1} \wedge \cdots \wedge s_{j_q} \right) = (xy) s_{i_1} \wedge \cdots \wedge s_{j_q}.$$

If one of the forms φ, ψ is of degree zero, then we write $\varphi\psi$ instead of $\varphi \wedge \psi$.

Suppose that $a = (a_1, \ldots, a_n)$ is a commuting system of morphisms of the K-module X. Then we denote by $K^\bullet(a, X)$ the cochain complex consisting of the spaces $K^p(a, X) = \Lambda^p(s, X)$ and of the K-module homomorphisms $\delta^p \colon K^p(a, X) \to K^{p+1}(a, X)$ given by

$$\delta^p(x) = \left(\sum_{i=1}^{n} a_i s_i \right) \wedge x,$$

where $\wedge$ stands for the exterior product between $\Lambda(s, \operatorname{End}_K(X))$ and $\Lambda(s, X)$. We call $K^\bullet(a, X)$ the *cochain Koszul complex induced by* a. The *Koszul complex* $K_\bullet(a, X)$ *induced by* a is by definition the complex consisting of the K-modules $K_p(a, X) = \Lambda^p(s, X)$ and the K-module homomorphisms $\delta_p \colon K_p(a, X) \to K_{p-1}(a, X)$, given by

$$\delta_p(x s_i) = \sum_{\rho=1}^{p} (-1)^{\rho-1} (a_{i_\rho} x) s_{i_1} \wedge \cdots \wedge \hat{s}_{i_\rho} \wedge \cdots \wedge s_{i_p},$$

as boundaries. The coboundary and boundary maps defined above will sometimes simply be denoted by δ_a or δ. Sometimes we simply write a to refer to the coboundary or boundary induced by the commuting tuple a. We denote by

$$H^p(a, X) = \operatorname{Ker} \delta^p / \operatorname{Im} \delta^{p-1} \qquad \text{and} \qquad H_p(a, X) = \operatorname{Ker} \delta_p / \operatorname{Im} \delta_{p+1}$$

the cohomology, respectively, homology groups of the above complexes.

There is a close relationship between the cochain Koszul complex and the Koszul complex induced by a. For an ordered p-tuple $i = (i_1, \ldots, i_p)$ of integers $1 \le i_1 < \cdots < i_p \le n$, let us denote by $i' = (i'_1, \ldots, i'_{n-p})$ the ordered $(n - p)$-tuple consisting of those integers $1 \le i'_1 < \cdots < i'_{n-p} \le n$ not appearing among the components of i. We write $h^p \colon K_p(a, X) \to K^{n-p}(a, X)$ for the unique isomorphism of K-modules with

$$h^p(xs_i) = (-1)^{|i|-p} xs_{i'} \qquad (x \in X, 1 \le i_1 < \cdots < i_p \le n).$$

Here we use the notation $|i| = i_1 + \cdots + i_p$. Since the relation

$$h^{p-1} \circ \delta_p = \delta^{n-p} \circ h^p$$

holds for all p, the family $h = (h^p)$ induces K-module isomorphisms

$$h^p \colon H_p(a, X) \to H^{n-p}(a, X) \qquad (p \ge 0).$$

Let $a_1, \ldots, a_m, b_1, \ldots, b_n \in \operatorname{End}_K(X)$ be commuting module homomorphisms, and let $a = (a_1, \ldots, a_m)$, $b = (b_1, \ldots, b_n)$, and

$$(a, b) = (a_1, \ldots, a_m, b_1, \ldots, b_n).$$

We denote by $s = (s_1, \ldots, s_m)$, $t = (t_1, \ldots, t_n)$ systems of indeterminates corresponding to the tuples a and b. The (cochain) Koszul complex of the composed system (a, b) can be constructed in a standard way from the (cochain) Koszul complexes of its components a and b. We confine ourselves to the cochain version. Let us denote by $K^{p,q}$ the set of all forms of bidegree (p, q) in (s, t) with coefficients in X, and let us consider the double complex $\mathcal{K} = (K^{p,q})_{p,q \ge 0}$ with rows and columns given by

$$a \colon K^{p,q} \to K^{p+1,q}, \qquad \varphi \mapsto \left(\sum_{i=1}^{m} a_i s_i \right) \wedge \varphi,$$

$$b \colon K^{p,q} \to K^{p,q+1}, \qquad \varphi \mapsto \left(\sum_{i=1}^{n} b_i t_i \right) \wedge \varphi.$$

Modulo the canonical identifications

$$\Lambda^r((s, t), X) = \bigoplus_{p+q=r} K^{p,q} \qquad (r \ge 0),$$

the cochain Koszul complex $K^\bullet((a, b), X)$ coincides with the total complex $K^\bullet = \operatorname{Tot}(\mathcal{K})$ of the double complex $\mathcal{K}$ (see Appendix 2). If $\varphi \in K^r$, then we denote by $\varphi_{p,q} \in K^{p,q}$, where $p + q = r$, the coefficients of φ relative to the above direct sum decomposition. For $r \geq 0$, the maps

$$j: H^r(a, H^0(b, X)) \to H^r((a, b), X),$$

$$[\varphi] \mapsto \left[(\varphi_{p,q})_{p+q=r}\right], \quad \text{where } \varphi_{r,0} = \varphi \text{ and } \varphi_{p,q} = 0 \text{ for } p \neq r,$$

and

$$k: H^{r+n}((a, b), X) \to H^r(a, H^n(b, X)),$$

$$\left[(\varphi_{p,q})_{p+q=r+n}\right] \mapsto [\varphi_{r,n} + \operatorname{Im} b],$$

respectively, are well-defined module homomorphisms. They are usually called the *edge homomorphisms* of the double complex $\mathscr{K}$.

In the following X, Y, and Z will be fixed K-modules.

Definition 2.2.1 A commuting tuple $a = (a_1, \ldots, a_n) \in \operatorname{End}_K(X)^n$ will be called *non-singular* if $H^p(a, X) = 0$ for each p.

Because of the remarks preceding Definition 2.2.1 we could as well have used the Koszul complex instead of the cochain Koszul complex to define the notion of non-singularity. If σ is a permutation of the set $\{1, \ldots, n\}$ and $a_\sigma = (a_{\sigma(1)}, \ldots, a_{\sigma(n)})$, then a is non-singular if and only if a_σ is non-singular. To see this, it suffices to observe that the complexes $K^\bullet(a, X)$ and $K^\bullet(a_\sigma, X)$ are isomorphic via the morphisms $\sigma^p: K^p(a, X) \to K^p(a_\sigma, X)$ defined by

$$\sigma^p\left(xs_{i_1} \wedge \cdots \wedge s_{i_p}\right) = xs_{\sigma^{-1}(i_1)} \wedge \cdots \wedge s_{\sigma^{-1}(i_p)}.$$

Thus the notion of nonsingularity is independent of the order of elements in the system a.

For later use the reader should note that the first and the last cohomology groups of the complex $K^\bullet(a, X)$ are given by

$$H^0(a, X) = \bigcap_{i=1}^n \operatorname{Ker} a_i,$$

$$H^n(a, X) = X / \sum_{i=1}^n a_i X.$$

In the following we summarize some basic results concerning the notion of non-singularity.

Lemma 2.2.2 *Let $a_1, \ldots, a_n, b_1, \ldots, b_m$ be commuting morphisms of the K-module X. Set $a = (a_1, \ldots, a_n)$ and $b = (b_1, \ldots, b_m)$. If a is non-singular, then so is (a, b).*

Proof It suffices to consider the case where $m = 1$. Let $a_1, \ldots, a_{n+1}$ in $\operatorname{End}_K(X)$ be commuting module homomorphisms, and let $a = (a_1, \ldots, a_n)$, $\tilde{a} = (a_1, \ldots, a_{n+1})$. We obtain a short exact sequence

$$0 \to K^\bullet(a, X) \xrightarrow{s} K^\bullet(\tilde{a}, X) \xrightarrow{t} K^\bullet(a, X) \to 0$$

of cochain maps of degree 1 and 0, respectively, by setting

$$s(xs_i) = xs_i \wedge s_{n+1}, \quad t\left(\sum_{1 \le i_1 < \cdots < i_p \le n+1} x_i s_i \right) = \sum_{1 \le i_1 < \cdots < i_p \le n} x_i s_i.$$

The induced long exact sequence of cohomology (Lang 1984, Chapter IV, §2) is of the form

$$0 \longrightarrow H^0(\tilde{a}, X) \xrightarrow{\;i\;} H^0(a, X)$$

$$\xrightarrow{a_{n+1}} H^0(a, X) \xrightarrow{\;s\;} H^1(\tilde{a}, X) \xrightarrow{\;t\;} H^1(a, X)$$

$$\xrightarrow{a_{n+1}} H^1(a, X) \xrightarrow{\;s\;} H^2(\tilde{a}, X) \xrightarrow{\;t\;} \cdots$$

$$\cdots \cdots \cdots \cdots \cdots \cdots \cdots \cdots \cdots \cdots \cdots$$

$$\xrightarrow{a_{n+1}} H^n(a, X) \xrightarrow{\;s\;} H^{n+1}(\tilde{a}, X) \longrightarrow 0,$$

where i denotes the inclusion map and $H^p(a, X) \xrightarrow{a_{n+1}} H^p(a, X)$ is the cohomology map induced by the componentwise action of a_{n+1} on the representing p-forms. Since this sequence is exact, the non-singularity of a implies the non-singularity of $\tilde{a}$.   □

Let X and Y be K-modules. Let $a = (a_1, \ldots, a_n) \in \operatorname{End}_K(X)^n$ and $b = (b_1, \ldots, b_n) \in \operatorname{End}_K(Y)^n$ be commuting tuples of module homomorphisms. A K-module homomorphism $d \colon X \to Y$ is said to *intertwine* the tuples a and b if

$$da_i = b_i d \qquad (i = 1, \ldots, n).$$

The next result allows one to compare the non-singularity of tuples related by a short exact sequence of intertwining module homomorphisms.

Lemma 2.2.3 *Let $0 \to X \xrightarrow{j} Y \xrightarrow{q} Z \to 0$ be a short exact sequence of modules. Let $a \in \operatorname{End}_K(X)^n$, $b \in \operatorname{End}_K(Y)^n$, and $c \in \operatorname{End}_K(Z)^n$ be commuting tuples of module homomorphisms such that j intertwines a and b, and q intertwines b and c. If any two of the tuples a, b, and c are non-singular, then so is the third.*

Proof The componentwise action of j and q induces a short exact sequence

$$0 \to K^{\bullet}(a, X) \xrightarrow{j} K^{\bullet}(b, Y) \xrightarrow{q} K^{\bullet}(c, Z) \to 0$$

of complexes of K-modules. Again the induced long exact cohomology sequence

$$\cdots \to H^p(a, X) \to H^p(b, Y) \to H^p(c, Z)$$
$$\to H^{p+1}(a, X) \to H^{p+1}(b, Y) \to H^{p+1}(c, Z)$$
$$\to \cdots$$

can be used in an obvious way to prove the assertion.   □

If $a = (a_1, \ldots, a_n) \in \operatorname{End}_K(X)^n$ is a commuting system such that there are solutions $b_1, \ldots, b_n$ of the equation $a_1 b_1 + \cdots + a_n b_n = 1_X$ in the commutant

$(a)'$ of $\{a_1, \ldots, a_n\}$, then a is non-singular. This follows as a special application of the next result.

Lemma 2.2.4 *Let* $a = (a_1, \ldots, a_n)$ *be a commuting tuple of morphisms of the module* X. *If* $b_1, \ldots, b_n$ *belong to the commutant of* a *and* $c = a_1 b_1 + \cdots + a_n b_n$, *then* c: $H^p(a, X) \to H^p(a, X)$, *induced by the componentwise action of* c, *is the zero operator for each* p.

Proof Let β_p: $K^p(a, X) \to K^{p-1}(a, X)$ be the module homomorphism defined by

$$\beta_p\big(xs_{i_1} \wedge \cdots \wedge s_{i_p}\big) = \sum_{\rho=1}^{p} (-1)^{\rho-1} b_{i_\rho} x s_{i_1} \wedge \cdots \wedge \hat{s}_{i_\rho} \wedge \cdots \wedge s_{i_p}.$$

Then $(\delta_a^{p-1}\beta_p + \beta_{p+1}\delta_a^p)(xs_i) = cxs_i$ holds for each tuple i of strictly increasing integers $1 \le i_1 < \cdots < i_p \le n$ and each $x \in X$. Hence the assertion follows. $\square$

2.3 CAUCHY–WEIL SYSTEMS

In the following K will be a commutative, unital ring and X, Y, and Z will be modules over K.

Definition 2.3.1 Let u^0: $X \to Y$ be a morphism of K-modules, and let (u_{ij}) be an $m \times n$ matrix of commuting elements in $\mathrm{End}_K(Y)$. We denote by $s = (s_1, \ldots, s_n)$ and $t = (t_1, \ldots, t_m)$ systems of indeterminates, and define

$$u(s_j) = u_{1j}t_1 + \cdots + u_{mj}t_m \in \Lambda^1(t, \mathrm{End}_K(Y))$$

for $j = 1, \ldots, n$. The *special transformation* determined by u^0 and the matrix (u_{ij}) is the morphism u: $\Lambda(s, X) \to \Lambda(t, Y)$ (of degree 0) of graded K-modules defined by

$$u\Big(\sum_{1 \le i_1 < \cdots < i_p \le n} x_i s_i \Big) = \sum_{1 \le i_1 < \cdots < i_p \le n} u^0(x_i) u(s_{i_1}) \wedge \cdots \wedge u(s_{i_p}). \quad (2.3.1)$$

In the above definition, we regard Y as a right $\mathrm{End}_K(Y)$-module via $y \cdot v = v(y)$ for $y \in Y$ and $v \in \mathrm{End}_K(Y)$. Since the coefficients of the matrix (u_{ij}) were assumed to be commuting, there is no associativity problem in defining the right-hand side of (2.3.1). By the definition of exterior products, we obtain the following useful formula:

$$u\big(xs_{j_1} \wedge \cdots \wedge s_{j_p}\big) = u^0(x) u(s_{j_1}) \wedge \cdots \wedge u(s_{j_p})$$

$$= \sum_{i_1, \ldots, i_p = 1}^{m} u_{i_1 j_1} \cdots u_{i_p j_p} u^0(x) t_{i_1} \wedge \cdots \wedge t_{i_p}$$

$$= \sum_{1 \le i_1 < \cdots < i_p \le m} \big(\det(u_{i_\mu j_\nu})_{1 \le \mu, \nu \le p} u^0(x) \big) t_{i_1} \wedge \cdots \wedge t_{i_p}.$$

Lemma 2.3.2   *Let $u: \Lambda(s, X) \to \Lambda(t, Y)$ be as in the preceding definition. Let $a \in \operatorname{End}_K(X)^n$ and $b \in \operatorname{End}_K(Y)^m$ be tuples of commuting morphisms of X and Y, respectively. Suppose that the components of b commute with u_{ij} for all i, j and that the diagram*

$$
\begin{array}{ccc}
X & \xrightarrow{\ \delta_a\ } & \Lambda^1(s, X) \\
{\scriptstyle u^0}\downarrow & & \downarrow{\scriptstyle u} \\
Y & \xrightarrow{\ \delta_b\ } & \Lambda^1(t, Y)
\end{array}
$$

commutes. Then u is a cochain map from $K^\bullet(a, X)$ into $K^\bullet(b, Y)$.

Proof   For $x \in X$ and $1 \le j_1 < \cdots < j_p \le n$, we obtain

$$
\begin{aligned}
u \circ \delta_a\big(x s_{j_1} \wedge \cdots \wedge s_{j_p}\big) &= \sum_{j=1}^n u^0(a_j x) u(s_j) \wedge u(s_{j_1}) \wedge \cdots \wedge u(s_{j_p}) \\
&= (\delta_b \circ u^0(x)) \wedge u(s_{j_1}) \wedge \cdots \wedge u(s_{j_p}) \\
&= \left(\left(\sum_{i=1}^m b_i t_i\right) u^0(x)\right) \wedge u(s_{j_1}) \wedge \cdots \wedge u(s_{j_p}) \\
&= \left(\sum_{i=1}^m b_i t_i\right) \wedge \big(u^0(x) u(s_{j_1}) \wedge \cdots \wedge u(s_{j_p})\big) \\
&= \delta_b \circ u\big(x s_{j_1} \wedge \cdots \wedge s_{j_p}\big).
\end{aligned}
$$

Here we have used the assumption that the components of b commute with the coefficients of the matrix (u_{ij}). □

The central notion in Taylor's construction of an analytic functional calculus for several commuting operators is that of a Cauchy–Weil system. We use a slightly modified version, which was introduced by Frunză (1975b).

Let $X_0 \subset X$ be a submodule of a given K-module X. To simplify the notation, we shall denote by $\operatorname{End}_K(X \mid X_0)$ the set of all morphisms u of the K-module X with the property that $u(X_0) \subset X_0$.

Definition 2.3.3   Suppose that $X_0 \subset X$ is a submodule of the K-module X. Let $a_1, \ldots, a_n, d_1, \ldots, d_m \in \operatorname{End}_K(X \mid X_0)$ be commuting module homomorphisms. Set $a = (a_1 \cdots a_n)$ and $d = (d_1, \ldots, d_m)$. Then (X, X_0, a, d) is a *Cauchy–Weil system* if the tuple (d, a) is non-singular on the quotient module X/X_0.

Suppose that (X, X_0, a, d) is a Cauchy–Weil system as above. Our next aim is to construct a family $R_a: H^p(d, X) \to H^{p+n}(d, X_0)$ $(p \ge 0)$ of module homomorphisms. For this purpose, we fix systems of indeterminates $s = (s_1, \ldots, s_n)$ and $t = (t_1, \ldots, t_m)$ corresponding to the tuples a and d, respectively. We define a cochain map $s: K^\bullet(d, X) \to K^\bullet((d, a), X)$ of degree n

by setting $s(\varphi) = \varphi \wedge s_1 \wedge \cdots \wedge s_n$ for $\varphi \in \Lambda(t, X)$. Using the long exact sequence of cohomology corresponding to the short exact sequence

$$0 \to X_0 \xrightarrow{\ i\ } X \xrightarrow{\ q\ } X/X_0 \to 0,$$

one sees that the inclusion map $i\colon X_0 \to X$ induces module isomorphisms $i\colon H^p((d, a), X_0) \to H^p((d, a), X)$. Finally, let $\pi\colon \Lambda((t, s), X_0) \to \Lambda(t, X_0)$ be the canonical projection, that is, the special transformation determined by $\pi^0 = 1_{X_0}$ and the $m \times (m + n)$ matrix (π_{ij}) with $\pi_{ij} = 1_{X_0}$ if $i = j$ and $\pi_{ij} = 0$ otherwise. Then $\pi\colon K^\bullet((d, a), X_0) \to K^\bullet(d, X_0)$ is a cochain map by Lemma 2.3.2.

Definition 2.3.4 Let (X, X_0, a, d) be a Cauchy–Weil system. Then the maps $R_a\colon H^p(d, X) \to H^{p+n}(d, X_0)$ defined for $p \geq 0$ as the composition of the module homomorphisms

$$H^p(d, X) \xrightarrow{\ s\ } H^{p+n}((d, a), X) \xrightarrow{\ i^{-1}\ } H^{p+n}((d, a), X_0) \xrightarrow{\ \pi\ } H^{p+n}(d, X_0)$$

are called the *Cauchy–Weil maps relative to the system* (X, X_0, a, d).

In the following we summarize some basic properties of Cauchy–Weil maps. First, we derive a criterion for the compatibility of the Cauchy–Weil maps with a given special transformation.

Proposition 2.3.5 *Let* (X, X_0, a, d) *and* (X', X_0', a', d') *be Cauchy–Weil systems with* $a = (a_1, \ldots, a_n)$, $d = (d_1, \ldots, d_m)$, *and* $a' = (a_1', \ldots, a_n')$, $d' = (d_1', \ldots, d_m')$. *We denote by* s, t, s' *and* t' *the corresponding tuples of indeterminates. Consider a special transformation* $u\colon \Lambda(t, X) \to \Lambda(t', X')$ *determined by a morphism* $u^0\colon X \to X'$ *and a commuting matrix* $(u_{ij}) \in \mathrm{End}_K(X' \mid X_0')^{(m', m)}$. *Suppose that* u *is a cochain map from* $K^\bullet(d, X)$ *into* $K^\bullet(d', X')$ *such that:*

(i) *the components of* (d', a') *commute with the coefficients of* (u_{ij});
(ii) $u^0(X_0) \subset X_0'$;
(iii) $u^0 a_i = a_i' u^0$ *for all* i.

Then for each $p \geq 0$, *the diagram*

$$
\begin{array}{ccc}
H^p(d, X) & \xrightarrow{\ R_a\ } & H^{p+n}(d, X_0) \\
\big\downarrow{\scriptstyle u} & & \big\downarrow{\scriptstyle u} \\
H^p(d', X') & \xrightarrow{\ R_{a'}\ } & H^{p+n}(d', X_0')
\end{array}
$$

commutes.

Proof Let $v\colon \Lambda((t, s), X) \to \Lambda((t', s'), X)$ be the special transformation with

$v^0 = u^0$, $v(t_j) = u(t_j)$ for $1 \leq j \leq m$ and $v(s_j) = s'_j$ for $1 \leq j \leq n$. Then the assertion follows from the observation that

$$K^\bullet(d, X) \xrightarrow{\ s\ } K^\bullet((d, a), X) \xleftarrow{\ i\ } K^\bullet((d, a), X_0) \xrightarrow{\ \pi\ } K^\bullet(d, X_0)$$

with vertical maps u, v, v, u respectively, to

$$K^\bullet(d', X') \xrightarrow{\ s'\ } K^\bullet((d', a'), X') \xleftarrow{\ i\ } K^\bullet((d', a'), X'_0) \xrightarrow{\ \pi\ } K^\bullet(d', X'_0)$$

is a commuting diagram of cochain maps.   □

Next we study what happens to the Cauchy–Weil maps if in a Cauchy–Weil system (X, X_0, a, d) the tuple a is replaced by its image under a matrix transformation.

Proposition 2.3.6 *Consider Cauchy–Weil systems (X, X_0, a, d) and (X, X_0, b, d), where a and b are tuples of length n and d is a tuple of length m. Suppose that there is a matrix $(u_{ij}) \in \operatorname{End}_K(X \mid X_0)^{(n, n)}$ such that the coefficients u_{ij} commute with one another and with the components of (a, d) and such that*

$$b_i = \sum_{j=1}^n u_{ij} a_j \qquad (1 \leq i \leq n).$$

Then for each $p \geq 0$ the componentwise action of the morphism $\det(u_{ij})$ induces a cohomology map $\det(u_{ij})$: $H^p(d, X) \to H^p(d, X)$ such that

$$R_b \circ \det(u_{ij}) = R_a.$$

Proof Let v: $\Lambda((t, s), X) \to \Lambda((t, s), X)$ be the special transformation with $v^0 = 1_X$ and $v(t_j) = t_j$, $v(s_j) = \sum_{i=1}^n u_{ij} s_i$ for $1 \leq j \leq n$. We write v_0 for the corresponding map obtained by the same definition, but with X replaced by X_0. The proof follows from the observation that

$$K^\bullet(d, X) \xrightarrow{\ s\ } K^\bullet((d, a), X) \xleftarrow{\ i\ } K^\bullet((d, a), X_0) \xrightarrow{\ \pi\ } K^\bullet(d, X_0)$$

with vertical maps $\det(u_{ij})$, v, v_0, id respectively, to

$$K^\bullet(d, X) \xrightarrow{\ s\ } K^\bullet((d, b), X) \xleftarrow{\ i\ } K^\bullet((d, b), X_0) \xrightarrow{\ \pi\ } K^\bullet(d, X_0)$$

is a commutative diagram of cochain maps.   □

Combining Lemma 2.2.4 with the definition of the Cauchy–Weil maps, we obtain the following result.

Proposition 2.3.7 *Suppose that (X, X_0, a, d) is a Cauchy–Weil system, and that $c \in \operatorname{End}_K(X)$ belongs to the ideal generated by the components of (d, a) in the commutant of (d, a) (formed in $\operatorname{End}_K(X)$). Then, for each $p \geq 0$, the composition*

$$H^p(d, X) \xrightarrow{\ c\ } H^p(d, X) \xrightarrow{\ R_a\ } H^{p+n}(d, X_0),$$

where the first operator is the one induced by the componentwise action of c, is the zero operator.

Proof The assertion follows from Lemma 2.2.4 and from the fact that the diagram

$$
\begin{array}{ccc}
H^p(d,X) & \xrightarrow{\ c\ } & H^p(d,X) \\
{\scriptstyle s}\downarrow & & \downarrow{\scriptstyle s} \\
H^{p+n}((d,a),X) & \xrightarrow{\ c\ } & H^{p+n}((d,a),X)
\end{array}
$$

commutes. $\square$

Roughly speaking, the next result shows that the Cauchy–Weil maps $R_{(a,b)}$ with respect to a composed system (a,b) can be computed as the composition of the Cauchy–Weil maps of the single tuples a and b.

Theorem 2.3.8 *Let $X_0 \subset X_1$ be submodules of the K-module X, and let $a_1,\ldots,a_n$, $b_1,\ldots,b_{n'}$, $d_1,\ldots,d_m$ be commuting morphisms of X that leave X_0 and X_1 invariant. Set $a=(a_1,\ldots,a_n)$, $b=(b_1,\ldots,b_{n'})$, and $d=(d_1,\ldots,d_m)$. Suppose that (X,X_1,a,d) and (X_1,X_0,b,d) are Cauchy–Weil systems. Then $(X,X_0,(a,b),d)$ is a Cauchy–Weil system and $R_{(a,b)}=R_b \circ R_a$ holds on $H^p(d,X)$ for each p.*

Proof Since the tuple (d,a,b) is non-singular on the first and the last space of the canonical short exact sequence

$$
0 \to X_1/X_0 \to X/X_0 \to X/X_1 \to 0,
$$

it is non-singular on X/X_0 (Lemma 2.2.2 and Lemma 2.2.3). Hence $(X,X_0,(a,b),d)$ is a Cauchy–Weil system.

Let s, s', and t be tuples of indeterminates associated with the systems a, b, and d, respectively. The identity $R_{(a,b)}=R_b \circ R_a$ follows from the commutativity of the cochain diagram

$$
\begin{array}{ccccccccc}
K(d,X) & \xrightarrow{\ s\wedge s'\ } & K((d,c),X) & \xleftarrow{\ i\ } & K((d,c),X_1) & \xleftarrow{\ i\ } & K((d,c),X_0) & \xrightarrow{\ \pi\ } & K(d,X_0) \\
{\scriptstyle s}\downarrow & \nearrow{\scriptstyle s'} & & \nearrow{\scriptstyle s'} & & \searrow{\scriptstyle \pi_s} & & \searrow{\scriptstyle \pi_s} & \uparrow{\scriptstyle \pi_{s'}} \\
K((d,a),X) & \xleftarrow{\ i\ } & K((d,a),X_1) & \xrightarrow{\ \pi_s\ } & K(d,X_1) & \xrightarrow{\ s'\ } & K((d,b),X_1) & \xleftarrow{\ i\ } & K((d,b),X_0).
\end{array}
$$

To abbreviate the notation we have written K instead of $K^\bullet$ and c instead of (a,b). $\square$

Let (X,X_0,a,d) and $(X,X_0,a,(b,d))$ be Cauchy–Weil systems, where a, b, and d are tuples of length n, m, and n, respectively. We denote by s, s', and t corresponding systems of indeterminates, and we set $c=(d,a)$. Let

$$
\mathscr{K}=\mathscr{K}(b,d), \quad \mathscr{L}=\mathscr{K}(b,c), \quad \mathscr{L}_0=\mathscr{K}(b|X_0,c|X_0), \quad \mathscr{K}_0=\mathscr{K}(b|X_0,d|X_0)
$$

be the double complexes induced by the indicated pairs (cf. the beginning of Section 2.2). The maps

$$
K^{p,q} \xrightarrow{\ s\ } L^{p,q+n}, \qquad L_0^{p,q} \xrightarrow{\ i\ } L^{p,q}, \qquad L_0^{p,q} \xrightarrow{\ \pi\ } K_0^{p,q},
$$

where $s(\varphi) = \varphi \wedge s_1 \wedge \cdots \wedge s_n$, the map i is induced by the inclusion $X_0 \to X$, and π is the restriction of the canonical projection $\Lambda((s', t, s), X_0) \to \Lambda((s', t), X_0)$, determine morphisms of double complexes

$$\mathscr{K} \xrightarrow{s} \mathscr{L} \xleftarrow{i} \mathscr{L}_0 \xrightarrow{\pi} \mathscr{K}_0,$$

that is, they commute with the row and column differentials in the double complexes $\mathscr{K}, \mathscr{L}, \mathscr{L}_0, \mathscr{K}_0$, respectively. The induced morphisms of the corresponding total complexes.

$$K^\bullet((b, d), X) \xrightarrow{s} K^\bullet((b, d, a), X) \xleftarrow{i} K^\bullet((b, d, a), X_0)$$
$$\xrightarrow{\pi} K^\bullet((b, d), X_0),$$

given by the componentwise actions of s, i and π, are precisely the maps occurring in the definition of the Cauchy–Weils maps

$$R_a: H^p((b, d), X) \to H^{p+n}((b, d), X_0) \qquad (p \geq 0). \qquad (2.3.2)$$

We consider the second homology groups

$$''H^{p,q}(\mathscr{K}) = \mathrm{Ker}(K^{p,q} \xrightarrow{d} K^{p,q+1})/\mathrm{Im}(K^{p,q-1} \xrightarrow{d} K^{p,q})$$

of the double complex $\mathscr{K}$, and use the corresponding notations for the double complexes $\mathscr{K}_0, \mathscr{L}, \mathscr{L}_0$. The induced complexes $(''H^{\bullet,0}(\mathscr{K}), b)$ and $(''H^{\bullet,n}(\mathscr{K}_0), b)$ are isomorphic to the cochain Koszul complexes $K^\bullet(b, H^0(d, X))$ and $K^\bullet(b, H^n(d, X_0))$, respectively. The composition

$$''H^{p,0}(\mathscr{K}) \xrightarrow{s} ''H^{p,n}(\mathscr{L}) \xrightarrow{i^{-1}} ''H^{p,n}(\mathscr{L}_0) \xrightarrow{\pi} ''H^{p,n}(\mathscr{K}_0)$$

can be identified with the morphism

$$\Lambda^p(s', H^0(d, X)) \xrightarrow{\oplus R_a} \Lambda^p(s', H^n(d, X_0)),$$

where $R_a: H^0(d, X) \to H^n(d, X_0)$ is the Cauchy–Weil map with respect to the Cauchy–Weil system (X, X_0, a, d). In particular, the componentwise action of R_a yields a morphism $R_a: K^\bullet(b, H^0(d, X)) \to K^\bullet(b, H^n(d, X_0))$ of complexes of K-modules. We denote by $H(R_a)$ the corresponding map between the cohomology groups of both complexes.

Proposition 2.3.9 *The diagram*

$$
\begin{array}{ccc}
H^p(b, H^0(d, X)) & \xrightarrow{H(R_a)} & H^p(b, H^n(d, X_0)) \\
\downarrow{\scriptstyle j} & & \uparrow{\scriptstyle k} \\
H^p((b, d), X) & \xrightarrow{R_a} & H^{p+n}((b, d), X_0),
\end{array}
$$

where j and k are the edge homomorphisms of the double complexes $\mathscr{K}$ and $\mathscr{K}_0$, respectively, is commutative.

Proof We fix an arbitrary element $[x] = [\sum_{|j|=p} x_j s'_j]$ in $H^p(b, H^0(d, X))$. If $y = (y_{r,s})_{r+s=p+n} \in L^{p+n}$ is defined by

$$y_{p,n} = sx \qquad \text{and} \qquad y_{r,s} = 0 \quad \text{for } r \neq p,$$

then $s \circ j[x] = [y]$ (for the definition of j and k see Section 2.2).

Let us denote by $L^\bullet, L_0^\bullet$ the total complexes of the double complexes $\mathscr{L}$ and $\mathscr{L}_0$, respectively. We consider in detail the problem of finding an element $[z] \in H^{p+n}(L_0^\bullet)$ with

$$i[z] = [y]. \tag{2.3.3}$$

By assumption the last complex in the short exact sequence

$$0 \to K^\bullet((b,c), X_0) \xrightarrow{i} K^\bullet((b,c), X) \xrightarrow{q} K^\bullet((b,c), X/X_0) \to 0$$

is exact. Hence there is an element $u = (u_{r,s})$ in L^{p+n-1} with

$$\delta_{(b,c)}(qu) = qy. \tag{2.3.4}$$

But then $z = y - \delta_{(b,c)}u \in L_0^{p+n}$ solves equation (2.3.3).

Let us describe a way of constructing a solution u of equation (2.3.4). For this purpose, we regard $K^\bullet(X/X_0, (b,c))$ as the total complex of the double complex $\mathscr{M}$ composed of the complexes $K^\bullet(X/X_0, b)$ and $K^\bullet(X/X_0, c)$. We write ∂' and ∂'' for the row and column differentials of $\mathscr{M}$. Since the columns of $\mathscr{M}$ are exact, for each element $\varphi \in L^{p,n-1}$ with $\partial''(q\varphi) = q(sx)$, there is a solution u of (2.3.4) with $u_{p,n-1} = \varphi$ and $u_{r,s} = 0$ for $r < p$. Recall that, up to the sign $(-1)^p$, the pth column of the double complex $\mathscr{M}$ is the direct sum

$$\bigoplus_{1 \le j_1 < \cdots < j_p \le m} K^\bullet(X/X_0, c).$$

Since (X, X_0, a, d) is a Cauchy–Weil system, for $1 \le j_1 < \cdots < j_p \le m$, there are forms $v_j \in \Lambda^n((t,s), X_0)$, $w_j \in \Lambda^{n-1}((t,s), X)$ with

$$sx_j - iv_j = \delta_c w_j.$$

Via the identifications $L_0^{p,n} = \oplus_{|j|=p} \Lambda^n((t,s), X_0)$ and

$$L^{p,n-1} = \oplus_{|j|=p} \Lambda^{n-1}((t,s), X),$$

we regard $v = (v_j)_{|j|=p}$ and $w = (w_j)_{|j|=p}$ as elements of $L_0^{p,n}$ and $L^{p,n-1}$, respectively. Obviously, $\varphi = (-1)^p w$ is a solution of the equation

$$\partial''(q\varphi) = q(sx).$$

Therefore we can choose a solution $u \in L^{p+n-1}$ of (2.3.4) with

$$u_{p,n-1} = \varphi, \qquad u_{r,s} = 0 \quad \text{for } r < p.$$

To complete the proof, it suffices to observe that both $H(R_a)[x]$ and $k \circ R_a \circ j[x]$ are represented by the form $\sum_{|j|=p}[\pi v_j]s_j'$, where π is the canonical projection from $\Lambda((t,s), X_0)$ onto $\Lambda(t, X_0)$. $\qquad\square$

The following is an application of Proposition 2.3.9 which will be needed in the sequel.

Corollary 2.3.10   Let $a_1, \ldots, a_n, b_1, \ldots, b_n, d_1, \ldots, d_n \in \mathrm{End}_K(X \mid X_0)$ be commuting module homomorphisms, and let $a = (a_1, \ldots, a_n)$, $b = (b_1, \ldots, b_n)$, $d = (d_1, \ldots, d_n)$, and $a - b = (a_1 - b_1, \ldots, a_n - b_n)$. Suppose that (X, X_0, a, d) and (X, X_0, b, d) are Cauchy–Weil systems. Then the maps R_a and R_b induce identical module homomorphisms

$$H^p(a - b, H^0(d, X)) \xrightarrow{\;H(R_a) = H(R_b)\;} H^p(a - b, H^n(d, X_0))$$

for each $p \geq 0$.

Proof   By Proposition 2.3.9 it suffices to show that the Cauchy–Weil maps R_a and R_b formed with respect to

$$(X, X_0, a, (a - b, d)) \quad \text{and} \quad (X, X_0, b, (a - b, d)),$$

coincide as maps from $H^p((a - b, d), X)$ into $H^{p+n}((a - b, d), X_0)$.

Let $u \colon K^\bullet((a - b, d, a), X) \to K^\bullet((a - b, d, b), X)$ be the morphism of complexes given by the special transformation defined by

$$u^0 = 1_x, \quad u(s_i^{a-b}) = s_i^{a-b} - s_i^b, \quad u(s_i^d) = s_i^d, \quad u(s_i^a) = s_i^b,$$

where the upper index indicates the system to which the indeterminates are associated. To conclude the proof it suffices to check that

$$\begin{array}{ccccccc}
K^\bullet((a-b,d),X) & \xrightarrow{\;s\;} & K^\bullet((a-b,d,a),X) & \xleftarrow{\;i\;} & K^\bullet((a-b,d,a),X_0) & \xrightarrow{\;\pi\;} & K^\bullet((a-b,d),X_0) \\
\downarrow{\scriptstyle id} & & \downarrow{\scriptstyle u} & & \downarrow{\scriptstyle u} & & \downarrow{\scriptstyle id} \\
K^\bullet((a-b,d),X) & \xrightarrow{\;s\;} & K^\bullet((a-b,d,b),X) & \xleftarrow{\;i\;} & K^\bullet((a-b,d,b),X_0) & \xrightarrow{\;\pi\;} & K^\bullet((a-b,d),X_0)
\end{array}$$

is a commutative diagram of cochain maps. □

2.4   THE CAUCHY–WEIL INTEGRAL

Let X be a Fréchet space, and let $L(X)$ be the algebra of all continuous linear operators on X. We now equip $L(X)$ with the locally convex topology given by the uniform convergence on all bounded subsets of X. Let $U \subset \mathbb{C}^n$ be open, and let $a = (a_1, \ldots, a_n) \in \mathcal{O}(U, L(X))^n$ be a tuple of commuting operator functions, that is, suppose that $a_i(z) a_j(z) = a_j(z) a_i(z)$ for all $z \in U$ and all $i, j = 1, \ldots, n$. We denote by $\mathcal{E}(U, X)$ the space of all X-valued C^∞-functions on U and by $\mathcal{D}(U, X)$ its subspace consisting of all functions $f \in \mathcal{E}(U, X)$ with compact support in U. We shall show that, if the tuple a is in a suitable sense non-singular outside a compact subset of U, then $(\mathcal{E}(U, X), \mathcal{D}(U, X), a, \bar{\partial})$ is a Cauchy–Weil system. Here the components of a are regarded as multiplication operators on $\mathcal{E}(U, X)$ and $\bar{\partial}$ is the commuting tuple with components

$$\bar{\partial}_i \colon \mathcal{E}(U, X) \to \mathcal{E}(U, X), \qquad f \to (\partial / \partial \bar{z}_i) f.$$

We use the Cauchy–Weil map

$$R_a \colon \mathcal{O}(U, X) \to H^n(\bar{\partial}, \mathcal{D}(U, X))$$

relative to this system to define the Cauchy–Weil integral. More precisely, if $f \in \mathcal{O}(U, X)$ and if $R_a(f) \in H^n(\bar{\partial}, \mathcal{D}(U, X))$ is represented by an n-form with coefficient $g \in \mathcal{D}(U, X)$, then we set

$$\int R_a(f) = C_n \int g \, dz,$$

where the integral is the Lebesgue integral and C_n is a suitable constant. The fact that this definition is independent of the choice of g follows as an easy application of Stokes' theorem.

To deduce the basic properties of the Cauchy–Weil integral, we study a concept that is slightly more general. Let $U \subset \mathbb{C}^{n+m}$ be an open set. We write the elements of $\mathbb{C}^{n+m}$ in the form (z, w) with $z \in \mathbb{C}^n$, $w \in \mathbb{C}^m$, we denote by $\pi: \mathbb{C}^{n+m} \to \mathbb{C}^m$ the projection of $\mathbb{C}^{n+m}$ onto its last m coordinates, and we set $W = \pi U$.

Definition 2.4.1   A subset σ of U is called *z-compact* if it is closed in U and if $\sigma \cap (\mathbb{C}^n \times L)$ is compact for each compact subset L of W.

We shall need the following elementary properties of z-compact sets.

Lemma 2.4.2   *Let U be an open subset of $\mathbb{C}^{n+m}$.*

(a) *Suppose that $\sigma \subset U$ is z-compact. Then there is an open neighbourhood V of σ in U such that the closure of V in U is z-compact.*

(b) *Suppose that $f \in \mathcal{E}(U \setminus \sigma, X)$ for some z-compact subset σ of U. Then there is a larger z-compact set σ_1 in U and a function $g \in \mathcal{E}(U, X)$ with $f = g$ on $U \setminus \sigma_1$.*

Proof   (a) By using an exhaustion of W by relatively compact open subsets, one can construct a countable family $(V_i)_{i \geq 1}$ of open, relatively compact subsets of U such that $\sigma \subset \bigcup_{i \geq 1} V_i$ and such that, for each compact set L in W, we have

$$V_i \cap (\mathbb{C}^n \times L) = \varnothing$$

for almost all i. Set $V = \bigcup_{i \geq 1} V_i$. Then, for each compact set L in W, there is an index k with

$$\bar{V}^U \cap (\mathbb{C}^n \times L) \subset \bigcup_{i=1}^{k} \bar{V}_i.$$

(b) We choose open subsets V, V_1 of U with z-compact closure in U such that $\sigma \subset V \subset \bar{V}^U \subset V_1$. Let $(\theta_i)_{i \in I}$ be a locally finite C^∞-partition of unity subordinate to the open cover $U = V_1 \cup (U \setminus \bar{V}^U)$. Set $J = \{i \in I; \operatorname{supp}(\theta_i) \subset U \setminus \bar{V}^U\}$. Then $\theta = \sum_{i \in J} \theta_i \in \mathcal{E}(U)$ is a function with $\theta = 1$ on $U \setminus V_1$ and with $\theta = 0$ on V. Hence the desired function g can be defined by

$$g = \theta f \quad \text{on } U \setminus \sigma, \qquad g = 0 \quad \text{on } V. \qquad \square$$

We denote by $\mathscr{E}_z(U, X)$ the set of all functions $f \in \mathscr{E}(U, X)$ such that the support of f formed in U is a z-compact subset of U. The set

$$\mathscr{V} = \{V \subset U \text{ open}; \quad U \setminus V \text{ is } z\text{-compact in } U\}$$

is obviously directed downwards by inclusion.

Definition 2.4.3 Let $n, N \geq 1$ and $m \geq 0$ be natural numbers. Let $U \subset \mathbb{C}^{n+m}$ be open, and let $a = (a_1, \ldots, a_N) \in \mathscr{O}(U, L(X))^N$ be a tuple of commuting analytic operator functions. The *singularity set* σ_a of a is the complement in U of the largest open subset V of U with the property that $K^\bullet(a, \mathscr{O}_V^X)$ is exact.

The reader should note that the cases $\sigma_a = \varnothing$ or $\sigma_a = U$ are not excluded. Since the sheaf $\mathscr{O}_{\mathbb{C}^{n+m}}^X$ is acyclic on each pseudoconvex open set D in $\mathbb{C}^{n+m}$ (cf. the end of Section 2.1), it follows that, in the setting of Definition 2.4.3, the complex $K^\bullet(a, \mathscr{O}(D, X))$ is exact on each pseudoconvex open subset D of $U \setminus \sigma_a$ (Corollary II.4.3 in Bredon 1967).

Let us fix natural numbers $n, N, m \geq 1$, and a commuting tuple $a = (a_1, \ldots, a_N) \in \mathscr{O}(U, L(X))^N$ of analytic operator functions. We choose an arbitrary commutative ring $K \subset \mathscr{O}(U, L(X))$ containing the constant function 1_X and having the property that all elements of K commute with $a_1, \ldots, a_N$. We identify a with the induced tuple of multiplication operators $\mathscr{E}(U, X) \to \mathscr{E}(U, X)$, $f \mapsto a_i f$, and obtain in this way a commuting tuple of K-module homomorphisms. As before, we write the elements of $\mathbb{C}^{n+m}$ in the form (z, w), where $z \in \mathbb{C}^n$ and $w \in \mathbb{C}^m$. We define $\bar{\partial} = (\bar{\partial}_{z_1}, \ldots, \bar{\partial}_{z_n}, \bar{\partial}_{w_1}, \ldots, \bar{\partial}_{w_m})$ and use $d\bar{z} \oplus d\bar{w} = (d\bar{z}_1, \ldots, d\bar{z}_n, d\bar{w}_1, \ldots, d\bar{w}_m)$ as the associated tuple of indeterminates.

Proposition 2.4.4 *Suppose that $\sigma_a \subset U$ is z-compact. Then*

$$\left(\mathscr{E}(U, X), \mathscr{E}_z(U, X), a, \bar{\partial} \right)$$

is a Cauchy–Weil system of K-modules.

Proof The map

$$\mathscr{E}(U, X)/\mathscr{E}_z(U, X) \to \varinjlim_{V \in \mathscr{V}} \mathscr{E}(V, X), \quad [f] \mapsto [(f, U)]$$

is an isomorphism of K-modules. Note that the surjectivity of this map follows from Lemma 2.4.2. Since inductive limits preserve exactness, it is sufficient to prove that for each open set $V \subset U \setminus \sigma_a$, the cochain complex $K^\bullet(\bar{\partial} \oplus a, \mathscr{E}(V, X))$ is exact. Because the sheaf $\mathscr{E}_V^X$ of germs of X-valued C^∞-functions on V is acyclic (Section II.9 in Bredon 1967), we only have to check that the complex $K^\bullet(\bar{\partial} \oplus a, \mathscr{E}_V^X)$ is exact. But for each pseudoconvex open set $D \subset V$ the $\bar{\partial}$-sequence

$$0 \to \mathscr{O}(D, X) \to K^\bullet(\bar{\partial}, \mathscr{E}(D, X))$$

is exact, and hence (see Lemma A2.6)

$$H^i(\bar{\partial} \oplus a, \mathscr{E}(D, X)) \cong H^i(a, \mathscr{O}(D, X)) = 0 \qquad (i \geq 0).$$

Here we have used the remark following Definition 2.4.3. $\qquad\square$

In exactly the same way, one obtains a corresponding result for analytic operator functions with compact singularity sets.

Lemma 2.4.5 *Let $U \subset \mathbb{C}^n$ be open, and let $a \in \mathscr{O}(U, L(X))^N$ be a tuple of commuting operator functions on U. Suppose that $\sigma_a \subset U$ is compact. Then $(\mathscr{E}(U, X), \mathscr{D}(U, X), a, \bar{\partial})$ is a Cauchy–Weil system.* $\qquad\square$

As before, let $U \subset \mathbb{C}^{n+m}$ be open, and let $W = \pi U$. We define

$$\mathscr{W} = \{V \subset W \text{ open}; \ W \backslash V \text{ is compact}\}.$$

If $f \in \mathscr{O}(W, L(X))$, then the function $U \to L(X), (z, w) \mapsto f(w)$ is analytic on U, and we can consider the induced multiplication operator on $\mathscr{E}(U, X)$.

Before we give the definition of the Cauchy–Weil integral, we describe a third example of a Cauchy–Weil system.

Proposition 2.4.6 *Let $b \in \mathscr{O}(W, L(X))^M$ be a tuple of commuting operator functions, and let $K \subset \mathscr{O}(U, L(X))$ be a commutative ring with $1_X \in K$ and such that the elements of K commute with the components of b (regarded as functions on U). Suppose that $\sigma_b \subset W$ is compact. Then $(\mathscr{E}_z(U), \mathscr{D}(U), b, \bar{\partial})$ is a Cauchy–Weil system*

Proof The map

$$\varphi: \mathscr{E}_z(U, X) / \mathscr{D}(U, X) \to \varinjlim_{V \in \mathscr{W}} \mathscr{E}_z(U \cap (\mathbb{C}^n \times V), X), \qquad [f] \mapsto [(f, U)]$$

is an isomorphism of K-modules. We indicate only a proof of the surjectivity. Let $f \in \mathscr{E}_z(U \cap (\mathbb{C}^n \times V), X)$ for some set $V \in \mathscr{W}$. We choose compact sets K_1 and K_2 in W with $W \backslash V \subset \text{Int } K_1 \subset K_1 \subset \text{Int } K_2$ and a function $\theta \in \mathscr{E}(W)$ with $\theta = 0$ on K_1 and $\theta = 1$ on $W \backslash K_2$. Then the function $g \in \mathscr{E}(U, X)$ with $g = 0$ on $U \cap (\mathbb{C}^n \times \text{Int}(K_1))$ and

$$g(z, w) = \theta(w) f(z, w) \qquad ((z, w) \in U \cap (\mathbb{C}^n \times V))$$

belongs to $\mathscr{E}_z(U, X)$ and is mapped by φ onto the equivalence class of f.

To conclude the proof, we show that for each open set $D \subset \mathbb{C}^n \times (W \backslash \sigma_b)$ the complex $K^\bullet(\bar{\partial}_w \oplus b, \mathscr{E}_z(D, X))$ is in fact exact. If $D = P \times Q$ is an open polydisc with $Q \subset W \backslash \sigma_b$, then

$$K^\bullet(\bar{\partial}_w \oplus b, \mathscr{E}(D, X)) = 1_{\mathscr{E}(P, X)} \otimes K^\bullet(\bar{\partial}_w \oplus b, \mathscr{E}(Q, X))$$

is exact. Since the sheaf $\mathscr{E}^X_{\mathbb{C}^{m+n}}$ is fine, we find that the complex

$$K^\bullet\left(\bar\partial_w \oplus b, \mathscr{E}(D, X)\right)$$

is exact for each open set $D \subset \mathbb{C}^n \times (W \backslash \sigma_b)$. Using the inductive limit representation of $\mathscr{E}(D, X)/\mathscr{E}_z(D, X)$ obtained in the proof of Proposition 2.4.4, we obtain that $\bar\partial_w \oplus b$ is non-singular on $\mathscr{E}(D, X)/\mathscr{E}_z(D, X)$ for each such set D. Now it suffices to apply Lemma 2.2.3. $\square$

Let $U \subset \mathbb{C}^n$ be open, and let $a = (a_1, \ldots, a_n) \in \mathscr{O}(U, L(X))^n$ be a commuting tuple of operator functions on U such that $\sigma_a \subset U$ is compact. Let $K \subset \mathscr{O}(U, L(X))$ be a commutative ring with $1_X \in K$ such that K commutes with the components of a. Then $(\mathscr{E}(U, X), \mathscr{D}(U, X), a, \bar\partial)$ is a Cauchy–Weil system of K-modules and the Cauchy–Weil map

$$R_a: \mathscr{O}(U, X) \to H^n(\bar\partial, \mathscr{D}(U, X))$$

is a K-module homomorphism. Because of Stokes' theorem, the map

$$\int : H^n(\bar\partial, \mathscr{D}(U, X)) \to X, \qquad [g\, d\bar z] \mapsto \left(-\frac{1}{\pi}\right)^n \int_U g\, dz,$$

where the integration is performed with respect to the $(2n)$-dimensional Lebesgue measure on U, is well defined and $\mathbb{C}$-linear.

Definition 2.4.7   Let $a = (a_1, \ldots, a_n) \in \mathscr{O}(U, L(X))^n$ be a tuple of commuting operator functions on an open set U in $\mathbb{C}^n$ such that $\sigma_a \subset U$ is compact. The $\mathbb{C}$-linear map

$$\mathscr{O}(U, X) \to X, \qquad f \mapsto \int R_a(f)$$

is the *Cauchy–Weil integral with respect to a on U*.

The Cauchy–Weil integral of f is sometimes also denoted by $\int R_a(f) dz$ or $\int_U R_a(f) dz$.

Let $A \subset L(X)$ be the algebra of all operators that commute with $a_i(z)$ for each i and each $z \in U$. Since R_a is an A-module homomorphism, so is the Cauchy–Weil integral. Another elementary property of the Cauchy–Weil integral, which follows immediately from what we proved before, is the following. If $f = a_1 g_1 + \cdots + a_n g_n$ for some functions $g_1, \ldots, g_n \in \mathscr{O}(U, X)$, then $\int R_a(f) = 0$. The reason for this is that, by Proposition 2.3.7, $R_a(a_i f) = 0$ for $i = 1, \ldots, n$.

To prove the continuity of the Cauchy–Weil integral, we study first the case where the argument $f \in \mathscr{O}(U, X)$ depends continuously on an additional parameter. For this purpose, let Λ be a compact Hausdorff space. We equip the space $C(\Lambda, X)$ of all continuous X-valued functions on Λ with its canonical Fréchet space topology given by the semi-norms

$$\|f\|_p = \sup_{\lambda \in \Lambda} p(f(\lambda)),$$

where p runs through all continuous semi-norms on the Fréchet space X.

Proposition 2.4.8 *Let $a \in \mathcal{O}(U, L(X))^n$ be a commuting tuple of operator functions on an open set U in $\mathbb{C}^n$ such that $\sigma_a \subset U$ is compact. Let Λ be a compact Hausdorff space. Then, for each function $f \in C(\Lambda, \mathcal{O}(U, X))$, the map*

$$\Lambda \to X, \qquad \lambda \mapsto \int R_a(f(\lambda))$$

is continuous.

Proof Let $f \in C(\Lambda, \mathcal{O}(U, X))$ and $\lambda \in \Lambda$ be fixed. We define $Y = C(\Lambda, X)$ and denote by $\delta_\lambda : Y \to X$ the point evaluation at λ. We obtain a function $f' \in \mathcal{O}(U, Y)$ by setting $f'(z)(\lambda) = f(\lambda)(z)$.

The operator functions $a'_i : U \to L(Y)$ defined by

$$a'_i(z)f = a_i(z) \circ f \qquad (f \in Y)$$

form a commuting tuple $a' \in \mathcal{O}(U, L(Y))^n$. For each open polydisc D in $U \setminus \sigma_a$ the complex $K^\bullet(a', \mathcal{O}(D, Y)) \cong C(\Lambda, K^\bullet(a, \mathcal{O}(D, X)))$ is exact by Theorem 2.1.2. We have thus shown that $\sigma_{a'} \subset \sigma_a$. Let us choose a function $g \in \mathscr{D}(U, Y)$ with $R_{a'}(f') = [g\,d\bar{z}]$, where $R_{a'}$ is the Cauchy–Weil map with respect to the Cauchy–Weil system $(\mathscr{E}(U, Y), \mathscr{D}(U, Y), a', \bar{\partial})$. Proposition 2.3.5, with $\mathscr{E}(U, Y)$ as X, $\mathscr{E}(U, X)$ as X', $u^0(h) = \delta_\lambda \circ h$, and $(u_{ij}) = (\delta_{ij})$, gives rise to the identity

$$[(\delta_\lambda \circ g)\,d\bar{z}] = R_a(f(\lambda)).$$

By the definition of the Cauchy–Weil integral we obtain

$$\left(\int R_{a'}(f')\,dz \right)(\lambda) = \int R_a(f(\lambda))\,dz. \qquad \square$$

As an application of the last result we obtain the continuity of the Cauchy–Weil integral.

Corollary 2.4.9 *Let $a \in \mathcal{O}(U, L(X))^n$ be a tuple of commuting operator functions with $\sigma_a \subset U$ compact. Then the Cauchy–Weil integral*

$$\mathcal{O}(U, X) \to X, \qquad f \mapsto \int R_a(f)$$

is continuous.

Proof Suppose that (f_n) is a convergent sequence in $\mathcal{O}(U, X)$ with limit f, and that Λ denotes the one-point compactification of the positive integers. Then the function $F : \Lambda \to \mathcal{O}(U, X)$, defined by $F(n) = f_n$ for $n \in \mathbb{N}$ and $F(\infty) = f$, is continuous. Hence Proposition 2.4.8 implies that

$$\int R_a(f_n) \to \int R_a(f) \qquad \text{as } n \to \infty. \qquad \square$$

To deduce a Fubini-type formula for the Cauchy–Weil integral, we first extend its definition to a more general situation.

Let $U \subset \mathbb{C}^{n+m}$ be open, and let π_z, π_w be the projections of $\mathbb{C}^{n+m}$ onto its first n and last m coordinates. Suppose that $f \in \mathscr{E}_z(U, X)$. Then we obtain a function $F \in \mathscr{E}(\mathbb{C}^n \times W, X)$ by setting

$$F(z,w) = f(z,w) \quad \text{for} \quad (z,w) \in U, \qquad F(z,w) = 0 \quad \text{for} \quad (z,w) \notin U.$$

Furthermore, the reader may easily check that for each compact set L in W there is a compact set K in $\pi_z(U)$ such that F vanishes on $(\mathbb{C}^n \setminus K) \times L$. In the following we shall use the same notation for f and its trivial extension F. For $f \in \mathscr{E}_z(U, X)$ and $\mathscr{D}(U, X)$, respectively, the function

$$\tilde{f}: W \to X, \qquad \tilde{f}(w) = \int_{\mathbb{C}^n} f(z,w)\,\mathrm{d}z$$

is well-defined and belongs to $\mathscr{E}(W, X)$ and $\mathscr{D}(W, X)$, respectively. To prove this, it suffices to show that $x' \circ \tilde{f}(w) = \int_{\mathbb{C}^n} x' \circ f(z,w)\,\mathrm{d}z$ defines a function in $\mathscr{E}(W)$ for each continuous linear form $x' \in X'$. Since, moreover, one may suppose that f vanishes outside $K \times W$ for a suitable compact set K, the assertion is reduced to a case where it is known to be true.

Obviously, we have the relations

$$\big((\partial/\partial \bar{w}_j)f\big)^{\sim} = (\partial/\partial \bar{w}_j)(\tilde{f}) \quad \text{for} \quad j = 1, \dots, m,$$

$$\big((\partial/\partial \bar{z}_j)f\big)^{\sim} = 0 \qquad\qquad \text{for} \quad j = 1, \dots, n.$$

For $n, p \in \mathbb{N}$, we define

$$C_n = \left(-\frac{1}{\pi}\right)^n, \qquad C_p^n = \left(\frac{1}{\pi}\right)^n (-1)^{(p+1)n}.$$

The reader will easily check that the maps

$$\rho: \Lambda^{n+p}(\mathrm{d}\bar{z} \oplus \mathrm{d}\bar{w}, \mathscr{E}_z(U, X)) \to \Lambda^p(\mathrm{d}\bar{w}, \mathscr{E}(W, X)),$$

$$f \mapsto C_p^n \sum_{|i|=p} \tilde{f}_{1 \dots n(n+i)}\,\mathrm{d}\bar{w}_i \quad \big(= \tilde{f}_{1 \dots n} \text{ if } p = 0\big),$$

where $n + i = (n + i_1, \dots, n + i_p)$, define a cochain map of degree $-n$ from $K^{\bullet}(\bar{\partial}_z \oplus \bar{\partial}_w, \mathscr{E}_z(U, X))$ into $K^{\bullet}(\bar{\partial}_w, \mathscr{E}(W, X))$. Everything remains true if $\mathscr{E}_z(U, X)$ and $\mathscr{E}(W, X)$ are replaced by $\mathscr{D}(U, X)$ and $\mathscr{D}(W, X)$, respectively. Moreover, if $b = (b_1, \dots, b_m) \in \mathscr{O}(W, L(X))^m$ is a tuple of commuting operator functions on W, then everything remains true with $\bar{\partial}_w$ replaced by $\bar{\partial}_w \oplus b$.

Definition 2.4.10 Let $U \subset \mathbb{C}^{n+m}$ be open, let $W = \pi_w U$, and let a in $\mathscr{O}(U, L(X))^n$ be a commuting tuple of operator functions such that $\sigma_a \subset U$ is z-compact. Then the maps $H^p(\bar{\partial}_z \oplus \bar{\partial}_w, \mathscr{E}(U, X)) \to H^p(\bar{\partial}_w, \mathscr{E}(W, X))$ defined for $p \geq 0$ by

$$f \mapsto \int R_a(f) = \rho \circ R_a(f),$$

where $R_a(f)$ is formed relative to the Cauchy–Weil system

$$\big(\mathscr{E}(U, X), \mathscr{E}_z(U, X), a, \bar{\partial}_z \oplus \bar{\partial}_w\big),$$

are called (*generalized*) *Cauchy–Weil integrals*.

For the generalized Cauchy–Weil integral we shall also use the notation $\int R_a(f)\,dz = \int R_a(f)$. The following is a Fubini-type theorem for the Cauchy–Weil integral.

Theorem 2.4.11 *Let $U \subset \mathbb{C}^{n+m}$ be an open set, and let $a \in \mathcal{O}(U, L(X))^n$ and $b \in \mathcal{O}(W, L(X))^m$, with $W = \pi_w U$, be tuples of commuting operator functions such that $\sigma_a \subset U$ is z-compact, $\sigma_b \subset W$ is compact, and*

$$a_i(z,w)b_j(w) = b_j(w)a_i(z,w) \qquad ((z,w) \in U, i = 1,\ldots,n, j = 1,\ldots,m).$$

Then $(\mathcal{E}(U,X), \mathcal{D}(U,X), (a,b), \bar{\partial}_z \oplus \bar{\partial}_w)$ is a Cauchy–Weil system and

$$\int R_{(a,b)}(f) = \int R_b\left(\int R_a(f)\right)$$

holds for each $f \in \mathcal{O}(U,X)$.

Proof Let us write $\bar{\partial}$ instead of $\bar{\partial}_z \oplus \bar{\partial}_w$. Our assumptions guarantee that $(\mathcal{E}(U,X), \mathcal{E}_z(U,X), a, \bar{\partial})$ and $(\mathcal{E}_z(U,X), \mathcal{D}(U,X), b, \bar{\partial})$ are Cauchy–Weil systems (Proposition 2.4.4 and Proposition 2.4.6). By Theorem 2.3.8 it also follows that $(\mathcal{E}(U,X), \mathcal{D}(U,X), (a,b), \bar{\partial})$ is a Cauchy–Weil system and that the diagram

$$\mathcal{O}(U,X) \xrightarrow{R_{(a,b)}} H^{n+m}(\bar{\partial}, \mathcal{D}(U,X))$$
$$\underset{R_a}{\searrow} \qquad \nearrow_{R_b}$$
$$H^n(\bar{\partial}, \mathcal{E}_z(U,X))$$

commutes. Let us suppose for a moment that we have proved the commutativity of the diagram

$$\begin{array}{ccc} H^n(\bar{\partial}, \mathcal{E}_z(U,X)) & \xrightarrow{R_b} & H^{n+m}(\bar{\partial}, \mathcal{D}(U,X)) \\ {\scriptstyle \rho}\downarrow & & \downarrow{\scriptstyle (-1)^{nm}p} \\ \mathcal{O}(W,X) & \xrightarrow{R_b} & H^m(\bar{\partial}_w, \mathcal{D}(W,X)). \end{array}$$

Then, if $f \in \mathcal{O}(U,X)$ and $R_{(a,b)}(f) = [g\,d\bar{z} \oplus d\bar{w}]$, the remaining assertion follows from the calculation

$$\int R_{a \oplus b}(f) = C_{n+m} \int_{\mathbb{C}^{n+m}} g(z,w)\,d(z,w)$$

$$= C_m \int_W \left((-1)^{nm} C_m^n \int_{\mathbb{C}^n} g(z,w)\,dz \right) dw$$

$$= \int R_b(\rho \circ R_a(f)) = \int R_b\left(\int R_a f\right).$$

Let us denote by $t = (t_1, \ldots, t_m)$ a tuple of indeterminates corresponding to the system b. To prove the commutativity of the above diagram, it suffices to observe that

$$K^\bullet(\bar{\partial}, \mathscr{E}_z(U, X)) \xrightarrow{t} K^\bullet(\bar{\partial} \oplus b, \mathscr{E}_z(U, X)) \xleftarrow{i} K^\bullet(\bar{\partial} \oplus b, \mathscr{D}(U, X)) \xrightarrow{\pi} K^\bullet(\bar{\partial}, \mathscr{D}(U, X))$$

$$\downarrow^\rho \qquad\quad \downarrow^{(-1)^{nm}\rho} \qquad\quad \downarrow^{(-1)^{nm}\rho} \qquad\quad \downarrow^{(-1)^{nm}\rho}$$

$$K^\bullet(\bar{\partial}_w, \mathscr{E}(W, X)) \xrightarrow{t} K^\bullet(\bar{\partial}_w \oplus b, \mathscr{E}(W, X)) \xleftarrow{i} K^\bullet(\bar{\partial}_w, \mathscr{D}(W, X)) \xrightarrow{\pi} K^\bullet(\bar{\partial}_w, \mathscr{D}(W, X))$$

is a commutative diagram of cochain maps. Thus the proof is complete.   □

Let U and a be as in Theorem 2.4.11. For $w \in W = \pi_w U$, we define $U_w = \{z \in \mathbb{C}^n; (z, w) \in U\}$ and $a_w(z) = (a_1(z, w), \ldots, a_n(z, w))(z \in U_w)$. Under suitable conditions, the function $\rho \circ R_a(f) \in \mathscr{O}(W, X)$ appearing in the last theorem can be computed at a given point $w \in W$ as the Cauchy–Weil integral of $f_w = f(\cdot, w)$ with respect to a_w.

Lemma 2.4.12 *Let U and a be as in Definition 2.4.10, and let $w \in W = \pi_w U$ be fixed. If $\sigma_{a_w} \subset U_w$ is compact, then for each $f \in \mathscr{O}(U, X)$, we have*

$$\left(\int R_a(f) \right)(w) = \int R_{a_w}(f_w),$$

where $f_w \in \mathscr{O}(U_w)$ is defined by $f_w(z) = f(z, w)$.

Proof Consider the Cauchy–Weil systems

$$\left(\mathscr{E}(U, X), \mathscr{E}_z(U, X), a, \bar{\partial}_z \oplus \bar{\partial}_w \right), (\mathscr{E}(U_w, X), \mathscr{D}(U_w, X), a_w, \bar{\partial}_z),$$

and the special transformation $u: K^\bullet(\bar{\partial}_z \oplus \bar{\partial}_w, \mathscr{E}(U, X)) \to K^\bullet(\bar{\partial}_z, \mathscr{E}(U_w, X))$ given by $u^0(f) = f_w$ and $u(d\bar{z}_j) = dz_j$, $u(d\bar{w}_j) = 0$. As an application of Proposition 2.3.5 we obtain

$$u \circ R_a = R_{a_w} \circ u.$$

Let $f \in \mathscr{O}(U, X)$, and let $R_a(f) = [g]$. Then

$$\rho \circ R_a(f)(w) = C_n \int_{\mathbb{C}^n} g_{1\ldots n}(z, w) dz = C_n \int_{U_w} g_{1\ldots n, w}(z) dz = \int R_{a_w}(f_w),$$

where $g_{1\ldots n}$ is the coefficient of g belonging to $d\bar{z}_1 \wedge \cdots \wedge d\bar{z}_n$.   □

Suppose that X is a Banach space. Then, in the setting of Lemma 2.4.12, the set σ_{a_w} is automatically compact. Indeed, by Lemma 2.1.3 and Theorem 2.1.8, we know that

$$\sigma_{a_w} = \{z \in U_w; K^\bullet(a(z, w), X) \text{ is not exact}\} = \pi_z(\sigma_a \cap (\mathbb{C}^n \times \{w\})).$$

Lemma 2.4.13 *Let U_1 and U_2 be two given open sets in $\mathbb{C}^n$, and let $a \in \mathscr{O}(U_1 \cup U_2, L(X))^n$ be a commuting tuple of operator functions such that σ_a is a compact subset of $U_1 \cap U_2$. Then*

$$\int R_{(a|U_1)}(f|U_1) = \int R_{(a|U_2)}(f|U_2) \qquad (f \in \mathscr{O}(U_1 \cup U_2, X)).$$

Proof It suffices to prove the result in the special case where $U_1 \subset U_2$. As in Proposition 2.4.4 or Proposition 2.4.6, one can show that there is a natural isomorphism

$$\mathscr{E}(U_2, X)/\mathscr{D}(U_1, X) \cong \varinjlim_V \mathscr{E}(V, X),$$

where the inductive limit is formed over all open sets $V \subset U_2$ with the property that $U_2 \setminus V \subset U_1$ is compact. Hence, as well as the Cauchy–Weil systems used in the definition of the Cauchy–Weil integrals of $f | U_1, f | U_2$, we can consider the Cauchy–Weil system $(\mathscr{E}(U_2, X), \mathscr{D}(U_1, X), a, \bar{\partial})$.

The diagram

$$
\begin{array}{ccc}
\mathscr{E}(U_2, X) & \xleftarrow{\ i\ } & \mathscr{D}(U_2, X) \\
{\scriptstyle id}\big\uparrow & & \big\uparrow{\scriptstyle j} \\
\mathscr{E}(U_2, X) & \xleftarrow{\ i\ } & \mathscr{D}(U_1, X) \\
{\scriptstyle r}\big\downarrow & & \big\downarrow{\scriptstyle id} \\
\mathscr{E}(U_1, X) & \xleftarrow{\ i\ } & \mathscr{D}(U_1, X),
\end{array}
$$

where j denotes the inclusion and r denotes the restriction, is commutative, and it follows from a double application of Proposition 2.3.5 that

$$R_{(a | U_2)}(f) = j \circ R_{(a | U_1)}(f | U_1)$$

for all $f \in \mathscr{E}(U_2, X)$. □

If a continuous linear operator $u \colon X \to Y$ between Fréchet spaces intertwines two commuting systems $a \in \mathscr{O}(U, L(X))^n$ and $b \in \mathscr{O}(U, L(Y))^n$, then it intertwines the corresponding Cauchy–Weil integrals. More precisely, we have the following result.

Lemma 2.4.14 *Let* $u \colon X \to Y$ *be a continuous linear operator between Fréchet spaces. Suppose that* $a \in \mathscr{O}(U, L(X))^n$ *and* $b \in \mathscr{O}(U, L(Y))^n$ *are tuples of commuting operator functions on an open set* U *in* $\mathbb{C}^n$ *such that* $\sigma_a \cup \sigma_b \subset U$ *is compact and* $u \circ a_i(z) = b_i(z) \circ u$ *holds for all* $z \in U, i = 1, \ldots, n$. *Then for each* $f \in \mathscr{O}(U, X)$, *we have*

$$u\left(\int R_a(f) \right) = \int R_b(uf).$$

Proof It suffices to apply Proposition 2.3.5 to the special transformation $\Lambda(d\bar{z}, \mathscr{E}(U, X)) \to \Lambda(d\bar{z}, \mathscr{E}(U, Y))$ given by $u^0 \colon \mathscr{E}(U, X) \to \mathscr{E}(U, Y)$, $f \mapsto u \circ f$, and the identity matrix $(u_{ij}) = (\delta_{ij})$. □

2.5 JOINT SPECTRUM AND ANALYTIC FUNCTIONAL CALCULUS

Throughout this section X will be a fixed Fréchet space. Let $T \in L(X)$ be a continuous linear operator. The singularity set σ_{z-T} of the operator function

$$z - T \colon \mathbb{C} \to L(X), \qquad z \mapsto z - T$$

is essentially what is usually called the (Waelbroeck-) spectrum of the operator T. Let us recall the definition of the spectrum $\sigma(T)$ of T in the sense of Waelbroeck. By definition, the *resolvent set* $\rho(T)$ of T consists of all points $w \in \hat{\mathbb{C}}$ in the extended complex plane for which there is an open neighbourhood U of w in $\hat{\mathbb{C}}$ such that:

(i) $z - T$ is invertible for all $z \in U \cap \mathbb{C}$;
(ii) $\{(z - T)^{-1}; z \in U \cap \mathbb{C}\} \subset L(X)$ is bounded.

The set $\sigma(T) = \hat{\mathbb{C}} \setminus \rho(T)$ is called the *Waelbroeck spectrum* of T. For z in $\rho(T) \cap \mathbb{C}$, we use the standard notation $R(z, T) = (z - T)^{-1}$. If $\infty \in \rho(T)$, then we set $R(\infty, T) = 0$.

Lemma 2.5.1 *With the above notation, we have $\sigma_{z-T} = \sigma(T) \cap \mathbb{C}$.*

Proof The inclusion $\sigma_{z-T} \subset \sigma(T) \cap \mathbb{C}$ follows from the fact that the resolvent $z \mapsto R(z, T)$ is analytic and yields, for each open set $U \subset \rho(T) \cap \mathbb{C}$, an inverse for the operator $\mathscr{O}(U, X) \to \mathscr{O}(U, X)$, $f \mapsto (z - T)f$.

If, conversely, the last operator is bijective for some open set U in $\mathbb{C}$, then for each $x \in X$, there is a unique function $\tilde{x} \in \mathscr{O}(U, X)$ with $(z - T)\tilde{x}(z) = x$ on U, and for $w \in U$, the operator $w - T$ is invertible with inverse

$$R_w \colon X \to X, x \mapsto \tilde{x}(w).$$

Obviously, for each compact subset K of U, the family $(R_w)_{w \in K}$ is pointwise bounded, hence equicontinuous, and therefore bounded in $L(X)$. $\square$

As usual we call a continuous linear operator $T \in L(X)$ *regular* if $\sigma(T) \subset \mathbb{C}$. For a regular operator $T \in L(X)$ the map

$$\Phi \colon \mathscr{O}(\sigma(T)) \to L(X), \qquad f \mapsto f(T) = \frac{1}{2\pi i} \int_\Gamma f(z) R(z, T) \, dz,$$

where Γ is a path surrounding $\sigma(T)$ in the domain of f, defines a continuous algebra homomorphism with $\Phi(1) = 1_X$ and $\Phi(z) = T$. Furthermore, this map satisfies the analytic spectral mapping theorem, that is,

$$\sigma(f(T)) = f(\sigma(T))$$

holds for all $f \in \mathscr{O}(\sigma(T))$. The map Φ is called the *analytic functional calculus* of T. The above properties of the analytic functional calculus for regular Fréchet space operators can be proved in the same way as in the Banach space case (see, for instance, Chapter III.3 in Vasilescu 1982).

We base the definition of the multidimensional analytic functional calculus for commuting tuples of regular Fréchet space operators on a version of Stokes' integral formula for rectangles in $\mathbb{C}$.

Lemma 2.5.2 *Let $U \subset \mathbb{C}$ be open, and let $R = \{z \in \mathbb{C}; \ a < \operatorname{Re} z < b, \ c < \operatorname{Im} z < d\} \Subset U$ be an open rectangle. Then for each $f \in C^1(U, X)$, we have*

$$2\mathrm{i} \int_R \bar{\partial} f \, \mathrm{d}z = \int_{\partial R} f \, \mathrm{d}z,$$

where ∂R is the positively oriented boundary of R.

Proof Obviously, we have

$$2\mathrm{i} \int_R \bar{\partial} f \, \mathrm{d}z = -\int_a^b \left(\int_c^d \frac{\partial f}{\partial y}(x + \mathrm{i}y) \mathrm{d}y \right) \mathrm{d}x + \mathrm{i} \int_c^d \left(\int_a^b \frac{\partial f}{\partial x}(x + \mathrm{i}y) \mathrm{d}x \right) \mathrm{d}y$$

$$= \int_a^b f(x + \mathrm{i}c) \mathrm{d}x + \mathrm{i} \int_c^d f(b + \mathrm{i}y) \mathrm{d}y - \int_a^b f(a + b - x + \mathrm{i}d) \, \mathrm{d}x$$

$$- \mathrm{i} \int_c^d f(a + \mathrm{i}(c + d - y)) \mathrm{d}y = \int_{\partial R} f(z) \, \mathrm{d}z. \qquad \square$$

Let $T \in L(X)$ be regular, let R be an open rectangle containing $\sigma(T)$, and let $f \in \mathcal{O}(U, X)$ be an analytic function defined on an open neighbourhood U of $\bar{R}$. If $g \in \mathscr{E}(U, X)$ is a function that is zero near $\sigma(T)$ and equal to $R(z, T)f(z)$ outside a compact subset of R, then $R_{z - T}(f) = [-(\bar{\partial}g)\mathrm{d}\bar{z}]$, where $R_{z - T}$ denotes the Cauchy–Weil map relative to the Cauchy–Weil system $(\mathscr{E}(U, X), \mathscr{D}(U, X), z - T, \bar{\partial})$. Hence we obtain

$$\int R_{z - T}(f) = \frac{1}{\pi} \int_U \bar{\partial}g \, \mathrm{d}z = \frac{1}{2\pi \mathrm{i}} \int_{\partial R} R(z, T)f(z) \, \mathrm{d}z.$$

We use these observations as a motivation for the following definitions. Let $a = (a_1, \ldots, a_n) \in L(X)^n$ be a commuting tuple of regular operators on a Fréchet space X. Then we write $z - a = (z_1 - a_1, \ldots, z_n - a_n)$ for the tuple of commuting operator functions on $\mathbb{C}^n$ with components given by

$$\mathbb{C}^n \to L(X), \quad z \mapsto z_i - a_i \quad (1 \leq i \leq n).$$

Obviously, the inclusion $\sigma_{z - a} \subset \sigma(a_1) \times \cdots \times \sigma(a_n)$ holds.

Definition 2.5.3 Let $a = (a_1, \ldots, a_n) \in L(X)^n$ be a commuting tuple of regular operators on X. The compact subset of $\mathbb{C}^n$ defined by

$$\sigma(a) = \sigma_{z - a}$$

is called the *(Taylor-) spectrum* of a. Let $f \in \mathcal{O}(U)$ for some open neighbourhood U of $\sigma(a)$. Then for each $x \in X$, we define

$$f(a)x = \int_U R_{z - a}(fx) \mathrm{d}z.$$

Here $R_{z - a}$ denotes the Cauchy–Weil map with respect to the Cauchy–Weil system $(\mathscr{E}(U, X), \mathscr{D}(U, X), z - a, \bar{\partial})$.

Suppose that X is a Banach space. Then by Lemma 2.1.3 and Theorem 2.1.8 (equivalence of (iii) and (iv)) we know that

$$\sigma(a) = \{z \in \mathbb{C}^n; \ K^\bullet(z-a, X) \text{ is not exact}\}.$$

This is the original definition of $\sigma(a)$ as given in Taylor (1970a).

Let us come back to the Fréchet space case. Our first observation is that the Taylor spectrum satisfies the projection property.

Theorem 2.5.4 *Let $a' = (a_1, \ldots, a_{n+1}) \in L(X)^{n+1}$ be a commuting tuple of regular operators on X, and let $a = (a_1, \ldots, a_n)$. Let π denote the projection of $\mathbb{C}^{n+1}$ onto its first n coordinates. Then $\sigma(a) = \pi\sigma(a')$.*

Proof We write the elements of $\mathbb{C}^{n+1}$ in the form (z, w), where $z \in \mathbb{C}^n$ and $w \in \mathbb{C}$.

Suppose that $D = P \times Q, P \subset \mathbb{C}^n, Q \subset \mathbb{C}$, is an open polydisc such that $P \cap \sigma(a) = \varnothing$. Then

$$K^\bullet(z-a, \mathcal{O}(D, X)) = K^\bullet(z-a, \mathcal{O}(P, X)) \otimes 1_{\mathcal{O}(Q)}$$

is exact. This proves the inclusion $\pi\sigma(a') \subset \sigma(a)$.

To prove the opposite inclusion, we fix a point $v \in \mathbb{C}^n \setminus \sigma(a')$. Then for each $w \in \mathbb{C}$ there are open neighbourhoods P_w of v and Q_w of w such that $K^\bullet(z' - a', \mathcal{O}(W, X))$ is exact for each open polydisc $W \subset P_w \times Q_w$. We choose an open disc $Q \supset \sigma(a_{n+1})$, a path Γ that surrounds $\sigma(a_{n+1})$ in Q, and points $w_1, \ldots, w_r \in \mathbb{C}$ such that $Q \subset Q_{w_1} \cup \cdots \cup Q_{w_r}$. Then $P = P_{w_1} \cap \cdots \cap P_{w_r}$ is an open neighbourhood of v such that $K^\bullet(z' - a', \mathcal{O}^X_{P \times Q})$ is exact. Since $\mathcal{O}^X_{\mathbb{C}^{n+1}}$ is acyclic on pseudoconvex open sets, the complex $K^\bullet(z' - a', \mathcal{O}(W \times Q, X))$ is exact for each open polydisc $W \subset P$.

Let $W \subset P$ be an open polydisc. We complete the proof by showing that $K^\bullet(z - a, \mathcal{O}(W, X))$ is exact. For $f \in \mathcal{O}(W \times Q, X)$,

$$\tilde{f}(z) = \frac{1}{2\pi i} \int_\Gamma R(w, a_{n+1}) f(z, w) dw \qquad (z \in W)$$

defines a function $\tilde{f} \in \mathcal{O}(W, X)$. We obtain a cochain map

$$\rho: K^\bullet(z' - a', \mathcal{O}(W \times Q, W)) \to K^\bullet(z - a, \mathcal{O}(W, X))$$

of degree -1 by setting $\rho(fs_{i_1} \wedge \cdots \wedge s_{i_{p+1}}) = \tilde{f}s_{i_1} \wedge \cdots \wedge s_{i_p}$ if $i_{p+1} = n+1$ and $\rho(fs_{i_1} \wedge \cdots \wedge s_{i_{p+1}}) = 0$ otherwise. Suppose that $f \in \Lambda^p(s, \mathcal{O}(W, X))$ satisfies $(z - a)f = 0$. Then $f' = f \wedge s_{n+1}$, regarded as an element in

$$\Lambda^{p+1}(s', \mathcal{O}(W \times Q, X)),$$

satisfies $(z' - a')f' = 0$. Hence there is a form $g \in \Lambda^p(s', \mathcal{O}(W \times Q, X))$ with $(z' - a')g = f$. But then

$$f = \rho f' = (z - a)\rho g,$$

and the proof is complete. $\qquad\qquad\qquad\qquad\qquad\qquad\qquad\qquad\square$

Using the fact that the non-singularity of a tuple of commuting module homomorphisms is stable with respect to permutations of its components, one can prove more general versions of Theorem 2.5.4 (cf. Theorem 3.2 in Taylor 1970a). Since the Waelbroeck spectrum of a single continuous Fréchet space operator on a space $X \neq 0$ is not empty, Theorem 2.5.4 yields the same result in the multidimensional case.

Corollary 2.5.5 *Suppose that $a = (a_1, \ldots, a_n)$ is a commuting tuple of regular operators on a Fréchet space $X \neq 0$. Then $\sigma(a) \neq \varnothing$.* $\square$

We now turn our attention to the analytic functional calculus. If $f \in \mathcal{O}(U)$ is defined in a sufficiently large neighbourhood of $\sigma(a)$, then one can give an alternative description of $f(a)x$.

Lemma 2.5.6 *Let $a \in L(X)^n$ be a commuting tuple of regular operators on X, and let $R = R_1 \times \cdots \times R_n \supset \sigma(a)$ be an open polyrectangle in $\mathbb{C}^n$. If $f \in \mathcal{O}(U)$ is analytic on an open neighbourhood U of $\overline{R}$ and if $x \in X$, then*

$$f(a)x = \left(\frac{1}{2\pi i}\right)^n \int_{\partial R_n} \cdots \int_{\partial R_1} f(z) R(z_1, a_1) \cdot \cdots \cdot R(z_n, a_n) x \, dz_1 \cdots dz_n.$$

Proof Because of Lemma 2.4.13 we may suppose that $U = U_1 \times \cdots \times U_n$.

For $n = 1$ the assertion has been proved above. Suppose that the assertion is true for $n - 1$, where $n \geq 2$. Then, using Theorem 2.4.11 and Lemma 2.4.12, we obtain

$$f(a)x = \int R_{(z'-a', z_n - a_n)}(fx) = \int R_{z_n - a_n}\left(\int R_{z'-a'}(fx)\right)$$

$$= \int R_{z_n - a_n}\left(\left(\frac{1}{2\pi i}\right)^{n-1} \int_{\partial R_{n-1}} \cdots \int_{\partial R_1}\right.$$

$$\left. \times f(z) R(z_1, a_1) \cdots R(z_{n-1}, a_{n-1}) x \, dz_1 \cdots dz_{n-1}\right)$$

$$= \left(\frac{1}{2\pi i}\right)^n \int_{\partial R_n} \cdots \int_{\partial R_1} f(z) R(z_1, a_1) \cdots R(z_n, a_n) x \, dz_1 \cdots dz_n.$$

Here we have written $z' - a'$ for $(z_1 - a_1, \ldots, z_{n-1} - a_{n-1})$, regarded as a tuple of operator functions on U. Thus the proof is complete. $\square$

Our next aim is to show that Definition 2.5.3 yields a solution of the analytic functional calculus problem. More precisely, we shall show that the Cauchy–Weil integral used in Definition 2.5.3 defines a continuous algebra homomorphism $\mathcal{O}(\sigma(a)) \to L(X)$ that extends the polynomial functional calculus of a and satisfies a suitable spectral mapping theorem.

Theorem 2.5.7 *Let $a \in L(X)^n$ be a commuting tuple of regular Fréchet space operators, and let U be an open neighbourhood of $\sigma(a)$. Then*

$$\Phi: \mathcal{O}(U) \to L(X), \qquad \Phi(f)x = f(a)x$$

defines a continuous algebra homomorphism with values in the bicommutant $(a)''$ of $\{a_1, \ldots, a_n\}$ in $L(X)$ such that $\Phi(1) = 1_X$ and $\Phi(z_i) = a_i$ for $i = 1, \ldots, n$.

Proof Obviously, $f(a)x$ is $\mathbb{C}$-linear in f and x. By Corollary 2.4.9 the map Φ is well defined and continuous. The remarks following Definition 2.4.7 show that Φ has values in the bicommutant of a. In view of Lemma 2.4.13 and Lemma 2.5.6 it is clear that $\Phi(1) = 1_X$ and $\Phi(z_i) = a_i$ for $i = 1, \ldots, n$.

Let $f, g \in \mathcal{O}(U)$ and $x \in X$. By Theorem 2.4.11 and Lemma 2.4.12,

$$f(a)g(a)x = \int R_{w-a}\left(f(w)\int R_{z-a}g(z)x\right)$$

$$= \int R_{(z-a,w-a)}(f(w)g(z)x).$$

By Proposition 2.3.6 the last integral remains unchanged if $(z - a, w - a)$ is replaced by $(z - w, w - a)$. Indeed, the second system is obtained by multiplying the first tuple by the matrix (u_{ij}) of determinant 1 defined by $u_{ij} = 1$ if $i = j$, $u_{ij} = -1$ if $j = n + i$, and $u_{ij} = 0$ otherwise. If $z - w$ is regarded as an element in $\mathcal{O}(U \times U, L(X))^n$, then $\sigma_{z-w} = \{(z, z); z \in U\}$ is z-compact in $U \times U$. Again using Theorem 2.4.11 and Lemma 2.4.12, we obtain

$$f(a)g(a)x = \int R_{w-a}\left(f(w)\int R_{z-w}(g(z)x)\right).$$

To complete the proof it suffices to observe that, by Lemma 2.5.6 and Cauchy's integral formula,

$$\int R_{z-w}(g(z)x) = g(w)x. \qquad \square$$

Lemma 2.4.13 together with the last theorem show that Definition 2.5.3 in fact yields a continuous algebra homomorphism

$$\mathcal{O}(\sigma(a)) \to L(X), \qquad f \mapsto f(a)$$

with $1(a) = 1_X$ and $z_i(a) = a_i (1 \le i \le n)$, where $\mathcal{O}(\sigma(a))$ is equipped with its natural inductive limit topology. In the case where $n = 1$, such a map is easily seen to be unique. In Chapter 5 we shall return to the uniqueness question for the multidimensional case.

The next result is a straightforward application of Lemma 2.4.14.

Lemma 2.5.8 *Let $u: X \to Y$ be a continuous linear operator between Fréchet spaces. Suppose that $a \in L(X)^n$ and $b \in L(Y)^n$ are commuting tuples of regular operators such that $ua_i = b_i u$ for $i = 1, \ldots, n$. If $f \in \mathcal{O}(U)$ is analytic on an open neighbourhood U of $\sigma(a) \cup \sigma(b)$, then*

$$uf(a) = f(b)u. \qquad \square$$

As an elementary application of Theorem 2.5.7 one obtains that, for $f \in \mathscr{O}(\sigma(a))$, the spectral inclusion $\sigma(f(a)) \subset f(\sigma(a))$ holds. Our final aim in this section is to prove a much more general spectral mapping theorem.

Proposition 2.5.9 *Let $a_1, \ldots, a_n$, $b_1, \ldots, b_n$, and $c_1, \ldots, c_m \in L(X)$ be commuting regular operators on a Fréchet space X. We set $a = (a_1, \ldots, a_n)$, $b = (b_1, \ldots, b_n)$, $c = (c_1, \ldots, c_m)$, and $a - b = (a_1 - b_1, \ldots, a_n - b_n)$. Suppose that $f \in \mathscr{O}(U)$ is analytic on an open neighbourhood U of $\sigma(a) \cup \sigma(b)$. Then the componentwise action of $f(a) - f(b)$ induces the zero operator on $H^p((a - b, c), X)$ for each p.*

Proof Let $x = \sum_{|i|=p} x_i s_i \in \Lambda^p(s, X)$ be a form with $(a - b)x = 0$. Then $f = \sum_{|i|=p} (fx_i) s_i \in \Lambda^p(s, \mathscr{O}(U, X))$ is a form with $(a - b)f = 0$. Since by Corollary 2.3.10 the Cauchy–Weil maps R_{z-a} and R_{z-b} induce the same operator

$$H^p(a - b, \mathscr{O}(U, X)) \to H^p(a - b, H^n(\bar{\partial}, \mathscr{D}(U, X))),$$

there is a $(p - 1)$-form $\theta = \sum_{|i|=p-1} [\theta_i] s_i$ with coefficients in $H^n(\bar{\partial}, \mathscr{D}(U, X))$ such that

$$\sum_{|i|=p} (R_{z-a} - R_{z-b})(fx_i) s_i = (a - b)\theta.$$

By the definition of the Cauchy–Weil integral we obtain that

$$\sum_{|i|=p} (f(a) - f(b)) x_i s_i = \sum_{|i|=p} \left(\int R_{z-a}(fx_i) - \int R_{z-b}(fx_i) \right) s_i$$

$$= (a - b) \sum_{|i|=p-1} \left(c_n \int_U \theta_i \, dz \right) s_i.$$

Thus we have shown that $f(a) - f(b)$ acts as the zero operator on $H^p(a - b, X)$ for each p.

For c as in the proposition we set $\tilde{a} = (a, c)$ and $\tilde{b} = (b, 0)$. If $\tilde{f} \in \mathscr{O}(U \times \mathbb{C}^m)$ is defined by $\tilde{f}(z, w) = f(z)$, then by Theorem 2.4.11 and Lemma 2.4.12, we know that $\tilde{f}(\tilde{a}) = f(a)$, $\tilde{f}(\tilde{b}) = f(b)$. Hence the general case is reduced to the special case for which the result has been proved above.   □

Let $a = (a_1, \ldots, a_n) \in L(X)^n$ be a commuting tuple of regular operators on a Fréchet space X, and let $f \in \mathscr{O}(U)$ be analytic on an open neighbourhood U of $\sigma(a)$. Let us fix a natural number $r \geq 1$, and open polydiscs $P \subset \mathbb{C}^n$, $Q \subset \mathbb{C}^r$ with $\bar{P} \subset U$. We define $D = P \times Q$, and we write the variables in $P \times Q$ in the form (z, w) with $z \in P, w \in Q$. Then $Y = \mathscr{O}(D, X)$ is a Fréchet space and $a_1', \ldots, a_n', b_1, \ldots, b_n \in L(Y)$, defined by

$$a_i'(f) = a_i f, \qquad b_i(f) = z_i f,$$

are commuting regular operators on Y. Set $a' = (a'_1, \ldots, a'_n)$ and $b = (b_1, \ldots, b_n)$. Then $\sigma(a') \cup \sigma(b) \subset U$. Indeed, for each open polydisc W in $\mathbb{C}^n$ with $W \cap (\sigma(a) \cup \bar{P}) = \varnothing$, the complexes

$$K^\bullet(\lambda - a', \mathcal{O}(W, Y)) \cong K^\bullet(\lambda - a, \mathcal{O}(W, X)) \otimes 1_{\mathcal{O}(D)},$$

$$K^\bullet(\lambda - z, \mathcal{O}(W, Y)) \cong K^\bullet(\lambda - z, \mathcal{O}(W \times D)) \otimes 1_X,$$

where λ denotes the variable in W, are exact. An elementary application of Lemma 2.5.8 yields that

$$f(a')(g) = f(a)g \qquad \text{and} \qquad f(b)(g)(z, w) = f(z)g(z, w).$$

Proposition 2.5.9 shows that whenever $c \in L(Y)^m$ is a commuting tuple of regular operators on Y such that the components of c commute with the components of (a', b), then $f(z)$ and $f(a)$ induce the same operator on $H^p((z - a, c), \mathcal{O}(D, X))$ for all p.

Theorem 2.5.10 *Let $a = (a_1, \ldots, a_n)$ be a commuting tuple of regular operators on a Fréchet space X. Suppose that $f \in \mathcal{O}(U)^m$ for some open neighbourhood U of $\sigma(a)$. Then $\sigma(f(a)) = f(\sigma(a))$.*

Proof Because of the projection property (Theorem 2.5.4) it suffices to show that

$$\sigma(a, f(a)) = \{(z, f(z)); z \in \sigma(a)\}.$$

To do this, we check for given points $z_0 \in \sigma(a)$ and $w_0 \in \mathbb{C}^m$ that $(z_0, w_0) \notin \sigma(a, f(a))$ if and only if $w_0 \neq f(z_0)$. We shall prove this equivalence by induction on the length m of the tuple f.

Let $m \in \mathbb{Z}^+$. We suppose either that $m = 0$ or that $m \geq 1$ is such that the equivalence has been proved for all tuples f of length m. In the case $m = 0$ the reader should simply cancel all terms in the following definitions and constructions that refer to m.

We fix points $z_0 \in \sigma(a)$, $u_0 \in \mathbb{C}^m$, $v_0 \in \mathbb{C}$ and functions g in $\mathcal{O}(U)^m$, h in $\mathcal{O}(U)$, and we define $w_0 = (u_0, v_0)$, $f = (g, h)$. We consider open polydiscs $P \Subset U$, $Q \subset \mathbb{C}^m$, and $V \subset \mathbb{C}$. We set $D = P \times Q \times V$ and write the variables of $\mathbb{C}^{n+m+1}$ in the form $(z, u, v) = (z, w)$ with $z \in \mathbb{C}^n$, $u \in \mathbb{C}^m$, $v \in \mathbb{C}$, and $w = (u, v)$. The components of the tuples $\alpha = (z - a, u - g(a))$, and $\tilde{\alpha} = (z - a, w - f(a))$ will be regarded as multiplication operators on the Fréchet space $Y = \mathcal{O}(D, X)$. By the proof of Lemma 2.2.2 there is a long exact sequence of cohomology

$$0 \to H^0(\tilde{\alpha}, Y) \qquad \to H^0(\alpha, Y)$$

$$\xrightarrow{v - h(a)} H^0(\alpha, Y) \to H^1(\tilde{\alpha}, Y) \qquad \to H^1(\alpha, Y)$$

$$\xrightarrow{v - h(a)} H^1(\alpha, Y) \to H^2(\tilde{\alpha}, Y) \qquad \to \cdots$$

$$\cdots\cdots\cdots\cdots\cdots\cdots\cdots\cdots\cdots\cdots\cdots\cdots\cdots\cdots\cdots\cdots\cdots\cdots\cdots$$

$$\xrightarrow{v - h(a)} H^{n+m}(\alpha, Y) \to H^{n+m+1}(\tilde{\alpha}, Y) \to 0.$$

As explained in the section preceding Theorem 2.5.10, $h(a)$ and $h(z)$ induce the same operator on $H^p(\alpha, Y)$.

Let us suppose now that $w_0 \neq f(z_0)$. If $m \geq 1$ and $u_0 \neq g(z_0)$, then by the induction hypothesis $(z_0, u_0) \notin \sigma(a, g(a))$. Hence, for $P \times Q$ contained in a sufficiently small neighbourhood of (z_0, u_0), we know that $H^p(\alpha, Y) = 0$ for each $p \geq 0$. This follows from the remark following Definition 2.4.3 and the results concerning the preservation of exactness under tensor products proved in Appendix 1. If $v_0 \neq h(z_0)$, then for $P \times V$ contained in a sufficiently small neighbourhood of (z_0, v_0), the function $v - h(z)$ is invertible on D, and hence the operators

$$v - h(z) \colon H^p(\alpha, Y) \to H^p(\alpha, Y) \qquad (p \geq 0)$$

are isomorphisms. In both cases, it follows that, for D contained in a sufficiently small neighbourhood of (z_0, w_0), the cohomology groups $H^p(\tilde{\alpha}, Y)$ vanish for all p. Therefore, $(z_0, w_0) \notin \sigma(a, f(a))$.

Conversely, let us suppose that $(z_0, w_0) \notin \sigma(a, f(a))$ and that $v_0 = h(z_0)$. Let $D = P \times Q \times V \subset \mathbb{C}^{n+m+1}$ be an open polydisc as before. But this time we suppose in addition that D is centred at (z_0, w_0) and is disjoint from $\sigma(a, f(a))$. It suffices to prove that $K^\bullet(\alpha, \mathcal{O}(W, X))$ is exact for each open polydisc W in $\mathbb{C}^{n+m}$ with $W \subset (P \cap h^{-1}(V)) \times Q$. Suppose that this has been done. Then it follows that $(z_0, u_0) \notin \sigma(a, g(a))$. For $m = 0$, this is not possible. Hence, in this case, the condition that $(z_0, w_0) \notin \sigma(a, f(a))$ implies that $v_0 \neq h(z_0)$. For $m \geq 1$, the induction hypothesis would imply that $u_0 \neq g(z_0)$. In both cases we would obtain that $w_0 \neq f(z_0)$. Thus the proof would be complete.

Let us fix an open polydisc $W \subset (P \cap h^{-1}(V)) \times Q$. To prove that α is non-singular on $\mathcal{O}(W, X)$, we form the long exact cohomology sequence described above, but this time with $Y = \mathcal{O}(D, X)$ replaced by the Fréchet space $Z = \mathcal{O}(W \times V, X)$. Since $H^p(\tilde{\alpha}, Z) = 0$ for all p, we conclude that the maps

$$H^p(\alpha, Z) \xrightarrow{\;v - h(z)\;} H^p(\alpha, Z) \qquad (p \geq 0)$$

are isomorphisms. Let $p \geq 0$, and let $\varphi \in K^p(\alpha, \mathcal{O}(W, X))$ be a form with $\alpha\varphi = 0$. We regard the coefficients of φ as functions on $W \times V$ that are independent of the last variable. Thus φ becomes a form in $K^p(\alpha, Z)$ with $\alpha\varphi = 0$. Hence we can choose forms $\psi \in K^p(\alpha, Z)$ and $\eta \in K^{p-1}(\alpha, Z)$ with

$$\varphi = (v - h(z))\psi + \alpha\eta$$

on $W \times V$. By composing the coefficients of the form on the right-hand side with the analytic function

$$W \to W \times V, \qquad (z, u) \mapsto (z, u, h(z)),$$

one finally obtains that φ is exact with respect to the complex $K^\bullet(\alpha, \mathcal{O}(W, X))$. Thus the proof is complete. $\qquad\square$

2.6 ANALYTIC SPECTRAL MAPPING THEOREMS
FOR JOINT SPECTRA

In the Banach space case, one can use Gelfand theory to prove the analytic
spectral mapping theorem for certain natural subspectra of the Taylor spectrum. Let X be a complex Banach space, and let A be a closed, commutative
subalgebra of $L(X)$ containing the identity operator. We denote by $c(A)$ the
set of all tuples $a = (a_1, \ldots, a_n)$ in A of arbitrary finite length. For systems a
and b in $c(A)$ of length n and m, we set $(a, b) = (a_1, \ldots, a_n, b_1, \ldots, b_m)$, and
we denote by p, q the corresponding projections of $\mathbb{C}^{n \times m}$ onto its first n and
last m coordinates.

A *spectral system* on A is a rule which assigns to each element $a = (a_1, \ldots, a_n) \in c(A)$ a closed subset $s(a)$ of $\mathbb{C}^n$ such that for a and b in $c(A)$:

(i) $ps(a, b) = s(a), \qquad qs(a, b) = s(b)$;
(ii) $s(a) \subset \sigma_A(a)$.

Here $\sigma_A(a)$ is the usual joint spectrum of a with respect to the Banach
algebra A. There is a one-to-one correspondence between the spectral
systems on A and the closed subsets of the character space Δ_A of A. To
make this relation precise, let us fix a spectral system s on A. For each given
tuple $a = (a_1, \ldots, a_n) \in c(A)$, the Gelfand transform $\hat{a} = (\hat{a}_1, \ldots, \hat{a}_n)$ is a
continuous function on Δ_A with values in $\mathbb{C}^n$, and

$$\Delta_a = \Delta_a(s) = (\hat{a})^{-1}(s(a))$$

is a closed subset of Δ_A.

Proposition 2.6.1 *Let s be a spectral system on A. Then*

$$\Delta = \Delta(A, s) = \bigcap(\Delta_a; a \in c(A))$$

is the unique closed subset of Δ_A satisfying

$$s(a) = \hat{a}(\Delta) \qquad (a \in c(A)). \tag{2.6.1}$$

*Conversely, for each closed subset Δ of Δ_A, the last formula defines a spectral
system on A.*

Proof Suppose that s is a spectral system on A, and let $\Delta = \Delta(A, s)$ be
defined as above.

Suppose that $a \in c(A)$ and that $z \in s(a)$. Then for each element $b \in c(A)$,
there is a tuple $w \in s(b)$ and a functional λ in Δ_A such that $(z, w) = \widehat{(a, b)}(\lambda)$.
Hence

$$s(a) \subset \hat{a}(\Delta_b) \qquad (a, b \in c(A)).$$

Let $a \in c(A)$ and $z \in s(a)$ be arbitrary. The family $\mathcal{M}$ of all sets $\Delta_b(b \in c(A))$
is directed downwards by inclusion, and for each $M \in \mathcal{M}$, there is a functional
$\lambda_M \in M$ with $\hat{a}(\lambda_M) = z$. By compactness the net $(\lambda_M)_M$ has a convergent

subnet. But if λ is the limit of such a convergent subnet, then $\lambda \in \Delta$ and $\hat{a}(\lambda) = z$. Thus we have shown that

$$s(a) = \hat{a}(\Delta) \qquad (a \in c(A)).$$

To prove the claimed uniqueness, let us suppose that Δ_1 and Δ_2 are two closed subsets of Δ_A satisfying (2.6.1) and that $\lambda \in \Delta_1$. If $\lambda \notin \Delta_2$, then by continuity and compactness one could choose open sets $U_1, \ldots, U_n$ in Δ_A and a tuple $a = (a_1, \ldots, a_n) \in c(A)$ with

$$\Delta_2 \subset U_1 \cup \cdots \cup U_n \qquad \text{and} \qquad \hat{a}(\lambda) \notin \hat{a}(U_1 \cup \cdots \cup U_n).$$

Since this is impossible, we conclude that $\Delta_1 \subset \Delta_2$. $\qquad\qquad\square$

Suppose that s is a spectral system on A, and that B is a closed, unital subalgebra of A. Then, as an elementary application of the last result, one obtains the identity

$$\Delta(B, s|B) = \Delta(A, s)|B.$$

The next result shows that for any solution of the analytic functional calculus problem, regardless of its definition, the validity of the analytic spectral mapping theorem is equivalent to its compatibility with the Gelfand transformation.

Theorem 2.6.2 *Let s be a spectral system on A, and let a be an n-tuple in $c(A)$ such that there is an algebra homomorphism*

$$\Phi \colon \mathscr{O}(s(a)) \to A, f \mapsto \Phi(f)$$

with $\Phi(1) = I$ and $\Phi(z_i) = a_i (1 \le i \le n)$. Then the following conditions are equivalent:

(i) $s(\Phi(f)) = f(s(a)) \qquad (f \in \mathscr{O}(s(a))^k, k \ge 1)$;
(ii) $f(\hat{a}(\lambda)) = \widehat{\Phi(f)}(\lambda) \qquad (f \in \mathscr{O}(s(a)), \lambda \in \Delta(A, s))$.

Proof As usual we shall write $f(a)$ instead of $\Phi(f)$. Because of the remark following Proposition 2.6.1 we may suppose that the operators $f(A)$ with $f \in \mathscr{O}(s(a))$ form a dense subset of A.

Let us suppose that condition (i) holds. By Proposition 2.6.1 it suffices to show that the set

$$\Delta = \left\{ \lambda \in \Delta(A, s); f(\hat{a}(\lambda)) = \widetilde{f(a)}(\lambda) \qquad \text{for each} \quad f \in \mathscr{O}(s(a)) \right\}$$

satisfies condition 2.6.1. For this purpose, we assign to each $\lambda \in \Delta(A, s)$ the functional $\Lambda \in \Delta_A$ uniquely determined by

$$\Lambda(f(a)) = f(\hat{a}(\lambda)) \qquad (f \in \mathscr{O}(s(a))).$$

To show that $\Lambda \in \Delta$, it suffices to check that $\Lambda \in \Delta(A, s)$. But for a given $b \in c(A)$, there is a sequence $(f^k(a))_k$ such that

$$b = \lim_{k \to \infty} f^k(a)$$

componentwise. Let us fix k for the moment. Since $z = f^k(\hat{a}(\lambda))$ is contained in $s(f^k(a))$, there is a point $w \in s(b)$ with $(z, w) \in s(f^k(a), b)$. Hence, with respect to the maximum norm, we have

$$\mathrm{dist}(z, s(b)) \leq \|f^k(a) - b\|.$$

Since also

$$\|z - \hat{b}(\Lambda)\| \leq \|f^k(a) - b\|,$$

we conclude that $\hat{b}(\Lambda) \in s(b)$. Thus we have shown that $\Lambda \in \Delta$.

To prove (ii), it suffices to find, for each $b \in c(A)$ and each point $w \in s(b)$, a functional $\lambda \in \Delta(A, s)$ such that $w = \hat{b}(\Lambda)$. If $(f^k(a))_k$ is a sequence converging to b, then exactly as above it follows that

$$\mathrm{dist}(w, s(f^k(a))) \leq \|f^k(a) - b\| \qquad (k \in \mathbb{N}).$$

Hence there is a sequence (z_k) in $s(a)$ with $w = \lim_{k \to \infty} f^k(z_k)$. By passing to a suitable subsequence we may suppose that $z = \lim_{k \to \infty} z_k$ exists. Condition (i) implies that the sequence (f^k) is uniformly convergent on $s(a)$, in particular that it is equicontinuous. To conclude the proof of this implication, it suffices to observe that

$$w = \lim_{k \to \infty} f^k(\hat{a}(\lambda)) = \Lambda(b)$$

for each $\lambda \in \Delta(A, s)$ with $z = \hat{a}(\lambda)$.

Suppose, conversely, that Φ commutes with the Gelfand transform as explained in (ii). Then, for each $f \in \mathcal{O}(s(a))^k$, we have

$$s(f(a)) = \widehat{f(a)}(\Delta) = f(\hat{a}(\Delta)) = f(s(a))$$

with $\Delta = \Delta(A, s)$. $\qquad\qquad\qquad\qquad\qquad\qquad\qquad\qquad\qquad\qquad\square$

Suppose that s is a spectral system on A. Then a *subspectrum* of s is a spectral system τ on A with the property that

$$\tau(a) \subset s(a) \qquad (a \in c(A)).$$

The analytic spectral mapping theorem is inherited by subspectra.

Corollary 2.6.3 *Suppose that s, a, and Φ satisfy the conditions of Theorem 2.6.2. Then for each subspectrum τ of s, we have*

$$f(\tau(a)) = \tau(f(a)) \qquad \left(f \in \mathcal{O}(s(a))^k, k \geq 1\right).$$

Proof It suffices to observe that $\Delta(A, \tau) \subset \Delta(A, s)$. $\qquad\qquad\qquad\qquad\square$

In the following we shall apply the above results to concrete subspectra of the Taylor spectrum. Let $a \in L(X)^n$ be a commuting tuple on X. The *essential spectrum* $\sigma_e(a)$ of a is the set of those points $z \in \mathbb{C}^n$ with the property that

$$\dim H^p(z - a, X) = \infty$$

for at least one $p = 0, \ldots, n$. Before we prove spectral mapping theorems for concrete subspectra of the Taylor spectrum and the essential Taylor spectrum, we recall a standard construction that can be used to reduce the latter case to the previous one.

For a given Banach space X, let X^∞ be the Banach space of all bounded sequences in X with the supremum norm, and denote by X^{pc} the subspace of X^∞ consisting of all bounded sequences (x_k) in X with the property that each subsequence of (x_k) has a convergent subsequence. A diagonal argument shows that X^{pc} is a closed subspace of X^∞. We denote by X^q the quotient X^∞ / X^{pc}, a Banach space under the quotient norm. Suppose that $T \colon X \to Y$ is a continuous linear operator between Banach spaces. Then $T^\infty \colon X^\infty \to Y^\infty, (x_k) \mapsto (Tx_k)$, $T^{pc} = T^\infty | X^{pc}$, and $T^q = T^\infty / X^{pc}$ are continuous linear operators.

Proposition 2.6.4 *Let $T \colon X \to Y$ be a continuous linear operator between Banach spaces. Then $T^q \colon X^q \to Y^q$ is onto if and only if $\dim(Y/\operatorname{Im} T) < \infty$.*

Proof Suppose that T^q is onto. Let us assume that $\operatorname{Im} T$ is not closed. Then the adjoint T' has non-closed range, and one can choose a bounded sequence (u_k) in Y' with $\lim_{k \to \infty} T'u_k = 0$ such that the vectors $z_k = u_k + \operatorname{Ker} T'$ in $Y'/\operatorname{Ker} T'$ have norm one. Since (z_k) cannot have a convergent subsequence, the norm r of the induced element in $(Y'/\operatorname{Ker} T')^q$ is positive. By passing to a suitable subsequence of (u_k), we may suppose that $\|z_k - z_l\| > r/2$ for $k \neq l$. Since $Y'/\operatorname{Ker} T' \cong (\operatorname{Im} T)'$, for each pair k, l of indices with $k \neq l$ there is a vector $x_{kl} \in X$ with $\|Tx_{kl}\| \leq 1$ and

$$|\langle u_k - u_l, Tx_{kl} \rangle| > r/2.$$

The assumption that T^q is onto allows us to choose a bounded family (a_{kl}) in X and a compact subset M of Y such that

$$Tx_{kl} - Ta_{kl} \in M \qquad (k \neq l).$$

By the theorem of Arzela–Ascoli applied to the space $C(M)$, we may suppose (by passing to a suitable subsequence) that, in addition, $\|u_k - u_l\|_{\infty, M} < r/4$ for all k and l. But then

$$|\langle T'u_k - T'u_l, a_{kl} \rangle| > r/4 \qquad (k \neq l).$$

This is clearly not possible.

If $\operatorname{Im} T$ were not of finite codimension in Y, then a repeated application of the Riesz lemma would yield a bounded sequence (y_k) in Y with $\operatorname{dist}(y_k - y_l, TX) > 1/2$ for $k \neq l$. But then the equivalence class of (y_k) in Y^q would not belong to $\operatorname{Im} T^q$.

Conversely, let TX be of finite codimension in Y. To prove the surjectivity of T^q, we choose a finite-dimensional subspace N of Y with $Y = TX + N$ and observe that, for each bounded sequence (y_k) in Y, there are bounded sequences (x_k) in X and (z_k) in N with $y_k = Tx_k + z_k$ for each k. $\qquad \square$

The last result can be used to show that the functor $X \rightarrow X^q$ preserves the exactness of finite complexes of Banach spaces.

Lemma 2.6.5 *Let $X \xrightarrow{S} Y \xrightarrow{T} Z$ be a sequence of continuous linear operators between Banach spaces. Then the following conditions are equivalent*:

(*i*) *Im T is closed and* $\dim(\operatorname{Ker} T/\operatorname{Im} S) < \infty$;
(*ii*) *Im T is closed and* $\operatorname{Ker} T^q = \operatorname{Im} S^q$;
(*iii*) *Im S is closed and* $\operatorname{Ker} T^q = \operatorname{Im} S^q$.

Proof Since epimorphisms between Banach spaces allow the lifting of bounded and precompact sequences, the complex

$$0 \rightarrow (\operatorname{Ker} T)^q \xrightarrow{i^q} Y^q \xrightarrow{T^q} (\operatorname{Im} T)^q \rightarrow 0$$

is exact whenever Im T is closed. Hence the equivalence of (i) and (ii) follows from Proposition 2.6.4.

To conclude the proof, we show that (iii) implies (ii). By the first part the sequence

$$0 \rightarrow X/\operatorname{Ker} S \xrightarrow{\hat{S}} Y \rightarrow Y/\operatorname{Im} S \rightarrow 0$$

remains exact when applying the q-functor. Hence the assumption implies that the second map in

$$Y^q/\operatorname{Im}(S^q) \xrightarrow{\sim} (Y/\operatorname{Im} S)^q \xrightarrow{(T/\operatorname{Im} S)^q} Z^q$$

is injective. Therefore, it suffices to show that each continuous linear Banach space operator $R \colon X \rightarrow Y$ for which R^q is injective has closed range. This follows by exactly the same argument that was applied to T' in the beginning of the proof of Proposition 2.6.4. $\square$

Let $a \in L(X)^n$ be a commuting tuple on a Banach space X. For $k \geq 0$ we define $\sigma^{\delta,k}(a)$ as the complement of the set of those points $z \in \mathbb{C}^n$ for which

$$H^p(z-a, X) = 0 \qquad (n-k \leq p \leq n),$$

and $\sigma^{\pi,k}(a)$ as the complement of the set of those points for which

$$H^p(z-a, X) = 0 \qquad (0 \leq p \leq k)$$

and $H^{k+1}(z-a, X)$ is separated in its natural quotient topology. Analogously, we define $\sigma_e^{\delta,k}(a)$ as the complement of the set of those points $z \in \mathbb{C}^n$ for which

$$\dim H^p(z-a, X) < \infty \qquad (n-k \leq p \leq n),$$

and $\sigma_e^{\pi,k}(a)$ as the complement of the set of those points $z \in \mathbb{C}^n$ for which

$$\dim H^p(z-a, X) < \infty \qquad (0 \leq p \leq k)$$

and $H^{k+1}(z-a, X)$ is separated. The sets $\sigma^{\delta,k}(a)$ and $\sigma^{\pi,k}(a)$ are called the *defect* and *approximate point spectra* of degree k for a. The *defect* and *approximate point spectrum* of a are defined as

$$\sigma_\pi(a) = \sigma^{\pi,0}(a) \qquad \text{and} \qquad \sigma_\delta(a) = \sigma^{\delta,0}(a).$$

Via the topological identification $\Lambda^p(s, X)' \cong \Lambda^p(s, X')$, given by

$$\left\langle \sum_{|i|=p} x_i s_i, \sum_{|i|=p} u_i s_i \right\rangle = \sum_{|i|=p} \langle x_i, u_i \rangle,$$

the dual sequence of $K_\bullet(a, X)$ is the cochain complex $K^\bullet(a', X')$, and $K_\bullet(a', X')$ is dual to $K^\bullet(a, X)$. There are canonical topological identifications (Section 2.2)

$$H_p(a, X) \cong H^{n-p}(a, X) \quad \text{and} \quad H_p(a', X') \cong H^{n-p}(a', X') \qquad (p \geq 0).$$

Hence elementary duality theory shows that

$$\sigma^{\pi, k}(a) = \sigma^{\delta, k}(a') \qquad \text{and} \qquad \sigma^{\delta, k}(a) = \sigma^{\pi, k}(a') \qquad (k \geq 0).$$

In particular, we obtain that $\sigma(a) = \sigma(a')$. By Lemma 2.1.3, the defect and approximate point spectra form closed subsets of the Taylor spectrum.

Theorem 2.6.6 *Let A be a closed, commutative subalgebra of $L(X)$ with $I \in A$. Then, for $k \geq 0$, the rule assigning to each $a \in c(a)$ the set $\sigma^{\delta, k}(a)$ (or the set $\sigma^{\pi, k}(a)$) defines a spectral system on A.*

Proof In view of Lemma 2.2.4 it is obvious that $\sigma(a) \subset \sigma_A(a)$ for each $a \in c(A)$.

It remains to prove that the defect spectra satisfy the projection property. Since for each n-tuple $a \in c(A)$ and each permutation π of the index set $\{1, \ldots, n\}$ the cochain Koszul complexes of a and $(a_{\pi(1)}, \ldots, a_{\pi(n)})$ are isomorphic, it suffices to prove that

$$\sigma^{\delta, k}(a) = p\sigma^{\delta, k}(a, b) \qquad (k \geq 0) \tag{2.6.2}$$

in the special case where $b \in c(A)$ is a tuple of length 1. In view of the long exact cohomology sequence described in Lemma 2.2.2, the right-hand side of (2.6.2) is obviously contained in the left-hand side. To prove the opposite inclusion, let us suppose that

$$H^p((a, z - b), X) = 0 \qquad (n + 1 - k \leq p \leq n + 1, z \in \mathbb{C}).$$

Then, for each $z \in \mathbb{C}$, the map

$$H^p(a, X) \xrightarrow{z-b} H^p(a, X)$$

given by the componentwise action of $z - b$ is bijective for $n + 1 - k \leq p \leq n$, and surjective for $p = n - k$. Therefore, to complete the proof, it suffices to apply the following lemma. □

Lemma 2.6.7 *Suppose that*

$$\begin{array}{ccc} X & \xrightarrow{d} & Y \\ s \downarrow & & \downarrow T \\ X & \xrightarrow{d} & Y \end{array}$$

is a commutative diagram of continuous linear operators between Banach spaces such that

$$dX + (z - T)Y = Y$$

for each $z \in \mathbb{C}$. Then the operator d is onto.

Proof By dualizing the above diagram, one obtains that

$$(d', z - T'): Y' \to X' \oplus Y' \qquad (2.6.3)$$

is bounded from below for each $z \in \mathbb{C}$. Since T' restricted to $\operatorname{Ker} d'$ has empty approximate point spectrum, the operator d' is injective. By applying the q-functor to (2.6.3) one also obtains that $(d')^q$ is injective. Lemma 2.6.5 implies that d' has closed range. But then also the operator d has closed range, and the proof is complete.   □

Using the fact that the kernel of an epimorphism $\varphi: \mathcal{O}^X \to \mathcal{O}^Y$ between Banach-free analytic sheaves on a Stein space is acyclic, which will be proved in Chapter 4 (see Proposition 4.5.7), one can give an alternative proof of the projection property for the defect spectra which proceeds exactly as the corresponding proof for the projection property of the Taylor spectrum given in Section 2.5.

Let $\Phi: \mathcal{O}(\sigma(a)) \to L(X)$ be Taylor's analytic functional calculus as constructed in Section 2.5, and let A be the closure of the image of Φ. Then Corollary 2.6.3 is applicable and yields the following result.

Corollary 2.6.8 *For each $k \geq 0$, the analytic spectral mapping theorem as formulated in part (i) of Theorem 2.6.2 is valid for the spectra $\sigma^{\delta,k}(a)$ and $\sigma^{\pi,k}(a)$ with respect to Φ.*   □

By an iterative application of Lemma 2.6.5, one obtains that, for each finite complex of Banach spaces

$$0 \to X^0 \xrightarrow{\delta^0} X^1 \xrightarrow{\delta^1} \cdots \xrightarrow{\delta^{n-1}} X^n \to 0$$

and each integer $k \geq 0$, the condition that

$$\dim H^p(X^\bullet) < \infty \qquad (n - k \leq p \leq n)$$

is equivalent to the fact that $H^p((X^\bullet)^q) = 0$ for $n - k \leq p \leq n$. Analogously, if δ^k has closed range, then

$$\dim H^p(X^\bullet) < \infty \qquad (0 \leq p \leq k)$$

if and only if $H^p((X^\bullet)^q) = 0$ for $0 \leq p \leq k$.

Corollary 2.6.9 *For each integer $k \geq 0$ the spectra $\sigma_e^{\delta,k}$ and $\sigma_e^{\pi,k}$ satisfy the analytic spectral mapping theorem with respect to Φ.*

Proof Since for each commuting tuple $a \in L(X)^n$ on a Banach space X the identities $\sigma_e^{\delta,k}(a) = \sigma^{\delta,k}(a^q)$ and $\sigma_e^{\pi,k}(a) = \sigma^{\pi,k}(a^q)$ hold, the assertion follows as an application of Corollary 2.6.3 and Theorem 2.6.6. $\square$

We denote by $K(X,Y)$ the space of all compact operators between two Banach spaces X and Y, and we write $\mathscr{C}(X,Y)$ for the quotient space $L(X,Y)/K(X,Y)$. For a single operator $T \in L(X)$ the spectrum and essential spectrum of T coincide with the spectrum of the operator L_T of left multiplication with T on $L(X)$ and on $\mathscr{C}(X)$, respectively. Whether the corresponding results for commuting systems are true or not seems not to be known. But, even for a single operator T on X, the parts of $\sigma(T)$ considered in Corollary 2.6.8 are in general different from the corresponding parts of $\sigma(L_T)$.

To study these phenomena, we call a complex of Banach spaces

$$0 \to X^0 \xrightarrow{d^0} X^1 \xrightarrow{d^1} \cdots \xrightarrow{d^{n-1}} X^n \to 0$$

split (respectively, *Fredholm*) in degree $p(0 \le p \le n)$ if there are operators

$$X^{p-1} \xleftarrow{h} X^p \xleftarrow{h} X^{p+1}$$

such that $I_{X^p} = dh + hd$ (respectively, $I_{X^p} - (dh + hd) \in K(X^p)$). To keep the notation simple, we have omitted, and shall omit in general, the indices of the differential d. If Y is a Banach space, then $L(Y, X^\bullet)$ and $L(X^\bullet, Y)$ will denote the complexes induced by left and right multiplication with the differentials forming the complex $(X^\bullet, d^\bullet)$. Similarly, we define the complexes $\mathscr{C}(Y, X^\bullet)$ and $\mathscr{C}(X^\bullet, Y)$. Let $a \in L(X)^n$ be a fixed commuting tuple on a Banach space X.

Lemma 2.6.10 *Let k be an integer with $0 \le k \le n$.*

(a) The following conditions are equivalent:

   (i) $(X^\bullet, d^\bullet)$ is split (respectively, Fredholm) in degree $p = n - k, \ldots, n$;
   (ii) $L(Y, X^\bullet)$ (respectively, $\mathscr{C}(Y, X^\bullet)$) is exact in degree $p = n - k, \ldots, n$ for each Banach space Y.

If $(X^\bullet, d^\bullet) = K^\bullet(a, X)$, then (i) and (ii) are equivalent to:

   (iii) $L(X, X^\bullet)$ (respectively, $\mathscr{C}(X, X^\bullet)$) is exact in degree $p = n - k, \ldots, n$.

(b) The following conditions are equivalent:

   (i) $(X^\bullet, d^\bullet)$ is split (respectively, Fredholm) in degree $p = 0, \ldots, k$;
   (ii) $L(X^\bullet, Y)$ (respectively, $\mathscr{C}(X^\bullet, Y)$) is exact in degree $p = 0, \ldots, k$ for each Banach space Y.

If $(X^\bullet, d^\bullet) = K^\bullet(a, X)$, then (i) and (ii) are equivalent to:

   (iii) $L(X^\bullet, X)$ (respectively, $\mathscr{C}(X^\bullet, X)$) is exact in degree $p = 0, \ldots, k$.

Proof We indicate the main ideas for the non-trivial parts of the proof of the Fredholm case of part (b). All other cases can be proved in a similar way.

Suppose that $\mathscr{C}(X^{\bullet}, Y)$ is exact in degree $p = 0, \ldots, k$ for each Banach space Y, and that, for some $r < k$, operators

$$X^{r-1} \xleftarrow{\ h\ } X^r \xleftarrow{\ h\ } X^{r+1}$$

have been found with the property that

$$hd \in I_{X^r} - dh + K(X^r).$$

Then $(I_{X^{r+1}} - dh)d \in K(X^r, X^{r+1})$. Hence there is an operator h in $L(X^{r+2}, X^{r+1})$ satisfying $I_{X^{r+1}} - dh \in hd + K(X^{r+1})$. Thus a finite induction can be used to prove that (ii) implies (i). The reverse implication is obviously true.

To prove the equivalence with condition (iii) in the case of Koszul complexes, the reader should note that, for arbitrary Banach spaces Y and Z, the exactness of $\mathscr{C}(X^{\bullet}, Y)$ and $\mathscr{C}(X^{\bullet}, Z)$ in degree p implies the exactness of $\mathscr{C}(X^{\bullet}, Y \oplus Z) \cong \mathscr{C}(X^{\bullet}, Y) \oplus \mathscr{C}(X^{\bullet}, Z)$. $\qquad\square$

For a commuting tuple $a \in L(X)^n$ we define $sp^{\delta, k}(a)$ (respectively, $sp_e^{\delta, k}(a)$) as the complement of the set of those points $z \in \mathbb{C}^n$ for which $K^{\bullet}(z - a, X)$ is split (respectively, Fredholm) in degree $p = n - k, \ldots, n$. Correspondingly, we define $sp^{\pi, k}(a)$ (respectively, $sp_e^{\pi, k}(a)$) as the complement of the set of those points $z \in \mathbb{C}^n$ for which $K^{\bullet}(z - a, X)$ is split (respectively, Fredholm) in degree $p = 0, \ldots, k$. We call $sp(a) = sp^{\delta, n}(a)$ and $sp_e(a) = sp_e^{\delta, n}(a)$, respectively, the *split* and *essential split spectrum* of a.

Corollary 2.6.11 *For $k \geq 0$ we have the identities*:

$$sp^{\delta, k}(a) = \sigma^{\delta, k}(L_a, L(X)) \quad and \quad sp^{\pi, k}(a) = \sigma^{\delta, k}(R_a, L(X)),$$
$$sp_e^{\delta, k}(a) = \sigma^{\delta, k}(L_a, \mathscr{C}(X)) \quad and \quad sp_e^{\pi, k}(a) = \sigma^{\delta, k}(R_a, \mathscr{C}(X)).$$

Proof We consider only the Fredholm case. Since $K^{\bullet}(L_a, \mathscr{C}(X))$ is isomorphic to $\mathscr{C}(X, K^{\bullet}(a, X))$, the first part of Lemma 2.6.10 implies that

$$sp_e^{\delta, k}(a) = \sigma^{\delta, k}(L_a, \mathscr{C}(X)).$$

On the other hand, the Koszul complex $K_{\bullet}(R_a, \mathscr{C}(X))$ is isomorphic to the complex $\mathscr{C}(K^{\bullet}(a, X), X)$. By using the relation between the Koszul complex and the cochain Koszul complex described at the beginning of Section 2.2, one obtains that

$$sp_e^{\pi, k}(a) = \sigma^{\delta, k}(R_a, \mathscr{C}(X)). \qquad\square$$

We obtain the spectral mapping theorems for the split spectra by applying Corollary 2.6.3 to the algebra homomorphism $\Psi \colon \mathscr{O}(sp(a)) \to L(X), f \mapsto f(a)$, induced by Taylor's analytic functional calculus.

Corollary 2.6.12   *Each of the following spectra*

$$sp^{\delta,k}, \quad sp^{\pi,k}, \quad sp_e^{\delta,k}, \quad sp_e^{\pi,k} \qquad (0 \le k \le n)$$

satisfies the analytic spectral mapping theorem with respect to Ψ.

Proof   An obvious application of Lemma 2.5.8 yields that $f(L_a) = L_{f(a)}$ for each function $f \in \mathcal{O}(sp(a))$. Let A be the norm closure of the algebra $\psi(\mathcal{O}(sp(A)))$ in $L(X)$. By Corollary 2.6.11 all the above spectra are subspectra of the spectral system induced on A by the split spectrum. Hence the assertion follows as an application of Corollary 2.6.3.   □

We conclude this section with a result which implies that there is no difference between split spectra and spectra on Hilbert spaces.

Lemma 2.6.13   *Let*

$$0 \to X^0 \xrightarrow{d^0} X^1 \xrightarrow{d^1} \cdots \xrightarrow{d^{n-1}} X^n \to 0$$

be a complex of Banach spaces, and let $k \ge 0$ be an integer.

(*a*) *The following conditions are equivalent*:

   (*i*) $(X^\bullet, d^\bullet)$ *is split (respectively, Fredholm) in degree $p = k, \ldots, n$;*
   (*ii*) $H^p(X^\bullet) = 0$ *(respectively,* $\dim H^p(X^\bullet) < \infty$) *for $p = k, \ldots, n$ and* $\operatorname{Ker} d^p$ *is a complemented subspace of X^p for $p = k-1, \ldots, n$.*

(*b*) *The following conditions are equivalent*:

   (*i*) $(X^\bullet, d^\bullet)$ *is split (respectively, Fredholm) in degree $p = 0, \ldots, k$;*
   (*ii*) $H^p(X^\bullet) = 0$ *(respectively,* $\dim H^p(X^\bullet) < \infty$) *for $p = 0, \ldots, k$ and* $\operatorname{Im} d^p$ *is a complemented subspace of X^{p+1} for $p = 0, \ldots, k$.*

Proof   We prove only the Fredholm case of part (b). To simplify the notation, we suppress all indices on the occurring operators.

Let $h \in L(X^1, X^0)$ be an operator with $hd = I - K$ for suitable $K \in K(X^0)$. We choose closed subspaces L and M of X^0 with

$$X^0 = \operatorname{Ker}(I - K) \oplus L = \operatorname{Im}(I - K) \oplus M,$$

and we define $S \in L(X^0)$ by

$$S(hdx + y) = x \qquad \text{if} \quad x \in L, y \in M.$$

Then $p = I - hdS$ is the projection of X^0 onto M along $\operatorname{Im}(hd)$. Since $Sp = 0$, the operator $P = dSh \in L(X^{-1})$ is a projection. Because $dx = dShdx = Pdx$ for $x \in L$ and because dX^0/dL is finite dimensional, we have

$$\dim(dX^0/\operatorname{Im} P) < \infty. \tag{2.6.4}$$

In particular, it follows that dX^0 is a closed complemented subspace of X^1.

The identity $ShP = Sh$ implies that $h(\operatorname{Ker} P) \subset \operatorname{Ker} S = M$. Thus

$$\dim(\operatorname{Ker} P/\operatorname{Ker} h) < \infty, \tag{2.6.5}$$

and we can choose a projection $Q \in L(X^1)$ such that

$$QX^1 = \operatorname{Ker} h,$$

$$\dim \operatorname{Ker} Q/(\operatorname{Ker} d \cap \operatorname{Ker} Q) < \infty. \tag{2.6.6}$$

We set $P_1 = P$ and $Q_1 = Q$. Suppose that there is a second operator h in $L(X^2, X^1)$ with $dh + hd = I - K_1$ for a suitable operator $K_1 \in K(X^1)$. Then the relation $(Q_1 h)d = I - Q_1 K_1$ holds on $\operatorname{Ker} h \subset X^1$, and, according to the first part of the proof, there are projections P_2 and Q_2 on X^2 with

$$\dim d(\operatorname{Ker} h)/\operatorname{Im} P_2 < \infty \tag{2.6.7}$$

as well as

$$\dim \operatorname{Ker} P_2/\operatorname{Ker}(Q_1 h) < \infty,$$

$$Q_2 X_2 = \operatorname{Ker}(Q_1 h),$$

$$\dim \operatorname{Ker} Q_2/(\operatorname{Ker} d \cap \operatorname{Ker} Q_2) < \infty.$$

Because of (2.6.5) the quotient

$$dX^1/d(\operatorname{Ker} h) = d(\operatorname{Ker} P_1)/d(\operatorname{Ker} h)$$

is finite dimensional, and because of (2.6.7), this shows that dX^1 is a closed complemented subspace of X^2. If, in addition, an equation of the form

$$dh + hd = I - K_2 \quad (K_2 \in K(X^2))$$

holds on X^2, then by (2.6.6) we know that

$$d(Q_1 h) + hd = I - C$$

holds on X^2 with a suitable operator $C \in K(X^2)$. Thus the identity $(Q_2 h)d = I - Q_2 C$ holds on the kernel of $Q_1 h$, and one can repeat the process.

On the other hand, each compact operator $K \in K(X^p)$ satisfying an equation of the form $dh + hd = I - K$ on X^p induces the identity operator on $H^p(X^\bullet, d^\bullet)$. Consequently, if this space is Hausdorff, then it is necessarily finite dimensional.

Conversely, suppose that the assumptions described in (ii) are satisfied. For $p = 0, \ldots, k$, one can choose closed subspaces N^p and L^p of X^p such that

$$X^p = \operatorname{Ker} d^p \oplus L^p \quad \text{and} \quad \operatorname{Ker} d^p = \operatorname{Im} d^{p-1} \oplus N^p.$$

Then $M^p = N^p \oplus L^p$ is a closed subspace of X^p with $X^p = \operatorname{Im} d^{p-1} \oplus M^p$. Let M^{k+1} be a topologically direct complement of $\operatorname{Im} d^k$ in X^{k+1}. We define $h: X^{p+1} \to X^p$ for $0 \le p \le k$ by

$$h(d^p x + y) = x \quad \text{if} \quad x \in L^p \text{ and } y \in M^{p+1}.$$

Then, for each of these indices, the operator $I - (dh + hd)$ is the projection of X^p onto N^p along Im $d^{p-1} \oplus L^p$. □

In Chapter 10 we shall use the results of this section to study Fredholm complexes of Banach spaces, and the Fredholm theory for systems of Banach space operators, in a more detailed and systematic way.

2.7 REFERENCES AND COMMENTS

Results on parametrized complexes of Banach spaces have been used by Taylor (1970a) to deduce the fundamental properties of the joint spectrum introduced in the same paper. For surjective operators between Banach spaces depending continuously on a parameter, a result of the form stated in Theorem 2.1.2 has been proved by Bartle and Graves (1952). A discussion of this and related results can be found in Mantlik (1988). Elementary examples show that in the Fréchet space case Theorem 2.1.2 may fail without the condition that the given complex is locally uniformly exact. A version of Lemma 2.1.5 for surjective operators depending analytically on a parameter (i.e. the case where $\beta = 0$) was proved by Gleason (1964), while Lemma 2.1.5 in its present form can be found in Taylor (1970a).

The analytic functional calculus for finite systems of elements in a commutative Banach algebra was introduced in papers of Shilov (1953), Waelbroeck (1954), and Arens and Calderon (1955). The definitive form of this functional calculus, including the uniqueness result mentioned in the introduction of Chapter 2, can be found in Bourbaki (1967).

Our construction of an analytic functional calculus for systems of commuting operators follows closely the original work of Taylor (1970a,b). The slight simplification of the notion of a Cauchy–Weil system (Definition 2.3.3) is due to Frunză (1975b). In Taylor's original articles the analytic functional calculus is developed for commuting systems of Banach space operators. The definition of the joint spectrum (Definition 2.5.3) for systems of commuting operators on Fréchet spaces was suggested by Taylor (1972b). The extension of Taylor's constructive solution of the analytic functional calculus problem to the case of commuting regular Fréchet space operators is due to Putinar (1979). In particular, the proof of the projection property (Theorem 2.5.4) and of the spectral mapping theorem (Theorem 2.5.10) for the Fréchet space case can be found in Putinar (1979). An alternative description of Taylor's constructive solution of the analytic functional calculus problem for Fréchet space operators is given in Vasilescu (1982). In the Hilbert space case, more canonical integral representations of Taylor's analytic functional calculus have been obtained in Vasilescu (1978a) by forming surface integrals with respect to operator-valued integral kernels of Martinelli type (see also Curto 1988).

An axiomatic approach to the notion of joint spectrum for commuting

Banach space operators has been given by Słodkowski and Zelazko (1974). The generalized defect and approximate point spectra defined in Section 2.6 have been introduced by Słodkowski (1977) who, in the same paper, proved the polynomial spectral mapping theorem for these spectra. Other closely related polynomial spectral mapping theorems have been obtained by Harte (1972). The analytic spectral mapping theorem for the essential Taylor spectrum is due to Fainshtein (1980), while the analytic spectral mapping theorem for the generalized defect and approximate point spectra and their essential and split versions has been proved by Eschmeier (1987a). The discussion of these topics in Section 2.6 closely follows Eschmeier (1987a). In particular, the observation that for an arbitrary, abstract analytic functional calculus, the validity of a spectral mapping theorem is equivalent to its compatibility with the Gelfand transform (Theorem 2.6.2 and Corollary 2.6.3) can be found there. The result on the representability of spectral systems (Proposition 2.6.1), and some consequences, can also be found in Curto (1988).

3
Topological homology

The homology of topological algebras and of topological modules over topological algebras has been developed relatively late and rather at a tangent to the main directions of research in homological algebra. However, this gap has been considerably reduced in recent times, when the interest in the Hochschild cohomology of operator algebras has increased, partly due to the elaboration of cyclic cohomology and of non-commutative geometries.

The present chapter is devoted to a basic and rather narrow part of modern topological homology; its content was dictated by the aims of this book, and only secondly by the subject itself. To provide only a few hints of the development of the homology theories which will concern us, we must begin by mentioning the contribution of Kamowitz (1962). It appears that Kamowitz was the first to remark that Hochschild cohomology theory makes perfect sense in the category of Banach modules over (commutative) Banach algebras. This algebraic point of view soon led to the solution of some concrete problems concerning, in the first place, derivations, and secondly extensions or perturbations of Banach algebras (Guichardet 1966, Johnson 1969, Helemskii 1964, and Kadison and Ringrose 1971). Later this field expanded considerably, especially its applications to C^*- and W^*-algebras (see the monographs by Johnson 1972, Guichardet 1980, and Helemskii 1986). Recent progress in cohomology theories with internal symmetries, such as the cyclic or dihedral cohomologies is the latest and the most spectacular application of topological Hochschild cohomology (see Connes 1985, 1994 and Karoubi 1987).

One of the main tools used in this book is a topological module tensor product relative to a Fréchet algebra. The homological algebra of this tensor product controls operations such as the base change, the transversality, and the intersection multiplicity of two modules. It was J. L. Taylor who gave a new perspective on (local) spectral theory by transforming the basic problems, and most of the statements of this field, into homological terms. He developed for this purpose a relative homology theory which is based on the topological module tensor product mentioned previously (see Taylor 1972a). Quite surprisingly, the same topological homology was discovered by Kiehl and Verdier (1971), and independently by Helemskii (1970). Kiehl and Verdier were motivated by some technical questions from the deformation theory of complex spaces, while Helemskii was interested in applications to

the theory of commutative Banach algebras. There is a quite natural explanation for the similarity between these theories, disregarding their distinct
applications. The functors introduced independently in these different areas
represent the homological counterpart of the geometric notions of localization and transversality.

The present chapter is devoted to an elaboration of the abstract setting in
which the homological constructions of Taylor, Helemskii, and Kiehl and
Verdier can naturally be formulated. Once the algebraic framework has been
established, we shall discuss a series of applications of these objects in
subsequent chapters, beginning in Chapter 4. Our selection contained in the
present chapter definitely does not exhaust all topological homology theories,
or their various applications.

For the convenience of the reader, a few notational conventions, the
standard terminology, and some facts from homological algebra are recalled
in Appendix 2. The needed facts concerning topological vector spaces, and
especially nuclear spaces and topological tensor products, are briefly recalled
in Appendix 1.

3.1   TENSOR PRODUCTS RELATIVE TO
A FRÉCHET ALGEBRA

One of the geometric motivations for studying topological tensor products is
the following question. Let M and M' be complex manifolds. Given the
spaces of analytic (continuous, differentiable or measurable) functions on M
and M', describe the corresponding space of functions on the direct product
$M \times M'$. To give a simple example, we consider $M = D$ and $M' = D'$ to be
open discs in the complex plane centered at zero. Let $\mathscr{O}(D)$ and $\mathscr{O}(D')$
denote the spaces of analytic functions defined on D and D', respectively.
Then any analytic function f on the bidisc $D \times D'$ is represented by a double
power series

$$f(z,w) = \sum_{m,n=0}^{\infty} a_{mn} z^m w^n,$$

which is absolutely convergent on all compact subsets of $D \times D'$. On the
other hand, the elements of $\mathscr{O}(D)$ and $\mathscr{O}(D')$ are simple power series in z
and w, respectively. Clearly the algebraic tensor product of complex vector
spaces $\mathscr{O}(D) \otimes_{\mathbb{C}} \mathscr{O}(D')$ is dense in $\mathscr{O}(D \times D')$ relative to a system of semi-
norms which defines the completed projective tensor product $\mathscr{O}(D) \hat{\otimes} \mathscr{O}(D')$
(see Appendix 1 for terminology). Thus the solution of the above question in
this case is

$$\mathscr{O}(D \times D') \cong \mathscr{O}(D) \hat{\otimes} (D').$$

The main problem to be discussed in this section has a similar origin. The
question is: how does one compute the space of analytic functions $\mathscr{O}(X \times_S Y)$

defined on the fibre product of analytic maps between analytic spaces X, Y, and S. Although we shall postpone the complete answer to this question until Chapter 4 (where it is shown that $\mathcal{O}(X \times_S Y) = \mathcal{O}(X) \,\hat{\otimes}_{\mathcal{O}(S)}\, \mathcal{O}(Y)$), the topological operation which forms the correct substitute for the projective tensor product will be analysed in detail below.

To be more specific, the geometric fibre product of complex spaces

$$
\begin{array}{ccc}
X \times_S Y & \longrightarrow & Y \\
\downarrow & & \downarrow{\scriptstyle g} \\
X & \xrightarrow{\;f\;} & S,
\end{array}
$$

where $X \times_S Y = \{(x, y) \in X \times Y; f(x) = g(y)\}$, gives rise to a dual diagram of Fréchet $\mathcal{O}(S)$-modules

$$
\begin{array}{ccc}
\mathcal{O}(X \times_S Y) & \longleftarrow & \mathcal{O}(Y) \\
\uparrow & & \uparrow{\scriptstyle g^*} \\
\mathcal{O}(X) & \xleftarrow{\;f^*\;} & \mathcal{O}(S),
\end{array}
$$

and the universal object $\mathcal{O}(X \times_S Y)$ can be computed from the $\mathcal{O}(S)$-modules $\mathcal{O}(X)$ and $\mathcal{O}(Y)$ as a relative topological tensor product,

$$
\mathcal{O}(X \times_S Y) = \mathcal{O}(X) \,\hat{\otimes}_{\mathcal{O}(S)}\, \mathcal{O}(Y).
$$

This transcendental operation is the analogue of the algebraic tensor product over an algebra, which is well known in algebraic geometry as the function space of a fibred product. To be more specific, by transcendental we mean here a non-algebraic operation which requires completion with respect to a certain topology. This geometric point of view, and the terminology which arises from it, will be explained in Chapter 4.

We start with the basic definition of a Fréchet algebra.

Definition 3.1.1 A *Fréchet algebra A with identity* is a Fréchet space A endowed with a continuous bilinear (multiplication) map

$$
A \times A \to A, \qquad (a, b) \mapsto a \cdot b,
$$

which satisfies the usual axioms of associativity, distributivity, and possesses a unit element.

If in addition this multiplication is commutative, then the algebra A is called a *commutative Fréchet algebra*.

We shall denote by 1 the unit element of an algebra A.

Definition 3.1.2 Let A be a commutative Fréchet algebra. A *Fréchet A-module* is a Fréchet space E together with a continuous bilinear map

$$
A \times E \to E, \qquad (a, x) \mapsto ax,
$$

which satisfies the usual axioms turning E into an algebraic A-module.

Examples of Fréchet A-modules are *free modules* $A^n = A \otimes \mathbb{C}^n$ ($n \in \mathbb{N}$), or, more generally, *topologically free A-modules* $A \hat{\otimes} V$, where V is a Fréchet space. If V is in addition a nuclear space (see Appendix 1), then $A \hat{\otimes} V$ will be called a *nuclearly free A-module*.

Besides these artificial examples of Fréchet A-modules it is important to bring forward at the very beginning some natural examples which will appear frequently in this book. Let Ω be a bounded open set in $\mathbb{C}^n$, and let $\mathcal{O}(\Omega)$ denote the algebra of analytic functions defined on Ω. Then $\mathcal{O}(\Omega)$ is a Fréchet algebra with identity, and each of the following spaces is a Fréchet $\mathcal{O}(\Omega)$-module via the usual pointwise multiplication of functions: $\mathscr{C}^r(\Omega)$, $L^p_{\mathrm{loc}}(\Omega)$, $\mathrm{Lip}_\alpha(\Omega)$, $I(V) = \{f \in \mathcal{O}(\Omega); f|_V = 0\}$. Here $\mathscr{C}^r$ is the space of r times continuously differentiable functions ($r \in \mathbb{N}$), L^p_{loc} is the space of locally p-integrable functions with respect to the $2n$-dimensional Lebesgue measure on $\mathbb{C}^n$ ($1 \le p \le \infty$), Lip_α denotes the space of locally Lipschitz functions of order α ($0 < \alpha < 1$), and V is a closed subset of Ω.

Definition 3.1.3 Let A be a Fréchet algebra with identity, and let M, N be Fréchet A-modules. Then $\mathrm{Hom}_A(M, N)$ is the vector space of continuous A-linear maps between M and N.

As expected, the free objects defined above are projective, at least in a relative sense.

Lemma 3.1.4 *Let* $F \xrightarrow{g} G \to 0$ *be an exact sequence of Fréchet A-modules which is split over* $\mathbb{C}$. *Then for every topologically free A-module* $A \hat{\otimes} V$, *the sequence* $\mathrm{Hom}_A(A \hat{\otimes} V, F) \to \mathrm{Hom}_A(A \hat{\otimes} V, G) \to 0$ *remains exact.*

Proof We first note that, for every Fréchet A-module F, the following natural isomorphism holds:

$$\mathrm{Hom}_A(A \hat{\otimes} V, F) \cong \mathrm{Hom}_\mathbb{C}(V, F). \qquad (3.1.1)$$

Indeed, the map $\mathrm{Hom}_A(A \hat{\otimes} V, F) \to \mathrm{Hom}_\mathbb{C}(V, F)$, $f \mapsto f(1 \otimes \cdot)$, has the bilateral inverse $\mathrm{Hom}_\mathbb{C}(V, F) \to \mathrm{Hom}_A(A \hat{\otimes} V, F)$, $g \mapsto 1_A \otimes g$, where the last map is regarded as an operator with values in F.

By hypothesis, the surjective morphism $g: F \to G$ admits a continuous $\mathbb{C}$-linear right inverse $h: G \to F$, that is, $gh = id_G$, where id_G denotes the identity map on G. Hence the sequence

$$\mathrm{Hom}_\mathbb{C}(V, F) \to \mathrm{Hom}_\mathbb{C}(V, G) \to 0$$

is still (split) exact. $\square$

Though the category of Fréchet A-modules is not abelian (because, for instance, the cokernel of a morphism of Fréchet A-modules need not be well defined), Lemma 3.1.4 provides sufficiently many projective objects relative to a suitable admissible class of exact sequences, to ensure that the derived functors of a relative tensor product of Fréchet A-modules exists (see Appendix 2 for the terminology). To begin with, one needs to establish the existence of projective resolutions for every object in our category. We can actually prove even more, namely that there are canonical projective resolutions with a series of additional symmetry properties.

Proposition 3.1.5 *Every Fréchet A-module E has a topologically free resolution which is split over $\mathbb{C}$. Any two such resolutions are homotopically equivalent.*

Proof   Let E be a Fréchet A-module. Then the natural multiplication map $m(a, e) = ae$ $(a \in A, e \in E)$ factors through $A \mathbin{\hat{\otimes}_\pi} E$:

$$
\begin{array}{ccc}
A \times E & \xrightarrow{\ m\ } & E\,. \\[2pt]
& \searrow_{\,n} \nearrow_{n} & \\[2pt]
& A \mathbin{\hat{\otimes}_\pi} E &
\end{array}
$$

Moreover, the canonical map n has a $\mathbb{C}$-linear right inverse $h: E \to A \mathbin{\hat{\otimes}_\pi} E$, $h(x) = 1 \otimes x$.

By repeating this construction for $\mathrm{Ker}(n)$, and so on, we find a $\mathbb{C}$-split, topologically free resolution of E:

$$
\cdots \to L_1 \to L_0 \to E \to 0.
$$

The second assertion in the statement is a routine application of the projectivity property of topologically free A-modules (see Appendix 2).   □

However, there exists a canonical resolution for every Fréchet A-module, based on the following construction of the so-called Bar complex.

Definition 3.1.6   Let A be a Fréchet algebra with unit, and let E, F be Fréchet A-modules. The *Bar complex* $B_\bullet^A(E, F)$ consists of the spaces

$$
B_0^A(E, F) = E \mathbin{\hat{\otimes}_\pi} F,
$$

$$
B_n^A(E, F) = E \underbrace{\mathbin{\hat{\otimes}_\pi} A \mathbin{\hat{\otimes}_\pi} \cdots \mathbin{\hat{\otimes}_\pi} A}_{n \text{ times}} \mathbin{\hat{\otimes}_\pi} F \qquad (n \geq 1),
$$

and the boundary operators

$$
d_n: B_n^A(E, F) \to B_{n-1}^A(E, F) \qquad (n \geq 1)
$$

defined by

$$
d_n(e \otimes a_1 \otimes \cdots \otimes a_n \otimes f) = a_1 e \otimes \cdots \otimes a_n \otimes f + (-1)^n e \otimes a_1 \otimes \cdots \otimes a_n f
$$

$$
+ \sum_{j=1}^{n-1} (-1)^j e \otimes \cdots \otimes a_j a_{j+1} \otimes \cdots \otimes f.
$$

The verification of the identity $d_n d_{n+1} = 0$ $(n \geq 1)$ is a simple computation. The main advantage of the Bar complex is its naturality in all of its parameters. Furthermore, the Bar complex yields canonical resolutions for Fréchet A-modules in the following way.

Lemma 3.1.7 *Let A be a Fréchet algebra with unit and let E be a Fréchet A-module. The complex $B_\bullet^A(A, E)$ is a topologically free resolution of E which is split over $\mathbb{C}$.*

Proof It suffices to remark that the $\mathbb{C}$-linear maps

$$h_n \colon B_n^A(A, E) \to B_{n+1}^A(A, E),$$

$$h_n(a_0 \otimes a_1 \otimes \cdots \otimes a_n \otimes e) = 1 \otimes a_0 \otimes a_1 \otimes \cdots \otimes e$$

are continuous, and satisfy the splitting conditions

$$d_{n+1} h_n + h_{n-1} d_n = id_n \qquad (n \in \mathbb{N}),$$

where id_n denotes the identity operator on $B_n^A(A, E)$ and h_{-1}, d_0 are by definition equal to zero. □

Note that the commutativity of A was not used in the previous arguments. Henceforth we shall keep the commutativity assumption for simplicity, although it will not always be necessary.

Let E and F be two Fréchet A-modules, The homology spaces of the Bar complex $B_\bullet^A(E, F)$ have an intrinsic meaning, which is explained in what follows.

Definition 3.1.8 Let A be a commutative Fréchet algebra with identity. With each pair of Fréchet A-modules E and F one associates the locally convex spaces

$$E \hat{\otimes}_A F = H_0(B_\bullet^A(E, F)),$$

$$\text{Tôr}_n^A(E, F) = H_n(B_\bullet^A(E, F)) \qquad (n \in \mathbb{N}).$$

In other terms, we define

$$E \hat{\otimes}_A F = \text{Coker}\Big(d_1 \colon E \hat{\otimes}_\pi A \hat{\otimes}_\pi F \to E \hat{\otimes}_\pi F\Big),$$

$$d_1(e \otimes a \otimes f) = ae \otimes f - e \otimes af$$

and

$$\text{Tôr}_n^A(E, F) = \text{Ker}\, d_n / \text{Im}\, d_{n+1},$$

and we equip these spaces with their (not necessarily Hausdorff) quotient topology.

Lemma 3.1.9 *Let A be a Fréchet algebra with identity. For any Fréchet*

A-module E and Fréchet space V, the following isomorphism of Fréchet A-modules holds:

$$E \,\hat{\otimes}_A \left(A \,\hat{\otimes}_\pi V \right) \cong E \,\hat{\otimes}_\pi V. \qquad (3.1.2)$$

Proof The sequence

$$E \,\hat{\otimes}_\pi A \,\hat{\otimes}_\pi \left(A \,\hat{\otimes}_\pi V \right) \xrightarrow{d_1} E \,\hat{\otimes}_\pi \left(A \,\hat{\otimes}_\pi V \right) \xrightarrow{p} E \,\hat{\otimes}_\pi V \to 0,$$

where $p(e \otimes a \otimes v) = ae \otimes v$, is exact and $\mathbb{C}$-split. Indeed, the continuous $\mathbb{C}$-linear operators

$$h: E \,\hat{\otimes}_\pi V \to E \,\hat{\otimes}_\pi A \,\hat{\otimes}_\pi V, \qquad h(e \otimes v) = e \otimes 1 \otimes v$$

and

$$k: E \,\hat{\otimes}_\pi A \,\hat{\otimes}_\pi V \to E \,\hat{\otimes}_\pi A \,\hat{\otimes}_\pi A \,\hat{\otimes}_\pi V, \qquad k(e \otimes a \otimes v) = e \otimes a \otimes 1 \otimes v$$

satisfy the relations

$$ph = id, \qquad hp - d_1 k = id. \qquad \square$$

Exactly as in the algebraic case, the invariants Tôr can be computed with the help of any free resolution. This follows from the next result.

Proposition 3.1.10 *Let A be a commutative unital Fréchet algebra, and let $L_\bullet \to F \to 0$ be a $\mathbb{C}$-split, topologically free resolution of the Fréchet A-module F. For every Fréchet A-module E, we have isomorphisms of locally convex spaces*

$$\mathrm{T\hat{o}r}_n^A(E, F) \cong H_n\left(E \,\hat{\otimes}_A L_\bullet \right) \qquad (n \in \mathbb{N}).$$

Proof In view of the above isomorphism (3.1.2), we have $B_\bullet^A(E, F) \cong E \,\hat{\otimes}_A B_\bullet^A(A, F)$. According to Proposition 3.1.5, the complexes $B_\bullet^A(A, F)$ and $L_\bullet$ are homotopically equivalent in the category of Fréchet A-modules. It follows that the complexes $B_\bullet^A(E, F)$ and $E \,\hat{\otimes}_A L_\bullet$ are homotopically equivalent, but this time only in the category of locally convex spaces. This suffices to ensure that their homology spaces are topologically isomorphic. $\square$

Corollary 3.1.11 *Let E be a Fréchet A-module, and let V denote a Fréchet space. Then*

$$\mathrm{T\hat{o}r}_n^A\left(E, A \,\hat{\otimes}_\pi V \right) = \begin{cases} E \,\hat{\otimes}_\pi V & (n = 0) \\ 0 & (n > 0). \end{cases}$$

Proof It suffices to observe that the Fréchet A-module $A \,\hat{\otimes}_\pi V$ is topologically free. $\square$

In the same way, one proves the following result.

Corollary 3.1.11' *Let E be a Fréchet A-module, and let V denote a Fréchet space. Then*

$$\mathrm{T\hat{o}r}_n^A\left(A \mathbin{\hat{\otimes}_\pi} V, E\right) = \begin{cases} E \mathbin{\hat{\otimes}_\pi} V & (n = 0) \\ 0 & (n > 0). \end{cases} \qquad \square$$

The next theorem establishes the characteristic property of the functors Tôr.

Theorem 3.1.12 *Let A be a commutative Fréchet algebra with identity, and let $0 \to E_1 \to E_2 \to E_3 \to 0$ be an exact sequence of Fréchet A-modules. For any Fréchet A-module F, the sequence of locally convex spaces*

$$\cdots \mathrm{T\hat{o}r}_1^A(E_1, F) \to \mathrm{T\hat{o}r}_1^A(E_2, F) \to \mathrm{T\hat{o}r}_1^A(E_3, F)$$

$$\to E_1 \mathbin{\hat{\otimes}_A} F \to E_2 \mathbin{\hat{\otimes}_A} F \to E_3 \mathbin{\hat{\otimes}_A} F \to 0$$

is exact provided at least one of the following conditions is satisfied:

(a) *E_2 is a nuclear Fréchet space;*
(b) *the sequence $0 \to E_1 \to E_2 \to E_3 \to 0$ is $\mathbb{C}$-split;*
(c) *there exists a $\mathbb{C}$-split, nuclearly free resolution of F;*
(d) *A and F are nuclear Fréchet spaces.*

Proof Let $L_\bullet$ be a $\mathbb{C}$-split topologically free resolution of the A-module F. The only assertion to be proved is the exactness of the short sequence of complexes

$$0 \to E_1 \mathbin{\hat{\otimes}_A} L_\bullet \to E_2 \mathbin{\hat{\otimes}_A} L_\bullet \to E_3 \mathbin{\hat{\otimes}_A} L_\bullet \to 0. \qquad (3.1.3)$$

The induced long exact sequence of homology spaces (Lemma 1 in Appendix 2) is isomorphic to the sequence occurring in the statement of the theorem.

By taking into account the isomorphism (3.1.2) and the exactness properties of the tensor multiplication with nuclear spaces (cf. Appendix 1), it suffices to know either that E_1, E_2, E_3 are nuclear Fréchet spaces (condition (a)), or that $L_\bullet$ is nuclearly free (conditions (c) and (d)). Of course, condition (b) also implies that (3.1.3) is an exact sequence of complexes. $\square$

Corollary 3.1.13 *Let E and F be two Fréchet A-modules, and let $L_\bullet$ denote a topologically free resolution of E (not necessarily $\mathbb{C}$-split).*

Suppose that at least one of the following conditions is satisfied:

(α) *$L_\bullet$ consists of nuclear Fréchet spaces;*
(β) *A and F are nuclear Fréchet spaces;*
(γ) *$L_\bullet$ is $\mathbb{C}$-split.*

Then

$$\mathrm{T\hat{o}r}_n^A(E, F) \cong H_n\left(L_\bullet \mathbin{\hat{\otimes}_A} F\right) \qquad (n \in \mathbb{N})$$

as locally convex spaces.

Proof If one decomposes the exact sequence $L_\bullet \to E \to 0$ into short exact sequences

$$0 \to Z_n \to L_n \to Z_{n-1} \to 0 \qquad (n \geq 0)$$

where $Z_n = \text{Ker}(d_n : L_n \to L_{n-1})$ $(n \geq 0)$, $L_{-1} = E$, and $Z_{-1} = E$, then each of the conditions (α), (β), or (γ) guarantees the existence of an induced long exact Tôr-sequence. Consequently, by taking into account Corollary 3.1.11′, one finds topological isomorphisms

$$\text{Tôr}_j^A(Z_{n-1}, F) \cong \text{Tôr}_{j-1}^A(Z_n, F)$$

for any $j > 1$ and $n \geq 0$. In particular, for $n \geq 1$, we have

$$\text{Tôr}_n^A(E, F) \cong \text{Tôr}_1^A(Z_{n-2}, F).$$

Using the exact sequences

$$0 \to \text{Tôr}_1^A(Z_{n-2}, F) \to Z_{n-1} \,\hat{\otimes}_A\, F \to L_{n-1} \,\hat{\otimes}_A\, F \qquad (n \geq 1)$$

and

$$L_{n+1} \,\hat{\otimes}_A\, F \to L_n \,\hat{\otimes}_A\, F \to Z_{n-1} \,\hat{\otimes}_A\, F \to 0 \qquad (n \geq 0),$$

one can identify (even topologically) $\text{Tôr}_1^A(Z_{n-2}, F)$ with $H_n(L_\bullet \,\hat{\otimes}_A\, F)$ for $n \geq 1$. The second sequence with $n = 0$ yields the topological identification $H_0(L_\bullet \,\hat{\otimes}_A\, F) \cong E \,\hat{\otimes}_A\, F$. $\qquad \square$

This corollary shows, among other things, that

$$\text{Tôr}_n^A(E, F) \cong \text{Tôr}_n^A(F, E) \qquad (n \in \mathbb{N})$$

for any pair of Fréchet A-modules E and F.

At this point the basic properties of the derived functors of the relative tensor product $(\hat{\otimes}_A)$ have been established. Further, one may treat the inherent algebraic questions, for instance flatness, projectivity, (global) projective dimension, and so on, in a similar way. We shall not pursue these directions at this abstract level. They are well represented by the recent results of the Russian school (see the book by Helemskii, 1986 for a synthesis). We shall return to these questions later in the particular case of Fréchet modules over algebras of analytic functions.

The following technical notion will be of central importance for the applications to be given later.

Definition 3.1.14 Let A be a Fréchet algebra with identity. Two Fréchet A-modules E and F are called *transversal* if $E \,\hat{\otimes}_A\, F$ is a Hausdorff locally convex space and $\text{Tôr}_n^A(E, F) = 0$ for $n \geq 1$.

Suppose that E and F are transversal over A. Then we denote this by $E \perp_A F$.

As an application of the long exact Tôr-sequence described above, one obtains the next result.

Proposition 3.1.15 *Let F be a Fréchet A-module, and suppose that $0 \to E_1 \xrightarrow{j} E_2 \xrightarrow{q} E_3 \to 0$ is an exact sequence of Fréchet A-modules such that the spaces F and E_i satisfy at least one of the conditions specified in Theorem 3.1.12. Suppose that $E_3 \perp_A F$. Then $E_1 \perp_A F$ if and only if $E_2 \perp_A F$.*

Proof Under any of the conditions described in Theorem 3.1.12, one can choose a topologically free, $\mathbb{C}$-split resolution $L_\bullet$ of F such that

$$0 \to E_1 \hat{\otimes}_A L_\bullet \xrightarrow{j \otimes 1} E_2 \hat{\otimes}_A L_\bullet \xrightarrow{q \otimes 1} E_3 \hat{\otimes}_A L_\bullet \to 0$$

becomes a short exact sequence between complexes of Fréchet spaces. The assertion follows as an application of the induced long exact homology sequence (see Theorem 3.1.12).

If $E_2 \perp_A F$ and $E_3 \perp_A F$, then it follows immediately that $\text{Tôr}_k^A(E_1, F) = 0$ for all $k \geq 1$. Since the map

$$E_1 \hat{\otimes}_A F \xrightarrow{j \otimes 1} E_2 \hat{\otimes}_A F$$

is continuous and injective with values in a Hausdorff space, the space $E_1 \hat{\otimes}_A F$ is Hausdorff.

If $E_1 \perp_A F$ and $E_3 \perp_A F$, then $\text{Tôr}_k^A(E_2, F) = 0$ for $k \geq 1$, and it remains to check that $E_2 \hat{\otimes}_A F$ is a Hausdorff locally convex space, or equivalently, that the map

$$E_2 \hat{\otimes}_A L_1 \to E_2 \hat{\otimes}_A L_0$$

has closed range. To simplify the notation, we denote all boundary maps by δ.

Let (u_n) be a sequence in $E_2 \hat{\otimes}_A L_1$ such that (δu_n) converges to zero in $E_2 \hat{\otimes}_A L_0$. By using the hypothesis for E_3 and the above short exact sequence of complexes, we can choose a sequence (v_n) in $E_2 \hat{\otimes}_A L_2$ such that $(q \otimes 1(u_n - \delta v_n))$ tends to zero in $E_3 \hat{\otimes}_A L_1$. Hence there is a sequence (w_n) in $E_1 \hat{\otimes}_A L_1$ with the property that

$$\lim_{n \to \infty} (u_n - \delta v_n - j \otimes 1 w_n) = 0.$$

But then (δw_n) tends to zero in $E_1 \hat{\otimes}_A L_0$. Using the hypothesis for E_1, we can choose a sequence (t_n) in $E_1 \hat{\otimes}_A L_2$ such that $(u_n - \delta(v_n - j \otimes 1 t_n))$ converges to zero in $E_2 \hat{\otimes}_A L_1$. Thus the proof of Proposition 3.1.15 is complete.   □

A repeated use of Proposition 3.1.15 yields the next result.

Corollary 3.1.16 *Let E and F be Fréchet A-modules, and let $0 \to E \to L_\bullet$ be a finite resolution to the right such that $L_\bullet$, A, and F satisfy at least one of the conditions (α), (β), (γ) from Corollary 3.1.13. Suppose that $L_j \perp_A F$ for $j \geq 0$. Then $E \perp_A F$.*   □

Moreover, if F is a Fréchet A-module and if

$$L_{p+1} \to L_p \to L_{p-1} \to \cdots \to L_0 \to 0$$

is an exact complex of Fréchet A-modules such that $L_\bullet$, A, and F satisfy one of the conditions (α), (β), (γ) and such that $L_j \perp_A F$ $(0 \le j \le p+1)$, then

$$L_{p+1} \hat{\otimes}_A F \to L_p \hat{\otimes}_A F \to L_{p-1} \hat{\otimes}_A F \to \cdots \to L_0 \hat{\otimes}_A F \to 0$$

is an exact complex of Fréchet spaces.

Lemma 3.1.17 *Let $f: L^\bullet \to M^\bullet$ be a morphism of complexes of Fréchet A-modules that are bounded to the right. Let $K^\bullet$ be a complex of Fréchet A-modules that is bounded to the right such that*

$$L^p \perp_A K^q \qquad and \qquad M^p \perp_A K^q$$

for all p, $q \in \mathbb{Z}$. Suppose that one of the following conditions holds:

(i) *A and all K^q are nuclear $(q \in \mathbb{Z})$;*
(ii) *L^p and M^p are nuclear for all p.*

Suppose that $f \hat{\otimes}_A 1_{K^q}$ is a quasi-isomorphism for each $q \in \mathbb{Z}$. Then

$$f \hat{\otimes}_A 1 : \mathrm{Tot}\big(L^\bullet \hat{\otimes}_A K^\bullet\big) \to \mathrm{Tot}\big(M^\bullet \hat{\otimes}_A K^\bullet\big)$$

is a quasi-isomorphism. In particular, this is true if $f: L^\bullet \to M^\bullet$ is a quasi-isomorphism.

Proof For each q, the complexes $C^\bullet(f) \hat{\otimes}_A 1_{K^q}$ and $C^\bullet(f \hat{\otimes}_A 1_{K^q})$ are canonically isomorphic as complexes of Fréchet spaces. Hence by the remark following Corollary 3.1.16, we know that, if f is a quasi-isomorphism, then so is $f \hat{\otimes}_A 1_{K^q}$ for each $q \in \mathbb{Z}$.

In the latter case, all rows in the bounded double complex $C^\bullet(f) \hat{\otimes}_A K^\bullet$ are exact. Hence the associated total complex is exact. But this total complex is isomorphic to the mapping cone of the morphism

$$f \hat{\otimes}_A 1 : \mathrm{Tot}\big(L^\bullet \hat{\otimes}_A K^\bullet\big) \to \mathrm{Tot}\big(M^\bullet \hat{\otimes}_A K^\bullet\big). \qquad \square$$

If the third complex $K^\bullet$ degenerates to a single Fréchet A-module K, then the problem dealt with in the last lemma is simply the question of whether a given quasi-isomorphism between complexes of Fréchet A-modules remains a quasi-isomorphism, when forming the module tensor product with the identity on K. We shall need results of the above type in Chapters 5 and 9.

3.2 INVERSE LIMITS

The second important functor to be studied is the limit of an inverse system of locally convex spaces. To give only one motivation for the importance of this subject, let us recall the classical *Mittag–Leffler problem*: find a meromorphic function on a domain Ω in $\mathbb{C}$ with prescribed polar expansions in a discrete subset of Ω.

The standard solution to this problem uses an approximation device as follows. Consider a relatively compact exhaustion $(\Omega_n)_{n=0}^{\infty}$ of Ω: $\overline{\Omega}_n \subset \Omega_{n+1}$, $\bigcup_{n=0}^{\infty} \Omega_n = \Omega$, such that the restriction maps $\mathscr{O}(\Omega_{n+1}) \to \mathscr{O}(\Omega_n)$ have dense range. Then solve the problem on each Ω_n by elementary means, inductively choosing the solutions on Ω_{n+1} and Ω_n so that they are close together on Ω_{n-1}. Then these functions will converge on Ω to the desired meromorphic function. This proof depends on the existence of the topological isomorphism $\mathscr{O}(\Omega) \cong \varprojlim_n \mathscr{O}(\Omega_n)$, and on the fact that the structural maps in the inverse system $\mathscr{O}(\Omega_0) \leftarrow \mathscr{O}(\Omega_1) \leftarrow \mathscr{O}(\Omega_2) \leftarrow \cdots$ have dense range. Exhaustions like this are also extremely useful in the theory of functions of several complex variables.

Next we shall investigate this phenomenon at an abstract level. The main reference here is Palamodov (1971). For the sake of simplicity, we confine ourselves to the case of Fréchet spaces.

Let $(I, \leq)$ be a partially ordered countable set, and let $(E_\alpha)_{\alpha \in I}$ be an *inverse system* of Fréchet spaces, that is, a family of Fréchet spaces E_α ($\alpha \in I$) together with a system of continuous linear maps

$$i_\alpha^\beta: E_\alpha \to E_\beta \qquad (\alpha \geq \beta)$$

such that

$$i_\alpha^\alpha = id, \qquad i_\beta^\gamma i_\alpha^\beta = i_\alpha^\gamma$$

for $\alpha \geq \beta \geq \gamma$.

Then the *projective limit* $E = \varprojlim_\alpha (E_\alpha)$ is, by definition, the closed subspace of the topological product $\prod_{\alpha \in I} E_\alpha$ consisting of all families $(x_\alpha)_{\alpha \in I}$ with the property that $i_\alpha^\beta(x_\alpha) = x_\beta$ for all $\alpha \geq \beta$. Equipped with the relative topology of the product space, E is a Fréchet space. Let $\pi_\alpha: E \to E_\alpha$ be the canonical projection for each α. Then by definition $i_\alpha^\beta \pi_\alpha = \pi_\beta$ for all $\alpha \geq \beta$.

Whenever $(G, (g_\alpha))$ is another system consisting of a Fréchet space G and maps $g_\alpha: G \to E_\alpha$ satisfying $i_\alpha^\beta g_\alpha = g_\beta$ for all $\alpha \geq \beta$, then there is a unique continuous linear map $g: G \to E$ with the property that $g_\alpha = \pi_\alpha \circ g$ for all α. Moreover, up to canonical isomorphisms, the projective limit is the only solution of this universal problem (see, for instance, Robertson and Robertson 1964 or Floret and Wloka 1968).

Our aim is to study the exactness properties of the functor $\varprojlim$. From the observation that $\varprojlim E_\alpha$ is precisely the algebraic inverse limit endowed with a certain specified topology, we obtain the first elementary result.

Lemma 3.2.1 *The topological inverse limit is a left-exact functor.* $\square$

In other terms, for every short exact sequence of inverse systems (in the natural sense)

$$0 \to E_\alpha \to F_\alpha \to G_\alpha \to 0 \qquad (\alpha \in I),$$

the complex

$$0 \to \varprojlim E_\alpha \to \varprojlim F_\alpha \to \varprojlim G_\alpha$$

is exact.

To prove the existence of the derived functors of the inverse limit functor, we need some more terminology; and we make the assumption that $I = \mathbb{N}$ with the natural order (see Nöbeling 1961 and Palamodov 1971).

Definition 3.2.2 An inverse system $(E_\alpha)_{\alpha \geq 1}$ is called *free*, with Fréchet spaces $G_1, G_2, \ldots$ as generators, if $E_\alpha = \prod_{j=1}^\alpha G_j$ and if the structural maps are the canonical projections.

In spite of the terminology, a free inverse system has an injectivity property rather than the expected projectivity behaviour.

Lemma 3.2.3 *Every inverse system of Fréchet spaces $(E_\alpha)_{\alpha \geq 1}$ possesses a free resolution to the right of the form*

$$0 \to (E_\alpha) \xrightarrow{u} (L_\alpha) \xrightarrow{v} (L_{\alpha-1}) \to 0. \qquad (3.2.1)$$

Proof Let $L_\alpha = \prod_{j=1}^\alpha E_j$ be the free inverse system with generators $E_1, E_2, \ldots$. Let us define

$$u(x) = (i_\alpha^1 x, \ldots, i_\alpha^\alpha x) \qquad (x \in E_\alpha)$$

and

$$v(x_1, \ldots, x_\alpha) = (x_1 - i_2^1 x_2, \ldots, x_{\alpha-1} - i_\alpha^{\alpha-1} x_\alpha) \qquad (x_j \in E_j).$$

Then it is straightforward to verify the exactness of the short sequences

$$0 \to E_\alpha \to L_\alpha \to L_{\alpha-1} \to 0 \qquad (\alpha \geq 1)$$

where $L_0 = 0$. $\qquad \square$

By a *morphism* $u: E \to F$ of *countable inverse systems* $E = (E_\alpha)_{\alpha \geq 1}$ and $F = (F_\alpha)_{\alpha \geq 1}$ of Fréchet spaces we mean a family $u = (u_\alpha)_{\alpha \geq 1}$ of continuous linear maps $u_\alpha: E_\alpha \to F_\alpha$ that commute with the structural maps of E and F in the obvious sense. A morphism $u = (u_\alpha)$ is called a *monomorphism* if all the components u_α are topological monomorphisms.

Since the category of locally convex spaces possesses sufficiently many injective objects, a familiar algebraic argument can be used to show that the category of inverse systems of locally convex spaces also has sufficiently many injective objects (see Palamodov 1971 and for terminology Appendix 2).

More precisely, the Banach spaces of the form $l^\infty(\Lambda)$, where Λ is an arbitrary index set, are injective objects in the category of locally convex spaces. Indeed, whenever we have a topological monomorphism $j: X \to Y$ and a continuous linear map $f: X \to l^\infty(\Lambda)$, then the Hahn–Banach theorem implies the existence of a continuous linear map $g: Y \to l^\infty(\Lambda)$ satisfying $f = g \circ j$.

Moreover, every separated locally convex space X is contained in a topological product of such spaces. Namely, the map

$$X \to \prod_{U \in \mathcal{U}} l^\infty(U^\circ), \qquad x \mapsto (\langle x, \lambda \rangle_{\lambda \in U^\circ})_{U \in \mathcal{U}}$$

defines a topological monomorphism. Here $\mathcal{U}$ is a basis of neighbourhoods of $0 \in X$, and U° denotes the polar of an element $U \in \mathcal{U}$ (Appendix 1). Since topological products of injective locally convex spaces are injective, we have just shown that each separated locally convex space is contained in an injective space of the same category.

By an *injective object* in the category of all countable inverse systems of Fréchet spaces, we mean an inverse system $I = (I_\alpha)_{\alpha \geq 1}$ of Fréchet spaces with the property that, for each monomorphism $j: E \to F$ of inverse systems and each morphism $f: E \to I$, there exists a morphism $g: F \to I$ such that the following diagram commutes:

$$E \xrightarrow{\;\;j\;\;} F .$$
$$\underset{f}{\searrow} \;\; \underset{I}{\nearrow} g$$

The category of countable inverse systems of Fréchet spaces possesses sufficiently many injective objects, that is, for each countable inverse system E of Fréchet spaces, there is an injective system I and a suitable monomorphism $j: E \to I$. To prove this assertion, the reader should note that each free inverse system with injective Fréchet spaces as generators is injective, and that each countable inverse system of Fréchet spaces can be embedded in a free inverse system with injective generators (see Lemma 3.2.3).

One can use the above remarks exactly as in the algebraic case (see Appendix 2) to construct, for each countable inverse system E of Fréchet spaces, an injective resolution

$$0 \to E \to I^0 \to I^1 \to I^2 \to \cdots$$

and to define the derived functors of $\varprojlim$ as

$$\varprojlim_\alpha{}^{(n)}(E) = H^n\left(\varprojlim_\alpha I^\bullet\right).$$

Standard arguments show that this definition is independent of the choice of the injective resolution.

Our main result in this section can be proved without the language of derived functors, but it has a natural homological interpretation.

Theorem 3.2.4   *Consider a short exact sequence*

$$0 \to (E_\alpha) \xrightarrow{\ u\ } (F_\alpha) \xrightarrow{\ v\ } (G_\alpha) \to 0$$

of countable inverse systems of Fréchet spaces. If all structural maps $i_\alpha^\beta \colon E_\alpha \to E_\beta$
($\alpha \geq \beta \geq 1$) have dense range, then the induced sequence

$$0 \to \varprojlim (E_\alpha) \to \varprojlim (F_\alpha) \to \varprojlim (G_\alpha) \to 0$$

of inverse limits remains exact.

Proof   We start by choosing in each space F_α a translation-invariant metric
ρ_α generating the topology of F_α such that all the structural maps belonging
to (F_α) are contractive. To see that this is possible, choose in each F_α a
translation-invariant metric d_α generating the topology of F_α and set

$$\rho_\alpha(x, y) = \max_{\beta \leq \alpha} d_\beta(i_\alpha^\beta(x), i_\alpha^\beta(y)) \qquad (x, y \in F_\alpha).$$

To simplify the notation, we shall write ρ instead of ρ_α.
   Let $(y_\alpha)_{\alpha \geq 1}$ be a sequence of elements $y_\alpha \in F_\alpha$ with

$$i_{\alpha+1}^\alpha(y_{\alpha+1}) - y_\alpha \in \operatorname{Im}(u)$$

for all $\alpha \geq 1$, and let $\varepsilon > 0$ be arbitrary. We define $y_1' = y_1$, and we choose
elements $k_1 \in E_1$, $x_2 \in E_2$ with

$$u(k_1) = i_2^1(y_2) - y_1', \qquad \rho(u i_2^1(x_2), u(k_1)) < \varepsilon/2.$$

Then for $y_2' = y_2 - u(x_2)$, one obtains the estimate

$$\rho(i_2^1(y_2'), y_1') = \rho(i_2^1(y_2) - u i_2^1(x_2), y_1') < \varepsilon/2.$$

Inductively one can define a sequence $(y_\alpha')_{\alpha \geq 1}$ of elements $y_\alpha' \in F_\alpha$ with
$y_\alpha' - y_\alpha \in \operatorname{Im} u$ and

$$\rho(i_{\alpha+1}^\alpha(y_{\alpha+1}'), y_\alpha') < \varepsilon/2^\alpha \qquad (\alpha \geq 1).$$

   Using the fact that the structural maps of the system (F_α) are contractive,
one easily deduces that $(i_{\alpha+n}^\alpha(y_{\alpha+n}'))_{n \geq 1}$ is a Cauchy sequence in F_α for each
fixed $\alpha \geq 1$. If y_α'' denotes its limit, then $\rho(y_1'', y_1) < \varepsilon$, and the relations

$$i_{\alpha+1}^\alpha(y_{\alpha+1}'') = y_\alpha'', \qquad y_\alpha'' - y_\alpha \in \operatorname{Im}(u)$$

hold for all $\alpha \geq 1$.
   To show that $v \colon \varprojlim_\alpha F_\alpha \to \varprojlim_\alpha G_\alpha$ is onto, it suffices to choose, for a given
element $(z_\alpha)_{\alpha \geq 1}$ in $\varprojlim_\alpha G_\alpha$, a sequence $(y_\alpha)_{\alpha \geq 1}$ of elements $y_\alpha \in F_\alpha$ with
$v(y_\alpha) = z_\alpha$ and then to apply the above construction to $(y_\alpha)_{\alpha \geq 1}$. $\square$

Remark 3.2.5   The last proof can be used to obtain other standard results for
inverse systems of Fréchet spaces. For later use, we indicate the following
particular cases.

(a) What was really shown in the proof of Theorem 3.2.4 is the following slightly stronger result. Whenever

$$(E_\alpha) \xrightarrow{\ u\ } (F_\alpha) \xrightarrow{\ v\ } (G_\alpha)$$

is an exact sequence of countable inverse systems of Fréchet spaces such that all structural maps in (E_α) have dense range, then for each element $(z_\alpha) \in \varprojlim G_\alpha$ with $z_\alpha \in \text{Im}(v)$ for all $\alpha \geq 1$, there is an element $(y_\alpha)_{\alpha \geq 1}$ in $\varprojlim F_\alpha$ with

$$(z_\alpha)_{\alpha \geq 1} = v(y_\alpha)_{\alpha \geq 1}.$$

(b) Let $(E_\alpha)_{\alpha \geq 1}$ be an inverse system of Fréchet spaces such that all structural maps have dense range, and let E denote its inverse limit. If $y_1 \in E_1$ is arbitrary, then the proof of Theorem 3.2.4 applied to the identity morphism $u = id$ on (E_α) and the sequence $(y_\alpha)_{\alpha \geq 1} = (y_1, 0, 0, \dots)$ shows that, for each $\varepsilon > 0$, there is an element (y''_α) in E with $\rho(y''_1, y_1) < \varepsilon$. Hence the canonical projection $E \to E_1$ has dense range. Of course, the same argument applies to all canonical projections $E \to E_\alpha \ (\alpha \geq 1)$.

(c) Let $v: (E_\alpha)_{\alpha \geq 1} \to (F_\alpha)_{\alpha \geq 1}$ be a morphism of countable inverse systems of Fréchet spaces such that all structural maps in $(E_\alpha)_{\alpha \geq 1}$ have dense range, and let us denote by E and F the corresponding inverse limits. We claim that, if all maps $v: E_\alpha \to F_\alpha$ have dense range, then $v: E \to F$ has dense range. By the theorem of Hahn–Banach it suffices to check that the only continuous linear form f on F with $f \circ v = 0$ is the trivial one. By the definition of the inverse limit topology (see Appendix 1 or §6.2.2 in Floret and Wloka 1968), there is an $\alpha \geq 1$ and a continuous semi-norm p on F_α such that $|f(x)| \leq p(\pi_\alpha x)$ holds for all $x \in F$. Here $\pi_\alpha: F \to F_\alpha$ is the canonical projection. Again by the theorem of Hahn–Banach, there is a continuous linear form f_α on F_α with

$$f = f_\alpha \circ \pi_\alpha.$$

Because $f_\alpha \circ v \circ \pi_\alpha = f \circ v = 0$, we conclude that $f_\alpha = 0$, and hence that $f = 0$.

As an application of Theorem 3.2.4, one can show that the higher derived functors of certain inverse systems of Fréchet spaces vanish, which in turn implies exactness results for sequences of inverse limits.

Corollary 3.2.6   *An inverse system of Fréchet spaces* $(E_\alpha)_{\alpha \geq 1}$ *is* $\varprojlim$ *-acyclic, that is,* $\varprojlim^{(n)}(E_\alpha) = 0$ *for* $n \geq 1$, *if all structural maps* $i_\alpha^\beta: E_\alpha \to E_\beta \ (\alpha \geq \beta)$ *have dense range.*

Proof   Let $E = (E_\alpha)_{\alpha \geq 1}$ and let $j: E \to I$ be a monomorphism of E into a free injective system $I = (I_\alpha)_{\alpha \geq 1}$ of Fréchet spaces. If $Q = I/\text{Im}(j)$ denotes the quotient system defined in an obvious sense, then the short exact sequence

$$0 \to E \to I \to Q \to 0$$

induces a long exact sequence of derived functors

$$0 \to \varprojlim{}^{(0)}(E) \to \varprojlim{}^{(0)}(I) \to \varprojlim{}^{(0)}(Q) \to$$
$$\to \varprojlim{}^{(1)}(E) \to \varprojlim{}^{(1)}(I) \to \varprojlim{}^{(1)}(Q) \to$$
$$\to \varprojlim{}^{(2)}(E) \to \cdots.$$

Since I is injective, the spaces $\varprojlim{}^{(n)}(I)$ vanish for $n \geq 1$. By Theorem 3.2.4, we know that $\varprojlim{}^{(1)}(E) = 0$. Now the proof can be concluded by an inductive argument using the fact that the structural maps of $(Q_\alpha)_{\alpha \geq 1}$ are surjective. $\square$

Since the structural maps in a free system are surjective, the long exact sequence of derived functors induced by the canonical free resolution described in Lemma 3.2.3 yields the following result.

Corollary 3.2.7 Let $(E_\alpha)_{\alpha \geq 1}$ be an inverse system of Fréchet spaces. Then $\varprojlim{}^{(n)}(E_\alpha) = 0$ for $n \geq 2$. $\square$

As a consequence we obtain the following characterization of $\varprojlim$-acyclic inverse systems.

Theorem 3.2.8 For an arbitrary inverse system $E = (E_\alpha)_{\alpha \geq 1}$ of Fréchet spaces, the following conditions are equivalent:

(i) E is $\varprojlim$-acyclic;
(ii) $\varprojlim{}^{(1)}(E) = 0$;
(iii) for each short exact sequence

$$0 \to E \to F \to G \to 0$$

of countable inverse systems of Fréchet spaces, the induced sequence

$$0 \to \varprojlim(E) \to \varprojlim(F) \to \varprojlim(G) \to 0$$

is exact;
(iv) the sequence of inverse limits

$$0 \to \varprojlim(E_\alpha) \to \varprojlim(L_\alpha) \to \varprojlim(L_{\alpha-1}) \to 0$$

induced by the canonical resolution (3.2.1) is exact. $\square$

To prove the missing implications, the reader should use the long exact sequence of derived functors induced by the short exact sequences occurring in the statement of the theorem. In the cited article of Palamodov (1971) one can find characterizations of acyclic inverse systems which improve our Theorem 3.2.4. In the applications it will be useful to have results not only for short exact sequences of inverse systems, but for exact sequences of arbitrary length.

Corollary 3.2.9 *Let* $(E^n, f^n)_{n \in \mathbb{Z}}$ *be an exact sequence of countable inverse systems of Fréchet spaces. Suppose that each E^n is* $\varprojlim$ *-acyclic* $(n \in \mathbb{Z})$. *Then the induced sequence*

$$\cdots \to \varprojlim (E^n) \to \varprojlim (E^{n+1}) \to \cdots$$

of inverse limits is exact.

Proof Using the derived long exact sequences corresponding to the short exact sequences of the form

$$0 \to \mathrm{Ker}(f^n) \to E^n \to \mathrm{Im}(f^n) \to 0,$$

one obtains that each of the inverse systems $\mathrm{Im}(f^n) = \mathrm{Ker}(f^{n+1})$ $(n \in \mathbb{Z})$ is $\varprojlim$ -acyclic. Therefore, the sequences

$$0 \to \varprojlim_{\alpha} (\mathrm{Ker}(f^n)) \to \varprojlim_{\alpha} (E^n) \to \varprojlim_{\alpha} (\mathrm{Im}(f^n)) \to 0$$

are exact, and the proof is complete.   □

In our applications we shall only consider inverse systems of the type studied in Theorem 3.2.4.

Definition 3.2.10 An inverse system of Fréchet spaces $(E_\alpha)_{\alpha \geq 1}$ with the property that all structural maps $i_\alpha^\beta \colon E_\alpha \to E_\beta$ $(\alpha \geq \beta \geq 1)$ have dense range will be called a *Mittag-Leffler system*.

Every Fréchet space E is isomorphic to the limit of an inverse system $(E_\alpha)_{\alpha \geq 1}$ of Banach spaces (see for instance Robertson and Robertson 1964). Since one can always replace the Banach spaces E_α by the closed subspaces $\overline{\pi_\alpha E_\alpha} \subset E_\alpha$, each Fréchet space is the limit of a Mittag-Leffler inverse system of Banach spaces.

If A is a commutative, unital Fréchet algebra and if E is a Fréchet A-module which is the inverse limit in the category of topological A-modules of a Mittag-Leffler system of Banach A-modules, then E will be said to possess the *Mittag-Leffler property*.

Proposition 3.2.11 *Let A be a commutative Fréchet algebra with identity, and let $(E_\alpha)_{\alpha \in \mathbb{N}}$ be a Mittag-Leffler inverse system of Fréchet A-modules. Suppose that F is a Fréchet A-module such that $E_\alpha \perp_A F$ for all α. Then $\left(\varprojlim_{\alpha} (E_\alpha) \right) \perp_A F$.*

Proof Let $L_\bullet \to F \to 0$ be a topologically free, $\mathbb{C}$-split resolution of the A-module F. By the results of Section 3.1, for each $\alpha \geq 1$, the induced sequence $E_\alpha \hat{\otimes}_A L_\bullet \to E_\alpha \hat{\otimes}_A F \to 0$ remains exact. In this way, we obtain an exact sequence

$$\cdots \to \left(E_\alpha \hat{\otimes}_A L_1 \right)_\alpha \to \left(E_\alpha \hat{\otimes}_A L_0 \right)_\alpha \to \left(E_\alpha \hat{\otimes}_A F \right)_\alpha \to 0$$

of morphisms of Mittag-Leffler inverse systems of Fréchet spaces. By Corollary 3.2.9 the induced sequence of inverse limits remains exact. By definition the spaces L_i $(i \geq 0)$ are of the form $L_i = A \hat{\otimes} M_i$ with suitable

Fréchet spaces M_i. Using the fact that the projective tensor product commutes with reduced inverse systems (see Appendix 1 or §41.6 in Köthe 1979), we obtain the identifications

$$\varprojlim_\alpha \left(E_\alpha \,\hat\otimes_A L_i \right) = \left(\varprojlim_\alpha E_\alpha \right) \hat\otimes_\pi M_i = \left(\varprojlim_\alpha E_\alpha \right) \hat\otimes_A L_i \qquad (i \geq 0).$$

Hence the sequence $(\varprojlim E_\alpha) \,\hat\otimes_A L_\bullet \to \varprojlim (E_\alpha \,\hat\otimes_A F) \to 0$ is exact, and the assertion follows from Corollary 3.1.13. □

Note that the preceding arguments also imply that, under the same assumptions,

$$\left(\varprojlim_\alpha E_\alpha \right) \hat\otimes_A F \cong \varprojlim_\alpha \left(E_\alpha \,\hat\otimes_A F \right)$$

holds as a topological isomorphism of Fréchet spaces.

A similar argument, using the compatibility of the projective tensor product with arbitrary products of locally convex spaces (§41.6 in Köthe 1979), shows that Proposition 3.2.11 remains true if countable Mittag-Leffler inverse limits are replaced by products (with arbitrary index set) of Fréchet A-modules.

3.3   DIRECT LIMITS AND DUALITY

In contrast to the case of inverse limits, the direct limit and the duality functor are exact. Here the only interesting problem from our homological point of view is the question of whether these functors preserve topological homomorphisms (see Appendix 1). Although this problem has a complete solution for general locally convex spaces (see Palamadov 1971), we shall only discuss the Fréchet space case.

Let $(I, \leq)$ be a partially ordered, countable set. An *inductive* or *direct system* of Fréchet spaces is a family $(E_\alpha)_{\alpha \in I}$ of Fréchet spaces endowed with structural maps $i_\alpha^\beta \colon E_\alpha \to E_\beta$ $(\alpha \leq \beta)$ satisfying

$$i_\alpha^\alpha = id, \quad i_\beta^\gamma i_\alpha^\beta = i_\alpha^\gamma \qquad (\alpha \leq \beta \leq \gamma).$$

The *direct limit* of such a system $(E_\alpha)_{\alpha \in I}$ is a locally convex space denoted by $\varinjlim_\alpha E_\alpha$ together with continuous linear maps $i_\alpha \colon E_\alpha \to \varinjlim_\alpha E_\alpha$ such that $i_\beta i_\alpha^\beta = i_\alpha$ $(\alpha \leq \beta)$ and such that the following universal property holds:

Whenever G is a locally convex space and $(g_\alpha)_{\alpha \in I}$ is a family of continuous linear maps $g_\alpha \colon E_\alpha \to G$ that commute with the structural maps of the direct system $(E_\alpha)_{\alpha \in I}$, then there is a unique continuous linear map g for which the following diagram commutes:

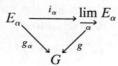

A solution of this universal problem always exists. For instance, one can define $\lim_{\overrightarrow{\alpha}} E_\alpha$ as the algebraic direct limit

$$E = \{(x, \alpha); \alpha \in I \text{ and } x \in E_\alpha\}/\sim$$

(where $(x, \alpha) \sim (y, \beta)$ if and only if there is an index $\gamma \geq \alpha, \beta$ with $i_\alpha^\gamma x = i_\beta^\gamma y$) equipped with the inductive locally convex topology relative to the mappings

$$i_\alpha: E_\alpha \to E, \quad x \to (x, \alpha)/\sim \qquad (\alpha \in I).$$

To give only two natural examples, let M be a complex manifold. For a fixed point x in M, the space $\mathscr{O}_x = \lim_{\overrightarrow{x \in U}} \mathscr{O}(U)$ is an inductive limit of Fréchet algebras with identity. The structural maps $\mathscr{O}(U) \to \mathscr{O}(V)$ are given by the restrictions of analytic functions from the open set U to the open set V ($x \in V \subset U \subset M$). Of course, only a fundamental system of neighbourhoods of x is needed for the computation of the above inductive limit. Secondly, let us consider an increasing sequence of compact subsets K_n of M such that $M = \bigcup_{n=1}^\infty K_n$. Let $\mathscr{D}_{K_n}(M)$ denote the space of C^∞-functions on M with support contained in K_n, and let $\mathscr{D}_{K_n}(M) \to \mathscr{D}_{K_{n+1}}(M)$ be the inclusion map. Then $\mathscr{D}(M) = \lim_{\overrightarrow{n}} \mathscr{D}_{K_n}(M)$ is the space of test functions on M. Its topological dual is the space of distributions defined on the manifold M.

Since $\lim_{\overrightarrow{\alpha}} E_\alpha$ as a vector space is the algebraic inductive limit of the system (E_α), we obtain the next result.

Lemma 3.3.1 *Let* $0 \to E_\alpha \xrightarrow{u} F_\alpha \xrightarrow{v} G_\alpha \to 0$ ($\alpha \in I$) *be an exact sequence of direct systems of Fréchet spaces. Then*

$$0 \to \lim_{\overrightarrow{\alpha}} E_\alpha \xrightarrow{u} \lim_{\overrightarrow{\alpha}} F_\alpha \xrightarrow{v} \lim_{\overrightarrow{\alpha}} G_\alpha \to 0 \qquad (3.3.1)$$

is an exact sequence.   □

The definition of the inductive limit topology (see §23.3.14 in Floret and Wloka 1968) shows that in the above setting the map $v: \lim_{\overrightarrow{\alpha}} F_\alpha \to \lim_{\overrightarrow{\alpha}} G_\alpha$ is a topological epimorphism. However, the map u need not be a topological monomorphism in spite of the fact that all the components u_α ($\alpha \in I$) are topological monomorphisms. The simplest criterion for testing whether a morphism of locally convex spaces is a topological monomorphism makes use of duality theory. Before formulating this result, let us make a few comments on non-Hausdorff locally convex spaces.

If E is an arbitrary (not necessarily separated) locally convex space, then the relative topology of E on the linear subspace $E_0 = \overline{\{0\}}$ is the trivial topology, that is, its only open sets are the empty set and E_0 itself. If p denotes any algebraic projection of E onto E_0, then the map

$$j: E \to E_0 \times \prod_{U \in \mathscr{U}} l^\infty(U^\circ), \quad x \mapsto (p(x), (\langle x, \lambda \rangle_{\lambda \in U^\circ})_U),$$

where $\mathcal{U}$ is a basis of neighbourhoods of $0 \in E$, defines a topological monomorphism. The topological product on the right-hand side is easily seen to be an injective object in the category of all locally convex spaces. Hence also in the non-separated case there are sufficiently many injective objects.

Lemma 3.3.2   *An injective continuous linear map* $u\colon E \to F$ *between (not necessarily separated) locally convex spaces is a topological monomorphism if and only if the induced map* $u^*\colon \mathrm{Hom}_\mathbb{C}(F, l^\infty(\Lambda)) \to \mathrm{Hom}_\mathbb{C}(E, l^\infty(\Lambda))$ *is onto for each index set* Λ.

Proof   If u is a topological monomorphism, then because of the injectivity property of $l^\infty(\Lambda)$ described in the preceding section, the map u^* is onto for each index set Λ.

To prove the converse, consider the topological embedding

$$j\colon E \to E_0 \times \prod_{U \in \mathcal{U}} l^\infty(U^\circ)$$

described above. Since $u^*\colon \mathrm{Hom}_\mathbb{C}(F, I) \to \mathrm{Hom}_\mathbb{C}(E, I)$ is onto for $I = E_0$ and $I = l^\infty(U^\circ)$ $(U \in \mathcal{U})$, the above map is clearly onto, when I denotes the topological product of all these spaces. Thus j admits a factorization of the form $j = f \circ u$ with a suitable continuous linear map f, and hence u is a topological monomorphism.   □

This result shows that the map u in (3.3.1) is a topological monomorphism if and only if the 'dual' sequence

$$0 \to \mathrm{Hom}_\mathbb{C}\left(\varinjlim_\alpha G_\alpha, l^\infty(\Lambda)\right) \xrightarrow{v^*} \mathrm{Hom}_\mathbb{C}\left(\varinjlim_\alpha F_\alpha, l^\infty(\Lambda)\right)$$

$$\xrightarrow{u^*} \mathrm{Hom}_\mathbb{C}\left(\varinjlim_\alpha E_\alpha, l^\infty(\Lambda)\right) \to 0$$

is exact for each index set Λ. Fortunately, in a few interesting cases the following observation allows one to check this criterion.

Lemma 3.3.3   *Let* $(E_\alpha)_{\alpha \in I}$ *be a direct system and let* G *be an arbitrary locally convex space. Then*

$$\mathrm{Hom}_\mathbb{C}\left(\varinjlim_\alpha E_\alpha, G\right) \cong \varprojlim_\alpha \mathrm{Hom}_\mathbb{C}(E_\alpha, G)$$

as vector spaces.   □

The proof is a direct consequence of the universal property of inductive limits described above.

Corollary 3.3.4   *For each inductive system of locally convex spaces* $(E_\alpha)_{\alpha \in I}$, *one has the algebraic isomorphism*

$$\left(\varinjlim_\alpha E_\alpha\right)' \cong \varprojlim_\alpha E_\alpha'.$$

□

The behaviour of the dual topologies with respect to this isomorphism is a more delicate question (see Séminaire Banach, 1972).

To provide only one possible application, we conclude our discussion with a representative (but far from the most general) result. Recall that an inductive system $(E_n)_{n=1}^{\infty}$ of Fréchet spaces is *strict* if all structural maps $E_n \to E_{n+1}$ are topological monomorphisms. In this case, the limit $\varinjlim_n E_n$ is a so-called strict (LF)-space.

Theorem 3.3.5 *Let* $0 \to E_n \xrightarrow{u} F_n \xrightarrow{v} G_n \to 0$ $(n \geq 0)$ *be an inductive system of exact sequences of Fréchet spaces. If* (G_n) *is a strict system, then the induced map* u: $\varinjlim_n E_n \to \varinjlim_n F_n$ *is a topological monomorphism.*

Proof For each index set Λ,

$$0 \to \mathrm{Hom}_{\mathbb{C}}(G_n, l^{\infty}(\Lambda)) \xrightarrow{v^*} \mathrm{Hom}_{\mathbb{C}}(F_n, l^{\infty}(\Lambda)) \xrightarrow{u^*} \mathrm{Hom}_{\mathbb{C}}(E_n, l^{\infty}(\Lambda)) \to 0$$

is an exact sequence of countable inverse systems of $\mathbb{C}$-vector spaces. Since the structural maps

$$(i_n^{n+1})^*: \mathrm{Hom}_{\mathbb{C}}(G_{n+1}, l^{\infty}(\Lambda)) \to \mathrm{Hom}_{\mathbb{C}}(G_n, l^{\infty}(\Lambda))$$

$(n \geq 0)$ are onto, an elementary algebraic argument similar to the one used in the proof of Theorem 3.2.4 allows us to deduce that the induced sequence of inverse limits is exact. □

A much deeper analysis is carried out by Palamodov (1971) for not necessarily strict inductive systems.

The duality functor, which has already occurred in the present section, admits a similar theory of derived functors. To be more precise, let (E', β) denote the strong dual of a Fréchet space E (see Appendix 1). The open mapping principle and the Hahn–Banach theorem yield the next result.

Proposition 3.3.6 *Let* $0 \to E \xrightarrow{u} F \xrightarrow{v} G \to 0$ *be an exact sequence of Fréchet spaces. Then the dual sequence*

$$0 \leftarrow (E', \beta) \xleftarrow{u'} (F', \beta) \xleftarrow{v'} (G', \beta) \leftarrow 0$$

is exact. □

However, the maps u' and v' in the above sequence need not be strong topological homomorphisms (see Palamodov, 1971). A criterion derived from Lemma 3.3.2 shows that u' and v' are topological homomorphisms if and only if the sequences

$$0 \to \mathrm{Hom}_{\mathbb{C}}(E'_{\beta}, l^{\infty}(\Lambda)) \xrightarrow{u_*} \mathrm{Hom}_{\mathbb{C}}(F'_{\beta}, l^{\infty}(\Lambda)) \xrightarrow{v_*} \mathrm{Hom}_{\mathbb{C}}(G'_{\beta}, l^{\infty}(\Lambda)) \to 0$$

are exact for each index set Λ.

Since it is clear that the map u_* is injective, the functor D_{Λ}: $E \mapsto \mathrm{Hom}_{\mathbb{C}}(E'_{\beta}, l^{\infty}(\Lambda))$ (where E is a Fréchet space) turns out to be exact to the left. Therefore the derived functors $D_\Lambda^{(p)}$ ($p \geq 0$) exist. The condition $D_\Lambda^{(1)} = 0$ will ensure that the morphisms u' and v' are topological homomorphisms with respect to the strong dual topologies. This condition is satisfied, for instance, when E is a so-called quasi-normable Fréchet space, or in particular, a Fréchet–Schwartz space (see Palamodov, 1971 for details).

3.4   ABSTRACT DERIVED FUNCTORS

In a similar way to the case of the inverse limit functor presented in Section 3.2, one can associate derived functors with each semi-exact functor defined on the category of Fréchet modules over a Fréchet algebra. This general construction is discussed in detail in the book of Helemskii (1986).

Let A be a Fréchet algebra with identity, and let A-**Mod** denote the category of Fréchet A-modules. An additive contravariant functor

$$F: A\text{-}\mathbf{Mod} \to \mathbf{Ab}$$

with values in the category of abelian groups is called *semi-exact to the right* if, for every exact and $\mathbb{C}$-split sequence of Fréchet A-modules

$$0 \to E' \to E \to E'' \to 0,$$

the sequence

$$F(E') \leftarrow F(E) \leftarrow F(E'') \leftarrow 0$$

is exact.

Since the category A-**Mod** has sufficiently many projective objects with respect to $\mathbb{C}$-split morphisms (see Lemma 3.1.4), the derived functors F^p ($p \geq 0$) of F exist, and they can be defined by

$$F^p(E) = H^p(F(B_A^\bullet(A, E))). \tag{3.4.1}$$

Indeed, following the well-known algebraic scheme, the complex $B_A^\bullet(A, E)$ is an admissible free resolution of the module E (see Appendix 2 and Lemma 3.1.7). Moreover, Proposition 3.1.2 shows that

$$F^p(E) = H^p(F(P^\bullet)),$$

whenever P is an admissible projective resolution of E.

Short exact sequences of Fréchet A-modules induce long exact sequences between the derived functors.

Proposition 3.4.1   *Let* $F: A\text{-}\mathbf{Mod} \to \mathbf{Ab}$ *be an additive, contravariant, semi-exact functor. To every short exact and $\mathbb{C}$-split sequence of Fréchet A-modules*

$$0 \to E' \to E \to E'' \to 0$$

corresponds a long exact sequence of the form

$$0 \to F(E'') \to F(E) \to F(E') \to F^1(E'') \to F^1(E) \to \cdots .$$

Proof   The short exact sequence of Bar resolutions

$$0 \to B_A^\bullet(A, E') \to B_A^\bullet(A, E) \to B_A^\bullet(A, E'') \to 0$$

splits (even A-linearly). Hence a long exact sequence as in the assertion exists. The fact that one can identify $F^0(E) \cong F(E)$ is a consequence of the semi-exactness of the functor F.   □

One treats the case of a covariant semi-exact functor similarly.

To give just one possible example, let G be a Fréchet A-module and consider the functor $\mathrm{Hom}_A(*, G)$. As in the algebraic case, this functor is exact to the right and contravariant. According to the last proposition, its derived functors exist. They are usually denoted by $\mathrm{E\hat{x}t}_A^p(*, G)$.

Though we shall not make any further use of these topological Ext-spaces, let us make a few remarks

(1) Let G be topological A-module. As usual, the condition

$$\mathrm{E\hat{x}t}_A^1(E, G) = 0$$

is equivalent to the relative projectivity of the module E (in the sense of Section 3.1). Therefore the projective dimension of a Fréchet A-module can be defined in terms of the $\mathrm{E\hat{x}t}$ spaces (see Helemskii 1970, 1986 for a detailed discussion).

(2) Taylor (1972a, b) computed the same invariants $\mathrm{E\hat{x}t}_A^p(*, G)$ by means of certain specific injective resolutions of G. The locally convex topologies of the components of these injective objects are essential in his approach (see Taylor, 1972a).

(3) The abelian groups

$$\mathcal{H}^p(A, E) = \mathrm{E\hat{x}t}_{A \hat{\otimes} A}^P(A, E) \qquad (p \geq 0)$$

are the continuous Hochschild cohomology groups of the $A \hat{\otimes} A$-module (that is A-bimodule) E. They appear in the work of Kamowitz (1962) and Johnson (1969) (see the introduction to this chapter).

(4) The group $\mathrm{E\hat{x}t}_A^p(E, G)$ parametrizes the family of $\mathbb{C}$-split extensions

$$0 \to G \to E_{p-1} \to \cdots \to E_1 \to E_0 \to E \to 0,$$

exactly as in the algebraic case. In particular the Yoneda pairing and all other algebraic properties of extensions are preserved in this topological setting.

(5) The determination of whether the invariants $\mathrm{E\hat{x}t}_A^p(E, G)$ vanish for certain (commutative or non-commutative) algebras A and specific values of p, E, G can be a highly non-trivial problem. For example, it is still an open question whether $\mathcal{H}^p(A, A) = 0$ $(p > 1)$ for a von Neumann algebra A. Similarly, it is not known for which C^*-algebras $A \subset L(H)$ one has $\mathcal{H}^1(A, L(H)) = 0$, where H is a Hilbert space (see Kadison and Ringrose 1971; Helemskii 1986).

3.5   REFERENCES AND COMMENTS

The books of Helemskii (1986), (1989) contain numerous and accurate historical remarks concerning the development of topological homology theory. In fact, the first monograph is a genetic approach to this field, with special emphasis on the theory of (commutative) Banach algebras. A categorial framework for the relative homological algebra, which is the prototype of the homology theory developed in this chapter, can be found in Chapter IX of the monograph of Mac Lane (1963).

Although the homological interpretation of the geometric notion of transversality as a relative flatness goes back to the founders of modern algebraic geometry (see Serre, 1965), the topological, homological transversality condition as expressed in Definition 3.1.14 appears only in *Séminaire de géométrie analytique* (1974). The notation '$M \perp_A N$' appears in Taylor (1972a), where it is also interpreted as a transversality condition.

The reader should be aware of the nuclearity or splitting conditions necessary for any application of the long exact sequence of Tôr spaces (see Theorem 3.1.12 and the following corollary). A very simple example shows that without these conditions the canonical long exact sequence of Tôr spaces may fail to exist. Take for instance $A = \mathbb{C}$, and consider an exact sequence of Banach spaces

$$0 \to E_1 \to E_2 \to E_3 \to 0.$$

Then for an arbitrary space F and for arbitrary completed topological tensor products ($\overline{\otimes}$), the sequence

$$E_1 \overline{\otimes} F \to E_2 \overline{\otimes} F \to E_3 \overline{\otimes} F \to 0$$

is not in general exact. Hence a long exact sequence of Tôr spaces cannot exist without further conditions.

A very general programme for formulating the theory of locally convex spaces in categorial and homological terms was initiated in Séminaire Banach (1972). However, this can only be seen as a useful translation of the fundamental principles of functional analysis into a different type of language. In general, the difficulties cannot be resolved by a mere translation into the language of homological algebra. From this point of view, the paper of Palamodov (1971) is more precise, and it contains some definite new results. For instance, in this article one finds characterizations of acyclic inverse systems of Fréchet spaces. For some recent advances in this direction, see Vogt (1989).

Besides the general functor Eẋt mentioned at the end of Section 3.4, there is a parallel theory of extensions of C^*-algebras originating in the work of Brown *et al.* (1973). This theory has found numerous applications in topology and in the theory of operator algebras. Moreover, it has drastically changed the latter field in the last two decades (see also Douglas, 1980; Kasparov, 1988). Several other generalized homology theories and their interactions with operator algebras are presently being studied. Two remarkable examples are algebraic K-theory and cyclic (co)homology.

4
Analytic sheaves

Algebraic and sheaf-theoretic methods have been used to solve a series of classical problems in the geometry of complex manifolds and in the theory of functions of several complex variables. They have also oriented the research related to several complex variables for a good few decades. Developed in parallel with Grothendieck's theory of schemes, the theory of complex analytic spaces represents nowadays the natural framework and the language in which these algebraic methods are formulated.

In the present section we bring forward some facts and terminology from the modern theory of complex analytic spaces and the corresponding sheaf cohomology theory. Far from attempting to achieve completeness, we discuss only those topics and proofs which will be relevant for the subsequent chapters. The reader is assumed to be acquainted with the basic theory of analytic (or algebraic) sheaves and with the terminology and methods of local commutative algebra. Our concise and selective exposition by no means supplies a rigorous introduction to the subject. We recommend the reader to consult the excellent monograph of Grauert and Remmert (1984) for more details.

4.1 STEIN SPACES AND STEIN ALGEBRAS

An *analytic space* in our terminology is a pair $(X, \mathcal{O}_X)$ consisting of a Hausdorff topological space X which is countable at infinity and a sheaf $\mathcal{O}_X$ of unital $\mathbb{C}$-algebras on X satisfying the following local requirement: every point $x \in X$ has an open neighbourhood U such that the pair $(U, \mathcal{O}_X | U)$ is isomorphic to a local model. By a *local model* we mean a pair $(Z, \mathcal{O}_Z)$, where Z is the set of common zeros of a finite family of analytic functions $f_1, \ldots, f_p \in \mathcal{O}(V)$ defined on an open set V in $\mathbb{C}^n$, that is,

$$Z = \{z \in V; f_1(z) = \cdots = f_p(z) = 0\},$$

and where

$$\mathcal{O}_Z = \left(\mathcal{O}_V / (f_1, \ldots, f_p) \mathcal{O}_V \right) | Z.$$

Here $\mathcal{O}_V$ denotes the sheaf of complex analytic functions on the open set V in $\mathbb{C}^n$. An analytic space $(X, \mathcal{O}_X)$ will sometimes simply be denoted by X. For the algebra of global sections of the sheaf $\mathcal{O}_X$ we write $\mathcal{O}_X(X)$ or simply $\mathcal{O}(X)$.

The stalks $\mathcal{O}_{X,x} = \varinjlim_{x \in U} \mathcal{O}(U)$ of the structural sheaf $\mathcal{O}_X$ of an analytic space are local algebras with residue field canonically isomorphic to $\mathbb{C}$.

Each complex manifold M is an analytic space if $\mathcal{O}_M$ is defined as the sheaf of analytic functions on M. The larger category of complex analytic spaces allows new operations which in general are not possible within the category of complex manifolds. For instance, the fibre product and the orbit space of certain (discrete) group actions are internal operations in the category of analytic spaces. In particular, consider the space $\mathbb{C}^2$ with complex coordinates (z, w) and identify two pairs (z, w) and $(ez, e^2 w)$, if e is a primitive root of order 3 of 1. Then the corresponding quotient space is an analytic space with a singular point at the equivalence class of $(0, 0)$. Sometimes analytic spaces are also called *complex singular spaces*.

The modern point of view which considers the structural sheaf together with the support variety seems to go back (implicitly) to Oka. This idea has certain clear advantages. For example, the refined theory of the singularities of an analytic set (i.e. the common zeros of a family of analytic functions) becomes in this way a question of local algebra rather than of complex analysis.

Another of the main tools in the theory of analytic spaces is the notion of analytic modules. Roughly speaking, an analytic module is the localization and abstraction of various familiar spaces of functions, differential forms, tensors, etc., encountered in global analysis or geometry.

An *analytic module* (or *analytic sheaf*) on an analytic space $(X, \mathcal{O}_X)$ is a sheaf of $\mathcal{O}_X$-modules defined on X. For example, each quotient of $\mathcal{O}_X$ modulo an ideal sheaf I in $\mathcal{O}_X$ is an analytic module. The analytic sheaves on a fixed analytic space form an abelian category with sufficiently many injective objects (here injective has in fact two meanings, one referring to the fibres, that is, a sheaf has fibres that are injective modules, and the other describes a sheaf which has the usual injectivity lifting property for global morphisms of analytic modules). The cohomology theory of analytic sheaves is based on the second definition of injectivity (see Grauert and Remmert 1984 for details). This theory reflects a number of intrinsic properties of the underlying space. To give only a single example, let Y be a closed analytic subspace of the space $(X, \mathcal{O}_X)$. The long exact sequence of cohomology associated with the short exact sequence

$$0 \to I \to \mathcal{O}_X \to \mathcal{O}_X / I \to 0,$$

where I is the ideal sheaf defining the space Y, yields obstructions to the surjectivity of the restriction map r

$$0 \to I(X) \to \mathcal{O}_X(X) \xrightarrow{\ r\ } (\mathcal{O}_X / I)(Y) \to H^1(X, I) \to \cdots .$$

Among the analytic $\mathcal{O}_X$-modules there is a distinguished class of sheaves, called coherent $\mathcal{O}_X$-modules, which have remarkable stability properties under algebraic and geometric operations (see Serre 1955; Grauert and Remmert 1984).

An analytic module $\mathscr{F}$ on the analytic space $(X, \mathscr{O}_X)$ is said to be *finitely generated* by sections $s_1, \ldots, s_m \in \mathscr{F}(X)$ if the morphism $\mathscr{O}_X^m \to \mathscr{F}$ induced by this m-tuple of sections is surjective.

Definition 4.1.1 An *analytic sheaf* $\mathscr{F}$ on the analytic space $(X, \mathscr{O}_X)$ is *coherent* if it is locally finitely generated and if, for every open subset $U \subset X$ and morphism of $\mathscr{O}_U$-modules $u: \mathscr{O}_U^p \to \mathscr{F} \mid U$, the kernel of u is a locally finitely generated $\mathscr{O}_U$-module.

In particular, a coherent analytic $\mathscr{O}_X$-module $\mathscr{F}$ admits locally finite presentations of length one, that is, each point $x \in X$ has an open neighbourhood U on which $\mathscr{F} \mid U$ can be identified with the cokernel of a morphism of $\mathscr{O}_U$-modules

$$\mathscr{O}_U^q \to \mathscr{O}_U^p \to \mathscr{F} \mid U \to 0. \tag{4.1.1}$$

A deep theorem of Oka simplifies the above definition of a coherent sheaf.

Theorem 4.1.2 (Oka) *Let $\mathscr{F}$ be an analytic sheaf on the analytic space $(X, \mathscr{O}_X)$. Then the following assertions are equivalent:*

(a) *$\mathscr{F}$ is a coherent $\mathscr{O}_X$-module;*
(b) *$\mathscr{F}$ admits locally finite presentations of length one;*
(c) *$\mathscr{F}$ admits locally finite-type $\mathscr{O}_X$-free resolutions to the left, of arbitrary finite length.*

Proof Obviously, assertion (b) follows from (a), and (c) implies (b). Since the kernel and the cokernel of a morphism between coherent analytic sheaves are coherent, one can reduce the proof of the remaining implications to the assertion that the structure sheaf $\mathscr{O}_{\mathbb{C}^n}$ is coherent for each positive integer n. This result is known as Oka's coherence theorem (for a complete proof see Grauert and Remmert 1984). $\square$

Note the difference between points (a) and (b) in Oka's theorem. While point (a) asserts that any morphism from a finite free module has a locally finitely generated kernel, point (b) states that only one finite presentation at each point of X suffices to ensure the coherence of $\mathscr{F}$. Returning to some more concrete examples, let us consider an open domain Ω in $\mathbb{C}^n$. If there exists an analytic function $f \in \mathscr{O}(\Omega)$ which cannot be holomorphically extended across any point of $\partial\Omega$, then Ω is said to be a *domain of holomorphy*.

Every open subset $\Omega \subset \mathbb{C}$ is a domain of holomorphy, but in higher dimensions this is no longer true. For example, the annular domain defined by $\Omega = \{z \in \mathbb{C}^2; 1 < |z| < 2\}$ is not a domain of holomorphy because of Hartogs' extension phenomenon (for details see Hörmander 1966).

There are several classical characterizations of domains of holomorphy.

Among them we recall Cartan–Thullen's theorem, which asserts that a domain in $\mathbb{C}^n$ is a domain of holomorphy if and only if it is holomorphically convex. This means that the holomorphically convex hull $\hat{K} = \{z \in \Omega; |f(z)| \leq \|f\|_{\infty, K}; f \in \mathscr{O}(\Omega)\}$ of every compact subset $K \subset \Omega$ is also compact in Ω. Various classical approximation, interpolation or extension problems for analytic functions can be solved in the affirmative on domains of holomorphy (see Shabat 1985; Grauert and Remmert 1984).

The necessity to extend the notion of a domain of holomorphy to analytic spaces has led to the following concept.

Definition 4.1.3 An analytic space $(X, \mathscr{O}_X)$ is said to be a *Stein space* if the algebra of global sections $\mathscr{O}_X(X)$ separates the points of X and if X is holomorphically convex with respect to the algebra $\mathscr{O}_X(X)$.

Most of the function theory which is valid on domains of holomorphy has its counterpart on abstract Stein spaces. However, the methods used in the case of singular spaces are quite different. They rely essentially on algebraic and sheaf theoretic language. The central result from this point of view is the following.

Theorem 4.1.4 (H. Cartan) *Let $(X, \mathscr{O}_X)$ be a Stein space. Then for every analytic coherent sheaf $\mathscr{F}$ on X, the following assertions hold:*

(A) *for every $x \in X$, the $\mathscr{O}_{X,x}$-module $\mathscr{F}_x$ is generated by restrictions of global sections in $\mathscr{F}(X)$;*
(B) *$H^p(X, \mathscr{F}) = 0$ for $p \geq 1$.* $\square$

In fact assertion (A) is redundant (more precisely (B) $\Rightarrow$ (A) on Stein spaces). The notation $H^\bullet(X, \mathscr{F})$ is used for the sheaf cohomology of $\mathscr{F}$ on the space X (for a complete proof of Cartan's theorem see Grauert and Remmert 1984).

The (*Zariski* or *embedding*) *dimension* of an analytic space $(X, \mathscr{O}_X)$ is the supremum of the Krull dimensions of the local rings $\mathscr{O}_{X,x}$

$$\dim X = \sup_{x \in X} (\dim \mathscr{O}_{X,x}),$$

that is, the supremum of the lengths of maximal chains of prime ideals in the local rings $\mathscr{O}_{X,x}$.

It can be shown that every finite-dimensional Stein space is isomorphic to a closed Stein subspace of $\mathbb{C}^N$ for a sufficiently large N (see for instance Hörmander 1966).

Since every closed analytic subspace of a Stein space is also Stein, an arbitrary analytic space can be covered by Stein open subspaces. But such a covering is acyclic, by virtue of Cartan's Theorem (B), with respect to any coherent analytic sheaf. This leads to a very useful method of computing the Čech cohomology by using Stein open coverings.

In what follows we recall the construction of the locally convex topology carried by the section spaces of a coherent analytic sheaf.

Let $\mathscr{F}$ be a coherent analytic sheaf on the analytic space $(X, \mathscr{O}_X)$. The locally finite presentations (4.1.1) give rise to exact sequences on sufficiently small Stein open subsets U

$$\mathscr{O}_X(U)^q \to \mathscr{O}_X(U)^p \to \mathscr{F}(U) \to 0.$$

If the set U is contained in a local model of the space X, then the space $\mathscr{O}_X(U)$ is a Fréchet nuclear algebra. Therefore $\mathscr{F}(U)$ is the quotient of a nuclear Fréchet space. To see that the topology on $\mathscr{F}(U)$ is separated, one considers the continuous map $\mathscr{F}(U) \to \prod_{x \in U} \mathscr{F}_x$ and uses the fact that each factor $\mathscr{F}_x$ is separated in the m_x-adic topology by virtue of Krull's lemma (see for details Grauert and Remmert 1984). The locally convex topology on the space $\mathscr{F}(U)$ does not depend on the chosen resolution because of the open mapping principle applied to the Fréchet space $\mathscr{F}(U)$ endowed with two comparable different topologies (see again Grauert and Remmert 1984).

By using Stein open coverings in a standard way we obtain the next well-known result.

Proposition 4.1.5 *Let $\mathscr{F}$ be an analytic coherent sheaf on an analytic space $(X, \mathscr{O}_X)$. Then $\mathscr{F}(X)$ is a nuclear Fréchet $\mathscr{O}_X(X)$-module.*

Here we have used the terminology of the preceding chapter. Quite specifically, we mean that the product map

$$\mathscr{O}_X(X) \times \mathscr{F}(X) \to \mathscr{F}(X)$$

is jointly continuous.

In particular, $\mathscr{O}_X(X)$ is a unital Fréchet nuclear algebra. In the case of a Stein space this global structure determines the whole local structure. This makes the next definition legitimate.

Definition 4.1.6 A *Stein algebra* is a unital Fréchet algebra which is topologically isomorphic to $\mathscr{O}_X(X)$ for some Stein space $(X, \mathscr{O}_X)$.

Let $A = \mathscr{O}_X(X)$ be a Stein algebra. The spectrum of A (i.e. the space of continuous characters) is naturally identified with X. In other terms, the only continuous unital morphisms of algebras $\chi: A \to \mathbb{C}$ are of the form $\chi = \epsilon_x$, where $x \in X$ and $\epsilon_x(f) = f(x)$ ($f \in A$). The procedure of deriving the structural sheaf $\mathscr{O}_X$ from A is more involved. Similar questions will be discussed in the next two sections (see also Forster 1967).

Analogously, a morphism between two Stein algebras $\varphi: A \to A'$ is induced by a 'geometric' morphism of ringed spaces $\varphi^*: (X', \mathscr{O}_{X'}) \to (X, \mathscr{O}_X)$, and conversely any morphism of analytic spaces defines at the level of global section spaces a morphism of Stein algebras.

In conclusion, the category of Stein algebras is anti-equivalent to the category of Stein spaces.

Whether a unital morphism of algebras between two Stein algebras is necessarily continuous or not is still an important open question in the theory of automatic continuity (see Dales 1978).

There are various functional characterizations of Stein algebras among the Fréchet algebras (see Kramm 1980, 1984).

It is known that the global section space of a coherent $\mathscr{O}_X$-module on a Stein space $(X, \mathscr{O}_X)$ also determines its local structure. This question will be treated in detail in the subsequent sections.

4.2 ANALYTIC TRANSVERSALITY

The familiar notion of geometric transversality is extended in the following to an abstract cohomological setting. In particular we shall compute the global section spaces of certain fibre products of Stein spaces, or more generally, the sections of external tensor products of coherent sheaves. The main result below will turn out to be an important technical device at several crucial points in the book.

We recall that two vector subspaces E_1 and E_2 of a finite-dimensional complex vector space V are said to be *transversal* if

$$\dim E_1 + \dim E_2 = \dim V + \dim(E_1 \cap E_2).$$

Analogously, two closed complex submanifolds S_1 and S_2 of a complex manifold M *intersect transversally* at the point $x \in S_1 \cap S_2$ if the tangent spaces $T_x S_1$ and $T_x S_2$ are transversal as subspaces of $T_x M$. It is elementary to remark that the transversality property (for both subspaces and submanifolds) is stable under small perturbations, in the natural sense of the terms.

The transversality of two submanifolds S_1 and S_2 of M at the point $x \in S_1 \cap S_2$ has some important algebraic consequences (see Serre 1965). In order to motivate the subsequent definitions we recall only the following elementary example.

Let $z_1, \ldots, z_n$ be local coordinates of the manifold M at the point a $(z_1(a) = \cdots = z_n(a) = 0)$ such that S_1 and S_2 are given locally by the equations $z_1 = z_2 = \cdots = z_p = 0$ and $z_q = z_{q+1} = \cdots = z_n = 0$, respectively $(1 \le p \le q \le n)$. Let $\mathscr{O}_{M,a}$, $\mathscr{O}_{S_i,a}$ $(i = 1, 2)$ denote the local rings of holomorphic functions on the three complex manifolds, respectively. Then we have

$$\mathscr{O}_{S_1,a} = \mathscr{O}_{M,a}/(z_1, \ldots, z_p),$$

$$\mathscr{O}_{S_2,a} = \mathscr{O}_{M,a}/(z_q, \ldots, z_n),$$

and $(z_1, \ldots, z_n)$ is a regular sequence of generators of the maximal ideal of $\mathscr{O}_{M,a}$ (that is, for every $1 \le i \le n$, the multiplication with z_i is injective on the quotient space $\mathscr{O}_{M,a}/(z_1, \ldots, z_{i-1})$). Accordingly we obtain

$$\mathscr{O}_{S_1,a} \otimes_{\mathscr{O}_{M,a}} \mathscr{O}_{S_2,a} = \mathscr{O}_{M,a}/(z_1, \ldots, z_p) + (z_q, \ldots, z_n) = \mathscr{O}_{S_1 \cap S_2,a}.$$

Moreover, the derived functors of this tensor product can be computed with the help of the Koszul resolution

$$K_\bullet(z_1, \ldots, z_p; \mathcal{O}_{M,a}) \to \mathcal{O}_{S_1,a} \to 0$$

(see Serre 1965 or Appendix 2). Therefore we obtain

$$\mathrm{Tor}_j^{\mathcal{O}_{M,a}}(\mathcal{O}_{S_1,a}, \mathcal{O}_{S_2,a}) = H_j\big(K_\bullet(z_1, \ldots, z_p; \mathcal{O}_{S_2,a})\big) = 0$$

for $j \geq 1$, because $(z_1, \ldots, z_p)$ is a regular sequence for the local ring $\mathcal{O}_{S_2,a}$.

In conclusion, the transversality at $a \in M$ of the submanifolds S_1 and S_2 implies that

$$\mathrm{Tor}_j^{\mathcal{O}_{M,a}}(\mathcal{O}_{S_1,a}, \mathcal{O}_{S_2,a}) = \begin{cases} \mathcal{O}_{S_1 \cap S_2,a} & (j = 0) \\ 0 & (j > 0). \end{cases}$$

For singular analytic subspaces which intersect not necessarily transversally, the algebraic intersection number is defined in terms of the preceding Tors (see Serre 1965).

In the sequel we shall be concerned with a more general situation. Let

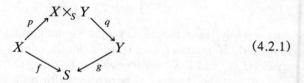

$$(4.2.1)$$

be the fibre product of two morphisms $f: X \to S$, $g: Y \to S$ of analytic spaces. Fix a point $(x, y) \in X \times_S Y$. Let us consider two open sets U in $\mathbb{C}^m$ and V in $\mathbb{C}^n$ such that there are holomorphic embeddings $\varphi: X_0 \to U$ and $\psi: Y_0 \to V$ defined on suitable open neighbourhoods X_0 of x and Y_0 of y, respectively. Consider the following two commuting diagrams

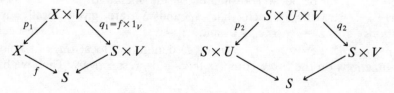

where p_1, p_2, q_2, and the non-specified maps are the canonical holomorphic projections.

Let $\mathcal{F}$ and $\mathcal{G}$ be coherent analytic sheaves on X and Y, respectively. Let $\mathcal{F}'$ and $\mathcal{G}'$ denote the extensions of $\mathcal{F}|_{X_0}$ and $\mathcal{G}|_{Y_0}$ to $S \times U$ and $S \times V$, respectively. We define $(s, u, v) = (f(x), \varphi(x), \psi(y))$. Suppose that $\mathcal{L}_\bullet$ is a complex of analytic sheaves on $S \times V$ which gives a free resolution of $\mathcal{G}'$ on a small open neighbourhood of (s, v).

Lemma 4.2.1 *With the above notations there exist isomorphisms*

$$\mathrm{Tor}_j^{(\mathscr{O}_{S\times U\times V})_{(s,u,v)}}\big((p_2^*\mathscr{F}')_{(s,u,v)},(q_2^*\mathscr{G}')_{(s,u,v)}\big)$$

$$\cong \mathrm{Tor}_j^{(\mathscr{O}_{S\times U})_{(s,u)}}\big(\mathscr{F}_{(s,u)},(q_2\mathscr{G}')_{(s,u,v)}\big)$$

$$\cong H_j\big((p_1^*\mathscr{F})_{(x,v)}\otimes_{\mathscr{O}_{X\times V,(x,v)}}(q_1^*\mathscr{L}_\bullet)_{(x,v)}\big)$$

$$\cong H_j\big(\mathscr{F}_x\otimes_{\mathscr{O}_{X,x}}(q_1^*\mathscr{L}_\bullet)_{(x,v)}\big)$$

of vector spaces for all $j\geq 0$.

Proof We have made use of the fact that each $\mathscr{O}_{S\times U\times V,(s,u,v)}$-module can be turned into an $\mathscr{O}_{S\times U,(s,u)}$-module using the natural ring homomorphism $\mathscr{O}_{S\times U,(s,u)}\to\mathscr{O}_{S\times U\times V,(s,u,v)}$ determined by p_2. Correspondingly, each $\mathscr{O}_{X\times V,(x,v)}$-module can also be regarded as an $\mathscr{O}_{X,x}$-module.

If one replaces in the last two diagrams the sets S,U,V by smaller open neighbourhoods S_0,U_0,V_0 of s,u,v, and if one replaces X by $(f|X_0\vee\varphi)^{-1}(S_0\times U_0)$, then the modules occurring in Lemma 4.2.1 can also be computed with the help of the induced diagrams. Without changing the notation, we shall therefore suppose that the maps $f\vee\varphi\colon X\to S\times U$ and $g\vee\psi\colon Y\to S\times V$ are global holomorphic embeddings, and that $\mathscr{L}_\bullet$ is a global resolution of $\mathscr{G}'$ on $S\times V$ by free analytic sheaves.

Since q_2 is a flat morphism, $q_2^*\mathscr{L}_\bullet$ becomes a free resolution of $q_2^*\mathscr{G}'$. We regard

$$p_2^*\mathscr{F}'\otimes_{\mathscr{O}_{S\times U\times V}}q_2^*\mathscr{L}_\bullet$$

$$=\big(\bar p_2^1(\mathscr{F}')\otimes_{\bar p_2^1(\mathscr{O}_{S\times U})}\mathscr{O}_{S\times U\times V}\big)\otimes_{\mathscr{O}_{S\times U\times V}}q_2^*\mathscr{L}_\bullet$$

$$\cong\bar p_2^1(\mathscr{F}')\otimes_{\bar p_2^1(\mathscr{O}_{S\times U})}q_2^*\mathscr{L}_\bullet$$

as an isomorphism between complexes of $\bar p_2^1(\mathscr{O}_{S\times U})$-modules. For the stalks at (s,u,v) we obtain induced isomorphisms of complexes of $\mathscr{O}_{S\times U,(s,u)}$-modules

$$(p_2^*\mathscr{F}')_{(s,u,v)}\otimes_{\mathscr{O}_{S\times U\times V,(s,u,v)}}(q_2^*\mathscr{L}_\bullet)_{(s,u,v)}\cong\mathscr{F}_{(s,u)}\otimes_{\mathscr{O}_{S\times U,(s,u)}}(q_2^*\mathscr{L}_\bullet)_{(s,u,v)}.$$

Since p_2 is a flat morphism, the last complex computes the spaces occurring in the second place of the isomorphism chain described in the statement of Lemma 4.2.1. Thus we have shown that the first isomorphism in Lemma 4.2.1 holds as an isomorphism between $\mathscr{O}_{S\times U,(s,u)}$-modules.

Similarly, the last isomorphism follows from the identification (of complexes of $\bar p_1^1(\mathscr{O}_X)$-modules)

$$p_1^*\mathscr{F}\otimes_{\mathscr{O}_{X\times V}}q_1^*\mathscr{L}_\bullet=\big(\bar p_1^1(\mathscr{F})\otimes_{\bar p_1^1(\mathscr{O}_X)}\mathscr{O}_{X\times V}\big)\otimes_{\mathscr{O}_{X\times V}}q_1^*\mathscr{L}_\bullet$$

$$\cong\bar p_1^1(\mathscr{F})\otimes_{\bar p_1^{-1}(\mathscr{O}_X)}q_1^*\mathscr{L}_\bullet.$$

In particular, the last isomorphism in Lemma 4.2.1 holds as an isomorphism of $\mathscr{O}_{X,x}$-modules.

To complete the proof, we show that the spaces in the second and the last position are isomorphic. Let us denote by $\mathscr{J}$ the kernel of the morphism $\widetilde{f \vee \varphi}$: $\mathscr{O}_{S \times U} \to f \vee \varphi(\mathscr{O}_X)$. Since $\mathscr{F}'_{(s,u)}$ is annihilated by the ideal $\mathscr{J}_{(s,u)} \subset \mathscr{O}_{S \times U,(s,u)}$, the stalk $\mathscr{F}'_{(s,u)}$ becomes a module over $A = \mathscr{O}_{S \otimes U,(s,u)}/\mathscr{J}_{(s,u)}$, and we obtain isomorphisms

$$\mathscr{F}'_{(s,u)} \otimes_{\mathscr{O}_{S \times U,(s,u)}} (q_2^* \mathscr{L}_j)_{(s,u,v)} \cong \mathscr{F}'_{(s,u)} \otimes_A \left[(q_2^* \mathscr{L}_j)_{(s,u,v)}/\mathscr{J}_{(s,u)}(q_2^* \mathscr{L}_j)_{(s,u,v)} \right]$$

of A-modules. If we regard the last tensor product as an $\mathscr{O}_{X,x}$-module via the identification $A \cong \mathscr{O}_{X,x}$ induced by $f \vee \varphi$, then this module is isomorphic to

$$\mathscr{F}_x \otimes_{\mathscr{O}_{X,x}} \tilde{\mathscr{L}}_j,$$

where $\tilde{\mathscr{L}}_j$ denotes the second factor regarded as an $\mathscr{O}_{X,x}$-module. The holomorphic embedding $h = (f \vee \varphi) \times 1_V$: $X \times V \to (S \times U) \times V$ factors through the closed complex subspace of $S \times U \times V$ determined by the ideal $\mathrm{Ker}(\tilde{h}) = \bar{p}_2^1(\mathscr{J})\mathscr{O}_{S \times U \times V}$. The right vertical in the commutative diagram

$$
\begin{array}{ccc}
\mathscr{O}_{S \times U,(s,u)}/\mathscr{J}_{(s,u)} & \xrightarrow{\;\;p_2\;\;} & \mathscr{O}_{S \times U \times V,(s,u,v)}/\mathscr{J}_{(s,u)}\mathscr{O}_{S \times U \times V,(s,u,v)} \\
{\scriptstyle f \vee \varphi}\Big\downarrow {\scriptstyle \sim} & & {\scriptstyle \sim}\Big\downarrow {\scriptstyle h} \\
\mathscr{O}_{X,x} & \xrightarrow{\;\;p_1\;\;} & \mathscr{O}_{X \times V,(x,v)}
\end{array}
$$

induces an isomorphism $\tilde{\mathscr{L}}_j \to (q_1^* \mathscr{L}_j)_{(x,v)}$ of $\mathscr{O}_{X,x}$-modules. Thus for each $j \geq 0$, we obtain an isomorphism

$$\mathscr{F}'_{(s,u)} \otimes_{\mathscr{O}_{S \times U,(s,u)}} (q_2^* \mathscr{L}_j)_{(s,u,v)} \cong \mathscr{F}_x \otimes_{\mathscr{O}_{X,x}} (q_1^* \mathscr{L}_j)_{(x,v)}$$

of $\mathscr{O}_{X,x}$-modules. It is elementary to check that in this way one really obtains an isomorphism of complexes. $\square$

Definition 4.2.2 With the above notation, the sheaves $\mathscr{F}$ and $\mathscr{G}$ are said to be *transversal over S at the point* $(x,y) \in X \times_S Y$ if the spaces listed in Lemma 4.2.1 vanish for $j > 0$. The sheaves $\mathscr{F}$ and $\mathscr{G}$ are said to be *transversal over S* if they are transversal over S at each point of $X \times_S Y$.

In fact, we shall not need the preceding definition in its full generality. First note that the transversality relation is symmetric because the first space in Lemma 4.2.1 is symmetric in $\mathscr{F}$ and $\mathscr{G}$. Also the transversality condition is local, and it does not depend on the local embeddings of X and Y, because $S \times U$ is not involved in the last space of Lemma 4.2.1.

Proposition 4.2.3 *Let* $f: X \to S$ *be a morphism of analytic spaces, and let* $\mathscr{F}$ *be a coherent* $\mathscr{O}_X$-*module which is* $\mathscr{O}_S$-*flat. Then for every morphism* $g: Y \to S$ *and coherent* $\mathscr{O}_Y$-*module* $\mathscr{G}$, $\mathscr{F}$ *and* $\mathscr{G}$ *are transversal over S.*

Proof Use the last space in Lemma 4.2.1. $\square$

In applications only the case of a Stein space will be interesting. Recall that we defined a global transversality notion for Fréchet modules over Fréchet algebras in Chapter 3. The relationship between the above local analytic transversality and the global transversality is established by the following important theorem.

Theorem 4.2.4 *Let* (4.2.1) *be the fibre product of two morphisms acting between Stein spaces, and let $\mathscr{F}$ and $\mathscr{G}$ be coherent modules on X and Y, respectively.*

Suppose that $\mathscr{F}$ and $\mathscr{G}$ are transversal over S. Then $\mathscr{F}(X) \perp_{\mathscr{O}(S)} \mathscr{G}(Y)$, and

$$\mathscr{F}(X) \hat{\otimes}_{\mathscr{O}(S)} \mathscr{G}(Y) \cong \left(p^*\mathscr{F} \otimes_{\mathscr{O}_{X\times_S Y}} q^*\mathscr{G} \right)(X \times_S Y).$$

Proof The proof will be divided into several steps, corresponding to increasing degrees of generality for the space Y and the sheaf $\mathscr{G}$.

(1) Suppose first that $Y = S \times V$, where $V \subset \mathbb{C}^m$ is a Stein open subset and $\mathscr{G} = \mathscr{O}_{S\times V}$. Then $X \times_S Y = X \times V$ and $\mathscr{G}(Y) = \mathscr{O}(S \times V) = \mathscr{O}(S) \hat{\otimes} \mathscr{O}(V)$. Consequently, we have

$$\text{Tôr}_j^{\mathscr{O}(S)}(\mathscr{F}(X), \mathscr{G}(Y)) = \text{Tôr}_j^{\mathscr{O}(S)}(\mathscr{F}(X), \mathscr{O}(S) \hat{\otimes} \mathscr{O}(V))$$

$$\cong \begin{cases} \mathscr{F}(X) \hat{\otimes} \mathscr{O}(V) & (j = 0) \\ 0 & (j \geq 1), \end{cases}$$

as an application of Corollary 3.1.11.

Here it is evident that the sheaves $\mathscr{F}$ and $\mathscr{G}$ are transversal in the local sense of Definition 4.2.2.

(2) The case where $Y = S \times V$ and $\mathscr{G} = \mathscr{O}_{S\times V}^r$ can easily be deduced from (1).

(3) Suppose next that $Y = S \times V$ is as in case (1), but $\mathscr{G}$ is a coherent $\mathscr{O}_Y$-module which has an infinite free resolution of finite type.

Let $\mathscr{L}_\bullet \to \mathscr{G} \to 0$ be such a resolution. Then according to Lemma 4.2.1, $p^*\mathscr{F} \otimes_{\mathscr{O}_{X\times_S Y}} q^*\mathscr{L}_\bullet$ is a resolution of the coherent sheaf $p^*\mathscr{F} \otimes_{\mathscr{O}_{X\times_S Y}} q^*\mathscr{G}$. Moreover, $q^*\mathscr{L}_\bullet(X \times_S Y)$ is a free resolution of the Fréchet $\mathscr{O}(X \times_S Y)$-module $q^*\mathscr{G}(X \times_S Y)$. Hence

$$\text{Tôr}_j^{\mathscr{O}(X\times_S Y)}(p^*\mathscr{F}(X \times_S Y), q^*\mathscr{G}(X \times_S Y))$$

$$\cong H_j\left(\left(p^*\mathscr{F} \otimes_{\mathscr{O}_{X\times_S Y}} q^*\mathscr{L}_\bullet \right)(X \times_S Y) \right)$$

for every $j \geq 0$.

The complex $(p^*\mathscr{F} \otimes_{\mathscr{O}_{X\times_S Y}} q^*\mathscr{L}_\bullet)(X \times_S Y)$ is exact in positive degrees by Cartan's Theorem (B), and its zeroth homology space coincides with $(p^*\mathscr{F} \otimes_{\mathscr{O}_{X\times_S Y}} q^*\mathscr{G})(X \times_S Y)$ as desired.

(4) Suppose that Y is a relatively compact Stein open subset of a larger Stein space to which the sheaf $\mathscr{G}$ extends.

Then there exists an embedding of a neighbourhood of Y into $S \times V$, relative to S, such that the trivial extension $\mathcal{G}'$ of the sheaf $\mathcal{G}$ to $S \times V$ satisfies the conditions of case (3).

(5) The general case needs a projective limit argument. Quite specifically, one chooses a sequence of Stein open sets $Y_1 \subset Y_2 \subset \cdots$ in Y with $\bigcup_{j \geq 1} Y_j = Y$ such that Y_j and $\mathcal{G}|_{Y_j}$ satisfy the assumptions of (4) and such that the restriction maps $\mathcal{O}(Y_{j+1}) \to \mathcal{O}(Y_j)$ $(j \geq 1)$ have dense range. Then the maps $\mathcal{O}(Y_{j+1})^r \to \mathcal{O}(Y_j)^r$, and hence also the induced maps $\mathcal{G}(Y_{j+1}) \to \mathcal{G}(Y_j)$, have dense range. Thus $\{\mathcal{G}(Y_j)\}_{j \geq 1}$ is a Mittag-Leffler inverse system of Fréchet $\mathcal{O}(X \times_S Y)$-modules, and Proposition 3.2.11 yields the conclusion of Theorem 4.2.4.

It remains to show how one can construct an exhaustion as above. For instance, let $\varphi_j \colon Y \to \mathbb{C}^{n_j}$ be morphisms of Stein spaces which are embeddings on some Stein open sets $Y_j' \subset Y$, such that Y_j' is relatively compact in Y_{j+1}'. Let $P_j \subset \mathbb{C}^{n_j}$ be bounded open polydiscs such that $\varphi_j^{-1}(P_j) = Y_j \Subset Y_j'$, $Y_j \subset Y_{j+1}$, and $Y = \bigcup_{j \geq 1} Y_j$. Fix an integer j and denote by

$$\varphi = (\varphi_j, \varphi_{j+1}) \colon Y \to \mathbb{C}^{n_j + n_{j+1}}$$

the diagonal morphism which yields an embedding of Y_{j+1}'. Then

$$\varphi^{-1}(\mathbb{C}^n \times P_{j+1}) = Y_{j+1},$$

$$\varphi^{-1}(P_j \times P_{j+1}) = Y_j,$$

and the restriction maps $\mathcal{O}(\mathbb{C}^n \times P_{j+1}) \to \mathcal{O}(Y_{j+1})$, $\mathcal{O}(P_j \times P_{j+1}) \to \mathcal{O}(Y_j)$ are onto by Cartan's Theorem (B). By Runge's approximation theorem, it follows that the restriction $\mathcal{O}(Y_{j+1}) \to \mathcal{O}(Y_j)$ has dense range (see Grauert and Remmert 1984 for details concerning this familiar argument in the theory of functions of several complex variables).

This completes the proof of Theorem 4.2.4.   □

Corollary 4.2.5 *Let $f \colon X \to S$ be a morphism of Stein spaces, and let $\mathcal{F}$ be a coherent $\mathcal{O}_X$-module. Suppose that $U \subset S$ is a Stein open subset. Then $\mathcal{O}(U) \perp_{\mathcal{O}(S)} \mathcal{F}(X)$ and $\mathcal{O}(U) \hat{\otimes}_{\mathcal{O}(S)} \mathcal{F}(X) \cong \mathcal{F}(f^{-1}(U))$.*

Proof Apply Theorem 4.2.4 by noticing that the inclusion map $(U, \mathcal{O}_S|_U) \to (S, \mathcal{O}_S)$ is flat.   □

This corollary shows in particular (when $S = X$ and $f = \mathrm{id}_X$) how the global $\mathcal{O}(X)$-Fréchet module structure on $\mathcal{F}(X)$ determines the sheaf $\mathcal{F}$. This observation forms the starting point for the next section.

Corollary 4.2.6 *Let $f \colon X \to S$ be a morphism of Stein spaces. Let $\mathcal{F}$ be a coherent $\mathcal{O}_X$-module which is flat over S, and let $s \in S$ be a fixed point. Then $\mathcal{F}(X) \perp_{\mathcal{O}(S)} \mathcal{O}_{S,s}/m$, and $\mathcal{F}(X) \hat{\otimes}_{\mathcal{O}(S)} \mathcal{O}_{S,s}/m \cong \mathcal{F}_s/m\mathcal{F}_s$, where m denotes the maximal ideal of the local ring $\mathcal{O}_{S,s}$.*   □

4.3 QUASI-COHERENT ANALYTIC SHEAVES

The last part of the preceding section contained a direct method of constructing a coherent sheaf $\mathscr{F}$ from its global section space $\mathscr{F}(X)$ on Stein spaces
X. However, this procedure is not specific for coherent analytic sheaves. The
class of Fréchet analytic modules which satisfy the same transversality condition will be investigated in the present section. Sheaves of this type appeared
in the work of Ramis and Ruget (1974) as a tool in certain analytic duality
questions. The same analytic sheaves have since been used, implicitly or
explicitly, in a variety of other problems.

To give a motivation for the following definition, let $\mathscr{F}$ be a coherent
$\mathscr{O}_X$-module on a Stein space X. Then $\mathscr{F}$ has finite-type free presentations in
neighbourhoods of each given compact set $K \subset X$

$$\mathscr{O}^q|_K \to \mathscr{O}^p|_K \to \mathscr{F}|_K \to 0,$$

but this is no longer true for the whole space X. Let us consider for instance
a sheaf $\mathscr{F}$ supported by a discrete infinite subset $\{x_1, x_2, \ldots\}$ of X, and such
that $\dim_{\mathbb{C}}(\mathscr{F}_{x_j}/m_{x_j}\mathscr{F}_{x_j}) = j$ $(j \geq 1)$. Then a presentation as above cannot exist
globally on X, since $\dim_{\mathbb{C}}(\mathscr{O}_{x_j}^p/m_{x_j}\mathscr{O}_{x_j}^p) = p$ and since $\mathscr{F}_{x_j}/m_{x_j}\mathscr{F}_{x_j}$ would have
to be a quotient of $\mathscr{O}_{x_j}^p/m_{x_j}\mathscr{O}_{x_j}^p$ $(j \geq 1)$.

However, for some global problems, one needs to replace the sheaf $\mathscr{F}$ by a
free presentation or resolution on the whole space X. There are essentially
two known ways to accomplish this. In both of them the class of coherent
sheaves is replaced by a more general category. In the first approach, due to
Forster and Knorr (1972), the analytic sheaves on X are replaced by simplicial systems of sheaves. The second method, which is due to Ramis and Ruget
(1974), will be discussed below. Roughly speaking, its main idea is to consider
infinite-dimensional free resolutions of the form

$$\cdots \to \mathscr{O} \hat{\otimes} \mathscr{O}(X) \hat{\otimes} \mathscr{F}(X) \to \mathscr{O} \hat{\otimes} \mathscr{F}(X) \to \mathscr{F} \to 0$$

obtained by localizing the Bar complex $B_{\bullet}^{\mathscr{O}(X)}(\mathscr{O}(X), \mathscr{F}(X))$ (see Corollary
4.2.5 and Lemma 3.1.4).

By a *sheaf $\mathscr{F}$ of Fréchet modules* on X, or simply an *analytic Fréchet sheaf*,
we mean an analytic sheaf $\mathscr{F}$ on X such that all section spaces $\mathscr{F}(U)$ $(U \subset X$
open) are Fréchet $\mathscr{O}(U)$-modules and such that the restriction maps $\mathscr{F}(U) \to$
$\mathscr{F}(V)$, $s \mapsto s \,|\, V$, are continuous for each pair of open sets U, V in X with
$V \subset U$. It suffices to have the last two conditions for U, V in the restricted
class of all Stein open subsets of X. Then the sheaf $\mathscr{F}$ can be turned
canonically into an analytic Fréchet sheaf.

A *(continuous) morphism* $\varphi: \mathscr{F} \to \mathscr{G}$ between analytic Fréchet sheaves is a
sheaf homomorphism such that the induced maps $\varphi: \mathscr{F}(U) \to \mathscr{G}(U)$ are
continuous for each open set U in X.

Definition 4.3.1 Let X be a Stein space. An analytic Fréchet sheaf $\mathscr{F}$ on X is *quasi-coherent* if, for every Stein open subset U of X, the natural restriction and multiplication map

$$\mathscr{O}(U) \otimes_{\mathscr{O}(X)} \mathscr{F}(X) \to \mathscr{F}(U)$$

is an isomorphism between Fréchet spaces and if $\mathscr{O}(U) \perp_{\mathscr{O}(X)} \mathscr{F}(X)$.

According to Proposition 3.1.10, an analytic Fréchet sheaf $\mathscr{F}$ is quasi-coherent if and only if the augmented Bar complex

$$B_\bullet^{\mathscr{O}(X)}(\mathscr{O}(U), \mathscr{F}(X)) \to \mathscr{F}(U) \to 0 \qquad\qquad (4.3.1)$$

is exact for each Stein open subset U of X. In analogy to the corresponding global definition (see Chapter 3) we call a sheaf of the form

$$(\mathscr{O} \,\hat{\otimes}\, E)(U) = \mathscr{O}(U) \,\hat{\otimes}\, E \qquad (U \subset X \text{ open})$$

a *topologically free $\mathscr{O}$-module*. In the discussion that follows, E will be a Fréchet space, but of course the preceding definition makes perfect sense for arbitrary locally convex spaces E.

A standard nuclearity argument shows that topologically free $\mathscr{O}$-modules are quasi-coherent. Indeed, the exact sequence of nuclear Fréchet spaces

$$B_\bullet^{\mathscr{O}(X)}(\mathscr{O}(U), \mathscr{O}(X)) \to \mathscr{O}(U) \to 0$$

remains exact when forming the tensor product with the identity operator on a Fréchet space E. Thus

$$B_\bullet^{\mathscr{O}(X)}(\mathscr{O}(U), (\mathscr{O} \,\hat{\otimes}\, E)(X)) \to (\mathscr{O} \,\hat{\otimes}\, E)(U) \to 0$$

is an exact complex for each Stein open subset U of X.

Similarly, let $\mathscr{U}$ denote a Stein open cover of the Stein space X. Since the augmented Čech complex of alternating chains

$$0 \to \mathscr{O}(X) \to \mathscr{C}^\bullet(\mathscr{U}, \mathscr{O})$$

is exact and consists of nuclear Fréchet spaces, the tensorized complex

$$0 \to (\mathscr{O} \,\hat{\otimes}\, E)(X) \to \mathscr{C}^\bullet(\mathscr{U}, \mathscr{O} \,\hat{\otimes}\, E)$$

remains exact. By passing to the inductive limit (with respect to $\mathscr{U}$), one finds that $H^q(X, \mathscr{O} \,\hat{\otimes}\, E) = 0$ for $q \geq 1$ (compare Chapter 1 in Grauert and Remmert 1984).

Summing up, we obtain the following result.

Lemma 4.3.2 *A topologically free $\mathscr{O}_X$-module defined on a Stein space X is quasi-coherent and acyclic.*   □

Let $\mathscr{F}$ be a quasi-coherent sheaf on a Stein space X. The exact complex (4.3.1) gives a global topologically free resolution of $\mathscr{F}$ at the level of sheaves,

$$B_\bullet^{\mathscr{O}(X)}(\mathscr{O}, \mathscr{F}(X)) \to \mathscr{F} \to 0.$$

The sheaves occurring in the Bar complex (see Section 3.1) are of the form $\mathscr{O} \hat{\otimes} \mathscr{O}(X) \hat{\otimes} \cdots \hat{\otimes} \mathscr{O}(X) \hat{\otimes} \mathscr{F}(X)$, and hence they are topologically free.

Since each coherent $\mathscr{O}$-module is quasi-coherent (Corollary 4.2.5), one obtains the first part of the following proposition.

Proposition 4.3.3   *Let X be a Stein space.*

(a) *Each coherent analytic sheaf on X has a global topologically free resolution to the left.*

(b) *Let $\mathscr{F}$ be a quasi-coherent sheaf on X. Then $\mathscr{F} \,|\, U$ is acyclic for every Stein open subset U of X.*

Proof   To prove part (b), let $\mathscr{F}$ be a quasi-coherent sheaf on a Stein space X, and let U be a Stein open subset of X. The Bar complex yields a topologically free resolution of $\mathscr{F} \,|\, U$:

$$\mathscr{L}_\bullet = (\mathscr{L}_n)_{n \geq 0} \to \mathscr{F} \,|\, U \to 0.$$

Suppose first that $\dim U < \infty$. Define $\mathscr{Z}_j = \mathrm{Ker}(\mathscr{L}_j \to \mathscr{L}_{j-1})$ for $j \in \mathbb{Z}$ where $\mathscr{L}_{-1} = \mathscr{F} \,|\, U$ and $\mathscr{L}_{-2} = \mathscr{L}_{-3} = \cdots = 0$. There are short exact sequences

$$0 \to \mathscr{Z}_j \to \mathscr{L}_j \to \mathscr{Z}_{j-1} \to 0 \qquad (j \geq 0).$$

Using the induced long exact cohomology sequences and Lemma 4.3.2, one obtains isomorphisms

$$H^k(U, \mathscr{Z}_{j-1}) \cong H^{k+n}(U, \mathscr{Z}_{j+n-1}) \qquad (j, n \geq 0, k \geq 1).$$

But $H^p(U, \mathscr{Z}_j) = 0$ for $p > \dim U$ and all j (compare Grauert and Remmert 1979). Hence all sheaves $\mathscr{Z}_j$, in particular $\mathscr{F} \,|\, U = \mathscr{Z}_{-1}$, are acyclic on U. Note that the induced sequence of section spaces $\Gamma(U, \mathscr{L}_\bullet) \to \Gamma(U, \mathscr{F}) \to 0$ remains exact.

Now let U be an arbitrary Stein open subset of X. Then one can choose a Stein open cover $\mathscr{U} = (U_i)_{i \in I}$ of U with $\dim U_i < \infty$ for all indices i. To reduce the assertion to the finite-dimensional case considered before, one can use a double complex of the form $\mathscr{C}^\bullet(\mathscr{U}, \mathscr{L}_\bullet)$, where $\mathscr{C}^\bullet(\mathscr{U}, \mathscr{L}_p)$ $(p \geq 0)$ is the alternating Čech complex with coefficients in $\mathscr{L}_p$ relative to the open cover $\mathscr{U}$.   □

Before proceeding any further, we need an elementary technical result.

Lemma 4.3.4   *Let $0 \to \mathscr{F} \to \mathscr{G} \to \mathscr{H} \to 0$ be an exact sequence of analytic Fréchet sheaves on a Stein space X such that the induced sequence of global section spaces $0 \to \mathscr{F}(X) \to \mathscr{G}(X) \to \mathscr{H}(X) \to 0$ is exact. If any two of the sheaves $\mathscr{F}$, $\mathscr{G}$, or $\mathscr{H}$ are quasi-coherent, then so is the third.*

Proof   Suppose that $\mathscr{F}$ and $\mathscr{G}$ are quasi-coherent. Let U be a Stein open subset of X. The long exact sequence of Tôr spaces (Theorem 3.1.12) yields the first exact row in the following commutative diagram

$$0 \to \text{Tôr}_1^{\mathscr{O}(X)}(\mathscr{O}(U), \mathscr{H}(X)) \to \mathscr{O}(U) \,\hat{\otimes}_{\mathscr{O}(X)}\, \mathscr{F}(X) \to \mathscr{O}(U) \,\hat{\otimes}_{\mathscr{O}(X)}\, \mathscr{G}(X) \to \mathscr{O}(U) \,\hat{\otimes}_{\mathscr{O}(X)}\, \mathscr{H}(X) \to 0$$

$$\begin{array}{ccccc} & \sim \downarrow \alpha & & \sim \downarrow \beta & & \downarrow \gamma \\ 0 \longrightarrow & \mathscr{F}(U) & \longrightarrow & \mathscr{G}(U) & \longrightarrow & \mathscr{H}(U) \to 0. \end{array}$$

The exactness of the lower horizontal sequence follows as an application of Proposition 4.3.3(b). Since α is an isomorphism, it follows that $\text{Tôr}_1^{\mathscr{O}(X)}(\mathscr{O}(U), \mathscr{H}(X)) = 0$. The long exact sequence of Tôr spaces, referred to above, also yields that $\text{Tôr}_j^{\mathscr{O}(X)}(\mathscr{O}(U), \mathscr{H}(X)) = 0$ for all $j \geq 2$. Elementary diagram chasing shows that the map γ is an isomorphism. Since U is an arbitrary Stein open subset of X, it follows that the sheaf $\mathscr{H}$ is quasi-coherent.

Similar arguments can be used to prove the assertion in the remaining two cases.   □

Corollary 4.3.5 *Let $\mathscr{L}_\bullet$ be a quasi-coherent, finite resolution to the right of the analytic Fréchet sheaf $\mathscr{F}$ over a Stein space X such that $0 \to \mathscr{F}(X) \to \mathscr{L}_\bullet(X)$ is an exact complex. Then $\mathscr{F}$ is quasi-coherent.*   □

We are now in a position to give first equivalent descriptions of the quasi-coherence property.

Theorem 4.3.6 *Let X be a finite-dimensional Stein space, and let $\mathscr{F}$ be a Fréchet $\mathscr{O}_X$-module. The following conditions are equivalent:*

(a) *$\mathscr{F}$ is quasi-coherent;*
(b) *$\mathscr{F}$ admits global topologically free resolutions to the left;*
(c) *$\mathscr{F}$ is acyclic on X and admits, locally on X, topologically free resolutions to the left.*

Proof In view of the remark preceding Proposition 4.3.3 and the proof of Proposition 4.3.3(b), it suffices to show that part (c) implies part (a).

To do this, fix a Stein open subset V of X such that on V there exists a topologically free resolution $\mathscr{L}_\bullet \to \mathscr{F}|V \to 0$. Since the complex $\mathscr{L}_\bullet(V) \to \mathscr{F}(V) \to 0$ is exact (see the argument given in the proof of Proposition 4.3.3(b)), the invariants $\text{Tôr}_p^{\mathscr{O}(X)}(\mathscr{F}(V), \mathscr{O}(U))$ $(p \geq 0)$ are isomorphic to the homology spaces of the complex $\mathscr{L}_\bullet(V) \hat{\otimes}_{\mathscr{O}(X)} \mathscr{O}(U)$ for each Stein open subset U of X. By virtue of Corollary 4.2.5, and an obvious tensor product argument, $\mathscr{L}_\bullet(V) \hat{\otimes}_{\mathscr{O}(X)} \mathscr{O}(U) \cong \mathscr{L}_\bullet(V \cap U)$. But $V \cap U$ is a Stein open set. Therefore the complex $\mathscr{L}_\bullet(V \cap U) \to \mathscr{F}(V \cap U) \to 0$ is exact.

In conclusion, we have proved that

$$\mathrm{T\hat{o}r}_j^{\mathscr{O}(X)}(\mathscr{F}(V),\mathscr{O}(U)) \cong \begin{cases} \mathscr{F}(V \cap U), & j = 0 \\ 0, & j \geq 1. \end{cases}$$

Next consider an open cover of X by Stein open sets $\mathscr{V} = (V_i)_{i \in I}$ such that the sheaf $\mathscr{F}$ admits topologically free resolutions on each of the sets V_i. Since the space X is finite dimensional, we may suppose that the cover $\mathscr{V}$ has finite-dimensional nerve, that is, there is an integer N such that any $N + 1$ distinct elements in the cover have empty intersection.

The preceding arguments show that the sheaf $\mathscr{F}$ is acyclic on every finite intersection $V_i \cap V_j \cap \cdots \cap V_k$. Thus according to Leray's theorem, its cohomology groups $H^p(X, \mathscr{F})$ can be computed by using the Čech complex of alternating chains $\mathscr{C}^\bullet(\mathscr{V}, \mathscr{F})$. As this cohomology vanishes in each positive degree, we obtain an exact sequence of Fréchet $\mathscr{O}(X)$-modules

$$0 \to \mathscr{F}(X) \to \mathscr{C}^0 \to \mathscr{C}^1 \to \cdots \to \mathscr{C}^N \to 0.$$

Since each component $\mathscr{C}^q$ of the Čech complex is a product space of the form $\prod_{i,j,\ldots,k} \mathscr{F}(V_i \cap V_j \cap \cdots \cap V_k)$, it follows that $\mathscr{C}^q \perp_{\mathscr{O}(X)} \mathscr{O}(U)$ and that $\mathscr{C}^q(\mathscr{V}, \mathscr{F}) \hat{\otimes}_{\mathscr{O}(X)} \mathscr{O}(U) \cong \mathscr{C}^q(\mathscr{V} \cap U, \mathscr{F})$ for every Stein open subset U of X. According to Corollary 3.1.16 and the following remarks, $\mathscr{F}(X) \perp_{\mathscr{O}(X)} \mathscr{O}(U)$, and the complex

$$0 \to \mathscr{F}(X) \hat{\otimes}_{\mathscr{O}(X)} \mathscr{O}(U) \to \mathscr{C}^\bullet(\mathscr{V} \cap U, \mathscr{F})$$

is exact. In particular, this shows that $\mathscr{F}(X) \hat{\otimes}_{\mathscr{O}(X)} \mathscr{O}(U) \cong \mathscr{F}(U)$ for any Stein open subset U of X.

This completes the proof of Theorem 4.3.6. □

Henceforth **Qcoh(X)** will denote the category of (Fréchet) quasi-coherent sheaves on the Stein space X. We write **FMod(A)** for the category of Fréchet modules over a unital Fréchet algebra A.

Proposition 4.3.7 *Let X be a Stein space. The assignment of global sections*

$$\Gamma(X, *) : \mathbf{Qcoh}(X) \to \mathbf{FMod}(\mathscr{O}(X)) \tag{4.3.2}$$

is an exact and completely faithful functor.

Proof Since quasi-coherent $\mathscr{O}$-modules are $\Gamma(X, *)$-acyclic, the functor $\Gamma(X, *)$ is exact.

Let $\mathscr{F}$ and $\mathscr{G}$ be two quasi-coherent $\mathscr{O}_X$-modules. It remains to verify that the canonical map

$$\mathrm{Hom}_{\mathbf{Qcoh}(X)}(\mathscr{F}, \mathscr{G}) \xrightarrow{s} \mathrm{Hom}_{\mathscr{O}(X)}(\mathscr{F}(X), \mathscr{G}(X))$$

has an inverse t. Let $\varphi \in \mathrm{Hom}_{\mathscr{O}(X)}(\mathscr{F}(X, \mathscr{G}(X))$, and let $t(\varphi)$ be the map induced by $id \otimes \varphi$ on $\mathscr{O} \hat{\otimes}_{\mathscr{O}(X)} \mathscr{F}(X) \cong \mathscr{F}$. Then a standard computation shows that the map $\varphi \to t(\varphi)$ defines a bilateral inverse for s. □

Corollary 4.3.8   *Let X be a finite-dimensional Stein space. A Fréchet $\mathcal{O}(X)$-module M belongs to the range of the functor (4.3.2) if and only if $M \perp_{\mathcal{O}(X)} \mathcal{O}(U)$ for every Stein open subset U of X.*

In this case, $\tilde{M} = \mathcal{O} \hat{\otimes}_{\mathcal{O}(X)} M$ is the unique quasi-coherent $\mathcal{O}_X$-module with the property that $\tilde{M}(X) \cong M$ as Fréchet $\mathcal{O}(X)$-modules.   □

A Fréchet $\mathcal{O}(X)$-module M which fulfils the transversality condition stated in Corollary 4.3.8 will be called a *quasi-coherent $\mathcal{O}(X)$-module*. Thus on a finite-dimensional Stein space X, we may speak without ambiguity of the quasi-coherent sheaf $\tilde{M}$ associated with a quasi-coherent $\mathcal{O}(X)$-module M. From now on, we shall switch freely between the global and local point of view, depending on the context.

For later use, we mention the following variant of Lemma 4.3.4 for quasi-coherent $\mathcal{O}(X)$-modules.

Lemma 4.3.9   *Let X be a Stein space, and let*

$$0 \to F \xrightarrow{\ j\ } G \xrightarrow{\ q\ } H \to 0$$

be a short exact sequence of Fréchet $\mathcal{O}(X)$-modules.

(a) *If G and H are quasi-coherent $\mathcal{O}(X)$-modules, then so is F.*
(b) *If F and H are quasi-coherent $\mathcal{O}(X)$-modules, then so is G.*

Proof   The assertion follows as an immediate application of Proposition 3.1.15.   □

In the setting of Lemma 4.3.9 the quasi-coherence of the $\mathcal{O}(X)$-modules F and G is, in general, not sufficient to ensure the quasi-coherence of the quotient module H. This will become obvious in Chapter 6, where we shall give a natural operator-theoretic interpretation of the quasi-coherence condition.

A remarkable property of quasi-coherent modules on Stein spaces, which distinguishes them from the coherent modules and in a certain sense complements their behaviour, is the compatibility with morphisms between analytic spaces. While coherent modules are preserved by pull-backs, quasi-coherent modules turn out to be invariant under direct images as well. This explains their role in the recent proofs of Grauert's direct image theorem (see Chapter 9).

Proposition 4.3.10   *Let $f: X \to Y$ be a morphism of Stein spaces, and let $\mathscr{F}$ be a quasi-coherent $\mathcal{O}_X$-module. Then $f_* \mathscr{F}$ is a quasi-coherent $\mathcal{O}_Y$-module.*

Proof   Recall that $f_* \mathscr{F}$ is the sheaf given by the presheaf $V \mapsto \mathscr{F}(f^{-1}(V))$ ($V \subset Y$ open).

We have to prove that, for any Stein open subset V of Y,

$$\mathrm{T\hat{o}r}_j^{\mathscr{O}(Y)}(\mathscr{F}(X), \mathscr{O}(V)) \cong \begin{cases} \mathscr{F}(f^{-1}(V)), & j = 0 \\ 0, & j \geq 1. \end{cases} \quad (4.3.3)$$

This is certainly true in the particular case that $\mathscr{F}$ is a topologically free $\mathscr{O}_X$-module (apply Corollary 4.2.5 and an obvious nuclearity argument).

In the general case, the Bar complex yields a topologically free resolution $\mathscr{L}_\bullet \to \mathscr{F} \to 0$ for $\mathscr{F}$ on X. Again because of Corollary 4.2.5, when applying the functor $* \hat{\otimes}_{\mathscr{O}(Y)} \mathscr{O}(V)$ to the complex $\mathscr{L}_\bullet(X)$, we obtain the complex $\mathscr{L}_\bullet(f^{-1}(V))$ whenever V is an arbitrary Stein open subset of Y. Moreover, since the components of the complex $\mathscr{L}_\bullet(X)$ are acyclic with respect to this functor, one obtains the isomorphisms

$$H_j(\mathscr{L}_\bullet(f^{-1}(V))) \cong \mathrm{T\hat{o}r}_j^{\mathscr{O}(Y)}(\mathscr{F}(X), \mathscr{O}(V)) \quad (j \geq 0).$$

On the other hand, since $f^{-1}(V)$ is Stein, the augmented complex $\mathscr{L}_\bullet(f^{-1}(V)) \to \mathscr{F}(f^{-1}(V)) \to 0$ is exact. Thus (4.3.3) also holds in the general case. $\qquad\qquad\Box$

Let us finally remark that, depending on the base space X, quasi-coherent sheaves may have finite topologically free resolutions. For instance, let X be a Stein open subset of $\mathbb{C}^n$. Then the diagonal $\Delta \subset X \times X$ is a complete intersection, the ideal of which is generated by the system $z - w = (z_1 - w_1, \ldots, z_n - w_n)$, where $(z_1, \ldots, z_n), (w_1, \ldots, w_n) \in X$. Consequently, the augmented Koszul complex

$$K_\bullet(z - w, \mathscr{O}(X \times X)) \to \mathscr{O}(\Delta) \to 0$$

is exact. Since Δ is canonically isomorphic to X via the second projection, we obtain the $\mathscr{O}(X)$-linear resolution

$$K_\bullet(z - w, \mathscr{O}(X) \hat{\otimes} \mathscr{O}(X)) \to \mathscr{O}(X) \to 0. \quad (4.3.4)$$

If M denotes a Fréchet $\mathscr{O}(X)$-module, then Corollary 3.1.11 yields the canonical topologically free resolution

$$K_\bullet(z - w, \mathscr{O}(X) \hat{\otimes} M) \to M \to 0. \quad (4.3.5)$$

In particular, if $\mathscr{F}$ is a quasi-coherent module on X, then the localization of (4.3.5) is

$$K_\bullet(z - w, \mathscr{O} \hat{\otimes} \mathscr{F}(X)) \to \mathscr{F} \to 0. \quad (4.3.6)$$

This simple resolution will appear in a series of applications to be discussed later in the book.

4.4 FRÉCHET ANALYTIC SOFT SHEAVES

The purpose of this section is to prove that, on Stein spaces, all Fréchet analytic soft sheaves are quasi-coherent. This result will provide a series of non-trivial examples of quasi-coherent modules, and besides, it will imply a useful quasi-coherence criterion.

Theorem 4.4.1   *A Fréchet analytic soft sheaf on a Stein space is quasi-coherent.*

Proof   We shall divide the proof into several steps.

(1) The reduction to the case $X = \mathbb{C}^n$. Suppose that X is a finite-dimensional Stein space, embedded as a closed analytic subspace in $\mathbb{C}^n$. If we extend $\mathscr{F}$ by 0 to the whole of $\mathbb{C}^n$, then the softness assumption is preserved, and it suffices to prove that the extension of $\mathscr{F}$ is quasi-coherent.

For a general space X, choose an open exhaustion of X by relatively compact Stein subspaces $(X_n)_{n=0}^{\infty}$, $X_n \subset X_{n+1} \subset \cdots \subset X$, such that the restriction maps $\mathscr{O}(X_{n+1}) \to \mathscr{O}(X_n)$ have dense range for $n \geq 0$. Because each Stein space X_n has finite dimension, our reduction enables us to assume that $\mathscr{F}|X_n$ is a quasi-coherent $\mathscr{O}_{X_n}$-module. In particular, we have the isomorphisms

$$\mathscr{O}(X_n) \, \hat{\otimes}_{\mathscr{O}(X_{n+1})} \, \mathscr{F}(X_{n+1}) \cong \mathscr{F}(X_n)$$

and

$$\mathscr{O}(X_{n+1}) \, \hat{\otimes}_{\mathscr{O}(X_{n+1})} \, \mathscr{F}(X_{n+1}) \cong \mathscr{F}(X_{n+1}).$$

Therefore $(\mathscr{F}(X_n))_{n=0}^{\infty}$ is a Mittag-Leffler system of Fréchet $\mathscr{O}(X)$-modules.

Let U be an arbitrary Stein open subset of X, and let us write $U_n = U \cap X_n$ $(n \geq 0)$. According to our reduction hypothesis, $\mathscr{O}(U_n) \perp_{\mathscr{O}(X_n)} \mathscr{F}(X_n)$. Moreover, for each n, we obtain an exact Bar complex

$$B_{\bullet}^{\mathscr{O}(X_n)}(\mathscr{O}(U_n), \mathscr{F}(X_n)) \to \mathscr{F}(U_n) \to 0.$$

The components of these complexes form Mittag-Leffler systems with respect to n. Hence the inverse limit complex remains exact (see Corollary 3.2.6 and Corollary 3.2.9). Thus we have shown that

$$B_{\bullet}^{\mathscr{O}(X)}(\mathscr{O}(U), \mathscr{F}(X)) \to \mathscr{F}(U) \to 0$$

is exact, and the assertion is reduced to the case $X = \mathbb{C}^n$.

(2) Suppose that $X = \mathbb{C}^n$. We shall show that $\mathscr{F}(\mathbb{C}^n)$ is a quasi-coherent $\mathscr{O}(\mathbb{C}^n)$-module. We have to check that $\mathscr{O}(U) \perp_{\mathscr{O}(\mathbb{C}^n)} \mathscr{F}(\mathbb{C}^n)$ for each Stein open set, or equivalently, that the Koszul resolution

$$K_{\bullet}(z - w, \mathscr{O}(U) \, \hat{\otimes} \, \mathscr{F}(\mathbb{C}^n))$$

has vanishing homology in positive degree and its zeroth homology space is separated (see the end of Section 4.3). Since the above complex is induced by a corresponding complex of acyclic sheaves

$$K_{\bullet}(z - w, \mathscr{O}_U^{\mathscr{F}(\mathbb{C}^n)}),$$

it suffices to check that this complex is exact in degree $q > 0$, and that each element η in the closure of the image of the boundary map

$$K_1(z - w, \mathscr{O}(U) \, \hat{\otimes} \, \mathscr{F}(\mathbb{C}^n)) \xrightarrow{\ \delta\ } K_0(z - w, \mathscr{O}(U) \, \hat{\otimes} \, \mathscr{F}(\mathbb{C}^n))$$

is locally exact, that is, can locally be written as $\eta = \delta\xi$ with a suitable form ξ of degree 1.

To prove this, let us fix an open polydisc U in $\mathbb{C}^n$, an integer q with $0 < q \leq n$, and a form $\eta \in K_q(z - w, \mathcal{O}(U) \,\hat{\otimes}\, \mathcal{F}(\mathbb{C}^n))$ with $\delta\eta = 0$. We choose open polydiscs V_0, V_1, V_2 with $V_0 \Subset V_1 \Subset V_2 \Subset U$ and define $E = \mathcal{F}(\mathbb{C}^n)$, $E_1 = \mathcal{F}_{\mathbb{C}^n \setminus V_1}(\mathbb{C}^n)$, $E_2 = \mathcal{F}_{\bar{V}_2}(\mathbb{C}^n)$. Here for a closed set F in $\mathbb{C}^n$, we write $\mathcal{F}_F(\mathbb{C}^n)$ for the set of all global sections with support in F. The softness of $\mathcal{F}$ implies that the map

$$E_1 \oplus E_2 \to E, \qquad (s, t) \mapsto s + t$$

is onto. Because E_1 and E_2 consist of sections with support contained in U, the augmented Koszul complexes

$$K_\bullet\big(z - w, \mathcal{O}(U) \,\hat{\otimes}\, E_2\big) \to E_2 \to 0$$

and

$$K_\bullet\big(z - w, \mathcal{O}(U) \,\hat{\otimes}\, (E_1 \cap E_2)\big) \to E_1 \cap E_2 \to 0$$

are exact (see Vasilescu 1982 for a detailed discussion of this argument).

But then the quotient Koszul complex

$$K_\bullet\big(z - w, \mathcal{O}(U) \,\hat{\otimes}\, (E_2/E_1 \cap E_2)\big) \to E_1/E_1 \cap E_2 \to 0$$

is exact as well. Using the fact that there is a canonical isomorphism $E_2/E_1 \cap E_2 \cong E/E_1$ of Fréchet $\mathcal{O}(\mathbb{C}^n)$-modules, we finally obtain the exactness of the complex

$$K_\bullet\big(z - w, \mathcal{O}(U) \,\hat{\otimes}\, (E/E_1)\big) \to E/E_1 \to 0.$$

Passing from η to the induced form with coefficients in $\mathcal{O}(U) \,\hat{\otimes}\, (E/E_1)$, one sees that there is an element $\xi \in K_{q+1}(z - w, \mathcal{O}(U) \,\hat{\otimes}\, E)$ with the property that $\eta - \delta\xi \in K_q(z - w, \mathcal{O}(U) \,\hat{\otimes}\, E_1)$. If $q = n$, then the last statement holds with $\xi = 0$.

We claim that there is an element $\zeta \in K_{q+1}(z - w, \mathcal{O}(V_0) \,\hat{\otimes}\, E)$ such that

$$(\eta - \delta\xi) \,|\, V_0 = \delta\zeta$$

if $q < n$, respectively, that $\eta \,|\, V_0 = 0$ if $q = n$. This assertion follows from the next general result.

Lemma 4.4.2 *Let V, W be disjoint open sets in $\mathbb{C}^n$. If V is Stein, then the complex*

$$K_\bullet\Big(z - w, \varinjlim_F \mathcal{O}(V, \mathcal{F}_F(\mathbb{C}^n))\Big),$$

where F runs through the closed subsets of W, is exact.

Proof   If, in addition, W is a Stein open set, then there are analytic functions $f_1, \ldots, f_n \in \mathcal{O}(V \times W)$ with the property that $1 = (z_1 - w_1)f_1(z, w) + \cdots + (z_n - w_n)f_n(z, w)$ on $V \times W$. Since $\mathcal{O}(V, \mathcal{F}_F(\mathbb{C}^n))$ is an $\mathcal{O}(V \times W)$-module for each closed set F in W, in this case we even obtain the exactness of $K_\bullet(z - w, \mathscr{O}(V, \mathscr{F}_F(\mathbb{C}^n)))$ for each fixed set F.

To prove the general result, we use, for given open sets V, W in $\mathbb{C}^n$, the abbreviation

$$\mathscr{J}(V, W) = \lim_{\overrightarrow{F}} \mathscr{O}(V, \mathscr{F}_F(\mathbb{C}^n)).$$

Let us now make the assumption that $W = W_1 \cup W_2$ is a union of two Stein open sets. If $F \subset U_1 \cup U_2$ is an open cover, then the softness of $\mathscr{F}$ implies that

$$\mathscr{F}_F(\mathbb{C}^n) \subset \mathscr{F}_{\bar{U}_1}(\mathbb{C}^n) + \mathscr{F}_{\bar{U}_2}(\mathbb{C}^n).$$

An elementary tensor product argument yields that

$$\mathscr{O}(V, \mathscr{F}_F(\mathbb{C}^n)) \subset \mathscr{O}\left(V, \mathscr{F}_{\bar{U}_1}(\mathbb{C}^n)\right) + \mathscr{O}\left(V, \mathscr{F}_{\bar{U}_2}(\mathbb{C}^n)\right).$$

If F is contained in W, then U_1, U_2 can be chosen in such a way that $\bar{U}_i \subset W$ ($i = 1, 2$), and the above observation is exactly what is required to prove that the canonical Mayer–Vietoris type sequence

$$0 \to \mathscr{J}(V, W_1 \cap W_2) \to \mathscr{J}(V, W_1) \oplus \mathscr{J}(V, W_2) \to \mathscr{J}(V, W) \to 0$$

is exact at the right end, while exactness at the beginning and in the middle is trivial. Since this short exact sequence induces a short exact sequence between the Koszul complexes induced by $z - w$, the assertion that $K_\bullet(z - w, \mathscr{J}(V, W))$ is exact is reduced to the case that W is Stein.

By induction one proves the assertion of Lemma 4.4.2 for all open sets W that are finite unions of Stein open sets. The general case follows by the observation that

$$\mathscr{J}(V, W) = \lim_{\overrightarrow{n}} \mathscr{J}(V, W_n),$$

whenever $(W_n)_{n=1}^\infty$ is an open cover of W by an increasing sequence of open subsets. $\square$

Let us return to the proof of Theorem 4.4.1. To conclude the proof of part (2), we still have to show that the limit of a convergent sequence (η_k) in the range of the operator

$$K_1(z - w, \mathscr{O}(U) \hat{\otimes} E) \xrightarrow{\delta} K_0(z - w, \mathscr{O}(U) \hat{\otimes} E)$$

is locally exact. Let η be the limit of such a sequence. If we replace E by E/E_1 in the last formula, then the resulting boundary operator has closed range. Hence there is a form $\xi \in K_1(z - w, \mathscr{O}(U) \hat{\otimes} E)$ such that $\eta - \delta\xi \in K_0(z - w, \mathscr{O}(U) \hat{\otimes} E_1)$. Now, as before, it suffices to apply Lemma 4.4.2.

(3) Since $\mathscr{F}(\mathbb{C}^n)$ is a quasi-coherent $\mathscr{O}(\mathbb{C}^n)$-module, we can consider the induced quasi-coherent analytic Fréchet sheaf $\tilde{\mathscr{F}}$ on $\mathbb{C}^n$. From Section 4.3 we recall that the section space of the sheaf $\tilde{\mathscr{F}}$ on a Stein open set U in $\mathbb{C}^n$ is given by

$$\tilde{\mathscr{F}}(U) = \mathscr{O}(U) \hat{\otimes}_{\mathscr{O}(\mathbb{C}^n)} \mathscr{F}(\mathbb{C}^n).$$

The canonical morphism

$$\mathscr{O}(U) \mathbin{\hat{\otimes}}_{\mathscr{O}(\mathbb{C}^n)} \mathscr{F}(\mathbb{C}^n) \to \mathscr{F}(U), \tag{4.4.1}$$

given by restriction and multiplication, induces a morphism $\varphi \colon \tilde{\mathscr{F}} \to \mathscr{F}$ of Fréchet $\mathscr{O}_{\mathbb{C}^n}$-modules. To conclude the proof of Theorem 4.4.1, we still have to check that the maps occurring in (4.4.1) are isomorphisms. To do this, we show that the resulting sheaf homomorphism φ is an isomorphism.

First, we observe that $\mathrm{supp}(\varphi \circ f) = \mathrm{supp}(f)$ for each global section f in $\tilde{\mathscr{F}}(\mathbb{C}^n)$. To prove the non-trivial inclusion, let D be an open polydisc with $\bar{D} \subset \mathbb{C}^n \setminus \mathrm{supp}(\varphi \circ f)$. Then Lemma 4.4.2 implies that there are analytic functions $h_1, \ldots, h_n \in \mathscr{O}(D) \mathbin{\hat{\otimes}} \mathscr{F}(\mathbb{C}^n)$ satisfying

$$\varphi \circ f = \sum_{j=1}^{n} (z_j - w_j) h_j(z) \qquad (z \in D).$$

Since $\varphi \colon \tilde{\mathscr{F}}(\mathbb{C}^n) \to \mathscr{F}(\mathbb{C}^n)$ is an isomorphism, we obtain the decomposition

$$f = \sum_{j=1}^{n} (z_j - w_j) \varphi^{-1}(h_j(z)) \qquad (z \in D).$$

The canonical map $\mathscr{O}(D) \mathbin{\hat{\otimes}} \tilde{\mathscr{F}}(\mathbb{C}^n) \to \tilde{\mathscr{F}}(D)$ intertwines the $\mathscr{O}(D) \mathbin{\hat{\otimes}} \mathscr{O}(\mathbb{C}^n)$-module structure of $\mathscr{O}(D) \mathbin{\hat{\otimes}} \tilde{\mathscr{F}}(\mathbb{C}^n)$ and the $\mathscr{O}(D)$-module structure of $\tilde{\mathscr{F}}(D)$ modulo the corresponding canonical map

$$\Delta \colon \mathscr{O}(D) \mathbin{\hat{\otimes}} \mathscr{O}(\mathbb{C}^n) \to \mathscr{O}(D), \qquad (\Delta f)(z) = f(z, z).$$

Therefore, the above decomposition of f immediately yields that $f|_D = 0$.

Secondly, to prove that φ is an isomorphism, let us fix a point $x \in \mathbb{C}^n$ and let us consider the induced map $\varphi_x \colon \tilde{\mathscr{F}}_x \to \mathscr{F}_x$. Since the right vertical map in the commutative diagram

$$
\begin{array}{ccc}
\tilde{\mathscr{F}}(\mathbb{C}^n) & \xrightarrow{\ \sim\ } & \mathscr{F}(\mathbb{C}^n) \\
\downarrow & & \downarrow \\
\tilde{\mathscr{F}}_x & \xrightarrow{\ \varphi_x\ } & \mathscr{F}_x
\end{array}
$$

is surjective, it follows that φ_x is surjective.

In order to prove that φ_x is injective, it is sufficient to know that, for each Stein open neighbourhood U of x and each $f \in \tilde{\mathscr{F}}(U)$, there is a decomposition $f = f_1 + f_2$ with $f_1, f_2 \in \tilde{\mathscr{F}}(U)$, $x \notin \mathrm{supp}(f_2)$, and $\mathrm{supp}(f_1) \subset U$. Namely, in this case, $\varphi(f) = 0$ implies that $x \notin \mathrm{supp}(\varphi(f_1)) = \mathrm{supp}(f_1)$, and hence that $x \notin \mathrm{supp}(f)$.

To decompose a given section $f \in \tilde{\mathscr{F}}(U)$, we choose a representative g of f in $\mathscr{O}(U) \mathbin{\hat{\otimes}} \mathscr{F}(\mathbb{C}^n)$ and open polydiscs $W \Subset V \Subset U$ with centre x. Using the softness of $\mathscr{F}$ we write g in the form $g = g_1 + g_2$ with suitable elements $g_1 \in \mathscr{O}(U) \mathbin{\hat{\otimes}} \mathscr{F}_V(\mathbb{C}^n)$, $g_2 \in \mathscr{F}_{\mathbb{C}^n \setminus V}(\mathbb{C}^n)$. Let $f = f_1 + f_2$, $f_i = [g_i]$ $(i = 1, 2)$ be the corresponding decomposition in $\tilde{\mathscr{F}}(U)$. Since for each open subset D of U the diagram

$$
\begin{array}{ccccc}
\mathscr{O}(U)\,\hat{\otimes}_{\mathscr{O}(\mathbb{C}^n)}\mathscr{F}(\mathbb{C}^n) & \xrightarrow{\;\sim\;} & \mathscr{O}(U)\,\hat{\otimes}_{\mathscr{O}(\mathbb{C}^n)}\tilde{\mathscr{F}}(\mathbb{C}^n) & \longrightarrow & \mathscr{O}(U)\,\hat{\otimes}_{\mathscr{O}(\mathbb{C}^n)}\tilde{\mathscr{F}}(\mathbb{C}^n) \\
\downarrow & & \downarrow & & \downarrow \\
\tilde{\mathscr{F}}(U) & \xrightarrow[id]{} & \tilde{\mathscr{F}}(U) & \xrightarrow[\text{restriction}]{} & \tilde{\mathscr{F}}(D)
\end{array}
$$

commutes, and since for $s \in \mathscr{F}(\mathbb{C}^n)$ the support of s and the support of the corresponding element $\varphi^{-1}(s) = [1 \otimes s] \in \tilde{\mathscr{F}}(\mathbb{C}^n)$ coincide, it follows that $f_2 \mid W = 0$ and that $f_1 \mid U \setminus \overline{V} = 0$.

Thus the proof of theorem 4.4.1 is complete. $\square$

There will be cases in the forthcoming applications where softness is easier to check than quasi-coherence. A simple abstract example is the following.

Corollary 4.4.3 *Let X be a Stein space, and let M be a Fréchet $\mathscr{O}(X)$-module. If there exists a Fréchet soft $\mathscr{O}$-module $\mathscr{F}$ on X such that $M \cong \mathscr{F}(X)$ as topological $\mathscr{O}(X)$-modules, then M is quasi-coherent and $\tilde{M} = \mathscr{F}$.* $\square$

To illustrate this result by a few particular cases, let μ be a positive measure on a Stein space X, and let $1 \leq p \leq \infty$ be fixed. Suppose that $L^p(X, \mu)$ is a Banach $\mathscr{O}(X)$-module, that is, all multiplication operators with sections in $\mathscr{O}(X)$ are bounded in the L^p-norm. Then this space is a quasi-coherent $\mathscr{O}(X)$-module. Since $L^p_{\mathrm{loc}}(\cdot, \mu)$ is a Fréchet soft $\mathscr{O}$-module with $L^p(X, \mu)$ as global section space, the associated sheaf is precisely

$$ L^p(X, \mu)^{\sim} = L^p_{\mathrm{loc}}(\cdot, \mu). $$

Similarly, one may consider examples involving Sobolev-type norms.

Another basic example of a quasi-coherent $\mathscr{O}_X$-module is the sheaf $\mathscr{E}$ of smooth functions on a Stein manifold X. Moreover, Corollary 4.4.3 shows that, on Stein manifolds, the sheaves of differentiable functions (of integral or fractional order) are quasi-coherent.

The importance of these examples is also reflected in the following criterion.

Theorem 4.4.4 *Let X be a finite-dimensional Stein space, and let $\mathscr{F}$ denote an analytic Fréchet $\mathscr{O}_X$-module. Then $\mathscr{F}$ is quasi-coherent if and only if $\mathscr{F}$ admits a finite resolution to the right by soft analytic Fréchet sheaves which remains a resolution at the level of global sections.*

Proof The condition is sufficient by Theorem 4.4.1 and Lemma 4.3.4. Suppose that $\mathscr{F}$ is a quasi-coherent $\mathscr{O}_X$-module. Let us choose a topologically free resolution $\mathscr{L}_\bullet \to \mathscr{F} \to 0$.

Since the space X was supposed to be finite dimensional, we may identify it with a closed analytic subspace of $\mathbb{C}^n$. Let $\mathscr{I}$ be a finite ideal in $\mathscr{O}_{\mathbb{C}^n}$ such that $X = \mathrm{supp}\,\mathscr{O}_{\mathbb{C}^n}/\mathscr{I}$.

By a well-known theorem of Malgrange (see Theorem 7.2.1 and Corollary 7.2.2), the coherent $\mathcal{O}_{\mathbb{C}^n}$-module $\mathcal{O}_X$ possesses the $\bar{\partial}$-resolution

$$0 \to \mathcal{O}_X \to \Lambda^0(\mathcal{E}/\mathcal{J}\mathcal{E}) \xrightarrow{\bar{\partial}} \Lambda^1(\mathcal{E}/\mathcal{J}\mathcal{E}) \xrightarrow{\bar{\partial}} \cdots \xrightarrow{\bar{\partial}} \Lambda^n(\mathcal{E}/\mathcal{J}\mathcal{E}) \to 0,$$

and $\mathcal{E}/\mathcal{J}\mathcal{E}$ is a sheaf of Fréchet $\mathcal{O}_X$-modules. Moreover, $\mathcal{E}/\mathcal{J}\mathcal{E}$ is a sheaf of $\mathcal{E}_{\mathbb{C}^n}$-modules. In particular, this quotient sheaf is soft.

If we multiply the above $\bar{\partial}$-resolution by $* \,\hat{\otimes}\, E$, where E is a Fréchet space, then the complex remains exact because the $\bar{\partial}$-complex consists of nuclear Fréchet spaces. In particular, each component $\mathcal{L}_q$ of the resolution $\mathcal{L}_\bullet$ of $\mathcal{F}$ fits into a generalized $\bar{\partial}$-complex

$$0 \to \mathcal{L}_q \to \mathcal{M}_q^\bullet \qquad (q \geq 0).$$

The naturality of the construction of $\mathcal{M}_q^\bullet$ shows that $\mathcal{M}_\bullet^\bullet$ is a double complex. By its very definition, this complex satisfies the condition

$$\mathcal{H}_q(\mathcal{H}^p(\mathcal{M}_\bullet^\bullet)) \cong \begin{cases} \mathcal{F}, & \text{if } p = q = 0 \\ 0, & \text{otherwise.} \end{cases}$$

Therefore, if $\mathcal{C}^\bullet$ denotes the simple complex associated with $\mathcal{M}_\bullet^\bullet$, then a familiar spectral sequence argument yields (with the canonical convention on the indices)

$$\mathcal{H}^p(\mathcal{C}^\bullet) \cong \begin{cases} \mathcal{F} & \text{if } p = 0 \\ 0 & \text{otherwise.} \end{cases} \qquad (4.4.2)$$

Note that the complex $\mathcal{C}^\bullet$ is bounded to the right and that it consists of Fréchet soft $\mathcal{O}_X$-modules. Let us define $\mathcal{B}^0 = \mathrm{Im}(\mathcal{C}^{-1} \to \mathcal{C}^0)$. Then relation (4.4.2) shows that the following two complexes of $\mathcal{O}_X$-modules

$$0 \to \mathcal{C}^0/\mathcal{B}^0 \to \mathcal{C}^1 \to \cdots \to \mathcal{C}^n \to 0,$$

$$\cdots \to \mathcal{C}^{-2} \to \mathcal{C}^{-1} \to \mathcal{B}^0 \to 0$$

are exact.

It remains to prove that $\mathcal{C}^0/\mathcal{B}^0$ is a Fréchet soft $\mathcal{O}_X$-module.

The exactness of the second sequence, together with the assumption that $\dim X < \infty$, implies the acyclicity of $\mathcal{B}^0$ on each open subset of X (see the proof of Proposition 4.3.3). The long exact cohomology sequence induced by the exact complex

$$0 \to \mathcal{B}^0 \to \mathcal{C}^0 \to \mathcal{C}^0/\mathcal{B}^0 \to 0$$

shows that $H^j(V, \mathcal{C}^0/\mathcal{B}^0) = 0$ for $j \geq 1$ and for each open subset $V \subset X$. But this is known to be equivalent to the softness of the sheaf $\mathcal{C}^0/\mathcal{B}^0$ (see Bredon 1967, Proposition II.15.1).

Finally, let us define $\mathcal{Z}^0 = \mathrm{Ker}(\mathcal{C}^0 \to \mathcal{C}^1)$. The acyclicity of the sheaf $\mathcal{F}$ on Stein open subsets, together with the softness of $\mathcal{B}^0$, show that $\mathcal{Z}^0$ is acyclic on all Stein open subsets of X.

In particular, this shows that $\mathcal{F}(U) = \mathcal{Z}^0(U)/\mathcal{B}^0(U)$ for every Stein open subset $U \subset X$. As $\mathcal{F}(U)$ and $\mathcal{Z}^0(U)$ are Fréchet spaces, it follows that $\mathcal{B}^0(U)$

is also a Fréchet space (apply Lemma A2.7 in Appendix 2 to the double complex $\mathcal{M}_\bullet^\bullet(U)$).

This completes the proof of Theorem 4.4.4.   □

It is worth mentioning that the length of the resolution obtained in the above proof is bounded by the dimension of the affine complex space into which X was embedded. Due to the existence of finite Koszul resolutions (see (4.3.6)) on affine complex spaces, we can state the following weakened version of Theorem 4.4.4.

First note that we may unambiguously call a Fréchet $\mathcal{O}(X)$-module M *soft* if there exists an analytic Fréchet soft sheaf $\mathcal{F}$ with $\mathcal{F}(X) = M$.

Corollary 4.4.5 *Let X be a Stein open subset of $\mathbb{C}^n$, and let M be a Fréchet $\mathcal{O}(X)$-module. Then M is quasi-coherent if and only if there exists an exact sequence of Fréchet $\mathcal{O}(X)$-modules*

$$0 \to M \to S^0 \to S^1 \to \cdots \to S^{n-1} \to C \to 0$$

such that each S^j $(j = 0, \ldots, n-1)$ is a Fréchet soft $\mathcal{O}(X)$-module.

Proof The necessity follows from Theorem 4.4.4.

Suppose that an exact complex as in the statement exists, and let U be a Stein open subset of X. We have to prove that $M \perp_{\mathcal{O}(X)} \mathcal{O}(U)$.

The long exact sequence of derived functors of the module tensor product $\bullet \, \hat{\otimes}_{\mathcal{O}(X)} \, \mathcal{O}(U)$ yields, when applied to the given resolution of M, topological isomorphisms

$$\mathrm{T\hat{o}r}_j^{\mathcal{O}(X)}(M, \mathcal{O}(U)) \cong \mathrm{T\hat{o}r}_{n+j}^{\mathcal{O}(X)}(C, \mathcal{O}(U)) \qquad (j \geq 1).$$

Using the Koszul resolution (4.3.5), one sees that the second term vanishes for $j \geq 1$.

Set $Z^1 = \mathrm{Im}(S^0 \to S^1)$ and $L_\bullet = K_\bullet(z - w, \mathcal{O}(X) \, \hat{\otimes} \, \mathcal{O}(U))$. Then $\mathrm{T\hat{o}r}_1^{\mathcal{O}(X)}(Z^1, \mathcal{O}(U)) \cong \mathrm{T\hat{o}r}_n^{\mathcal{O}(X)}(C, \mathcal{O}(U))$ is separated. The image of the map $M \, \hat{\otimes}_{\mathcal{O}(X)} \, L_1 \to M \, \hat{\otimes}_{\mathcal{O}(X)} L_0$ is isomorphic to a closed subspace of $S^0 \, \hat{\otimes}_{\mathcal{O}(X)} L_0$. Thus the proof is complete.   □

Corollary 4.4.6 *Let X be an open subset of $\mathbb{C}$. A Fréchet $\mathcal{O}(X)$-module is quasi-coherent if and only if it is a closed submodule of a Fréchet soft $\mathcal{O}(X)$-module.*   □

Corollary 4.4.7 *A Fréchet $\mathcal{O}_X$-module $\mathcal{F}$ on a Stein manifold X of dimension n is quasi-coherent if and only if it is acyclic on Stein open sets and there exist, locally on X, exact complexes of Fréchet $\mathcal{O}_X$-modules of the form*

$$0 \to \mathcal{F} \to \mathcal{S}^0 \to \mathcal{S}^1 \to \cdots \to \mathcal{S}^{n-1} \to \mathcal{C} \to 0, \qquad (4.4.3)$$

such that each $\mathcal{S}^j$ $(j = 0, \ldots, n-1)$ is a soft sheaf.

Proof The necessity is clear by Theorem 4.4.4.

Suppose that the sheaf $\mathcal{F}$ fulfills the conditions in the statement. Let U be a Stein open set in X on which a complex of the form (4.4.3) exists. Then Corollary 4.3.5 shows $\mathcal{F} \mid U$ is a quasi-coherent $\mathcal{O}_U$-module.

By Theorem 4.3.6 we conclude that $\mathcal{F}$ is a quasi-coherent $\mathcal{O}_X$-module.   □

The above quasi-coherence criteria will be applied to some concrete situations in Chapters 7 and 8. As a general conclusion, we may say that the quasi-coherence property for Fréchet analytic sheaves is equivalent to the existence of an abstract $\bar{\partial}$-resolution.

4.5 BANACH COHERENT ANALYTIC FRÉCHET SHEAVES

Unlike the classical notion of coherence, the definition of quasi-coherence, and the equivalent descriptions given in the previous section, are not in any obvious way of a local nature. However, there is an intermediate class of sheaves, between the coherent and quasi-coherent sheaves, which is closer to the classical concept of a coherent sheaf in this respect. This class was introduced and studied by Leiterer (1978), independently of the theory of quasi-coherent sheaves. Although we shall not use this notion in the remaining parts of the book, we give in this section a brief description of the main properties of this interesting class of analytic sheaves. Details can be found in the cited paper of Leiterer (1978).

Let $(X, \mathcal{O})$ be an analytic space, and let E be a Fréchet space. To simplify the notation, we write $\mathcal{O}^E = \mathcal{O} \hat{\otimes} E$. If E is a Banach space, then $\mathcal{O}^E$ will be called a *Banach-free $\mathcal{O}$-module* on X. Let E and F be Fréchet spaces. On a reduced complex space $(X, \mathcal{O})$ there is a one-to-one correspondence between the continuous morphisms $\varphi \colon \mathcal{O}^E \to \mathcal{O}^F$ and the global sections in $\mathcal{O}(X, L_b(E, F)) = \mathcal{O}(X) \hat{\otimes} L_b(E, F)$. Throughout this section, $(X, \mathcal{O})$ will denote a reduced analytic space.

Definition 4.5.1 A Fréchet $\mathcal{O}$-module $\mathcal{F}$ on X is called *Banach coherent* if, for each point $x \in X$ and natural number $n \geq 0$, there is an open neighbourhood U of x and a *Banach-free resolution for $\mathcal{F} \mid U$ of length n*, that is, an exact sequence of continuous morphisms

$$\mathcal{O}^{E_n} \mid U \to \cdots \to \mathcal{O}^{E_1} \mid U \to \mathcal{O}^{E_0} \mid U \to \mathcal{F} \mid U \to 0,$$

where $E_0, \ldots, E_n$ are suitable Banach spaces.

By Oka's Theorem each coherent analytic sheaf is Banach coherent. The main result of this section is contained in the following theorem.

Theorem 4.5.2 *Let $(X, \mathcal{O})$ be a Stein space. Then each Banach coherent (Fréchet) $\mathcal{O}$-module is quasi-coherent.*   □

In particular, a Banach coherent $\mathcal{O}$-module on an analytic space X is acyclic on all Stein open subsets of X. Instead of Theorem 4.5.2, we shall prove the following stronger result.

Theorem 4.5.3 *Let $\mathcal{F}$ be a Fréchet $\mathcal{O}$-module on a Stein space $(X, \mathcal{O})$. If $\mathcal{F}$ is Banach coherent, then $\mathcal{F}$ possesses a global l^1-free resolution*

$$\cdots \to \mathcal{O}^{l^1(I_1)} \to \mathcal{O}^{l^1(I_0)} \to \mathcal{F} \to 0,$$

where $I_0, I_1, \ldots$ are suitable index sets.

It is obvious that Theorem 4.5.3 implies Theorem 4.5.2 (see Theorem 4.3.6 and the first step in the proof of Theorem 4.4.1). The l^1-spaces are privileged among all Banach spaces, since they satisfy a natural projectivity condition (see Appendix 2).

Lemma 4.5.4 *Let E, F be Banach spaces, let $S \in L(E, F)$ be onto, and let $T \in \mathcal{O}(X) \hat{\otimes} L(l^1(I), F)$. Then there exists a lifting $\tilde{T}$ of T, that is, a map $\tilde{T} \in \mathcal{O}(X) \hat{\otimes} L(l^1(I), E)$ with $S\tilde{T} = T$.*

To prove this result, it suffices to recall that $l^1(I)$ is projective in the category of all Banach spaces and to apply an obvious tensor product argument.   □

Corollary 4.5.5 *Let $\mathcal{F}$ be a Fréchet $\mathcal{O}$-module on X. Suppose that $\mathcal{F}$ has a Banach-free resolution of length n on X. Then $\mathcal{F}$ has an l^1-free resolution of length n on X.*

Proof The result is proved by induction on n. If $n = 0$, then $\mathcal{F}$ is the quotient of a sheaf $\mathcal{O}^E$ with a suitable Banach space E. Since E is a quotient of an l^1-space, the result holds in this case.

Suppose that there exists a resolution of $\mathcal{F}$ of length n

$$\mathcal{O}^{E_n} \xrightarrow{d_n} \cdots \to \mathcal{O}^{E_1} \xrightarrow{d_1} \mathcal{O}^{E_0} \xrightarrow{d_0} \mathcal{F} \to 0.$$

The induction hypothesis yields an l^1-free resolution for $\operatorname{Ker} d_0$

$$\mathcal{O}^{l^1(I_{n-1})} \to \cdots \to \mathcal{O}^{l^1(I_0)} \to \operatorname{Ker} d_0 \to 0.$$

On the other hand, the Banach space E_0 admits an l^1-resolution

$$l^1(J_n) \to \cdots \to l^1(J_0) \to E_0 \to 0.$$

By recurrently applying Lemma 4.5.4, one finds a morphism of complexes

$$
\begin{array}{ccccccccccc}
0 & \to & 0 & \to & \mathcal{O}^{l^1(I_{n-1})} & \to & \cdots & \to & \mathcal{O}^{l^1(I_1)} & \to & \mathcal{O}^{l^1(I_0)} & \to & \operatorname{Ker} d_0 & \to & 0 \\
& & \downarrow{\scriptstyle \tau_n = 0} & & \downarrow{\scriptstyle \tau_{n-1}} & & & & \downarrow{\scriptstyle \tau_1} & & \downarrow{\scriptstyle \tau_0} & & \downarrow & & \\
0 & \to & \mathcal{O}^{l^1(J_n)} & \to & \mathcal{O}^{l^1(J_{n-1})} & \to & \cdots & \to & \mathcal{O}^{l^1(J_1)} & \to & \mathcal{O}^{l^1(J_0)} & \to & \mathcal{O}^{E_0} & \to & 0.
\end{array}
$$

Then the cone of the morphism $\tau^\bullet = (\tau_i)_{i=0}^n$ gives an l^1-free resolution of length n for $\mathscr{F}$.   □

Next we need an improved version of Lemma 4.5.4.

Lemma 4.5.6 *Let $\mathscr{F}$ be a Fréchet $\mathscr{O}$-module, let E be a Banach space, and let $s: \mathscr{O}^E \to \mathscr{F}$, $t: \mathscr{O}^{l^1(I)} \to \mathscr{F}$ be morphisms. If K is a compact subset of X such that $s: \mathscr{O}^E(K) \to \mathscr{F}(K)$ is onto, then there is an open neighbourhood U of K and a lifting $\tilde{\imath}: \mathscr{O}^{l^1(I)} | U \to \mathscr{O}^E | U$ of $t \,|\, U$, that is, a morphism such that the following diagram commutes:*

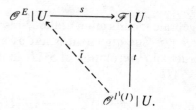

Proof Let $(U_n)_{n=0}^\infty$ be a basis of open neighbourhoods of K. The morphisms induced by t and s form diagrams

$$\mathscr{O}^E(U_n) \xrightarrow{\;s_n\;} \mathscr{F}(U_n)$$
$$\Big\uparrow{\scriptstyle t_n}$$
$$\mathscr{O}^{l^1(I)}(X).$$

The hypothesis implies that, for every $f \in \mathscr{O}^{l^1(I)}(X)$, there exists $n \geq 0$ such that $t_n(f) \in \mathrm{Im}(s_n)$. In other words, $\mathscr{O}^{l^1(I)}(X) = \bigcup_n t_n^{-1}(\mathrm{Im}\, s_n)$. Since $F = \mathscr{O}^{l^1(I)}(X)$ is a Fréchet space, one can show using Baire's theorem and the Banach–Schauder theorem (see Lemma 3.5 in Leiterer 1978 for details) that there is an integer m with $t_m^{-1}(\mathrm{Im}\, s_m) = F$ and such that $t_m^{-1}(s_m V)$ is open in F for each open set V in $\mathscr{O}^E(U_m)$.

In our concrete case, let U be an open neighbourhood of K such that $U \Subset U_m$. Then $V = \{f \in \mathscr{O}^E(U_m); \|f\|_{\infty, U} < 1\}$ is an open subset of $\mathscr{O}^E(U_m)$. Accordingly, there is a constant $c > 0$ such that

$$e_i \in c t_m^{-1}(s_m V) \qquad (i \in I),$$

where (e_i) denotes the canonical base of $l^1(I)$. Thus there are sections $f_i \in \mathscr{O}^E(U_m)$ with $s_m f_i = t_m e_i$ and $\|f_i\|_{\infty, U} < c$ $(i \in I)$.

Finally, observe that there is a well-defined map $\tilde{\imath} \in \mathscr{O}(U, L(l^1(I), E))$ with $\tilde{\imath}(z) e_i = f_i(z)$ for $i \in I$ and $z \in U$.   □

A first application of Lemma 4.5.6 is the following result.

Proposition 4.5.7 *Let $0 \to \mathscr{F} \xrightarrow{f} \mathscr{G} \xrightarrow{g} \mathscr{H} \to 0$ be an exact sequence of Fréchet $\mathscr{O}$-modules on X. If $\mathscr{G}$ and $\mathscr{H}$ are Banach coherent $\mathscr{O}$-modules, then so is $\mathscr{F}$.*

Proof Suppose first that $\mathscr{G} = \mathscr{O}^{l^1(I)}$. Fix $x \in X$ and a natural number n. By Corollary 4.5.5, we know that the sheaf $\mathscr{H}$ possesses an l^1-free resolution of length n on a suitable open neighbourhood U of x:

$$\mathscr{O}^{l^1(I_n)} | U \to \cdots \to \mathscr{O}^{l^1(I_1)} | U \longrightarrow \mathscr{O}^{l^1(I_0)} | U \xrightarrow{d_0} \mathscr{H} | U \to 0$$

$$0 \longrightarrow \mathscr{F} | U \xrightarrow{f} \mathscr{O}^{l^1(I)} | U \xrightarrow{g} \mathscr{H} | U \to 0. \tag{4.5.1}$$

It follows from Lemma 4.5.6 (applied with $K = \{x\}$) that, after shrinking U if necessary, there are morphisms α and β such that $d_0 \beta = g$ and $g \alpha = d_0$ (see diagram (4.5.1)).

Then a simple algebraic modification of the l^1-resolution of $\mathscr{H} | U$ produces an l^1-resolution of $\mathscr{F} | U$. The argument is left as an exercise to the reader.

In order to construct an l^1-resolution of $\mathscr{F} | U$ in the general case, one uses an l^1-resolution of $\mathscr{G} | U$

$$\mathscr{O}^{l^1(I_n)} | U \to \cdots \to \mathscr{O}^{l^1(I_0)} | U \xrightarrow{d_0} \mathscr{G} | U \to 0$$

$$0 \to \mathscr{F} | U \xrightarrow{f} \mathscr{G} | U \xrightarrow{g} \mathscr{H} | U \to 0,$$

and an l^1-resolution of $\mathrm{Ker}(g d_0)$. □

An essential step towards the proof of Theorem 4.5.3 is contained in the following proposition.

Proposition 4.5.8 *Let W be an open neighbourhood of a holomorphically convex compact subset K of X. Let $\mathscr{F}$ be a Banach coherent Fréchet $\mathscr{O}_W$-module. Then*:

(i) *there is an open neighbourhood U of K such that, for a suitable index set I, there is a resolution*

$$\mathscr{O}^{l^1(I)} | U \to \mathscr{F} | U \to 0$$

of length 0;

(ii) *for any Banach-free resolution $\mathscr{O}^E | U \to \mathscr{F} | U \to 0$ on a neighbourhood U of K, the map*

$$\mathscr{O}^E(K) \to \mathscr{F}(K) \to 0$$

is surjective.

Proof The proof is an adaptation of Cartan's original proof for what is now known as Cartan's lemma on invertible matrices (see Cartan 1940 or Grauert and Remmert 1979).

We may suppose, by shrinking X if necessary, that X is a closed analytic subspace of a Stein open subset of $\mathbb{C}^n$. Let $r_1(z), \ldots, r_{2n}(z)$ denote the real coordinates of a point $z \in X$. We denote by $(i)_k$ and $(ii)_k$ $(k = 0, \ldots, 2n)$ the statements of the proposition under the additional hypothesis that the coordinate functions r_{k+j} $(j = 1, \ldots, 2n - k)$ are constant on K. It is obvious that $(i)_0$ and $(ii)_0$ hold. We shall prove that the conditions $(i)_k$ and $(ii)_k$ imply the validity of the conditions $(i)_{k+1}$ and $(ii)_{k+1}$.

Suppose that the statements $(i)_k$ and $(ii)_k$ are true. Let K be a compact set as in the hypothesis of $(i)_{k+1}$. For $a, b \in \mathbb{R}$ with $a \leq b$, we define $K_{a,b} = \{z \in K; a \leq r_{k+1}(z) \leq b\}$.

If the real numbers a and b are close enough to each other, then assertion (i) is true for $K_{a,b}$. Hence, to prove $(i)_{k+1}$ it suffices to check that, if the sheaf $\mathscr{F}$ admits l^1-free resolutions on neighbourhoods of the compact sets $K_{a,b}$ and $K_{b,c}$, then the same is true on a neighbourhood of $K_{a,c}$.

Let

$$\mathscr{O}^{l^1(I)} | U \to \mathscr{F} | U \to 0$$

and

$$\mathscr{O}^{l^1(J)} | V \to \mathscr{F} | V \to 0$$

be resolutions on open neighbourhoods $U \supset K_{a,b}$ and $V \supset K_{b,c}$. By assumption $(ii)_k$, the horizontal maps in the diagram

$$\begin{array}{ccc} \mathscr{O}^{l^1(I)}(K_{b,b}) & \longrightarrow & \mathscr{F}(K_{b,b}) \longrightarrow 0 \\ \scriptstyle{\alpha}\downarrow\uparrow\scriptstyle{\beta} & & \downarrow \scriptstyle{id} \\ \mathscr{O}^{l^1(J)}(K_{b,b}) & \longrightarrow & \mathscr{F}(K_{b,b}) \longrightarrow 0 \end{array}$$

are surjective. According to Lemma 4.5.6, there are morphisms α and β which make the above diagram commutative. They define a holomorphic Banach bundle on $U \cup V$ with 1-cocycle on $U \cap V$ given by

$$\tau = \begin{pmatrix} 1 & 0 \\ \alpha & 1 \end{pmatrix} \begin{pmatrix} 1 & -\beta \\ 0 & 1 \end{pmatrix} \in GL(\mathscr{O}^{l^1(I) \oplus l^1(J)} | U \cap V).$$

Note that we may choose the neighbourhoods U, V and $U \cup V$ to be Stein. Then a generalization of Oka's principle to Banach bundles shows that the 1-cocycle τ is trivial (see Bungart 1964). Any resulting splitting of τ yields a surjective map

$$\mathscr{O}^{l^1(I) \oplus l^1(J)} | U \cup V \to \mathscr{F} | U \cup V \to 0,$$

as desired.

In order to prove $(ii)_{k+1}$, let

$$\mathscr{O}^E | U \xrightarrow{f} \mathscr{F} | U \to 0$$

be a Banach-free resolution of $\mathscr{F}$ on a neighbourhood of a compact set K as in $(ii)_{k+1}$.

Then Proposition 4.5.7 and condition $(i)_{k+1}$ yield a resolution for $\operatorname{Ker} f$. Repeating this argument, one obtains a decreasing sequence $(U_n)_{n \geq 0}$ of finite-dimensional Stein open neighbourhoods U_n of K such that, for every

$n \geq 0$, the sheaf $\mathscr{F} \mid U_n$ admits a Banach-free resolution of length n. The argument used in the proof of Proposition 4.3.3 shows that there is a natural number n_0 such that

$$H^1(U_n, \operatorname{Ker} f) = 0 \qquad (n \geq n_0).$$

Hence the map $\mathscr{O}^E(K) \xrightarrow{f} \mathscr{F}(K)$ is onto.

These are the principal points of the proof of Proposition 4.5.8 (for details see Leiterer 1978). □

Proof of Theorem 4.5.3 Let $\mathscr{F}$ be a Banach coherent Fréchet $\mathscr{O}$-module on a Stein space X. Let $(K_n)_{n \geq 0}$ be an exhaustion of X by holomorphically convex compact subsets. By Proposition 4.5.8, there are resolutions

$$\mathscr{O}^{l^1(I_n)} \xrightarrow{f_n} \mathscr{F} \mid U_n \to 0,$$

where $U_n \supset K_n$ are suitable open neighbourhoods $(n \geq 0)$.

Let I denote the disjoint union $I_0 \cup I_1 \cup \cdots$, and let $\pi_n : \mathscr{O}^{l^1(I)} \to \mathscr{O}^{l^1(I_n)}$ be the canonical projections $(n \geq 0)$. By Lemma 4.5.6 and Proposition 4.5.8, it follows that, after modifying the sets U_n if necessary, there are morphisms

$$\alpha_n : \mathscr{O}^{l^1(I_{n+1})} \mid U_n \to \mathscr{O}^{l^1(I_n)} \mid U_n$$

and

$$\beta_n : \mathscr{O}^{l^1(I_n)} \mid U_n \to \mathscr{O}^{l^1(I_{n+1})} \mid U_n$$

such that $f_{n+1} = f_n \alpha_n$ and $f_n = f_{n+1} \beta_n$.

Define $g_n = f_n \pi_n$ $(n \geq 0)$. We write $E_n = l^1(I_n)$ $(n \geq 0)$, and we regard each E_n as a closed subspace of $E = l^1(I)$. Then $\tau_n = (I + \alpha_n \pi_{n+1})(I - \beta_n \pi_n)$ defines an automorphism of the sheaf $\mathscr{O}^E \mid U_n$:

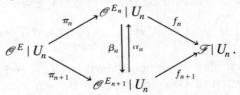

Moreover,

$$g_n \tau_n = f_n \pi_n (I + \alpha_n \pi_{n+1})(I - \beta_n \pi_n) = (f_n \pi_n + f_{n+1} \pi_{n+1})(I - \beta_n \pi_n)$$
$$= f_n \pi_n + f_{n+1} \pi_{n+1} - f_n \pi_n = f_{n+1} \pi_{n+1} = g_{n+1}. \tag{4.5.2}$$

Let $A_n \in \mathscr{O}(U_n) \hat{\otimes} L(E_{n+1}, E_n)$, $B_n \in \mathscr{O}(U_n) \hat{\otimes} L(E_n, E_{n+1})$, $T_n \in \mathscr{O}(U_n) \hat{\otimes} GL(E)$, and $P_N \in \mathscr{O}(U_n) \hat{\otimes} L(E, E_n)$ be the analytic operator functions corresponding to the morphisms α_n, β_n, τ_n, and π_n, respectively. Then $T_n = (I + A_n P_{n+1})(I - B_n P_n)$ $(n \geq 0)$.

Since K_n is a holomorphically convex compact set, the Oka–Weil approximation theorem shows that there are global sections $\tilde{A}_n \in \mathscr{O}(X) \hat{\otimes} L(E_{n+1}, E_n)$ and $\tilde{B}_n \in \mathscr{O}(X) \hat{\otimes} L(E_n, E_{n+1})$ approximating A_n and B_n uniformly on K_n as well as we require.

Let us define $\tilde{T}_n = (I + \tilde{A}_n P_{n+1})(I - \tilde{B}_n P_n)$ and $\tilde{S}_n = \tilde{T}_1 \tilde{T}_2 \cdots \tilde{T}_n$ for $n \geq 0$. After shrinking U_n if necessary, we may suppose that

$$\|\tilde{S}_n \tilde{T}_{n+1} \tilde{S}_{n+1}^{-1} - I\|_{\infty, U_n} < 2^{-n-1} \qquad (n \geq 0).$$

Consequently, the infinite product

$$R_n = \prod_{k=0}^{\infty} \tilde{S}_{n+k} \tilde{T}_{n+1+k} \tilde{S}_{n+1+k}^{-1}$$

converges uniformly on U_n, and it satisfies $\|R_n - I\|_{\infty, U_n} < 1$.

The automorphism $\omega_n \in \mathcal{O}(U_n) \,\hat{\otimes}\, GL(E)$ associated with $\tilde{T}_n \tilde{S}_n^{-1} R_n$ satisfies, in view of relation (4.5.2), the identity $g_n \omega_n = (g_{n+1} \omega_{n+1}) \,|\, U_n$. Hence there is a well-defined morphism $g \colon \mathcal{O}^E \to \mathcal{F}$ with $g = g_n$ on U_n $(n \geq 0)$. This morphism is surjective, since each g_n is surjective. A repeated application of Proposition 4.5.7 completes the proof.   □

We should point out that the conclusion of Theorem 4.5.3 is highly non-trivial even for coherent analytic sheaves on Stein spaces.

We conclude this chapter by mentioning a few open questions related to quasi-coherent or Banach coherent analytic sheaves.

Open problem 4.5.9 (Leiterer 1978)   Is the local existence of Banach-free resolutions of length one sufficient for a given Fréchet $\mathcal{O}$-module to be Banach coherent?

In other words, the question is of whether a Fréchet $\mathcal{O}_X$-module which locally admits resolutions

$$\mathcal{O}_U^E \to \mathcal{O}_U^F \to \mathcal{F} \,|\, U \to 0$$

with suitable Banach spaces E and F is Banach coherent. Note that for coherent analytic sheaves instead of Banach coherent analytic sheaves the answer is positive.

A closely related question is the following

Open problem 4.5.10   Let X be a Stein space, and let $0 \to \mathcal{F} \to \mathcal{G} \to \mathcal{H} \to 0$ be an exact sequence of Fréchet $\mathcal{O}_X$-modules. Is $\mathcal{F}$ a quasi-coherent sheaf whenever $\mathcal{G}$ and $\mathcal{H}$ have this property?

A positive answer to the similar problem raised for the weakened Banach coherence property indicated in Problem 4.5.9 would imply a positive answer to Problem 4.5.9.

Note that, in view of Lemma 4.3.4, only the vanishing of the cohomology group $H^1(X, \mathcal{F})$ must be shown in the setting of Problem 4.5.10 to settle this question in the affirmative. Moreover, a second reduction is possible. Indeed, using a holomorphic embedding, we may suppose that $X = \mathbb{C}^n$. In this case, the Koszul resolutions of $\mathscr{G}$ and $\mathscr{H}$

$$K_\bullet\left(z - w, \mathscr{O}_X \hat{\otimes} \mathscr{H}(X)\right) \longrightarrow \mathscr{H} \to 0$$

$$\uparrow$$

$$K_\bullet\left(z - w, \mathscr{O}_X \hat{\otimes} \mathscr{G}(X)\right) \longrightarrow \mathscr{G} \to 0$$

form a double complex of Fréchet-free $\mathscr{O}_X$-modules. Let $\mathscr{C}_\bullet$ be the associated total complex. Then

$$\mathscr{H}_q(\mathscr{C}_\bullet) = \begin{cases} \mathscr{F}, & q = 1 \\ 1, & q \neq 1. \end{cases}$$

Hence the sheaf $\mathscr{F}$ fits into the exact sequences

$$0 \to \mathscr{Z} \to \mathscr{C}_1 \to \mathscr{C}_0 \to 0,$$
$$0 \to \mathscr{B} \to \mathscr{Z} \to \mathscr{F} \to 0,$$
$$\cdots \to \mathscr{C}_3 \to \mathscr{C}_2 \to \mathscr{B} \to 0.$$

But $\mathscr{B}$ turns out to be a quasi-coherent $\mathscr{O}_X$-module. If $\mathscr{Z}$ were quasi-coherent, then according to Lemma 4.3.4, the sheaf $\mathscr{F}$ would have the same property.

In conclusion, we have reduced Problem 4.5.10 to the following question.

Open problem 4.5.10′ Let $u: E \to F$ be a morphism of Fréchet $\mathscr{O}(\mathbb{C}^n)$-modules. Consider the morphism of sheaves

$$\varphi: \left(\bigoplus_1^n \mathscr{O}^F \right) \oplus \mathscr{O}^E \to \mathscr{O}^F,$$

$$\varphi(f_1, \ldots, f_n, e) = (z_1 - w_1)f_1 + \cdots + (z_n - w_n)f_n + ue$$

where $z_1, \ldots, z_n$ are the coordinates of $\mathbb{C}^n$, and $w_1, \ldots, w_n$ denote their action on the module F.

Is φ onto at the level of global sections on $\mathbb{C}^n$ if φ is onto at the level of sheaves?

It is worthwhile to remark that pointwise surjectivity of the linear map φ does not suffice to ensure global surjectivity (see Mantlik 1988).

Another intriguing question related to the structure of analytic sheaves is the following.

Open problem 4.5.11 Does local Fréchet quasi-coherence imply global Fréchet quasi-coherence?

We have explicitly mentioned Fréchet quasi-coherence because the Banach-coherence property is local (see Definition 4.5.1).

In fact, an affirmative answer to Problem 4.5.10 would imply a positive solution to this problem. More precisely, this implication can be proved as a corollary of Theorem 4.4.4 as follows.

Let $\mathscr{F}$ be a Fréchet $\mathscr{O}_X$-module on a Stein space X. Suppose that $\mathscr{F}$ is locally quasi-coherent, that is, there exists a Stein open covering $\mathscr{U}$ of X with finite-dimensional nerve such that $\mathscr{F}|U$ is a quasi-coherent $\mathscr{O}_U$-module for every $U \in \mathscr{U}$. According to Theorem 4.4.4, for every finite intersection $V = U_0 \cap \cdots \cap U_q$ ($U_0,\ldots,U_q \in \mathscr{U}$), there exists a finite Fréchet soft resolution

$$0 \to \mathscr{F}|V \to \mathscr{S}_V^{\bullet}.$$

Moreover, the construction given in the proof of Theorem 4.4.4 shows that these resolutions can be selected in a functorial way.

Therefore there exists a Čech-type resolution of the sheaf $\mathscr{F}$ by complexes of soft Fréchet $\mathscr{O}_X$-modules

$$0 \to \mathscr{F} \to \bigoplus_{U \in \mathscr{U}} \mathscr{S}_U^{\bullet} \to \bigoplus_{U_0, U_1} \mathscr{S}_{U_0 \cap U_1}^{\bullet} \to \cdots.$$

By passing to the associated total complex, one obtains a finite resolution of $\mathscr{F}$ to the right by Fréchet soft $\mathscr{O}_X$-modules. The only missing property is the exactness of this resolution at the level of global section spaces. But this exactness would follow from a positive answer to Problem 4.5.10.

The last problem of this chapter is concerned with the definition of analytic transversality given in Section 4.2 (Definition 4.2.2).

Open problem 4.5.12 (Douady and Verdier)   Let $f: X \to S$ and $g: Y \to S$ be morphisms of Stein spaces, and let $\mathscr{F}$ and $\mathscr{G}$ be coherent analytic sheaves on X and Y, respectively. Are $\mathscr{F}$ and $\mathscr{G}$ analytically transversal over S whenever $\operatorname{Tor}_q^{\mathscr{O}_{S,s}}(\mathscr{F}_x, \mathscr{G}_y) = 0$ for every $q \geq 1$ and $s \in S$, $x \in X$, $y \in Y$ with $f(x) = g(y) = s$?

4.6   REFERENCES AND COMMENTS

Besides the references mentioned in Section 4.1, the recent volumes of the Soviet Mathematical Encyclopaedia devoted to 'several complex variables' contain excellent surveys on complex analytic geometry and on function theory in several complex variables. Within this series the theory of coherent analytic sheaves is extensively discussed and commented in Onishchik (1990). The reader interested in the algebraic aspects of complex analytic geometry will also find the compendium of Fischer (1976) useful. For the reader interested in the advanced aspects of the theory of analytic sheaves the collected works of Oka, Cartan, and the whole series of Séminaire Cartan are essential reading (for references see Oka 1984 and the introduction to Kashiwara and Schapira 1990).

A good exercise which complements Section 4.2 would be to prove Theorem 4.2.4 in the context of C^∞-modules on smooth C^∞-manifolds. While the analytic and smooth transversalities behave similarly, the transversality over function algebras such as C^k or Lip_α $(0 \le k < \infty, 0 < \alpha < 1)$ is more delicate because these spaces are not nuclear. The geometric counterpart of the analytic transversality developed in Section 4.2, namely a good intersection theory for complex spaces with singularities, is far from being complete. The most remarkable progress in this direction was made in the context of the theory of currents (see for details Chirka 1985).

The literature concerning quasi-coherent analytic Fréchet sheaves is not so plentiful, and is always related to concrete applications. We mention only the article of Ramis and Ruget (1974), where this class of sheaves was introduced, and a monograph on deformation theory by Bingener and Kosarew (1987) where this class of sheaves is used for other applications. We have to point out that the quasi-coherent analytic Fréchet sheaves do not coincide with the quasi-coherent sheaves used in algebraic geometry, although both classes of sheaves have many common properties. The proof of Theorem 4.3.6 is reproduced after Putinar (1986).

The idea of relating the softness and the quasi-coherence of an analytic sheaf appears implicitly in the early works on abstract spectral decompositions (see Bishop 1959, Colojoară and Foiaş 1968, Vasilescu 1982, and Chapter 1 of this monograph). This idea is made explicit in Putinar (1983a) and it is further developed in a series of papers, among which we mention Putinar (1986), (1990a), and Eschmeier and Putinar (1984).

The original motivation for the article of Leiterer (1978), in which the class of Banach coherent analytic Fréchet sheaves is introduced, was related to some factorization theorems for vector-valued meromorphic functions (see Leiterer 1978 and the references therein). A different motivation for considering this class of analytic modules is related to the following lifting problem for vector-valued analytic functions. Let $U \subset \mathbb{C}^n$ be a Stein open set, and let

$$A(z): E \to F \ (z \in U)$$

be an analytic family of bounded linear operators acting between two Banach spaces E and F. It follows from Theorem 4.5.2 and Proposition 4.5.7 (together with Lemma 2.1.5) that the induced map

$$A: \mathcal{O}(U) \,\hat{\otimes}\, E \to \mathcal{O}(U) \,\hat{\otimes}\, F$$

remains surjective if all the operators $A(z)$ $(z \in U)$ are surjective. For details see Leiterer (1978). A special case of this result was proved by Allan (1967).

5
Fréchet modules over Stein algebras

In Chapter 2 we constructed, for each finite commuting tuple $a \in L(E)^n$ of regular operators on a Fréchet space E, an extension of the global $\mathscr{O}(\mathbb{C}^n)$-functional calculus of a to an algebra homomorphism

$$\Phi: \mathscr{O}(\sigma(a)) \to L(E), \qquad f \mapsto f(a),$$

where $\sigma(a)$ is the Taylor spectrum of a. The analytic functional calculus was constructed with the help of a Cauchy–Weil type integral following Taylor (1970b). Among other results, we proved that the analytic functional calculus satisfies a general spectral mapping theorem of the form

$$\sigma(f(a)) = f(\sigma(a)) \big(f \in \mathscr{O}(\sigma(a))^k, k \in \mathbb{N} \big).$$

Soon after giving the above construction, Taylor (1972b) observed that an analytic functional calculus for a can also be defined by applying the topological homology theory developed in Chapter 3 to the Fréchet $\mathscr{O}(\mathbb{C}^n)$-module structure of E given by

$$\mathscr{O}(\mathbb{C}^n) \times E \to E, \qquad (f, x) \mapsto f(a)x.$$

For instance, using the fact that the augmented Koszul complex

$$K_\bullet(z - a, \mathscr{O}(\mathbb{C}^n, E)) \to E \to 0$$

yields a topologically free resolution of the Fréchet $\mathscr{O}(\mathbb{C}^n)$-module E, one obtains the result that the Taylor spectrum $\sigma(a)$ is the smallest closed set in $\mathbb{C}^n$ such that, for each Stein open subset U of its complement, the vanishing condition

$$\text{Tôr}_q^{\mathscr{O}(\mathbb{C}^n)}(\mathscr{O}(U), E) = 0 \quad (q \geq 0)$$

holds. If U is an open neighbourhood of $\sigma(a)$, then one can use the Koszul complex $K_\bullet(z - a, \mathscr{O}(U, E))$ or, more generally, the Bar resolution of the Fréchet $\mathscr{O}(\mathbb{C}^n)$-module E, together with the Čech complex of a suitable Stein open cover of U, to extend the $\mathscr{O}(\mathbb{C}^n)$-module structure of E to a Fréchet $\mathscr{O}(U)$-module structure (see Theorem 5.1.5 below).

At this stage it is not at all clear whether the above two constructions yield the same solution of the analytic functional calculus problem. That this is indeed the case was shown by Putinar (1983b). The idea is first to prove that the analytic functional calculus map defined via topological homology also satisfies the spectral mapping theorem indicated at the beginning, and then to observe that the validity of a spectral mapping theorem of this kind determines the analytic functional calculus map uniquely.

All the constructions necessary for the homological approach to the analytic functional calculus problem make perfect sense in the more general setting of analytic modules over finite-dimensional Stein spaces. Since, for the proof of the above-mentioned uniqueness result, it also seems to be necessary to depart from the category of open sets in Euclidean spaces, we work from the very beginning in this more general context.

5.1 SPECTRA OF ANALYTIC FRÉCHET MODULES

Let X be a Stein space. Throughout this section we suppose that X is of *finite (analytic) dimension*, that is, there is a natural number d such that

$$\dim_x X \le d$$

for all $x \in X$. Here $\dim_x X$ denotes the *analytic dimension* of X at x, that is, the minimal natural number k such that there is an open neighbourhood V of x and analytic sections $f_1, \ldots, f_k \in \mathcal{O}(V)$ with zero set $N(f_1, \ldots, f_k) = \{x\}$ (see Grauert and Remmert 1984, Chapter 5). We mention that the embedding dimension of an analytic space is in general larger than its analytic dimension (see Chapter 4 for a definition of the embedding dimension).

On a finite-dimensional Stein space each open cover of an open set possesses a refinement which is Stein and has finite-dimensional nerve. One can prove this assertion by using the fact that a Stein space of complex dimension d is triangulable and that the corresponding polyhedron has real dimension $2d$ (see for instance Lojasiewicz 1964). Or one can use for the proof of the same statement a topological embedding of a finite-dimensional Stein space into an affine space (see Grauert and Remmert 1979, p. 130).

Let M be a Fréchet $\mathcal{O}(X)$-module, and let $U \subset X$ be an open set in X. In the following we shall study the question of whether the $\mathcal{O}(X)$-module structure of M can be extended to a Fréchet $\mathcal{O}(U)$-module structure on M, and whether such extensions are unique. We start with an elementary uniqueness result.

Lemma 5.1.1 *Let $U \subset X$ be a Stein open subset.*

(a) *For any pair M, N of Fréchet $\mathcal{O}(U)$-modules, the restriction $r: \mathcal{O}(X) \to \mathcal{O}(U)$ induces a quasi-isomorphism*

$$r: B_\bullet^{\mathcal{O}(X)}(M, N) \to B_\bullet^{\mathcal{O}(U)}(M, N)$$

between complexes of Fréchet spaces. Here M and N are regarded as $\mathcal{O}(X)$-modules via the restriction map r.

(b) *If M is a Fréchet $\mathcal{O}(X)$-module, then there is at most one Fréchet $\mathcal{O}(U)$-module structure on M extending the $\mathcal{O}(X)$-module structure.*

Proof (a) By applying Corollary 4.2.5 to the inclusion map $i: U \to X$, we obtain $\mathscr{O}(U) \perp_{\mathscr{O}(X)} \mathscr{O}(U)$ and find that the diagonal map induces a topological isomorphism $\mathscr{O}(U) \hat{\otimes}_{\mathscr{O}(X)} \mathscr{O}(U) \xrightarrow{\sim} \mathscr{O}(U)$. Therefore the restriction map $r: \mathscr{O}(X) \to \mathscr{O}(U)$ induces a quasi-isomorphism

$$B_\bullet^{\mathscr{O}(X)}(\mathscr{O}(U), \mathscr{O}(U)) \to B_\bullet^{\mathscr{O}(U)}(\mathscr{O}(U), \mathscr{O}(U)).$$

Multiplying by $1_M \hat{\otimes}_{\mathscr{O}(U)} *$ and $* \hat{\otimes}_{\mathscr{O}(U)} N$, we obtain the quasi-isomorphism

$$r: B_\bullet^{\mathscr{O}(X)}(M, N) \to B_\bullet^{\mathscr{O}(U)}(M, N)$$

(see Lemma 3.1.17).

(b) The first part of the lemma applied to the pair of Fréchet $\mathscr{O}(U)$-modules $\mathscr{O}(U)$ and M shows that the canonical map

$$\mathscr{O}(U) \hat{\otimes}_{\mathscr{O}(X)} M \to \mathscr{O}(U) \hat{\otimes}_{\mathscr{O}(U)} M \cong M$$

is a topological isomorphism of Fréchet spaces. Modulo this identification the $\mathscr{O}(U)$-module structure of M is the free $\mathscr{O}(U)$-module structure of $\mathscr{O}(U) \hat{\otimes}_{\mathscr{O}(X)} M$. $\qquad\square$

Let E be a Fréchet space, and let $a = (a_1, \ldots, a_n) \in L(E)^n$ be a commuting tuple of regular operators on E. If Γ_i is the positively-oriented boundary of an open disc in $\mathbb{C}$ containing $\sigma(a_i)$ $(1 \le i \le n)$, then

$$\Phi: \mathscr{O}(\mathbb{C}^n) \to L(E),$$

$$f \mapsto (2\pi i)^{-n} \int_{\Gamma_1} \cdots \int_{\Gamma_n} f(z) R(z_1, a_1) \cdots R(z_n, a_n) \, dz_n \cdots dz_1$$

defines a continuous unital algebra homomorphism with $\Phi(z_i) = a_i$ for $i = 1, \ldots, n$. This can be checked exactly as in the one-dimensional case. But of course it also follows from the results proved in Section 2.5. Each such continuous algebra homomorphism is necessarily of the form

$$\Phi(f) = \sum_{k \in \mathbb{N}^n} a^k f^{(k)}(0)/k! \qquad (f \in \mathscr{O}(\mathbb{C}^n)).$$

If E is equipped with the Fréchet $\mathscr{O}(\mathbb{C}^n)$-module structure

$$\mathscr{O}(\mathbb{C}^n) \times E \to E, \qquad (f, x) \mapsto \Phi(f)x,$$

then according to Corollary 3.1.13 and the remarks at the end of Section 4.3, for each open set U in $\mathbb{C}^n$ and each $q \ge 0$, we have

$$\text{Tôr}_q^{\mathscr{O}(\mathbb{C}^n)}(\mathscr{O}(U), E) \cong H_q(K_\bullet(z - a, \mathscr{O}(U, E))).$$

Therefore, the Taylor spectrum of a is the smallest closed set σ in $\mathbb{C}^n$ with the property that, for each Stein open set $U \subset \mathbb{C}^n \setminus \sigma$ and for each $q \ge 0$,

$$\text{Tôr}_q^{\mathscr{O}(\mathbb{C}^n)}(\mathscr{O}(U), E) = 0.$$

In this last form the definition of the Taylor spectrum makes sense for an arbitrary Fréchet $\mathcal{O}(X)$-module M on an arbitrary Stein space X. To be more precise, let us denote by $B_q^{\mathcal{O}(X)}(\mathcal{O}, M)$ $(q \geq 0)$ the topologically free $\mathcal{O}$-module given by the presheaf

$$U \mapsto B_q^{\mathcal{O}(X)}(\mathcal{O}(U), M).$$

Then we obtain the following analogue of Corollary 2.1.9.

Lemma 5.1.2 *Let X be a Stein space of finite dimension, and let M be a Fréchet $\mathcal{O}(X)$-module. Then for a given Stein open set Ω in X, the following conditions are equivalent*:

  (i) *the complex $B_\bullet^{\mathcal{O}(X)}(\mathcal{O}(\Omega), M)$ is exact*;
 (ii) *the complex $B_\bullet^{\mathcal{O}(X)}(\mathcal{O}(U), M)$ is exact for each open subset U of Ω*;
(iii) *the complex $B_\bullet^{\mathcal{O}(X)}(\mathcal{O}, M)$ is exact on Ω*.

Proof If $B_\bullet^{\mathcal{O}(X)}(\mathcal{O}(\Omega), M)$ is an exact complex, then for each open subset $U \subset X$, the complex $B_\bullet^{\mathcal{O}(X)}(\mathcal{O}(U), M) = \mathcal{O}(U) \hat{\otimes}_{\mathcal{O}(\Omega)} B_\bullet^{\mathcal{O}(X)}(\mathcal{O}(\Omega), M)$ remains exact (see Section 3.1). Thus it suffices to show that condition (iii) implies condition (i).

To do this, let us consider more generally an arbitrary exact complex $\mathcal{L}_\bullet = (\mathcal{L}_n)_{n \geq 0}$ of acyclic $\mathcal{O}_\Omega$-modules on X. By defining $\mathcal{Z}_j = \text{Ker}(\mathcal{L}_j \to \mathcal{L}_{j-1})$ $(j \geq 0)$ with $\mathcal{L}_{-1} = 0$, we obtain a system of exact sequences

$$0 \to \mathcal{Z}_j \to \mathcal{L}_j \to \mathcal{Z}_{j-1} \to 0 \qquad (j \geq 1)$$

of $\mathcal{O}_\Omega$-modules. Using the induced long exact cohomology sequences, one obtains isomorphisms

$$H^k(\mathcal{Z}_{j-1}) \cong H^{k+n}(\mathcal{Z}_{j+n-1}) \qquad (j, k \geq 1, n \geq 0).$$

On the paracompact space Ω, these cohomology groups can be computed by means of alternating Čech cochains. Hence our assumption of finite dimensionality implies that all sheaves $\mathcal{Z}_j$ $(j \geq 0)$ are acyclic. In particular, the induced complex of section spaces $\Gamma(\Omega, \mathcal{L}_\bullet)$ is exact.   □

The preceding result shows that the following definition of the spectrum of a Fréchet $\mathcal{O}(X)$-module makes sense.

Definition 5.1.3 Let X be a Stein space of finite dimension, and let M be a Fréchet $\mathcal{O}(X)$-module. Then the *spectrum* $\sigma(X, M)$ of M is defined as the smallest closed set σ in X with the property that $\text{Tôr}_q^{\mathcal{O}(X)}(\mathcal{O}(U), M) = 0$ for each Stein open set $U \subset X \setminus \sigma$ and for each natural number $q \geq 0$.

The following observation shows that a reasonably defined spectrum cannot be much smaller than the one defined above.

Lemma 5.1.4 *Let M be a Fréchet $\mathcal{O}(X)$-module over a Stein space of finite dimension. If U is a Stein open subset of X such that the $\mathcal{O}(X)$-module structure of M extends to a Fréchet $\mathcal{O}(U)$-module structure, then $\sigma(X, M) \subset \overline{U}$.*

Proof If V is a Stein open set in X with $U \cap V = \emptyset$, then via the restriction map $\mathscr{O}(U \cup V) \to \mathscr{O}(U)$, the space M inherits a Fréchet $\mathscr{O}(U \cup V)$-module structure extending its $\mathscr{O}(X)$-module structure. By Lemma 5.1.1 and its proof, it follows that

$$0 = \operatorname{T\hat{o}r}_q^{\mathscr{O}(X)}(\mathscr{O}(U \cup V), M) \cong \operatorname{T\hat{o}r}_q^{\mathscr{O}(X)}(\mathscr{O}(U), M) \oplus \operatorname{T\hat{o}r}_q^{\mathscr{O}(X)}(\mathscr{O}(V), M)$$

for all $q \geq 1$, and that the horizontal maps in the commutative diagram

$$
\begin{array}{ccc}
\mathscr{O}(U) \mathbin{\hat{\otimes}}_{\mathscr{O}(X)} M \oplus \mathscr{O}(V) \mathbin{\hat{\otimes}}_{\mathscr{O}(X)} M = \mathscr{O}(U \cup V) \mathbin{\hat{\otimes}}_{\mathscr{O}(X)} M & \longrightarrow & M \\
\uparrow{\scriptstyle i} & & \uparrow{\scriptstyle id} \\
\mathscr{O}(U) \mathbin{\hat{\otimes}}_{\mathscr{O}(X)} M & \longrightarrow & M
\end{array}
$$

are topological isomorphisms. Hence $\mathscr{O}(V) \mathbin{\hat{\otimes}}_{\mathscr{O}(X)} M = 0$. $\qquad\square$

The following result gives a coordinate-free solution of the analytic functional calculus problem. The case $X = \mathbb{C}^n$ is due to Taylor (1972b).

Theorem 5.1.5 *If M is a Fréchet $\mathscr{O}(X)$-module over a Stein space of finite dimension, then for each open neighbourhood U of $\sigma(X, M)$, there is a Fréchet $\mathscr{O}(U)$-module structure on M extending its $\mathscr{O}(X)$-module structure.*

Proof Let $(U_i)_{i \in \mathbb{N}}$ and $(W_i)_{i \in \mathbb{N}}$ be Stein open covers with finite-dimensional nerve of U and $X \setminus \sigma$, respectively. We set $\mathscr{U} = (U_i)_{i \in \mathbb{N}}$ and $\mathscr{V} = (V_i)_{i \in \mathbb{N}} = (U_0, W_0, U_1, W_1, \ldots)$. Let $r: \mathscr{C}^{\bullet}(\mathscr{V}) \to \mathscr{C}^{\bullet}(\mathscr{U})$ be the morphism between the associated alternating analytic Čech complexes defined by

$$(rf)_{s_0 \cdots s_p} = f_{(2s_0) \cdots (2s_p)} \qquad (0 \leq s_0 < \cdots < s_p).$$

Obviously r is an epimorphism and its kernel is the complex $K^{\bullet}$ with components

$$K^p = \prod \mathscr{O}\left(V_{s_0} \cap \cdots \cap V_{s_p}\right),$$

where the topological product is formed over all index tuples $0 \leq s_0 < \cdots < s_p$ such that at least one of its components is odd.

We define $B^{\bullet} = (B^n)_{n \in \mathbb{Z}} = (B_{-n}^{\mathscr{O}(X)}(\mathscr{O}(X), M))_{n \in \mathbb{Z}}$, $C = K^{\bullet} \mathbin{\hat{\otimes}}_{\mathscr{O}(X)} B^{\bullet}$, $D = \mathscr{C}^{\bullet}(\mathscr{V}) \mathbin{\hat{\otimes}}_{\mathscr{O}(X)} B^{\bullet}$, $E = \mathscr{C}^{\bullet}(\mathscr{U}) \mathbin{\hat{\otimes}}_{\mathscr{O}(X)} B^{\bullet}$, and we consider the induced short exact sequence

$$0 \longrightarrow C \xrightarrow{\;i \otimes 1\;} D \xrightarrow{\;r \otimes 1\;} E \longrightarrow 0$$

of bounded double complexes. All columns of C are, up to sign, direct sums of complexes of the form $B_{\bullet}^{\mathscr{O}(X)}(\mathscr{O}(W), M)$, where W is a Stein open subset of $X \setminus \sigma(X, M)$. Hence all columns in C are exact, and the induced long exact sequence of cohomology yields topological isomorphisms

$$H^q(\operatorname{Tot} D) \to H^q(\operatorname{Tot} E) \qquad (q \in \mathbb{Z}).$$

Since all rows in D are exact except in degree 0, we obtain topological isomorphisms

$$H_q(B_{\bullet}^{\mathscr{O}(X)}(\mathscr{O}(X), M)) \cong H^{-q}(\operatorname{Tot} D) \qquad (q \in \mathbb{Z}).$$

For $q = 0$, the composition of the topological isomorphisms

$$M \xrightarrow{\sim} \mathcal{O}(X) \hat{\otimes}_{\mathcal{O}(X)} M \xrightarrow{\sim} H^0(\mathrm{Tot}\, D) \xrightarrow{\sim} H^0(\mathrm{Tot}\, E)$$

acts as the map which assigns to each element $m \in M$ the element

$$\left[(1|_{U_i})_{i \geq 0} \otimes m \oplus 0 \oplus 0 \oplus \cdots \oplus 0 \right]$$

in $H^0(\mathrm{Tot}\, E)$. Using the canonical Fréchet $\mathcal{O}(U)$-module structure of $H^0(\mathrm{Tot}\, E)$, one can turn M into a Fréchet $\mathcal{O}(U)$-module. In view of the above explicit form of the identification $M \cong H^0(\mathrm{Tot}\, E)$, it is clear that this $\mathcal{O}(U)$-module structure extends the $\mathcal{O}(X)$-module structure of M.   □

Our next aim is to show that the Fréchet $\mathcal{O}(X)$-module M in fact becomes an $\mathcal{O}(\sigma(X, M))$-module, and that the $\mathcal{O}(U)$-module structure of M defined in the last theorem is independent of the particular choice of the open cover $\mathcal{U}$.

Lemma 5.1.6 *Let X be a Stein space of finite dimension, and let M be a Fréchet $\mathcal{O}(X)$-module. If U and V are open neighbourhoods of $\sigma(X, M)$ with $V \subset U$ and if $\mathcal{U} = (U_i)_{i \in \mathbb{N}}$ and $\mathcal{V} = (V_j)_{j \in \mathbb{N}}$ are Stein open covers of U and V with finite-dimensional nerve such that $\mathcal{V}$ if finer than $\mathcal{U}$, then modulo the restriction map*

$$\mathcal{O}(U) \to \mathcal{O}(V), \qquad f \mapsto f \,|\, V$$

the $\mathcal{O}(V)$-module structure of M (defined with respect to $\mathcal{V}$) extends the $\mathcal{O}(U)$-module structure of M (defined with respect to $\mathcal{U}$).

Proof By assumption there is a map $\mathbb{N} \mapsto \mathbb{N}$, $j \mapsto i_j$, such that $V_j \subset U_{i_j}$ for all j. The associated canonical restriction map

$$r \colon \mathscr{C}^{\bullet}(\mathcal{U}) \to \mathscr{C}^{\bullet}(\mathcal{V})$$

is a morphism of complexes of Fréchet spaces that is compatible with the $\mathcal{O}(U)$-module structures of the spaces $\mathscr{C}^p(\mathcal{U})$ and the $\mathcal{O}(V)$-module structures of the spaces $\mathscr{C}^p(\mathcal{V})$. Then the morphism of double complexes

$$\mathscr{C}^{\bullet}(\mathcal{U}) \hat{\otimes}_{\mathcal{O}(X)} B^{\bullet} \xrightarrow{\ r \otimes 1\ } \mathscr{C}^{\bullet}(\mathcal{V}) \hat{\otimes}_{\mathcal{O}(X)} B^{\bullet},$$

where $B^{\bullet}$ is defined as in the preceding proof, induces a morphism between the corresponding total complexes such that the diagram

$$
\begin{array}{ccc}
M & \longrightarrow & H^0\!\left(\mathrm{Tot}\!\left(\mathscr{C}^{\bullet}(\mathcal{U}) \hat{\otimes}_{\mathcal{O}(X)} B^{\bullet}\right)\right) \\[2mm]
{\scriptstyle id}\big\downarrow & & \big\downarrow{\scriptstyle r \otimes 1} \\[2mm]
M & \longrightarrow & H^0\!\left(\mathrm{Tot}\!\left(\mathscr{C}^{\bullet}(\mathcal{V}) \hat{\otimes}_{\mathcal{O}(X)} B^{\bullet}\right)\right)
\end{array}
$$

commutes. Since the right vertical map respects the $\mathcal{O}(U)$- and $\mathcal{O}(V)$-module structures, respectively, the assertion follows.   □

If M is a Fréchet $\mathcal{O}(X)$-module over a Stein space X of finite dimension, then we shall denote, for each open neighbourhood U of $\sigma(X, M)$, the space M equipped with the Fréchet $\mathcal{O}(U)$-module structure defined above by M^U.

5.2 SPECTRAL MAPPING THEOREM

If $a = (a_1, \ldots, a_n) \in L(E)^n$ is a commuting tuple of regular operators on a Fréchet space E, then the constructions from Sections 5.1 yield a continuous analytic functional calculus map

$$\Phi \colon \mathcal{O}(\sigma(a)) \to L(E),$$

which extends the canonical Fréchet $\mathcal{O}(\mathbb{C}^n)$-module structure of E determined by a.

In Section 2.5 we used a Cauchy–Weil type integral formula to construct a solution of the same problem. Up to now it is not at all clear whether these two different methods yield the same analytic functional calculus.

One of the results of the present section is a uniqueness result which provides a positive answer to this question. Our first step is to prove a spectral mapping theorem for the module structures defined in Section 5.1.

Theorem 5.2.1 (Spectral mapping theorem) *Let X and Y be Stein spaces of finite dimension, and let M be a Fréchet $\mathcal{O}(X)$-module. If $X' \supset \sigma(X, M)$ is open and $f \colon X' \to Y$ is holomorphic, then*

$$\overline{f(\sigma(X, M))} = \sigma\left(Y, (M^{X'})^f\right),$$

where $(M^{X'})^f$ denotes $M^{X'}$ regarded as a Fréchet $\mathcal{O}(Y)$-module via the unital morphism of Fréchet algebras

$$\mathcal{O}(Y) \to \mathcal{O}(X'), \qquad g \mapsto g \circ f.$$

Proof The proof will be divided into two steps.

(1) We first suppose that $X' = X$. Let $V \subset Y$ be a Stein open set. If $f^{-1}(V) = \varnothing$, then it follows from Corollary 4.2.5 that the complex $B_\bullet^{\mathcal{O}(Y)}(\mathcal{O}(V), \mathcal{O}(X))$ is exact. If $f^{-1}(V) \neq \varnothing$, then again by Corollary 4.2.5, the morphism

$$B_\bullet^{\mathcal{O}(Y)}(\mathcal{O}(V), \mathcal{O}(X)) \to B_\bullet^{\mathcal{O}(X)}(\mathcal{O}(f^{-1}(V)), \mathcal{O}(X))$$

induced by f is a quasi-isomorphism between complexes of Fréchet $\mathcal{O}(X)$-modules. Applying the functor $_* \hat{\otimes}_{\mathcal{O}(X)} M$, we obtain the quasi-isomorphism (see Lemma 3.1.17)

$$B_\bullet^{\mathcal{O}(Y)}(\mathcal{O}(V), M^f) \to B_\bullet^{\mathcal{O}(X)}(\mathcal{O}(f^{-1}(V)), M)).$$

Using the fact that $f^{-1}(V)$ is a Stein open subset of X, the reader will easily deduce the assertion in this case.

(2) The general case is proved by factorizing f through its graph

$$\begin{array}{ccc} X' \times Y & \xrightarrow{\ i\ } & X \times Y \\ {\scriptstyle id \vee f}\Big\uparrow & & \Big\downarrow{\scriptstyle q} \\ X' & \xrightarrow{\ f\ } & Y \end{array}$$

We denote by p and q the holomorphic projections of $X \times Y$ onto X and Y, while i stands for the inclusion map and $id \vee f$ is the closed holomorphic embedding with components id and f. We write f in the form $f = q \circ f'$ with $f' = i \circ (id \vee f)$, and we set $Y' = X \times Y$.

Since the first part applies to the holomorphic map q and to the Fréchet $\mathscr{O}(Y')$-module $(M^{X'})^{f'}$, it suffices to prove that

$$f'\sigma(X,M) = \sigma\big(Y',(M^{X'})^{f'}\big).$$

The reader should observe that the left-hand side is automatically a closed subset of Y'.

Let $x \in X'$ be arbitrary. We choose a Stein open neighbourhood V of $f'(x)$ in Y' such that $(f')^{-1}V = pV \subset X$ is Stein. If $U \subset X'$ is a Stein open set, then the projection p induces a morphism

$$p^U : B_\bullet^{\mathscr{O}(X)}(\mathscr{O}(pV),\mathscr{O}(U)) \to B_\bullet^{\mathscr{O}(Y')}(\mathscr{O}(V),\mathscr{O}(U))$$

between complexes of Fréchet spaces. Here $\mathscr{O}(U)$ is turned into an $\mathscr{O}(Y')$-module via the map $f' : U \to Y'$. By applying Corollary 4.2.5 to

$$pV \times Y \xrightarrow{\ p\ } X,$$

to the Stein open set $U \subset X$, and to the coherent $\mathscr{O}_{pV \times Y}$-sheaf defined as the image sheaf of $\mathscr{O}_{pV}$ under the holomorphic embedding

$$id \vee (f|pV) : pV \to pV \times Y,$$

respectively, and by applying Corollary 4.2.5 to the map

$$U \xrightarrow{\ f'\ } Y',$$

to the Stein open set $V \subset Y'$, and to the sheaf $\mathscr{O}_U$, one obtains that p^U is a quasi-isomorphism.

We fix a Stein open cover $\mathscr{U} = (U_i)_{i \in \mathbb{N}}$ of X' with finite-dimensional nerve, and we denote by $K^\bullet$ the total complex of the double complex $\mathscr{C}^\bullet(\mathscr{U}) \, \hat{\otimes}_{\mathscr{O}(X)} \, B^\bullet$, where

$$B^\bullet = (B_{-n}^{\mathscr{O}(X)}(\mathscr{O}(X),M))_{n \le 0}.$$

To simplify the notation, we use the symbol p also for the morphism between the complexes

$$L^\bullet = (B_{-n}^{\mathscr{O}(X)}(\mathscr{O}(pV),\mathscr{O}(X')))_{n \le 0},$$
$$M^\bullet = (B_{-n}^{\mathscr{O}(Y')}(\mathscr{O}(V),\mathscr{O}(X')))_{n \le 0}$$

determined by the map $p: Y' \to X$. Then for each q, the morphism $p \hat{\otimes}_{\mathscr{O}(X')} 1_{K^q}$ is a direct sum of morphisms of the form

$$p^U \hat{\otimes} 1_{\mathscr{O}(X)\hat{\otimes} - \hat{\otimes}\mathscr{O}(X)\hat{\otimes}M} \qquad (U \subset X' \text{ Stein open subset}).$$

By Lemma 3.1.17,

$$p \hat{\otimes}_{\mathscr{O}(X')} 1: \operatorname{Tot}\left(L^{\bullet} \hat{\otimes}_{\mathscr{O}(X')} K^{\bullet}\right) \to \operatorname{Tot}\left(M^{\bullet} \hat{\otimes}_{\mathscr{O}(X')} K^{\bullet}\right) \qquad (5.2.1)$$

remains a quasi-isomorphism. On the other hand (see the proof of Theorem 5.1.5), we know that $K^{\bullet}$ is exact except in degree $q = 0$, where $H^0(K^{\bullet}) \cong M$ as Fréchet $\mathscr{O}(X')$-modules. Hence all columns in $L^{\bullet} \hat{\otimes}_{\mathscr{O}(X')} K^{\bullet}$ and $M^{\bullet} \hat{\otimes}_{\mathscr{O}(X')} K^{\bullet}$ are exact except in degree 0, where

$$H^0\left(L^p \hat{\otimes}_{\mathscr{O}(X')} K^{\bullet}\right) \cong B^{\mathscr{O}(X)}_{-p}(\mathscr{O}(pV), M),$$
$$H^0\left(M^p \hat{\otimes}_{\mathscr{O}(X')} K^{\bullet}\right) \cong B^{\mathscr{O}(Y')}_{-p}(\mathscr{O}(V), M). \qquad (5.2.2)$$

Hence we obtain induced topological isomorphisms

$$H_p(B^{\mathscr{O}(X)}_{\bullet}(\mathscr{O}(pV), M)) \to H_p(B^{\mathscr{O}(Y')}_{\bullet}(\mathscr{O}(V), M)) \qquad (5.2.3)$$

for all $p \geq 0$. Here the $\mathscr{O}(X')$-module M is turned into an $\mathscr{O}(Y')$-module via the map $f': X' \to Y'$.

Using the isomorphisms (5.2.3), one immediately obtains that

$$f'(\sigma(X, M)) = \sigma\left(Y', (M^{X'})^{f'}\right) \cap f'(X'). \qquad (5.2.4)$$

Indeed, for a given point $x \in X' \setminus \sigma(X, M)$, one can choose the Stein open neighbourhood V of $f'(x)$ in such a way that $pV \cap \sigma(X, M) = \varnothing$. But then $V \cap \sigma(Y', (M^X)^{f'}) = \varnothing$. Conversely, if $x \in X'$ is a point such that $f'(x)$ is not contained in the right-hand side of (5.2.4), then one can choose the Stein open neighbourhood V of $f'(x)$ in such a way that, in addition, $V \cap \sigma(Y', (M^X)^{f'}) = \varnothing$. Using the isomorphisms (5.2.3), one obtains in this case $pV \cap \sigma(X, M) = \varnothing$.

To conclude the proof, it suffices to show that

$$\sigma\left(Y', (M^{X'})^{f'}\right) \subset f'(X').$$

Since the natural extensions of the $\mathscr{O}(X)$-module structure of M are compatible with restrictions (see Lemma 5.1.6), it suffices to check that the left-hand side is contained in the closure of the right-hand side. But, if V is a Stein open set in Y' not intersecting $f'(X')$, then for each Stein open set $U \subset X'$, Corollary 4.2.5 applied to $f': U \to Y'$ yields that $B^{\mathscr{O}(Y')}_{\bullet}(\mathscr{O}(V), \mathscr{O}(U))$ is exact. Consequently, if one repeats the above constructions for this Stein open set, then the resulting complex $M^{\bullet} \hat{\otimes}_{\mathscr{O}(X')} K^{\bullet}$ has exact rows. In the same way as we demonstrated above, one obtains the isomorphisms (5.2.2). Hence $B^{\mathscr{O}(Y')}_{\bullet}(\mathscr{O}(V), M)$ is exact, and consequently $V \cap \sigma(Y', (M^{X'})^{f'}) = \varnothing$. $\square$

Applied to commuting systems of regular Fréchet space operators, the last result shows that the analytic functional calculus defined with the help of

Section 5.1 satisfies exactly the same spectral mapping theorem as the analytic functional calculus defined in Section 2.5. Indeed, if $a \in L(E)^n$ is a commuting system of regular Fréchet space operators and if

$$\Phi: \mathcal{O}(\sigma(a)) \to L(E)$$

denotes the analytic functional calculus map defined by extending the canonical Fréchet $\mathcal{O}(\mathbb{C}^n)$-module structure of E as explained in Section 5.1, then all resulting operators $\Phi(f)$ $(f \in \mathcal{O}(\sigma(a)))$ are regular, and hence the above spectral mapping theorem yields, for each m-tuple $f \in \mathcal{O}(U)^m$ of analytic functions defined on a neighbourhood U of $\sigma(a)$, the identity

$$\sigma(f(a)) = \sigma\left(\mathbb{C}^m, (E^U)^f\right) = f(\sigma(\mathbb{C}^n, E)) = f(\sigma(a)).$$

Our next aim is to show that, roughly speaking, the extensions constructed in Section 5.1 are compatible with morphisms of Fréchet $\mathcal{O}(X)$-modules. More precisely, we have the following result.

Lemma 5.2.2 *Let X be a Stein space of finite dimension, and let u: $M \to N$ be a morphism of Fréchet $\mathcal{O}(X)$-modules. If U is an open neighbourhood of $\sigma(X, M) \cup \sigma(X, N)$, then u remains a morphism between M and N with respect to the natural Fréchet $\mathcal{O}(U)$-module structures on both spaces.*

Proof Let $\mathcal{U}$ be a Stein open cover of U with finite-dimensional nerve. The map u induces a morphism

$$B_\bullet^{\mathcal{O}(X)}(\mathcal{O}(X), M) \xrightarrow{u} B_\bullet^{\mathcal{O}(X)}(\mathcal{O}(X), N)$$

of complexes of Fréchet $\mathcal{O}(X)$-modules. Let $B_M^\bullet$ and $B_N^\bullet$ be the same complexes written in cohomological form. Then the induced map

$$M \cong H^0\left(\mathrm{Tot}\left(\mathscr{C}^\bullet(\mathcal{U}) \,\hat{\otimes}_{\mathcal{O}(X)} B_M^\bullet\right)\right)$$

$$\downarrow 1 \otimes u$$

$$N \cong H^0\left(\mathrm{Tot}\left(\mathscr{C}^\bullet(\mathcal{U}) \,\hat{\otimes}_{\mathcal{O}(X)} B_N^\bullet\right)\right)$$

respects the Fréchet $\mathcal{O}(U)$-module structures. $\square$

Let $a \in L(E)^n$ be a commuting tuple of regular Fréchet space operators, and let

$$\Phi: \mathcal{O}(\sigma(a)) \to L(E), \qquad f \mapsto f(a)$$

be the analytic functional calculus of a determined by extending the canonical $\mathcal{O}(\mathbb{C}^n)$-module structure of E. If $f \in \mathcal{O}(\sigma(a))^m$ and $g \in \mathcal{O}(f(\sigma(a)))^k$ are given, then it is natural to ask whether

$$g(f(a)) = (g \circ f)(a) \tag{5.2.5}$$

holds, where the left-hand side is defined by extending the global Fréchet $\mathcal{O}(\mathbb{C}^m)$-module structure of E determined by $f(a)$ onto $\mathcal{O}(\sigma(f(a)))$. The next result give a positive answer to this question in a more general setting.

Theorem 5.2.3   *Let X be an n-dimensional Stein manifold. Let Y be a Stein space of finite analytic dimension, let $U \subset X$ and $V \subset Y$ be open, and let us consider a holomorphic map $f: U \to Y$ with $f(U) \subset V$. If M is a Fréchet $\mathcal{O}(X)$-module with $\sigma(X, M) \subset U$ and $\overline{f\sigma}(X, M) \subset V$, then $(M^U)^f$ is a Fréchet $\mathcal{O}(Y)$-module with $\sigma(Y, (M^U)^f) \subset V$ and*

$$((M^U)^f)^V = (M^U)^{(f:\, U \to V)}.$$

Proof   We first consider the case where f can be extended to a holomorphic map $F: X \to Y$. To prove the result in this case, we choose Stein open covers $\mathcal{U} = (U_i)_{i \in \mathbb{N}}$ of U and $\mathcal{V} = (V_i)_{i \in \mathbb{N}}$ of V with finite-dimensional nerve. Then $\mathcal{W} = f^{-1}(\mathcal{V}) \cap \mathcal{U}$ is a Stein open cover of U which refines $f^{-1}(\mathcal{V})$. We denote by $B_X^\bullet$ and $B_Y^\bullet$ the complexes

$$B_\bullet^{\mathcal{O}(X)}(\mathcal{O}(X), M) \quad \text{and} \quad B_\bullet^{\mathcal{O}(Y)}(\mathcal{O}(Y), (M^U)^f)$$

written as usual in their cohomological form, and we write

$$r: \mathscr{C}^\bullet(f^{-1}(\mathcal{V})) \to \mathscr{C}^\bullet(\mathcal{W})$$

for the restriction morphism. Then F and $r \otimes 1$ yield morphisms between the double complexes

$$K = \mathscr{C}^\bullet(\mathcal{V}) \hat{\otimes}_{\mathcal{O}(Y)} B_Y^\bullet,$$

$$L = \mathscr{C}^\bullet(f^{-1}(\mathcal{V})) \hat{\otimes}_{\mathcal{O}(X)} B_X^\bullet,$$

$$M = \mathscr{C}^\bullet(\mathcal{W}) \hat{\otimes}_{\mathcal{O}(X)} B_X^\bullet$$

such that the diagram

$$
\begin{array}{ccc}
M & \xrightarrow{\;\sim\;} & H^0(\mathrm{Tot}\, K) \\
\Big\downarrow{\scriptstyle id} & & \Big\downarrow{\scriptstyle F} \\
 & & H^0(\mathrm{Tot}\, L) \\
\Big\downarrow & & \Big\downarrow{\scriptstyle r \otimes 1} \\
M & \xrightarrow{\;\sim\;} & H^0(\mathrm{Tot}\, M)
\end{array}
$$

commutes. Here the horizontal maps are the isomorphisms defining the $\mathcal{O}(V)$-module structure of $(M^U)^f$ and the $\mathcal{O}(U)$-module structure of M, respectively (see the proof of Theorem 5.1.5).

If $m \in M$ and $h \in \mathcal{O}(V)$, then the element corresponding to hm in $H^0(\mathrm{Tot}\, K)$ is

$$\xi = [(h \,|\, V_i)_i \otimes m \oplus 0 \oplus \cdots \oplus 0].$$

But then

$$(r \otimes 1)F(\xi) = \left[((h \circ f) \,|\, f^{-1}(V_i) \cap U_j)_{i,j} \otimes m \oplus 0 \oplus \cdots \oplus 0 \right]$$

is precisely the element corresponding to $(h \circ f)m$ in $H^0(\mathrm{Tot}\, M)$.

To prove the assertion in the general case, let us denote by $\tilde{U}$ the holomorphic envelope of U (see Theorem 4.6 in Rossi 1963).

Thus $\tilde{U}$ is a Stein manifold of the same dimension as X such that U can be regarded as an open subset of $\tilde{U}$. Modulo this identification each holomorphic map $g: U \to Z$ into an arbitrary Stein space has a holomorphic extension $\tilde{g}: \tilde{U} \to Z$. In particular, there is a holomorphic extension $p: \tilde{U} \to X$ of the inclusion map $i: U \to X$. Finally, the restriction induces a topological isomorphism $\mathcal{O}(\tilde{U}) \xrightarrow{\sim} \mathcal{O}(U)$, $g \mapsto g \mid U$, of Fréchet algebras.

Via the last topological isomorphism we regard M as a Fréchet $\mathcal{O}(\tilde{U})$-module, and we claim that with respect to this module structure

$$\sigma(\tilde{U}, M) = \sigma(X, M). \tag{5.2.6}$$

To verify this, we choose a closed holomorphic embedding $g: \tilde{U} \to \mathbb{C}^N$, where N is a suitable integer (see Theorem 5.3.9 in Hörmander 1966). The $\mathcal{O}(\mathbb{C}^N)$-module structures on M, obtained from the $\mathcal{O}(U)$-module M via the map $g: U \to \mathbb{C}^N$, respectively, from the $\mathcal{O}(\tilde{U})$-module M via $g: \tilde{U} \to \mathbb{C}^N$, coincide. Moreover, since $\sigma(X, M)$ is closed in X, one easily deduces that it is also closed in $\tilde{U}$. Hence by the spectral mapping theorem (Theorem 5.2.1) it follows that

$$g\sigma(\tilde{U}, M) = \sigma(\mathbb{C}^N, M^g) = g\sigma(X, M).$$

Because of the identity (5.2.6), to prove the assertion in the general case, it suffices to replace X by $\tilde{U}$ and then to apply the first part of the proof. $\square$

In the setting of the last theorem, if $g: V \to Z$ is another holomorphic map into an arbitrary analytic space Z, then

$$\left(\left((M^U)^f\right)^V\right)^g = (M^U)^{g \circ f}$$

coincide as Fréchet $\mathcal{O}(Z)$-modules. Applying this observation to the case of regular commuting Fréchet space operators, one sees that formula (5.2.5) is valid under the conditions described in the paragraph preceding Theorem 5.2.3.

By similar methods one can prove the uniqueness result mentioned before.

Theorem 5.2.4 *Let M be a Fréchet $\mathcal{O}(X)$-module over an n-dimensional Stein manifold, and let $U \supset \sigma(X, M)$ be a fixed open set. We suppose that the $\mathcal{O}(X)$-module structure of M can be extended to a Fréchet $\mathcal{O}(U)$-module structure, and we denote by $\mathcal{M}$ the space M equipped with this $\mathcal{O}(U)$-module structure.*

Suppose that for each holomorphic map $f: U \to \mathbb{C}^k$ $(k \geq 1)$, the identity

$$\overline{f\sigma(X, M)} = \sigma(\mathbb{C}^k, \mathcal{M}^f) \tag{5.2.7}$$

holds. Then $\mathcal{M} = M^U$.

Proof We use the same notations as in the last proof. In particular, we regard $\mathcal{M}$ as a Fréchet $\mathcal{O}(\tilde{U})$-module, where $\tilde{U}$ is the holomorphic envelope of U. Applying the hypothesis (5.2.7) and the spectral mapping theorem to a suitable closed holomorphic embedding $g: \tilde{U} \to \mathbb{C}^N$ we obtain (precisely as in the preceding proof) that

$$\sigma(\tilde{U}, \mathcal{M}) = \sigma(X, M) \subset U.$$

Using Theorem 5.1.5, we can extend the Fréchet $\mathcal{O}(\tilde{U})$-module structure of $\mathcal{M}$ to a Fréchet $\mathcal{O}(U)$-module structure. But, since $\mathcal{O}(U) = \mathcal{O}(\tilde{U}) \mid U$, it follows that $\mathcal{M}^U$ is simply the original Fréchet $\mathcal{O}(U)$-module. Since this module structure extends the $\mathcal{O}(X)$-module structure, we obtain that

$$(\mathcal{M}^U)^{(p|U)} = M$$

as Fréchet $\mathcal{O}(X)$-modules. Applying Theorem 5.2.3 with $\tilde{U}$ instead of X, $p \mid U: U \to X$ instead of f, and $V = U$, we see that

$$\left((\mathcal{M}^U)^{(p|U)}\right)^U = \mathcal{M}^U$$

as Fréchet $\mathcal{O}(U)$-modules. Since the space on the left is M^U and the space on the right is $\mathcal{M}$, the proof is complete. $\square$

5.3   REFERENCES AND COMMENTS

The idea of studying linear operators via modules over function algebras is not new. For instance, the Jordan form of a matrix can be obtained in this way. A more recent development is a homological approach to the celebrated Sz.-Nagy–Foias dilation theory for contractions which is based on the study of certain Hilbert modules over the disc algebra (see Douglas and Paulsen 1989 for details).

For the study of commuting systems of operators, the language of modules is quite essential. For instance, operations with modules are more flexible and invariant than the corresponding operations with systems of operators. To give a simple example, let us mention that the idea to use Fréchet modules over Stein algebras and their spectra replaces the classical Arens and Calderon lemma, which is the basic tool in constructing the multivariable holomorphic functional calculus in commutative Banach algebras (see for instance Hörmander 1966, Chapter IV)

The presentation of Section 5.1 essentially follows Taylor (1972b). The actual form, in the context of Stein spaces, is only a slight modification of Taylor's original ideas, and it can be found in Putinar (1980, 1983b). The spectral mapping theorem and the uniqueness of the multivariable analytic functional calculus were proved in Putinar (1980, 1983b).

Beginning with Słodkowski (1977), several authors have studied abstract joint spectra which satisfy the spectral mapping theorem. Whether Taylor's joint spectrum is the smallest spectrum with this property is still an open question (for additional references see Harte 1988). The compatibility of the functional calculus with the Gelfand transform of an underlying Banach algebra is discussed in Eschmeier (1987a) and Putinar (1984d).

For various classical applications of the analytic functional calculus we recommend the monograph of Vasilescu (1982).

6
Bishop's condition (β) and invariant subspaces

The main idea of the homological approach to the analytic functional calculus problem is to identify a given commuting system $a \in L(X)^n$ of continuous linear Banach space operators with the $\mathcal{O}(\mathbb{C}^n)$-module structure

$$\mathcal{O}(\mathbb{C}^n) \times X \to X, \qquad (f, x) \mapsto f(a)x$$

determined by a, and then to extend this module structure to the algebra $\mathcal{O}(\sigma(a))$ of all germs of analytic functions defined near the spectrum $\sigma(a)$ of a, applying methods from topological homology theory.

The same idea can be used in a natural way to develop a multivariable local spectral theory for finite commuting systems of Banach space operators. For instance, in a generalization of the corresponding one-dimensional notion due to Bishop (1959), we say that a commuting tuple $a \in L(X)^n$ of Banach space operators satisfies *property* (β) if X is a quasi-coherent $\mathcal{O}(\mathbb{C}^n)$-module with respect to the $\mathcal{O}(\mathbb{C}^n)$-module structure determined by a. In this case, the associated quasi-coherent sheaf $\tilde{X} = \mathcal{O} \hat{\otimes}_{\mathcal{O}(\mathbb{C}^n)} X$ (see Section 4.3) is called the *canonical sheaf model* of a. From this point of view a single operator $T \in L(X)$ is decomposable in the sense of Foiaş (1963) if and only if it satisfies Bishop's property (β) and its canonical sheaf model is soft. Correspondingly, we shall call a commuting system $a \in L(X)^n$ of continuous linear Banach space operators *decomposable* if it satisfies the latter two conditions.

We shall show that this notion of decomposability is completely compatible with duality and that there is a natural, more operator-theoretic, characterization of decomposable systems. In the one-dimensional case Bishop's property (β) characterizes, up to similarity, those operators occurring as restrictions of decomposable operators. In the multidimensional case it turns out that Bishop's property (β) characterizes those commuting systems that possess a finite resolution by decomposable systems.

Bishop's condition (β) means by definition that, for each Stein open set U in $\mathbb{C}^n$, the transversality relation $\mathcal{O}(U) \perp_{\mathcal{O}(\mathbb{C}^n)} X$ holds. If the same condition holds with $\mathcal{O}(U)$ replaced by the Fréchet space $\mathcal{E}(U)$ of all smooth functions on U, then the given system a is said to possess *property* (β)$_{\mathcal{E}}$. Roughly speaking, property (β)$_{\mathcal{E}}$ characterizes those commuting systems that possess a finite resolution by generalized scalar systems. Even in the single-variable case there is a close relation between property $(\beta)_{\mathcal{E}}$ and the solvability of certain division problems for vector-valued distributions. As an application of our abstract operator-theoretical division results we shall prove some concrete division results for various spaces of distributions in Chapter 7.

In the final part of this chapter we prove the existence of joint invariant subspaces for commuting systems of Banach space operators that satisfy Bishop's property (β) and possess a sufficiently rich spectrum. In the case of a single operator one obtains as special cases invariant-subspace results due to Scott Brown, saying that on a Hilbert space all subnormal operators and all hyponormal operators with thick spectrum possess non-trivial invariant subspaces. Our invariant-subspace constructions are based on the Scott Brown technique as well as on the existence of canonical decomposable extensions for commuting systems with Bishop's property (β) and some explicit properties of these canonical extensions.

6.1 SPECTRAL DECOMPOSITIONS AND DUALITY

Let $a = (a_1, \ldots, a_n) \in L(X)^n$ be a commuting tuple of continuous linear operators on a complex Banach space X. Then X becomes a Banach $\mathcal{O}(\mathbb{C}^n)$-module with respect to the module structure

$$\mathcal{O}(\mathbb{C}^n) \times X \to X, \qquad (f, x) \mapsto f(a)x.$$

We shall say that the tuple a satisfies the *single-valued extension property* if, for each Stein open set U in $\mathbb{C}^n$,

$$\mathrm{T\hat{o}r}_p^{\mathcal{O}(\mathbb{C}^n)}(\mathcal{O}(U), X) = 0 \qquad \text{for} \quad p > 0.$$

Since the Koszul resolution

$$K_{\bullet}(z - w, \mathcal{O}(\mathbb{C}^n) \,\hat{\otimes}\, \mathcal{O}(\mathbb{C}^n)) \to \mathcal{O}(\mathbb{C}^n) \to 0$$

is a resolution of $\mathcal{O}(\mathbb{C}^n)$ by topologically free $\mathcal{O}(\mathbb{C}^n)$-modules, one obtains that the single-valued extension property is equivalent to the condition that, for each Stein open set U in $\mathbb{C}^n$, the Koszul complex $K_{\bullet}(z - a, \mathcal{O}(U, X))$ is exact in positive degrees.

If in addition, for each Stein open set U in $\mathbb{C}^n$,

$$\mathrm{T\hat{o}r}_0^{\mathcal{O}(\mathbb{C}^n)}(\mathcal{O}(U), X) = \mathcal{O}(U) \,\hat{\otimes}_{\mathcal{O}(\mathbb{C}^n)}\, X$$

is separated in its natural quotient topology, then a is said to satisfy *Bishop's property* (β). The reader should recall from Chapter 4 that the tuple a satisfies Bishop's property (β) if and only if the Banach $\mathcal{O}(\mathbb{C}^n)$-module X is quasi-coherent, or equivalently, if for each Stein open set U in $\mathbb{C}^n$, the transversality relation $\mathcal{O}(U) \perp_{\mathcal{O}(\mathbb{C}^n)} X$ holds.

In the case of a single operator $T \in L(X)$ on a complex Banach space X the single-valued extension property means that, for each open set U in $\mathbb{C}$, the multiplication operator

$$\mathcal{O}(U, X) \to \mathcal{O}(U, X), \qquad f \mapsto (z - T)f$$

is injective, while Bishop's property (β) means that this operator is injective with closed range. Thus, we are back to the original definitions (see Dunford and Schwartz 1971; Bishop 1959).

Suppose that $a \in L(X)^n$ satisfies the single-valued extension property. Then, following Putinar (1983a), we shall call the analytic sheaf $\mathcal{F}$ associated with the presheaf

$$U \mapsto F(U) = \mathcal{O}(U) \,\hat{\otimes}_{\mathcal{O}(\mathbb{C}^n)} X = \mathcal{O}(U, X) \Big/ \sum_{i=1}^{n} (z_i - a_i)\mathcal{O}(U, X),$$

where U runs through all open sets in $\mathbb{C}^n$, the *canonical sheaf model* of a. Since the Koszul complex

$$K_{\bullet}(z - a, \mathcal{O}^X) \to \mathcal{F} \to 0$$

provides a resolution of $\mathcal{F}$ by sheaves that are acyclic on Stein open sets in $\mathbb{C}^n$, the canonical sheaf model $\mathcal{F}$ is acyclic on Stein open sets. Hence on each Stein open set U in $\mathbb{C}^n$ the induced sequence of section spaces $K_{\bullet}(z - a, \mathcal{O}(U, X)) \to \mathcal{F}(U) \to 0$ is exact, and therefore the canonical map from $F(U)$ into the section space $\mathcal{F}(U)$ is an isomorphism.

Suppose that a satisfies Bishop's property (β). Then its canonical sheaf model $\mathcal{F}$ is, in a natural way, an analytic Fréchet sheaf on $\mathbb{C}^n$. Moreover, this sheaf is the unique quasi-coherent analytic Fréchet sheaf on $\mathbb{C}^n$ such that $\mathcal{F}(\mathbb{C}^n) \cong X$ as Fréchet $\mathcal{O}(\mathbb{C}^n)$-modules (see Chapter 4).

More generally, suppose that $a \in L(X)^n$ is a commuting tuple of continuous linear operators on a complex Banach space X. Then each analytic Fréchet sheaf $\mathcal{F}$ on $\mathbb{C}^n$ with $X \cong \mathcal{F}(\mathbb{C}^n)$ as topological $\mathcal{O}(\mathbb{C}^n)$-modules will be called a *sheaf model* for a.

Definition 6.1.1 A commuting tuple $a \in L(X)^n$ of Banach space operators will be said to possess a *Fréchet soft sheaf model* if a has a sheaf model which, in addition, is a soft sheaf.

Since each soft analytic Fréchet sheaf on $\mathbb{C}^n$ is quasi-coherent (Theorem 4.4.1), each commuting tuple a with a Fréchet soft sheaf model satisfies Bishop's property (β), and its sheaf model is isomorphic to the canonical sheaf model of a.

We recall some notions from local spectral theory. Suppose that $a \in L(X)^n$ is a commuting tuple of Banach space operators. Then, for each vector $x \in X$, we denote by $\sigma_a(x)$ the *local spectrum of a at x*, i.e. the smallest closed set in $\mathbb{C}^n$ on whose complement x can locally be written as

$$x = (z_1 - a_1)f_1(z) + \cdots + (z_n - a_n)f_n(z)$$

with suitable analytic X-valued functions $f_1, \ldots, f_n$. If $M \subset \mathbb{C}^n$ is arbitrary, then the a-invariant linear subspace of X defined by

$$X_a(M) = \{x \in X; \sigma_a(x) \subset M\}$$

is called the *spectral subspace* of a belonging to M.

Thus, if $\mathscr{F}$ is the canonical sheaf model of a commuting tuple a satisfying the single-valued extension property, then via the identification $X \cong \mathscr{F}(\mathbb{C}^n)$ the support of a vector $x \in X$, regarded as a global section in the sheaf $\mathscr{F}$, coincides with the local spectrum $\sigma_a(x)$ of a at x, and the spectral subspaces $X_a(M)$ correspond to the spaces $\mathscr{F}_M(\mathbb{C}^n)$ consisting of all global sections with support contained in M.

In particular, if $a \in L(X)^n$ is a commuting tuple with a Fréchet soft sheaf model $\mathscr{F}$, then the softness of $\mathscr{F}$ implies that, for each closed set F in $\mathbb{C}^n$ and each finite system $(U_i)_{i=1}^r$ of open sets in $\mathbb{C}^n$ with $F \subset U_1 \cup \cdots \cup U_r$, the decomposition

$$\mathscr{F}_F(\mathbb{C}^n) \subset \mathscr{F}_{\overline{U}_1}(\mathbb{C}^n) + \cdots + \mathscr{F}_{\overline{U}_r}(\mathbb{C}^n)$$

holds (Théorème 3.6.1 in Chapter II of Godement 1958), or equivalently, that

$$X_a(F) \subset X_a(\overline{U}_1) + \cdots + X_a(\overline{U}_r),$$

where the spectral subspaces occurring here are closed invariant subspaces for a.

The aim of this section is to show that the property of possessing a Fréchet soft sheaf model is completely compatible with duality. At the same time we shall obtain several alternative characterizations of tuples possessing a Fréchet soft sheaf model.

Definition 6.1.2   A commuting tuple $a \in L(X)^n$ of Banach space operators is said to be *decomposable* if

(i) a satisfies Bishop's property (β) and
(ii) $X_a(U_1 \cup U_2) = X_a(U_1) + X_a(U_2)$ for each pair of open sets U_1, U_2 in $\mathbb{C}^n$.

Obviously, condition (ii) implies the existence of corresponding decompositions for open covers of arbitrary finite length. In view of the above observations, it is clear that each commuting tuple with a Fréchet soft sheaf model is decomposable. If $a = (a_1, \ldots, a_n)$ is a commuting tuple of continuous linear operators on the complex Banach space X, then we denote by $a' = (a'_1, \ldots, a'_n)$ the commuting tuple consisting of the adjoint operators acting on the dual space X'.

Suppose that a satisfies the single-valued extension property, and that $\mathscr{F}$ denotes its canonical sheaf model. Then for each compact set K in $\mathbb{C}^n$, the identification $X_a(K) \cong \mathscr{F}_K(\mathbb{C}^n)$ turns the spectral subspace $X_a(K)$ into an $\mathscr{O}(K)$-module. This $\mathscr{O}(K)$-module structure, or equivalently, the algebra homomorphism

$$\Phi \colon \mathscr{O}(K) \to \operatorname{Hom}(X_a(K)), \qquad \Phi(f)x = fx$$

will be called the *local analytic functional calculus* of a. If a satisfies Bishop's property (β), then the spectral subspaces $X_a(K)$ are closed subspaces for all compact sets K, and the local analytic functional calculus is continuous in the following sense.

Lemma 6.1.3 *Let $a \in L(X)^n$ be a commuting tuple with Bishop's property (β). For each compact set K in $\mathbb{C}^n$, the local analytic functional calculus*

$$\Phi: \mathscr{O}(K) \to L(X_a(K))$$

is a well-defined continuous algebra homomorphism if $\mathscr{O}(K)$ is equipped with its canonical inductive limit topology.

Proof It suffices to check that, for each open neighbourhood U of K, the map

$$\mathscr{O}(U) \times X_a(K) \to X_a(K), \qquad (f, x) \mapsto fx$$

is separately continuous. For U and K as above, let us fix an open cover $(U_i)_{i=1}^r$ of K with Stein open sets U_i $(i = 1, \ldots, r)$. For each $x \in X_a(K)$ and each $f \in \mathscr{O}(U)$, the product fx is the unique vector $y = y_f$ in $X_a(K)$ satisfying

$$y \,|\, U_i - (f \,|\, U_i)x \in \sum_{j=1}^n (z_j - a_j)\mathscr{O}(U_i, X)$$

for each i. Since the space on the right is a closed subspace of $\mathscr{O}(U_i, X)$, the separate continuity follows by an application of the closed graph theorem. $\square$

Let us recall that a sheaf $\mathscr{F}$ on a topological space Ω is *c-soft* (Definition II.9.1 in Bredon 1967) if, for each compact subset K of Ω, the restriction map $\mathscr{F}(\Omega) \to \mathscr{F}(K)$, $s \mapsto s \,|\, K$, is onto. Furthermore, we shall say that $\mathscr{F}$ has *compact support* if there is a compact subset K of Ω such that the restriction of $\mathscr{F}$ to $\Omega \setminus K$ vanishes, that is, $\mathscr{F}_x = 0$ for all $x \in \Omega \setminus K$.

The following is our first duality result for tuples admitting spectral decompositions.

Theorem 6.1.4 *Suppose that a is decomposable. Then a' has a Fréchet soft sheaf model.*

Proof The sheaf model of a' is defined in a way similar to the construction of the sheaf of ordinary distributions. The proof is divided into several steps.

(1) For each open set U in $\mathbb{C}^n$, we regard $X_a(U)$ as the inductive limit

$$X_a(U) = \varinjlim_{K \Subset U} X_a(K),$$

where K ranges over all compact subsets of U, and each $X_a(K)$ carries its norm topology. Equipped with the inductive limit topology of the embeddings, $X_a(U)$ becomes a strict (LF)-space. Using the local analytic functional calculus, the space $X_a(U)$ becomes an $\mathscr{O}(U)$-module such that the multiplication

$$\mathscr{O}(U) \times X_a(U) \to X_a(U), \qquad (f, x) \mapsto fx$$

is at least separately continuous. The strong dual of $X_a(U)$ will be denoted by

$\mathscr{F}'(U)$. From the general theory of locally convex spaces (see §26.2.1 in Floret and Wloka 1968) it follows that

$$\mathscr{F}'(U) = \varprojlim_{K \in U} X_a(K)',$$

where each $X_a(K)'$ carries its norm topology. In particular, $\mathscr{F}'(U)$ is a Fréchet space. By dualizing the above module structure of $X_a(U)$, the space $\mathscr{F}'(U)$ becomes a Fréchet $\mathcal{O}(U)$-module. If $X_a(K)'$ is identified with $X'/X_a(K)^\perp$, then $\mathscr{F}'(U)$ consists of all families $u = (u_K/X_a(K)^\perp)_{K \in U}$ which satisfy $u_K - u_L \in X_a(K)^\perp$ for all compact sets K, L in U with $K \subset L$. The duality between $X_a(U)$ and $\mathscr{F}'(U)$ is given by $\langle x, u \rangle = \langle x, u_L \rangle$ if u is as above, $L \subset U$ is compact, and $x \in X_a(L)$. For $V \subset U \subset \mathbb{C}^n$ open, the restriction map

$$r_V^U : \mathscr{F}'(U) \to \mathscr{F}'(V), \qquad u \mapsto \left(u_K/X_a(K)^\perp\right)_{K \in V}$$

is the adjoint of the inclusion map $X_a(V) \to X_a(U)$. Instead of $r_V^U(u)$ we often write $u \mid V$.

(2) The family $\mathscr{F}'(U)$ ($U \subset \mathbb{C}^n$ open), together with the above restriction maps, form a presheaf of analytic Fréchet modules on $\mathbb{C}^n$. We claim that this presheaf is even a sheaf, i.e. satisfies the sheaf axioms.

Let $u \in \mathscr{F}'(U)$ and let (U_α) be an open cover of U such that $u \mid U_\alpha = 0$ for all α. Each $x \in X_a(U)$ is of the form $x = x_1 + \cdots + x_r$ with $x_i \in X_a(U_{\alpha_i})$ for suitable $\alpha_1, \ldots, \alpha_r$. It follows that

$$\langle x, u \rangle = \sum_{i=1}^r \langle x_i, u \mid U_{\alpha_i} \rangle = 0,$$

and hence that $u = 0$.

Secondly, let (U_α) be an open cover of U, and let $u_\alpha \in \mathscr{F}'(U_\alpha)$ satisfy $u_\alpha \mid U_\alpha \cap U_\beta = u_\beta \mid U_\alpha \cap U_\beta$ for all α, β. Induction on r shows that $\sum_{i=1}^r \langle x_i, u_{\alpha_i} \rangle = 0$, whenever $x_i \in X_a(U_{\alpha_i})$ for $i = 1, \ldots, r$ and $x_1 + \cdots + x_r = 0$. This observation allows us to define a linear form $u : X_a(U) \to \mathbb{C}$ by

$$x \mapsto \sum_{i=1}^r \langle x_i, u_{\alpha_i} \rangle \quad \text{if} \quad x = \sum_{i=1}^r x_i \quad \text{and} \quad x_i \in X_a(U_{\alpha_i}) \quad \text{for} \quad i = 1, \ldots, r.$$

To prove the continuity of u, it suffices to show that $u \mid X_a(K)$ is continuous for each compact set K in U. But for such a set K, there are $\alpha_1, \ldots, \alpha_r$ and compact sets $K_i \subset U_{\alpha_i}$ such that

$$X_a(K) \subset X_a(K_1) + \cdots + X_a(K_r).$$

By the open mapping principle, each sequence converging to zero in $X_a(K)$ can be written as a sum of sequences converging to zero in $X_a(K_i)$ ($i = 1, \ldots, r$). Therefore the continuity of $u \mid X_a(K_i)$ implies that of $u \mid X_a(K)$. By definition, $u \mid U_\alpha = u_\alpha$ for each α.

(3) We next prove that $\mathscr{F}'$ is a soft sheaf. Since $\mathscr{F}'$ has compact support, it

suffices to show that $\mathscr{F}'$ is c-soft. To do this, we prove that for arbitrary open sets U, V, W with compact closure in $\mathbb{C}^n$ such that $\overline{W} \subset V \subset \overline{V} \subset U$ and for each $u \in \mathscr{F}'(U)$, there is a section $u_0 \in \mathscr{F}'(U)$ with $u_0 | W = u | W$ and $\mathrm{supp}(u_0) \subset \overline{V}$. The action of u on $x \in X_a(W)$ is given by $\langle x, u \rangle = \langle x, u_{\overline{W}} \rangle$ if $u = (u_K / X_a(K)^{\perp})_{K \Subset U}$. The identification between spectral subspaces and section spaces allows one to deduce that

$$X_a(\mathbb{C}^n \setminus V) + X_a(\overline{W}) = X_a((\mathbb{C}^n \setminus V) \cup \overline{W}),$$

and that $X_a(\mathbb{C}^n \setminus V) \cap X_a(\overline{W}) = \{0\}$. An application of the open mapping principle shows that

$$X_a(\mathbb{C}^n \setminus V)^{\perp} + X_a(\overline{W})^{\perp} = \left(X_a(\mathbb{C}^n \setminus V) \cap X_a(\overline{W}) \right)^{\perp} = X'.$$

If we choose $\tilde{u}_0 \in X_a(\mathbb{C}^n \setminus V)^{\perp}$ with $u_{\overline{W}} - \tilde{u}_0 \in X_a(\overline{W})^{\perp}$, then

$$u_0 = \left(\tilde{u}_0 / X_a(K)^{\perp} \right)_{K \Subset U}$$

is an element in $\mathscr{F}'(U)$ that has the desired properties.

(4) The proof is completed by the observation that the map

$$J: X' \to \mathscr{F}'(\mathbb{C}^n), \qquad u \mapsto \left(u / X_a(K)^{\perp} \right)_{K \Subset \mathbb{C}^n}$$

is an isomorphism of Fréchet $\mathscr{O}(\mathbb{C}^n)$-modules. To see that this map is onto, note that $J(u_{\sigma(a)}) = u$ for all $u \in \mathscr{F}'(\mathbb{C}^n)$. $\qquad \square$

To prove the dual version of the last result, we need a criterion for the w^*-closedness of the spectral subspaces of a dual system.

Lemma 6.1.5 *Let $a \in L(X)^n$ be a commuting tuple of Banach space operators such that a' possesses the single-valued extension property. Suppose that $F \subset \mathbb{C}^n$ is closed and that $X'_{a'}(F)$ is a norm-closed subspace of X'. Then it is w^*-closed.*

Proof Let (u_j) be a net in $X'_{a'}(F)$ with w^*-limit $u \in X'$. By the theorem of Krein–Šmulian (see Corollary IV.6.4 in Schaefer 1966), we may assume that (u_j) is norm-bounded. For each relatively compact open polydisc D in $\mathbb{C}^n \setminus F$, the map

$$\mathscr{O}_b(D, X')^n \xrightarrow{\; z - a' \;} \mathscr{O}_b(D, X')$$

defines a continuous linear operator between Banach spaces. Here the subscript means that we consider the Banach space of all bounded analytic functions on D. Since the image of $z - a'$ contains $X'_{a'}(F)$, there is a bounded net (f_j) in $\mathscr{O}_b(D, X')^n$ such that $(z - a')f_j = u_j$ for all j. Since $\mathscr{O}(D)$ is a nuclear, in particular reflexive, Fréchet space, it follows that (see Appendix 1)

$$\mathscr{O}(D) \,\hat{\otimes}\, X' = (\mathscr{O}(D)' \,\hat{\otimes}\, X)'$$

topologically, if all dual spaces are equipped with their strong dual topology.

Because $\mathscr{O}(D)' \,\hat{\otimes}\, X$ is barrelled (see Corollary 1 of I.1.4 in Grothendieck 1955), bounded nets in $\mathscr{O}(D) \,\hat{\otimes}\, X'$ are equicontinuous, and hence possess weakly convergent subnets. Therefore, we may suppose that the components of (f_j) converge weakly to the components of a form $f \in (\mathscr{O}(D) \,\hat{\otimes}\, X')^n$. But weak convergence in $\mathscr{O}(D) \,\hat{\otimes}\, X'$ implies pointwise w^*-convergence in X'. Hence $u = (z - a')f$ on D. $\qquad\square$

This result enables us to interchange the roles of a and a' in the proof of Theorem 6.1.4.

Theorem 6.1.6 *Suppose that a' is decomposable. Then a has a Fréchet soft sheaf model.*

Proof The proof proceeds in much the same way as that of Theorem 6.1.4. To simplify the notation, we shall denote the spectral subspace of a' belonging to a set M in $\mathbb{C}^n$ by $X_{a'}(M)$ instead of $X'_{a'}(M)$.

(1) For each open set U in $\mathbb{C}^n$, we define the Fréchet space $\mathscr{F}(U)$ as the projective limit

$$\mathscr{F}(U) = \varprojlim_{K \Subset U} X/^\perp X_{a'}(K)$$

with respect to the canonical mappings $X/^\perp X_{a'}(L) \to X/^\perp X_{a'}(K)$, where $K \subset L \subset U$ are compact sets. By Lemma 6.1.5 and the bipolar theorem, the space $X_{a'}(K)$ can be regarded as the strong dual of the quotient $X/^\perp X_{a'}(K)$. The above projective limit is *reduced*, i.e. all projections

$$\mathscr{F}(U) \to X/^\perp X_{a'}(K) \qquad (K \subset U \text{ compact})$$

have dense range (see Remark 3.2.5). Therefore, we have the representation

$$\mathscr{F}(U)'_\tau = \varinjlim_{K \Subset U} X_{a'}(K)_\tau,$$

where τ denotes the topology of the Mackey duals (see Theorem IV.4.4 in Schaefer 1966). The inductive limit is formed with respect to the dual spectrum. Hence it can be identified with $X_{a'}(U)$ equipped with the inductive locally convex topology of the embeddings of $X_{a'}(K)_\tau$ into $X_{a'}(U)$, where K runs through all compact subsets of U. In the remainder of this proof, $X_{a'}(U)$ will always be equipped with this topology. Since $\mathscr{F}(U)$ is barrelled, it follows that (Appendix 1)

$$\mathscr{F}(U) = (\mathscr{F}(U)'_\tau)'_\beta = X_{a'}(U)'_\beta,$$

where β refers to the strong dual topology. The duality between $\mathscr{F}(U)$ and $X_{a'}(U)$ is given by $\langle u, x \rangle = \langle x_L, u \rangle$ if $u \in X_{a'}(L)$, $L \subset U$ is compact, and $x = (x_K/^\perp X_{a'}(K))_{K \Subset U}$. For open sets $V \subset U \subset \mathbb{C}^n$, the canonical restriction map

$$r_V^U : \mathscr{F}(U) \to \mathscr{F}(V), \qquad x \mapsto x \,|\, V$$

is the adjoint of the continuous inclusion mapping $X_{a'}(V) \to X_{a'}(U)$.

(2) The family $\mathscr{F}(U)$ $(U \subset \mathbb{C}^n$ open) forms, together with these restrictions, a presheaf of Fréchet spaces. This presheaf satisfies the sheaf axioms. The use of the Mackey topologies makes the proof of the second sheaf axiom more involved.

Suppose that (U_α) is an open cover of U and that $x_\alpha \in \mathscr{F}(U_\alpha)$ are given such that

$$x_\alpha \,|\, U_\alpha \cap U_\beta = x_\beta \,|\, U_\alpha \cap U_\beta$$

for all α and β. Then, exactly as in the proof of Theorem 6.1.4, one can define a linear form $x\colon X_{a'}(U) \to \mathbb{C}$. To prove the continuity of x it suffices to show that, for each compact set K in U, the kernel of x restricted to $X_{a'}(K)$ is closed in the Mackey, or equivalently, in the weak topology induced on $X_{a'}(K)$ by the duality $\langle X/^\perp X_{a'}(K), X_{a'}(K) \rangle$. Since this topology is the relative topology of the w^*-topology of X', it remains to verify that $\mathrm{Ker}(x) \cap X_{a'}(K)$ is a w^*-closed subspace of X'. Let (u_j) be a net in $\mathrm{Ker}(x) \cap X_{a'}(K)$ with w^*-limit u. Again by the theorem of Krein–Smulian, we may assume that $\|u_j\| \le 1$ for all j. By the open mapping principle, one can choose $\alpha_1, \dots, \alpha_r$, compact sets $K_i \subset U_{\alpha_i}$ for $1 \le i \le r$, and a constant $M > 0$ such that, for each $v \in X_{\alpha'}(K)$, there are $v_1, \dots, v_r$ with

$$v = v_1 + \cdots + v_r, \qquad v_i \in X_{a'}(K_i), \quad \|v_i\| \le M \|v\| \quad (i = 1, \dots, r).$$

In particular, each $u_j = u_{j1} + \cdots + u_{jr}$ for suitable $u_{ji} \in X_{a'}(K_i)$ with $\|u_{ji}\| \le M$. By the theorem of Alaoglu–Bourbaki and Lemma 6.1.5, we may suppose that, for each $i = 1, \dots, r$, the net $(u_{ji})_j$ converges with respect to the w^*-topology to an element $u^i \in X_{a'}(K_i)$. Then $u = \sum_{i=1}^r u^i$ and

$$\langle u, x \rangle = \sum_{i=1}^r \langle u^i, x_{\alpha_i} \rangle = \lim_j \sum_{i=1}^r \langle u_{ji}, x_{\alpha_i} \rangle = \lim_j \langle u_j, x \rangle = 0,$$

hence $u \in X_{a'}(K) \cap \mathrm{Ker}(x)$.

(3) $\mathscr{F}$ can be made into an analytic Fréchet sheaf on $\mathbb{C}^n$. If $U \subset \mathbb{C}^n$ is a Stein open set, then for each point $\lambda \in \mathbb{C}^n \setminus U$, there are analytic functions $f_1, \dots, f_n \in \mathscr{O}(U)$ such that

$$1 = \sum_{i=1}^n (\lambda_i - z_i) f_i(z) \quad (z \in U).$$

If K is a compact subset of U, then the local analytic functional calculus of a' yields solutions $c_1, \dots, c_n$ in the commutant of $a' \,|\, X_{a'}(K)$ of the equation

$$1_{X_{a'}(K)} = \sum_{i=1}^n (\lambda_i - a'_i) c_i.$$

Thus, by using Lemma 2.2.4 and duality (cf. Section 2.6), we obtain the spectral inclusion

$$\sigma(a, X/^\perp X_{a'}(K)) = \sigma(a', X_{a'}(K)) \subset U.$$

Hence, for fixed K, the analytic functional calculus yields a continuous algebra homomorphism $\mathscr{O}(U) \to L(X/^{\perp}X_{a'}(K))$. By Lemma 2.5.8 these functional calculi are compatible with the structural maps forming the inverse limit $\mathscr{F}(U)$. Consequently, for each Stein open set U in $\mathbb{C}^n$, the section space $\mathscr{F}(U)$ becomes a Fréchet $\mathscr{O}(U)$-module. Since, for any pair of Stein open sets $V \subset U$ in $\mathbb{C}^n$, the restriction $\mathscr{F}(U) \to \mathscr{F}(V)$ is continuous and compatible with the above module structures, there is a unique way to turn $\mathscr{F}$ into an analytic Fréchet sheaf such that on Stein open sets U in $\mathbb{C}^n$ the Fréchet $\mathscr{O}(U)$-module structure of $\mathscr{F}(U)$ is the one described above.

(4) The softness of $\mathscr{F}$ follows in the same way as the softness of $\mathscr{F}'$ in the proof of Theorem 6.1.4. One only has to know that $X = ^{\perp}X_{a'}(K) + ^{\perp}X_{a'}(L)$, whenever K and L are disjoint closed sets in $\mathbb{C}^n$. Let us denote the above spectral subspaces of a' by M and N, respectively. Then, for instance by Theorem IV.4.8 in Kato (1966), the fact that M and N are w^*-closed, and $M + N$ is norm-closed, implies that $^{\perp}M + ^{\perp}N \subset X$ is norm-closed. But then using the bipolar theorem one obtains that

$$X = ^{\perp}(M \cap N) = ^{\perp}\left[(^{\perp}M + ^{\perp}N)^{\perp}\right] = ^{\perp}M + ^{\perp}N.$$

(5) To conclude the proof, the reader should observe that
$$J: X \to \mathscr{F}(\mathbb{C}^n), \qquad x \mapsto (x/^{\perp}X_{a'}(K))_{K \in \mathbb{C}^n}$$
is an isomorphism of Fréchet $\mathscr{O}(\mathbb{C}^n)$-modules. $\qquad\qquad\square$

Since the existence of a Fréchet soft sheaf model for a commuting tuple implies its decomposability, the results obtained in Theorem 6.1.4 and Theorem 6.1.6 show that both conditions are equivalent, and that they are completely compatible with duality. In the following we use a special version of *Serre's duality principle* to add another equivalent characterization to this chain. As a preparation we need the following exactness result from the theory of locally convex spaces.

Lemma 6.1.7 *Let X be a Banach space, and let*
$$E \xrightarrow{\alpha} F \xrightarrow{\beta} G$$
be an exact sequence of nuclear Fréchet spaces. Suppose that $\mathrm{Im}(\beta)$ *is closed. Then the sequence*
$$G' \hat{\otimes} X \xrightarrow{\beta' \otimes 1} F' \hat{\otimes} X \xrightarrow{\alpha' \otimes 1} E' \hat{\otimes} X$$
is exact. Here all dual spaces are equipped with their strong topology.

Proof The reader should note first that E', F', and G' are nuclear. Here there is no reason to distinguish between the ε- and π-tensor product. The proof is based on the following factorization:

$$
\begin{array}{ccccc}
G' \hat{\otimes} X & \xrightarrow{\beta' \otimes 1} & F' \hat{\otimes} X & \xrightarrow{\alpha' \otimes 1} & E' \hat{\otimes} X \\
& {\scriptstyle \beta' \otimes 1} \searrow \quad \nearrow {\scriptstyle i \otimes 1} & \quad {\scriptstyle \alpha' \otimes 1} \searrow & \quad \nearrow {\scriptstyle i \otimes 1} & \\
& (\mathrm{Im}\,\beta') \hat{\otimes} X & & (\mathrm{Im}\,\alpha') \hat{\otimes} X. &
\end{array}
$$

First, since the inclusion maps denoted by i are topological monomorphisms, both of the maps denoted by $i \otimes 1$ are topological monomorphisms (see Theorem A1.6 in Appendix 1). In particular, the image of

$$i \otimes 1 \colon (\operatorname{Im} \beta') \mathbin{\hat{\otimes}} X \to F' \mathbin{\hat{\otimes}} X$$

is closed. Since nuclear Fréchet spaces are in particular (FM)-spaces, the map $\alpha' \colon F' \to \operatorname{Im} \alpha'$ is a homomorphism (see §33.6(1) in Köthe 1979). Therefore, it follows that

$$\operatorname{Ker}(\alpha' \otimes 1) = \overline{\operatorname{Ker}(\alpha') \otimes X}$$

(see Theorem A1.6 and the following remark in Appendix 1). Because of

$$\operatorname{Ker}(\alpha') = (\operatorname{Im} \alpha)^{\perp} = (\operatorname{Ker} \beta)^{\perp} = \operatorname{Im} \beta',$$

we conclude that

$$\operatorname{Ker}(\alpha' \otimes 1) = i \otimes 1\big((\operatorname{Im} \beta') \mathbin{\hat{\otimes}} X\big).$$

Using the algebraic identification $(F/\operatorname{Ker} \beta)' \cong \operatorname{Im}(\beta')$, the adjoint of the quotient map $F \to F/\operatorname{Ker} \beta$ becomes the inclusion map $\operatorname{Im}(\beta') \to F'$. Since the adjoint of a homomorphism between (FM)-spaces is a strong homomorphism, the relative topology and the strong dual topology of $\operatorname{Im}(\beta')$ (as the dual space of $F/\operatorname{Ker} \beta$) coincide. Therefore the proof of Lemma 6.1.7 is reduced to the case where $\alpha = 0$. In this case we have to check that

$$\beta' \otimes 1 \colon G' \mathbin{\hat{\otimes}} X \to F' \mathbin{\hat{\otimes}} X$$

is onto.

This is done in two steps. We first show that, for each bounded set C in F', there is a bounded set B in G' with $\beta'(B) = C$. To see this, note that the adjoint of the map $\beta \colon F \to \operatorname{Im} \beta$ becomes a topological isomorphism

$$G'/\operatorname{Ker} \beta' \xrightarrow{\ \beta'\ } F'$$

if both sides are equipped with their strong dual topology. Hence we can choose a strongly bounded subset $\tilde{B}$ of $G'/\operatorname{Ker} \beta'$ with $\beta'(\tilde{B}) = C$. Since $\tilde{B}$ is equicontinuous as a subset of $(\operatorname{Im} \beta)'$, the theorem of Hahn–Banach implies the existence of an equicontinuous set $B \subset G'$ with $B \mid \operatorname{Im} \beta = \tilde{B}$. But then $B \subset G'$ is bounded and $\beta'(B) = C$.

Finally, we prove the surjectivity of $\beta' \otimes 1$. Suppose that M is a bounded set in $F' \mathbin{\hat{\otimes}} X$. Then by a result of Grothendieck (see Corollary A1.11 in Appendix 1), there are closed, bounded, and absolutely convex sets C in F' and D in X such that M is contained in the image of the closed unit ball under the canonical map

$$(F')_C \mathbin{\hat{\otimes}_{\pi}} X_D \to F' \mathbin{\hat{\otimes}} X.$$

Here $(F')_C = \bigcup_{n \geq 1} nC \subset F'$ and $X_D = \bigcup_{n \geq 1} nD \subset X$ are equipped with the norm given by the Minkowski functionals of C and D, respectively (see Appendix 1). From the above considerations it follows that there is a closed, bounded, and absolutely convex set B in G' such that $\beta'(B) = C$. Since the diagram

$$
\begin{array}{ccc}
(G')_B \hat{\otimes}_\pi X_D & \longrightarrow & G' \hat{\otimes} X \\
{\scriptstyle \beta' \otimes 1} \downarrow & & \downarrow {\scriptstyle \beta' \otimes 1} \\
(F')_C \hat{\otimes}_\pi X_D & \longrightarrow & F' \hat{\otimes} X
\end{array}
$$

commutes and the left vertical map is onto, the set M is contained in the image of a bounded set in $G' \hat{\otimes} X$ under the right vertical map. $\qquad \square$

For future reference, the reader should observe that in the scalar case, that is, $X = \mathbb{C}$, the assertion of Lemma 6.1.7 is of course true for arbitrary Fréchet spaces E, F, and G without any nuclearity assumption (see the above proof). This observation is usually referred to as *Serre's duality lemma*. Furthermore, if

$$
0 \leftarrow E_0 \xleftarrow{d_1} E_1 \xleftarrow{d_2} \cdots \xleftarrow{d_r} E_r \leftarrow 0
$$

is an exact sequence of nuclear Fréchet spaces, then an iterative application of Lemma 6.1.7 shows that, for any Banach space X, the induced sequence

$$
0 \to E_0' \hat{\otimes} X \xrightarrow{d_1' \otimes 1} E_1' \hat{\otimes} X \xrightarrow{d_2' \otimes 1} \cdots \xrightarrow{d_r' \otimes 1} E_r' \hat{\otimes} X \to 0
$$

remains exact.

For a sheaf $\mathscr{F}$ of abelian groups on a topological space Ω, and any family of supports φ on Ω, we denote by $H_\varphi^p(\Omega, \mathscr{F})$ ($p \geq 0$) the *canonical cohomology groups* of $\mathscr{F}$ with respect to φ (see Section II.2 in Bredon 1967). The sheaf $\mathscr{F}$ is φ-*acyclic* if $H_\varphi^p(\Omega, \mathscr{F}) = 0$ for all $p > 0$. We write c for the family of supports consisting of all compact subsets of Ω, and we use the notation $\Gamma_c(\Omega, \mathscr{F})$ or $\mathscr{F}_c(\Omega)$ to refer to the space of all sections over Ω with compact support.

Proposition 6.1.8 *Let X be a Banach space. Then for each Stein open set U in $\mathbb{C}^n$, there are canonical algebraic isomorphisms*

$$
H_c^p(U, \mathscr{O}_{\mathbb{C}^n}^X) \cong \begin{cases} 0, & p \neq n, \\ \mathscr{O}(U)' \hat{\otimes} X, & p = n. \end{cases}
$$

Proof By applying Lemma 6.1.7 to the $\bar{\partial}$-resolution of $\mathscr{O}(U)$, we obtain an exact sequence of the form

$$
0 \to \mathscr{E}_{0,n}(U)' \hat{\otimes} X \xrightarrow{\bar{\partial}} \mathscr{E}_{0,n-1}(U)' \hat{\otimes} X \xrightarrow{\bar{\partial}} \cdots \xrightarrow{\bar{\partial}} \mathscr{E}_{0,0}(U)' \hat{\otimes} X
$$
$$
\to \mathscr{O}(U)' \hat{\otimes} X \to 0.
$$

For each $p = 0, \ldots, n$, there is a canonical topological isomorphism

$$
\mathscr{E}_{0,p}(U)' \hat{\otimes} X \cong L_b\big(\mathscr{E}_{0,p}(U), X\big),
$$

where the subscript b refers to the topology of uniform convergence on all bounded subsets of $\mathcal{E}_{0,p}(U)$ (see Treves 1967, p. 525, or Appendix 1). Moreover, the space $L_b(\mathcal{E}_{0,p}(U), X)$ can be regarded as the space of all sections with compact support over U in the sheaf $^X\mathcal{D}'_{0,n-p}$ of all $(n-p)$-forms of X-valued distributions on $\mathbb{C}^n$. Since

$$0 \to \mathcal{O}^X_{\mathbb{C}^n} \xrightarrow{i} {}^X\mathcal{D}'_{0,0} \xrightarrow{\bar{\partial}} {}^X\mathcal{D}'_{0,1} \xrightarrow{\bar{\partial}} \cdots \xrightarrow{\bar{\partial}} {}^X\mathcal{D}'_{0,n} \to 0$$

is a resolution of $\mathcal{O}^X_{\mathbb{C}^n}$ by soft sheaves, we obtain that

$$H^p_c\big(U, \mathcal{O}^X_{\mathbb{C}^n}\big) \cong H^p\big(\mathcal{E}_{0,n-\bullet}(U)' \hat{\otimes} X\big) \cong \begin{cases} 0, & p \neq n \\ \mathcal{O}(U)' \hat{\otimes} X, & p = n. \end{cases} \qquad \square$$

Let a be an n-tuple of commuting Banach space operators, and let U be a fixed Stein open set in $\mathbb{C}^n$. For each integer $p \geq 0$, we define $X_p = X^{\binom{n}{p}}$ and $\mathcal{O}_p = \mathcal{O}^{X_p}_U$. The Koszul complex $K_\bullet(z - a, \mathcal{O}^X_U)$ becomes a sequence of Fréchet analytic sheaves

$$\cdots \leftarrow \mathcal{O}_{p-1} \xleftarrow{z-a} \mathcal{O}_p \leftarrow \cdots .$$

We denote by $\mathcal{D}'_{p,q}$ the restriction of the sheaf $^{X_p}\mathcal{D}'_{0,q}$ to U, and we consider the resolution $0 \to \mathcal{O}_p \to \mathcal{D}'_{p,\bullet}$ used in the last proof. The Koszul complex induces morphisms of complexes

$$
\begin{array}{ccc}
0 \longrightarrow & \mathcal{O}_p & \longrightarrow \mathcal{D}'_{p,\bullet} \\
 & {\scriptstyle z-a}\downarrow & \quad\downarrow{\scriptstyle z-a} \\
0 \longrightarrow & \mathcal{O}_{p-1} & \longrightarrow \mathcal{D}'_{p-1,\bullet}
\end{array}
$$

If we identify as before $\Gamma_c(\mathcal{D}'_{p,\bullet}) \cong \mathcal{E}_{0,n-\bullet}(U)' \hat{\otimes} X_p$, then we obtain a commuting diagram.

$$
\begin{array}{ccccc}
H^n_c(\mathcal{O}_p) & \xleftarrow{\ \sim\ } & H^n(\mathcal{E}_{0,n-\bullet}(U)' \hat{\otimes} X_p) & \xrightarrow{\ \sim\ } & \mathcal{O}(U)' \hat{\otimes} X_p \\
{\scriptstyle z-a}\downarrow & & \downarrow{\scriptstyle z-a} & & \downarrow{\scriptstyle z-a} \\
H^n_c(\mathcal{O}_{p-1}) & \xleftarrow{\ \sim\ } & H^n(\mathcal{E}_{0,n-\bullet}(U)' \hat{\otimes} X_{p-1}) & \xrightarrow{\ \sim\ } & \mathcal{O}(U)' \hat{\otimes} X_{p-1}.
\end{array}
$$

The commutativity of the left rectangle follows from the naturality of the cohomology functor (see Theorem II.4.1 in Bredon 1967), while the commutativity of the right rectangle follows from the proof of Proposition 6.1.8.

Let us suppose that a satisfies the single-valued extension property. Then its canonical sheaf model $\mathcal{F}$ possesses the resolution

$$0 \leftarrow \mathcal{F} \leftarrow K_\bullet(z - a, \mathcal{O}^X).$$

We denote by $(\mathcal{O}_{p,q})_{0 \leq p \leq n, q \leq n}$ the double complex such that, for each $p = 0, \ldots, n$, the pth column is, up to the sign $(-1)^p$, the canonical flabby resolution of $\mathcal{O}_p$ written as

$$0 \to \mathcal{O}_p \to \mathcal{O}_{p,n} \to \mathcal{O}_{p,n-1} \to \cdots,$$

while the qth row is the complex induced by $K_\bullet(z - a, \mathscr{O}^X)$. Let $0 \to \mathscr{F} \to \mathscr{F}_n \to \mathscr{F}_{n-1} \to \cdots$ be the canonical flabby resolution of $\mathscr{F}$. Then all columns of the complex $\Gamma_c(\mathscr{O}_{\bullet,\bullet})$ are exact except in degree $q = 0$, where

$$H_0\big(\Gamma_c(\mathscr{O}_{p,\bullet})\big) = H_c^n(\mathscr{O}_p) \qquad (0 \le p \le n),$$

and all rows of $\Gamma_c(\mathscr{O}_{\bullet,\bullet})$ are exact except in degree $p = 0$, where

$$H_0\big(\Gamma_c(\mathscr{O}_{\bullet,q})\big) \cong \Gamma_c(U, \mathscr{F}_q) \qquad (q \le n).$$

Therefore, for each $q \ge 0$, we obtain the isomorphisms (Lemma A2.6 in Appendix 2)

$$H_c^q(U, \mathscr{F}) = H_{n-q}(\Gamma_c(U, \mathscr{F}_\bullet)) \cong H_{n-q}(H_c^n(\mathscr{O}_\bullet)).$$

Since the complex $H_c^n(\mathscr{O}_\bullet)$ is precisely the one induced by $K_\bullet(z - a, \mathscr{O}^X)$, we obtain, using the commutative diagram (6.1.1), the isomorphisms

$$H_c^q(U, \mathscr{F}) \cong H_{n-q}\big(K_\bullet(z - a, \mathscr{O}(U)' \,\hat{\otimes}\, X)\big) \qquad (q \ge 0), \qquad (6.1.2)$$

where the right-hand side has to be read as zero for $q > n$.

Using these observations, we obtain another characterization of decomposable n-tuples.

Theorem 6.1.9 *Suppose that $a \in L(X)^n$ is a commuting tuple such that both a and a' satisfy Bishop's property (β). Then a' has a Fréchet soft sheaf model.*

Proof Let $\mathscr{F}'$ be the canonical sheaf model of a'. If $0 \to \mathscr{F}' \to \mathscr{F}^\bullet$ is its canonical flabby resolution, then for each closed set A in $\mathbb{C}^n$, we obtain a short exact sequence of complexes of abelian groups

$$0 \to \Gamma_c(\mathscr{F}^\bullet | X \backslash A) \to \Gamma_c(\mathscr{F}^\bullet) \to \Gamma(\mathscr{F}^\bullet | A) \to 0.$$

To see this, the reader should note that, since $\mathscr{F}'$ has compact support, the same is true for all the sheaves $\mathscr{F}^i$, and that of course all the sheaves $\mathscr{F}^i$ are soft. In view of the induced long exact sequence of cohomology groups

$$0 \to \Gamma_c(\mathscr{F}' | X \backslash A) \to \Gamma(\mathscr{F}') \to \Gamma(\mathscr{F}' | A)$$
$$\to H_c^1(\mathscr{F}' | X \backslash A) \to H^1(\mathscr{F}') \to H^1(\mathscr{F}' | A)$$
$$\to \cdots$$

it is obvious that $\mathscr{F}'$ is soft if and only if $H_c^1(\mathscr{F}' | U) = 0$ for all open sets U in $\mathbb{C}^n$ (cf. Proposition II.15.1 in Bredon 1967).

The condition that a satisfies Bishop's property (β) means by definition that, for each Stein open set U in $\mathbb{C}^n$, the complex $K^\bullet(z - a, \mathscr{O}(U) \,\hat{\otimes}\, X)$ is exact in degree $q < n$ and has separated cohomology in degree $q = n$. Using Serre's duality lemma (see the remark following Lemma 6.1.7) and (6.1.2), we see that

$$H_c^q(U, \mathscr{F}') \cong H_{n-q}\big(K_\bullet(z - a', \mathscr{O}(U)' \,\hat{\otimes}\, X')\big) = 0 \qquad (q \ge 1).$$

Therefore, to complete the proof, it suffices to check that each sheaf of abelian groups on $\mathbb{C}^n$ that is c-acyclic on all Stein open sets in $\mathbb{C}^n$ is c-acyclic on all open sets U in $\mathbb{C}^n$. This follows from the next result.   □

Lemma 6.1.10 *Let X be a Hausdorff topological space, let $\mathcal{U} = (U_i)_{i \in \mathbb{N}}$ be an open cover of X with finite-dimensional nerve, and let $\mathcal{F}$ be a sheaf of abelian groups on X. Suppose that $\mathcal{F}$ is c-acyclic on all finite intersections of sets in $\mathcal{U}$. Then $\mathcal{F}$ is c-acyclic.*

Proof As usual we write $U_{i_0 \dots i_p} = U_{i_0} \cap \cdots \cap U_{i_p}$ for each tuple of indices $i_0, \dots, i_p$. For $p \geq 0$, the algebraic direct sum

$$\mathscr{C}_p(\mathcal{U}, \mathcal{F}_c) = \bigoplus_{0 \leq i_0 < \cdots < i_p} \mathcal{F}_c\big(U_{i_0 \dots i_p}\big),$$

consisting of all families with only finitely many non-zero components, is an abelian group. For each strictly increasing index tuple $i = (i_0, \dots, i_p)$ and each $f \in \mathcal{F}_c(U_i)$, we denote by $fs_i = fs_{i_0} \wedge \cdots \wedge s_{i_p}$ the element in $\mathscr{C}_p(\mathcal{U}, \mathcal{F}_c)$ with f as coefficient belonging to the indicated index tuple and all other coefficients equal to 0. For each open set U in X, we regard $\mathcal{F}_c(U)$ as a subspace of $\mathcal{F}(X)$ via trivial extension. Hence we may consider the augmented complex

$$0 \leftarrow \mathcal{F}_c(X) \xleftarrow{\; q \;} \mathscr{C}_0(\mathcal{U}, \mathcal{F}_c) \xleftarrow{\; \delta \;} \mathscr{C}_1(\mathcal{U}, \mathcal{F}_c) \xleftarrow{\; \delta \;} \cdots \qquad (6.1.3)$$

of abelian groups given by the boundary operators

$$\delta(fs_i) = \sum_{\rho=0}^{p} (-1)^\rho fs_{i_0} \wedge \cdots \wedge \hat{s}_{i_\rho} \wedge \cdots \wedge s_{i_p}$$

and the augmentation $q(f_i)_{i \geq 0} = \Sigma_{i \geq 0} f_i$.

The complex (6.1.3) can be regarded as the direct limit of all sequences of the form

$$0 \leftarrow \mathcal{F}_c(X) \xleftarrow{\; q \;} \mathscr{C}_{0n} \xleftarrow{\; \delta \;} \mathscr{C}_{1n} \xleftarrow{\; \delta \;} \cdots \xleftarrow{\; \delta \;} \mathscr{C}_{nn} \leftarrow 0, \qquad (6.1.4)$$

where for any pair of integers $0 \leq p \leq n$, the space $\mathscr{C}_{pn}$ is defined by

$$\mathscr{C}_{pn} = \bigoplus_{0 \leq i_0 < \cdots < i_p \leq n} \mathcal{F}_c\big(U_{i_0 \dots i_p}\big).$$

Using induction on n, one can show that, for each flabby sheaf $\mathcal{F}$ on X, all sequences of the form (6.1.4) are exact (cf. Lemma VII.3.9 in Bănică and Stănăşilă 1977). Hence for a flabby sheaf $\mathcal{F}$, the direct limit sequence (6.1.3) is exact.

Let $\mathcal{F}$ be an arbitrary sheaf on X, and let $0 \to \mathcal{F} \to \mathcal{F}^\bullet$ be its canonical flabby resolution. Then one can consider the double complex $\mathscr{C}_\bullet(\mathcal{U}, \mathcal{F}_c^\bullet)$ with rows $\mathscr{C}_q(\mathcal{U}, \mathcal{F}_c^\bullet)$ $(q \geq 0)$ induced by the above flabby resolution and columns given by

$$(-1)^p \mathscr{C}_\bullet(\mathcal{U}, \mathcal{F}_c^p) \qquad (p \geq 0).$$

As explained above, all columns are exact in all degrees, except in degree $q = 0$, where $H_0(\mathscr{C}_\bullet(\mathscr{U}, \mathscr{F}_c^p)) = \mathscr{F}_c^p(X)$. The assumption that $\mathscr{F}$ is a c-acyclic on all finite intersections of sets in $\mathscr{U}$ implies that all rows are exact in all degrees, except in degree $p = 0$, where $H^0(\mathscr{C}_q(\mathscr{U}, \mathscr{F}_c^\bullet)) = \mathscr{C}_q(\mathscr{U}, \mathscr{F}_c)$. The condition that $\mathscr{U}$ has finite-dimensional nerve, i.e. that there is an integer $N \geq 1$ such that all intersections of more than $N + 1$ sets of $\mathscr{U}$ are empty, ensures that $\mathscr{C}_\bullet(\mathscr{U}, \mathscr{F}_c^\bullet)$ has bounded diagonals.

Thus, again, our familiar double complex argument (Lemma A2.6 in Appendix 2) completes the proof:

$$H_c^p(X, \mathscr{F}) = H^p(\mathscr{F}_c^\bullet(X)) \cong H_{-p}(\mathscr{C}_\bullet(\mathscr{U}, \mathscr{F}_c)) = 0 \qquad (p \geq 1). \qquad \square$$

In the following theorem we gather together all previous results obtained in this section.

Theorem 6.1.11 *For a commuting tuple $a \in L(X)^n$ of continuous linear operators on a complex Banach space X, the following conditions are equivalent:*

(i) *a has a Fréchet soft sheaf model;*
(ii) *a' has a Fréchet soft sheaf model;*
(iii) *a is decomposable;*
(iv) *a' is decomposable;*
(v) *both a and a' satisfy Bishop's property (β).*

If one of these conditions holds, then the spectral subspaces of a and a' satisfy

$$X_{a'}(F) = X_a(\mathbb{C}^n \setminus F)^\perp, \qquad X_a(F) = {}^\perp X_{a'}(\mathbb{C}^n \setminus F)$$

for each closed set F in $\mathbb{C}^n$.

Proof As observed before, since the existence of a Fréchet soft sheaf model implies decomposability, the equivalence of the first four conditions follows from Theorems 6.1.4 and 6.1.6. Clearly, these conditions imply condition (v). The reverse implication is contained in Theorem 6.1.9.

It remains to prove the orthogonality relations for the spectral subspaces. For this purpose, let

$$J: X \to \mathscr{F}(\mathbb{C}^n), \qquad x \mapsto \left(x / {}^\perp X_{a'}(K)\right)_{K \Subset \mathbb{C}^n}$$

be the isomorphism defined in the proof of Theorem 6.1.6, and let $\mathscr{G}$ be the canonical sheaf model of a. Composing J with the corresponding isomorphism $X \cong \mathscr{G}(\mathbb{C}^n)$, one obtains an isomorphism $\varphi: \mathscr{F}(\mathbb{C}^n) \xrightarrow{\sim} \mathscr{G}(\mathbb{C}^n)$ of Fréchet $\mathscr{O}(\mathbb{C}^n)$-modules. Because of the quasi-coherence of $\mathscr{F}$ and $\mathscr{G}$ each such isomorphism is induced by a suitable isomorphism $\mathscr{F} \xrightarrow{\sim} \mathscr{G}$ of Fréchet analytic sheaves. In particular, the map φ preserves the support of each global section. Hence we obtain

$$X_a(F) = J^{-1}(\mathscr{F}_F(\mathbb{C}^n)) = {}^\perp X_{a'}(\mathbb{C}^n \setminus F)$$

for each closed set F in $\mathbb{C}^n$. In exactly the same way the dual relation is derived from Theorem 6.1.4. $\square$

In Theorems 6.1.4 and 6.1.6 we obtained alternative representations of the sheaf model of a decomposable tuple. We now formulate these results independently of the proofs given there.

Let a be a decomposable commuting n-tuple with associated Fréchet soft sheaf model $\mathscr{F}$, and let U be a fixed open set in $\mathbb{C}^n$. We identify X and $\mathscr{F}(\mathbb{C}^n)$ as Fréchet $\mathscr{O}(\mathbb{C}^n)$-modules, and we recall that via this identification we have $X_a(F) = \mathscr{F}_F(\mathbb{C}^n)$ for each closed set F in $\mathbb{C}^n$. For any pair of open sets V, W with $V \Subset W \Subset U$, the canonical isomorphism

$$X/X_a(\mathbb{C}^n \setminus V) \to \mathscr{F}_{\overline{W}}(\mathbb{C}^n)/X_a(\mathbb{C}^n \setminus V) \cap \mathscr{F}_{\overline{W}}(\mathbb{C}^n) \qquad (6.1.5)$$

induces a Fréchet $\mathscr{O}(U)$-module structure on $X/X_a(\mathbb{C}^n \setminus V)$ which is independent of the choice of W. Since the structural maps forming the inverse limit

$$\mathscr{G}(U) = \varprojlim_{V \Subset U} X/X_a(\mathbb{C}^n \setminus V)$$

are compatible with these $\mathscr{O}(U)$-module structures, the space $\mathscr{G}(U)$ becomes a Fréchet $\mathscr{O}(U)$-module. Combining the map (6.1.5) with the $\mathscr{O}(U)$-module homomorphism

$$\mathscr{F}_{\overline{W}}(\mathbb{C}^n)/X_a(\mathbb{C}^n \setminus V) \cap \mathscr{F}_{\overline{W}}(\mathbb{C}^n) \to \mathscr{F}(V), \qquad [s] \mapsto s \,|\, V,$$

one obtains morphisms $X/X_a(\mathbb{C}^n \setminus V) \to \mathscr{F}(V)$ which are independent of the choice of W and induce a continuous $\mathscr{O}(U)$-linear map

$$\varphi \colon \mathscr{G}(U) \to \mathscr{F}(U) = \varprojlim_{V \Subset U} \mathscr{F}(V).$$

We show that this map is an isomorphism by constructing its inverse. Note that, for U, V, and W as above, the softness of the sheaf $\mathscr{F}$ implies that $\mathscr{F}(U) = \mathscr{F}_{\mathbb{C}^n \setminus V}(U) + \mathscr{F}_{\overline{W}}(\mathbb{C}^n)$. The maps obtained as the composition of

$$\mathscr{F}(U) \xrightarrow{\;q\;} \mathscr{F}(U)/\mathscr{F}_{\mathbb{C}^n \setminus V}(U) \xrightarrow{\;\sim\;} \mathscr{F}_{\overline{W}}(\mathbb{C}^n)/\mathscr{F}_{\mathbb{C}^n \setminus V}(U) \cap \mathscr{F}_{\overline{W}}(\mathbb{C}^n)$$

$$\xrightarrow{\;\sim\;} X/X_a(\mathbb{C}^n \setminus V)$$

are compatible with the inverse spectrum forming $\mathscr{G}(U)$, and the induced map $\mathscr{F}(U) \to \mathscr{G}(U)$ is the inverse of φ.

Proposition 6.1.12 *Suppose that $a \in L(X)^n$ is a decomposable commuting tuple with Fréchet soft sheaf model $\mathscr{F}$. Then for each open set U in $\mathbb{C}^n$, there is a natural isomorphism of Fréchet $\mathscr{O}(U)$-modules*

$$\mathscr{F}(U) \cong \varprojlim_{V \Subset U} X/X_a(\mathbb{C}^n \setminus V).$$

                      $\square$

Our next aim is to describe a special, but important, class of decomposable n-tuples. As a preparation we need one more remark.

Let $a \in L(X)^n$ be a commuting n-tuple with the single-valued extension property, and let $U \subset \mathbb{C}^n$ be a Stein open set. Then each function $f \in \mathcal{O}(U, X)$ with the property that each point in U has an open neighbourhood V with $f|V \in \Sigma_{i=1}^n (z_i - a_i) \mathcal{O}(V, X)$ belongs to $\Sigma_{i=1}^n (z_i - a_i) \mathcal{O}(U, X)$. Indeed, under these conditions f defines a section in the image of the last sheaf homomorphism in the Koszul complex

$$0 \to \mathcal{O}_U^X \xrightarrow{z-a} (\mathcal{O}_U^X)^n \xrightarrow{z-a} \cdots \xrightarrow{z-a} (\mathcal{O}_U^X)^n \xrightarrow{z-a} \mathcal{O}_U^X \to 0.$$

But if the last sheaf on the right is replaced by this image sheaf, then the resulting sequence becomes an exact complex of acyclic sheaves on U.

Theorem 6.1.13 *Let $a \in L(X)^n$ be a commuting tuple of Banach space operators. Suppose that a possesses a continuous C^∞-functional calculus, i.e. that the analytic functional calculus of a extends to a continuous algebra homomorphism*

$$\psi : \mathscr{E}(\mathbb{C}^n) \to L(X).$$

Then a has a Fréchet soft sheaf model.

Proof Using C^∞-partitions of unity one shows that the map $E \colon \mathscr{F}(\mathbb{C}^n) = \{F; F = \bar{F} \subset \mathbb{C}^n\} \to \mathrm{Lat}(a)$, which assigns to each closed set F in $\mathbb{C}^n$ the closed a-invariant subspace

$$E(F) = \cap \{\mathrm{Ker}\, \psi(f); \quad f \in \mathscr{E}(\mathbb{C}^n) \text{ with } \mathrm{supp}(f) \cap F = \varnothing\},$$

satisfies the following axioms:

(i) $E(\varnothing) = \{0\}$, $E(\mathbb{C}^n) = X$;
(ii) $E(\cap_{i \in I} F_i) = \cap_{i \in I} E(F_i)$ for each family $(F_i)_{i \in I}$ of closed sets;
(iii) $E(F) \subset E(\bar{U}_1) + \cdots + E(\bar{U}_s)$ for each finite open cover $F \subset U_1 \cup \cdots \cup U_s$ of a closed set F;
(iv) $\sigma(a, E(F)) \subset F$ for all closed sets F.

The proof of these properties is based on the following observation. If Y is a closed linear subspace of X that is invariant under ψ, and if $f \in \mathscr{E}(\mathbb{C}^n)$ is a function with $\psi(f)|Y = 1_Y$, then $\sigma(a, Y) \subset \mathrm{supp}(f)$. For, if $\lambda \notin \mathrm{supp}(f)$, then using a C^∞-partition of unity with respect to an open cover of the form

$$\mathrm{supp}(f) \subset \bigcup_{i=1}^n \mathbb{C} \times \cdots \times \mathbb{C} \times (\mathbb{C} \setminus \bar{D}_r(\lambda_i)) \times \mathbb{C} \times \cdots \times \mathbb{C}$$

one can write $f = \Sigma_{i=1}^n (\lambda_i - z_i) f_i$ with suitable $f_i \in \mathscr{E}(\mathbb{C}^n)$. By applying ψ one obtains that $\lambda \notin \sigma(a, Y)$ (see Lemma 2.2.4).

The validity of (iv) follows directly from this remark, while condition (i) is trivially satisfied. Condition (iii) as well as condition (ii), for the case of finite intersections, follows by an obvious partition of unity argument.

To prove the general case of (ii), it suffices to verify that $\psi(f) = 0$ for each function $f \in \mathscr{E}(\mathbb{C}^n)$ with $\mathrm{supp}(f) \cap \sigma(a) = \varnothing$. But for f as above, one can choose a function $g \in \mathscr{E}(\mathbb{C}^n)$ with $(1 - g)f = 0$ and $\sigma(a, \mathrm{Ker}\,\psi(1 - g)) \subset \mathrm{supp}(g) \subset \rho(a)$, and an elementary application of Lemma 2.5.8 shows that the only a-invariant closed subspace Y of X with $\sigma(a, Y) \cap \sigma(a) = \varnothing$ is the trivial one $Y = \{0\}$.

To finish the proof, we show that each commuting tuple a for which there is a map E satisfying conditions (i) to (iv) is decomposable. Replacing the section spaces by the corresponding spaces formed with the help of the map E one can show, exactly as in part (2) of the proof of Theorem 4.4.1, that a satisfies Bishop's property (β).

To prove decomposability, it suffices to check that $E(F) = X_a(F)$ for all closed sets F. The inclusion $E(F) \subset X_a(F)$ is a consequence of (iv). To prove the opposite inclusion we choose, for each point $\lambda \in \mathbb{C}^n \setminus F$, open polydiscs $V \Subset U \Subset \mathbb{C}^n \setminus F$ with centre λ and observe that

$$\sigma(a, X/E(\mathbb{C}^n \setminus V)) = \sigma(a, E(\overline{U})/E(\overline{U} \setminus V)) \subset \overline{U}.$$

If $x \in X_a(F)$, then by the remark preceding Theorem 6.1.13, we know that $x \in \sum_{i=1}^n (z_i - a_i)\mathscr{O}(\overline{U}, X)$, and an application of the Cauchy–Weil integral associated with the system $a/E(\mathbb{C}^n \setminus V)$ to the vector $[x]$ yields that $[x] = 0$, i.e. $x \in E(\mathbb{C}^n \setminus V)$ (see Proposition 2.3.7). Using the intersection stability of E, we finally obtain the missing inclusion $X_a(F) \subset E(F)$.   □

A commuting tuple $a \in L(X)^n$ with a continuous C^∞-functional calculus is usually called a *generalized scalar n-tuple* (see Colojoară and Foiaş 1968 or Vasilescu 1982). A map $E: \mathscr{F}(\mathbb{C}^n) \to \mathrm{Lat}(a)$ satisfying the axioms (i) to (iv) is known as a *spectral capacity* for a. In this terminology, the last proof shows that generalized scalar tuples possess a spectral capacity, and that tuples possessing a spectral capacity have a Fréchet soft sheaf model. Moreover, it was shown that a spectral capacity, if it exists, is automatically given by the spectral subspaces of the system.

Replacing $\mathscr{E}(\mathbb{C}^n)$ by other function algebras admitting partitions of unity, one obtains other classes of decomposable systems. We should perhaps mention, although it is trivial, that commuting tuples consisting of normal Hilbert space operators are generalized scalar.

6.2 BISHOP'S CONDITION (β) AND DECOMPOSABLE RESOLUTIONS

In Chapter 4 we saw that, on a finite-dimensional Stein space, the quasi-coherent analytic Fréchet sheaves can be characterized as those analytic Fréchet sheaves which admit a finite resolution to the right by soft analytic Fréchet sheaves such that the induced sequence of global section spaces remains exact. In the present section we shall give an operator-theoretic interpretation of this result. In particular, we shall show that the Fréchet space constructions explained in Chapter 4 can be carried out within the category of Banach or Hilbert spaces. This observation, together with ideas originating in the work of S. Brown (1978), will be essential for the invariant subspace results to be derived later.

As before, let X be a complex Banach space, and let $a = (a_1, \ldots, a_n)$ be a commuting tuple of continuous linear operators on X. We fix an open polydisc D containing $\sigma(a)$ and we define for each multi-index $j \in \mathbb{N}^n$ the Sobolev type space

$$W^j(D) = \{f \in L^2(D); \bar{\partial}^k f \in L^2(D) \text{ for all } k \in \mathbb{N}^n \text{ with } 0 \le k \le j\}.$$

For each $f \in L^2(D)$ and each $k \in \mathbb{N}^n$, the derivative

$$\bar{\partial}^k f = (\partial/\partial \bar{z}_1)^{k_1} \cdots (\partial/\partial \bar{z}_n)^{k_n} f$$

can be formed in the sense of distributions. We write $\bar{\partial}^k f \in L^2(D)$ if this distribution can be represented by a function in $L^2(D)$. The space $W^j(D)$ is a Hilbert space with respect to the norm

$$\|f\| = \left(\sum_{0 \le k \le j} \|\bar{\partial}^k f\|_2^2 \right)^{1/2},$$

where $\| \; \|_2$ denotes the norm in $L^2(D)$.

If all components of j are non-zero, then there is a quite elementary way to build up an exact $\bar{\partial}$-sequence of Hilbert spaces starting with $W^j(D)$. We write $W^0 = W^j(D)$, and we consider for each integer q with $1 \le q \le n$ the Hilbert space direct sum

$$W^q = \bigoplus_{1 \le i_1 < \cdots < i_q \le n} W^{j - (e_{i_1} + \cdots + e_{i_q})}(D),$$

where $e_1, \ldots, e_n$ form the canonical basis of $\mathbb{C}^n$. If one regards the elements of W^q as q-forms in indeterminates $d\bar{z}_1, \ldots, d\bar{z}_n$, then the $\bar{\partial}$-sequence constructed in the usual way gives rise to a sequence of continuous linear operators between Hilbert spaces

$$0 \to L_a^2(D) \xrightarrow{i} W^0 \xrightarrow{\bar{\partial}} W^1 \xrightarrow{\bar{\partial}} \cdots \xrightarrow{\bar{\partial}} W^n \to 0. \qquad (6.2.1)$$

By definition, $L_a^2(D) = \{f \in W^0; \bar{\partial}f = 0\}$, and a suitable version of the classical Sobolev embedding lemma (Theorem A4.1 in Appendix 4) implies that $L_a^2(D)$ coincides with the usual Bergman space, i.e. with the space of all square-integrable analytic functions on D.

We first show that the definition of the Sobolev spaces $W^j(D)$ is compatible with Hilbertian tensor products, which will be denoted by $\tilde{\otimes}$.

Lemma 6.2.1 *Let $U \subset \mathbb{C}^m$ and $V \subset \mathbb{C}^n$ be open polydiscs. Then for each pair of multi-indices $k \in \mathbb{N}^m$ and $l \in \mathbb{N}^n$, the bilinear map*

$$W^k(U) \times W^l(V) \to W^{(k,l)}(U \times V), (f, g) \mapsto f \otimes g$$

induces an isometric isomorphism $W^k(U) \tilde{\otimes} W^l(V) \to W^{(k,l)}(U \times V)$.

Proof It is elementary to check that, for $f \in W^k(U)$ and $g \in W^l(V)$, the function $f \otimes g(z, w) = f(z)g(w)$ belongs to $W^{(k,l)}(U \times V)$, and satisfies

$$\bar{\partial}^{(\mu,\nu)}(f \otimes g) = (\bar{\partial}^\mu f) \otimes (\bar{\partial}^\nu g)$$

for all multi-indices $0 \le \mu \le k$, $0 \le \nu \le l$. To see that the induced linear map $W^k(U) \otimes W^l(V) \to W^{(k,l)}(U \times V)$ is an isometry, one can imitate the usual proof for the L^2-case.

It remains to prove that the induced isometry

$$\phi: W^k(U) \, \tilde{\otimes} \, W^l(V) \to W^{(k,l)}(U \times V)$$

has dense range. But this is obvious, since by Lemma A4.4 in Appendix 4 we know that $\mathscr{E}(\mathbb{C}^{n+m})$ is dense in the space on the right, while on the other hand, it is well known that $\mathscr{E}(\mathbb{C}^n) \otimes \mathscr{E}(\mathbb{C}^m) \subset \mathscr{E}(\mathbb{C}^{n \times m})$ is dense.   □

This result allows us to prove the exactness of the $\bar{\partial}$-sequence constructed above.

Lemma 6.2.2 *For each $j \in \mathbb{N}^n$ with $j_\nu \ge 1$ for all $\nu = 1, \ldots, n$, the $\bar{\partial}$-sequence* (6.2.1) *is exact.*

Proof If $U \subset \mathbb{C}$ is a bounded open set and $f \in L^2(U)$, then the convolution type integral

$$F(z) = \frac{1}{\pi} \int_U \frac{f(w)}{z - w} \, dw$$

defines a function $F \in L^2(U)$ with distributional $\bar{\partial}$-derivative $\bar{\partial} F = f$. The continuous linear operator $r: L^2(U) \to L^2(U)$, $f \mapsto F$, induces a right inverse for $\bar{\partial}: W^j(U) \to W^{j-1}(U)$. This settles the case $n = 1$.

The general case is proved by induction on n. To explain the induction step, we write $D = U \times V$ and $j = (k, l)$ with $U \subset \mathbb{C}^{n-1}$, $V \subset \mathbb{C}$, and $k \in \mathbb{N}^{n-1}$, $l \in \mathbb{N}$. We denote by $W_1^\bullet$ and $W_2^\bullet$ the $\bar{\partial}$-sequences starting with $W^k(U)$ and $W^l(V)$, respectively. Then using Lemma 6.2.1, one can identify the complex $W^\bullet$ with the total complex of the double complex

$$K^{\bullet, \bullet} = W_1^\bullet \, \tilde{\otimes}_\sigma \, W_2^\bullet.$$

Since by induction hypothesis and the case $n = 1$, all rows and columns of this double complex are exact in all degrees except 0, we conclude that the augmented $\bar{\partial}$-complex $0 \to L_a^2(D) \xrightarrow{\bar{\partial}} W^\bullet$ is exact.   □

To construct a decomposable resolution for a given tuple with Bishop's property (β) we replace, in the proof of Theorem 4.4.4, the Fréchet space $\bar{\partial}$-sequence by a $\bar{\partial}$-sequence consisting of Hilbert spaces, and we replace the Fréchet space resolution of the canonical sheaf model $\mathscr{F}$ of a by a Banach space resolution of $X = \mathscr{F}(\mathbb{C}^n)$. The first of these problems was solved above. Let us now turn to the second.

As before, let D be an open polydisc containing the spectrum $\sigma(a)$ of the commuting tuple $a \in L(X)^n$, where X is a given Banach space. In the following we shall write $\hat{\otimes}$ to denote the ε- or π-tensor product. In case X is a Hilbert space we also allow $\hat{\otimes}$ to be the Hilbertian tensor product. We denote by $z - a$ the n-tuple on $L_a^2(D) \hat{\otimes} X$ with components given by

$$L_a^2(D) \hat{\otimes} X \to L_a^2(D) \hat{\otimes} X, \qquad f \mapsto (z_i - a_i)f \quad (1 \leq i \leq n).$$

Because $L_a^2(D)$ is continuously embedded in $\mathcal{O}(D)$, there is a unique continuous linear map $\Phi: L_a^2(D) \hat{\otimes} X \to X$ with $\Phi(f \otimes x) = f(a)x$ for $f \in L_a^2(D)$ and $x \in X$.

Proposition 6.2.3   *The augmented cochain complex*

$$K^\bullet\left(z - a, L_a^2(D) \hat{\otimes} X\right) \xrightarrow{\Phi} X \to 0$$

is exact.

Proof   We first check that $K^\bullet(z - a, L_a^2(D) \hat{\otimes} X)$ is exact in degree $p \neq n$ and has separated cohomology in degree $p = n$. Using the notations of Section 2.6, we have to show that $0 \notin \sigma^{\pi, n-1}(z - a)$. Here z stands for the tuple induced by the multiplication with coordinate functions on $L_a^2(D) \hat{\otimes} X$, and $z - a$ is regarded as a tuple on the same space. Using the spectral mapping theorem (Corollary 2.6.8), one sees that

$$\sigma^{\pi, n-1}(z - a) = \{\lambda - \mu; (\lambda, \mu) \in \sigma^{\pi, n-1}(z, a)\}.$$

For $\lambda \in D$, the augmented cochain complex

$$K^\bullet(\lambda - a, L_a^2(D)) \xrightarrow{\delta_\lambda} \mathbb{C} \to 0 \qquad (\delta_\lambda(f) = f(\lambda))$$

is an exact sequence of Hilbert spaces. This can either be reduced, by an inductive argument, to the one-dimensional case as demonstrated in the proof of Lemma 6.2.2, or one can use the much more general results concerning the Gleason problem derived independently in Chapter 8. Forming the tensor product with the identity on X and using the projection property for $\sigma^{\pi, n-1}$ (Theorem 2.6.6), one obtains

$$(D \times \mathbb{C}^n) \cap \sigma^{\pi, n-1}(z, a) = \varnothing.$$

On the other hand, for $\lambda \notin D$, the analytic functional calculus of a yields operators $c_1, \ldots, c_n$ in the commutant of a such that $1_X = \sum_{i=1}^n (\lambda_i - a_i)c_i$. Therefore, by Lemma 2.2.4 and the projection property of the Taylor spectrum, we conclude that

$$(\mathbb{C}^n \times (\mathbb{C}^n \setminus D)) \cap \sigma(z, a) = \varnothing.$$

Thus, the first assertion has been verified.

To complete the proof, we choose an open polydisc U containing the closure of D and use the fact that the Fréchet-space complex

$$K^\bullet(z - a, \mathcal{O}(U, X)) \xrightarrow{\Phi_{an}} X \to 0,$$

where $\Phi_{an}\colon \mathscr{O}(U, X) \to X$ is the unique continuous linear map with $\Phi_{an}(f \otimes x) = f(a)x$ for $f \in \mathscr{O}(U)$ and $x \in X$, is exact (see the remarks at the end of Section 4.3). Namely, if $f \in L^2_a(D) \hat{\otimes} X$ satisfies $\Phi(f) = 0$, then using the density of the polynomials in $L^2_a(D)$ one can choose a sequence (f_k) in $\mathscr{O}(U, X)$ converging to f in $L^2_a(D) \hat{\otimes} X$. Hence $\lim_{k \to \infty} \Phi_{an}(f_k) = 0$, and there is a sequence (g_k) in $K^{n-1}(z - a, \mathscr{O}(U, X))$ such that $(f_k - (z - a)g_k)$ converges to zero in $\mathscr{O}(U, X)$. Then the sequence (G_k) induced by (g_k) in $K^{n-1}(z - a, L^2_a(D) \hat{\otimes} X)$ satisfies $f = \lim_{k \to \infty}(z - a)G_k$. $\square$

We shall say that a commuting tuple $a \in L(X)^n$ of Banach-space operators admits a *decomposable resolution* if there is a finite exact sequence

$$0 \to X \to X_0 \to X_1 \to \cdots \to X_r \to 0$$

of continuous linear operators between Banach spaces starting with the given space X, and a family of decomposable n-tuples $a_i \in L(X_i)^n$ $(0 \le i \le r)$ such that the above sequence intertwines the n-tuples, $a, a_0, \ldots, a_r$ component-wise.

The following proof is reproduced from Eschmeier (1992b).

Theorem 6.2.4   *Let $a \in L(X)^n$ be a commuting n-tuple of continuous linear operators on a complex Banach space X. Then a satisfies Bishop's property (β) if and only if a admits a decomposable resolution.*

Proof   A repeated application of part (a) of Lemma 4.3.9, starting on the right-hand side, shows that each commuting tuple of Banach-space operators with a decomposable resolution possesses Bishop's property (β).

Conversely, let us suppose that $a \in L(X)^n$ satisfies Bishop's property (β). We fix an open polydisc D containing $\sigma(a)$ as well as a multi-index $j \in \mathbb{N}^n$ with all components different from zero, and we form the exact $\bar{\partial}$-sequence

$$0 \to L^2_a(D) \xrightarrow{\;i\;} W^0 \xrightarrow{\;\bar{\partial}\;} W^1 \xrightarrow{\;\bar{\partial}\;} \cdots \xrightarrow{\;\bar{\partial}\;} W^n \to 0$$

as explained above. The spaces $W^q \hat{\otimes} X$ $(0 \le q \le n)$, where $\hat{\otimes}$ can be chosen as the ε- or π-tensor product, or if X is a Hilbert space, as the Hilbertian tensor product, become topological $\mathscr{E}(\mathbb{C}^n)$-modules if the multiplication is defined componentwise.

We consider the double complex

$$\mathscr{K} = (K^{p,q})_{p,q \ge 0} = (\Lambda^p(\sigma, W^q \hat{\otimes} X))_{p,q \ge 0},$$

where for each $q = 0, \ldots, n$, the qth row is given by $K^{\bullet}(z - a, W^q \hat{\otimes} X)$, while for $p = 0, \ldots, n$, the pth column consists of

$$(-1)^p \bigoplus_{\binom{n}{p}} W^{\bullet} \otimes 1_X.$$

With respect to componentwise multiplication the spaces occurring in the total complex $(X^\bullet, \delta^\bullet) = \text{Tot}(\mathscr{K})$ are topological $\mathscr{E}(\mathbb{C}^n)$-modules. The differentials $\delta^p \colon X^p \to X^{p+1}$ respect at least the $\mathscr{O}(\mathbb{C}^n)$-module structures. Since according to Lemma 6.2.2 all columns consist of homomorphisms and are exact in all degrees except in degree $q = 0$, there are topological isomorphisms (Lemma A2.7)

$$H^p(X^\bullet) \cong H^p\Big(K^\bullet\big(z - a, L_a^2(D) \,\hat{\otimes}\, X\big)\Big) \qquad (p \in \mathbb{Z})$$

with respect to the canonical quotient topologies of both sides. By Proposition 6.2.3, the cohomology groups on the right vanish for $p \neq n$, and for $p = n$ the analytic functional calculus of a induces the topological isomorphism

$$H^n\Big(K^\bullet\big(z - a, L_a^2(D) \,\hat{\otimes}\, X\big)\Big) \to X, \qquad [f] \mapsto \Phi(f).$$

Using these identifications, one obtains an exact sequence

$$0 \to X \xrightarrow{\ i\ } X^n/\text{Im } \delta^{n-1} \xrightarrow{\ \delta^n\ } X^{n+1} \xrightarrow{\ \delta^{n+1}\ } \cdots \xrightarrow{\ \delta^{2n-1}\ } X^{2n} \to 0 \quad (6.2.2)$$

of continuous linear operators between Banach spaces.

If the elements of X^n are written in the form $x_0 \oplus \cdots \oplus x_n$ with x_ν in $\Lambda^\nu(\sigma, W^{n-\nu} \hat{\otimes} X)$, then the map i acts as

$$i \colon X \to X^n/\text{Im } \delta^{n-1}, \qquad x \mapsto [0 \oplus \cdots \oplus 0 \oplus (1 \otimes x)].$$

The complex (6.2.2) intertwines the tuple a on X and the tuples M_z induced by the multiplication with the coordinate functions on all other spaces. Since for $i = 0, \ldots, 2n$ the $\mathscr{E}(\mathbb{C}^n)$-module structure of X^i yields a continuous C^∞-functional calculus for the tuple $M_z \in L(X^i)^n$, these tuples are decomposable by Theorem 6.1.13.

Repeated applications of part (a) of Lemma 4.3.9 to the short exact sequences obtained by decomposing the complex (6.2.2), starting on the right-hand side, show that M_z restricted to $\text{Ker } \delta^{n+1}$ satisfies Bishop's property (β). Then part (b) of the same lemma applied to

$$0 \to X \xrightarrow{\ i\ } X^n/\text{Im } \delta^{n-1} \xrightarrow{\ \delta^n\ } \text{Ker } \delta_{n+1} \to 0$$

implies property (β) for M_z on $X^n/\text{Im } \delta^{n-1}$.

To prove the decomposability of M_z on $X^n/\text{Im } \delta^{n-1}$, we use the exact sequence

$$0 \to X^0 \xrightarrow{\ \delta^0\ } X^1 \xrightarrow{\ \delta^1\ } \cdots \xrightarrow{\ \delta^{n-1}\ } X^n \xrightarrow{\ q\ } X^n/\text{Im } \delta^{n-1} \to 0. \quad (6.2.3)$$

This sequence intertwines the tuples induced by M_z on each of the occurring spaces. Since all these tuples satisfy Bishop's property (β), the induced sequence of canonical sheaf models remains exact (see the remark following Corollary 3.1.16). Because the sheaf models of M_z on X^i $(0 \leq i \leq 2n)$ are soft, to obtain the softness of the sheaf model of M_z on $X^n/\text{Im } \delta^{n-1}$ it suffices to decompose the complex (6.2.3) into short exact sequences, and then to use the fact that the quotient sheaf of a soft sheaf modulo a soft sheaf (on a paracompact space) remains soft (see Proposition II.9.7 in Bredon 1967). $\qquad\qquad\square$

If X is a Hilbert space, then choosing $\hat{\otimes}$ as the Hilbertian tensor product, the constructions in the preceding proof yield, for each commuting tuple on X satisfying Bishop's property (β), a finite resolution by decomposable Hilbert-space tuples.

The construction of the exact sequences (6.2.2) and (6.2.3) is possible for each commuting tuple $a \in L(X)^n$, independently of the fact whether a satisfies property (β) or not. In any case, we define $\hat{X} = X^n/\text{Im } \delta^{n-1}$, and denote by $\hat{a}$ the tuple induced by M_z on $\hat{X}$. If a satisfies Bishop's property (β), then we call $\hat{a}$, or $(X, a) \xrightarrow{i} (\hat{X}, \hat{a})$, the *canonical decomposable extension* of a. In the one-dimensional case, the above characterization of Bishop's property (β) is due to Albrecht and Eschmeier (1987). The multidimensional generalization was obtained by Putinar (1990a).

The invariant subspace constructions to be derived later depend on a special property of the dual of the canonical decomposable extension.

Let $a \in L(X)^n$ be a fixed commuting tuple satisfying Bishop's property (β). We construct the decomposable resolution as in the proof of Theorem 6.2.4. Since we shall need duality theory for tensor products, we use the ε-tensor product $\hat{\otimes} = \hat{\otimes}_\varepsilon$ as the underlying tensor product. To simplify the notations we define

$$Y = X', \quad b = a', \quad Z = (\hat{X})', \quad T = (\hat{a})'.$$

The adjoint of the map

$$X \to X^n, \qquad x \mapsto 0 \oplus \cdots \oplus 0 \oplus (1 \otimes x),$$

where X^n is as before the nth space in the total complex of the double complex $\mathscr{K}$, acts as

$$Q : (X^n)' = \bigoplus_{\nu=0}^{n} (K^{\nu, n-\nu})' \to Y, \qquad u_0 \oplus \cdots \oplus u_n \mapsto \mathbb{1}(u_n).$$

Modulo the isometric identification $(W^0 \hat{\otimes}_\varepsilon X)' \cong (W^0)' \hat{\otimes}_\pi Y$, the map $\mathbb{1}$ here is the unique continuous linear operator $(W^0)' \hat{\otimes}_\pi Y \to Y$ with

$$\mathbb{1}(u \otimes y) = u(1)y \qquad (u \in (W^0)', y \in Y).$$

The restriction q of Q onto $Z = (\hat{X})' = (\text{Im } \delta^{n-1})^\perp \subset (X^n)'$ is the adjoint of the embedding $i: X \to \hat{X}$, and it intertwines the tuple b on Y and the decomposable tuple T on Z (see Theorem 6.1.11 for the decomposability of T).

We shall call T, or $(Y, b) \xleftarrow{q} (Z, T)$, the *canonical decomposable lifting* of b. For our invariant subspace constructions we shall need the property that b has a decomposable lifting which allows the lifting of sequences of approximate eigenvectors. Our next aim is to show that the canonical decomposable lifting has this property provided the multi-index $j \in \mathbb{N}^n$ defining the space $W^0 = W^j(D)$ is chosen to be sufficiently large. Here D is as before an open polydisc containing the spectrum of the tuple a. More precisely, it suffices to choose j in such a way that $W^j(D)$ becomes a continuously embedded subspace of the Fréchet space $C(D)$ consisting of all continuous complex-valued functions on D. Here $C(D)$ is equipped with the topology of uniform convergence on all compact subsets of D. Because of the closed graph theorem it suffices to find a multi-index j such that $W^j(D)$ is contained in $C(D)$. By the analogue of the classical Sobolev embedding lemma, proved as Theorem A4.1 in Appendix 4, to ensure this it suffices to choose j in such a way that $j_\nu \geq 2n$ for $\nu = 1, \ldots, n$. From now one we shall suppose that this extra condition is satisfied. Under this hypothesis, we are able to prove the following lifting result for sequences of approximate eigenvectors.

Proposition 6.2.5 *Let* $T \in L(Z)^n$ *be the canonical decomposable lifting of the tuple* $b \in L(Y)^n$ *constructed as described in the preceding paragraph. Let* $\lambda \in D$ *be arbitrary. Then for each sequence* $(y_k)_{k \geq 1}$ *in* Y *with* $\lim_{k \to \infty}(\lambda - b)y_k = 0$, *there is a sequence* $(z_k)_{k \geq 1}$ *in* Z *with* $y_k = qz_k$ *for all k and* $\lim_{k \to \infty}(\lambda - T)z_k = 0$.

Proof   In the statement of the proposition $(\lambda - b)y_k$ (and similarly $(\lambda - T)z_k$) denotes the n-form resulting as the image of y_k under the map

$$\lambda - b: K_n(\lambda - b, Y) \to K_{n-1}(\lambda - b, Y).$$

For each $k \geq 1$, we define

$$u_k = 0 \oplus \cdots \oplus 0 \oplus (\delta_\lambda \otimes y_k) \in \bigoplus_{p=0}^n K'_{p, n-p} = (X^n)',$$

where $\delta_\lambda: W^j(D) \to \mathbb{C}$ denotes the continuous point evaluation at λ. Since

$$(\delta^{n-1})'(u_k) = 0 \oplus \cdots \oplus 0 \oplus ((z-b)\delta_\lambda \otimes y_k) = 0 \oplus \cdots \oplus 0 \oplus (\delta_\lambda \otimes (\lambda - b)y_k)$$

tends to zero in $\bigoplus_{\nu=0}^{n-1} K'_{\nu, n-1-\nu} = (X^{n-1})'$ as k tends to infinity, there is a sequence (v_k) in $Z = \mathrm{Ker}(\delta^{n-1})'$ such that $\lim_{k \to \infty}(u_k - v_k) = 0$.
    In view of

$$y_k - qv_k = Q(u_k - v_k) \xrightarrow{k} 0,$$

we can choose a sequence (w_k) in Z converging to zero such that $y_k = q(v_k + w_k)$ holds for all k. But then $(z_k) = (v_k + w_k)$ defines a sequence in Z with $y_k = qz_k$ for all k, and

$$(\lambda - T)z_k = (\lambda - M_z)((v_k - u_k) + w_k) \xrightarrow{k} 0. \qquad \square$$

To see that Proposition 6.2.5 describes a rather peculiar property of the canonical decomposable lifting, the reader should observe that, even in the case $n = 1$, for the most natural examples of decomposable liftings such a lifting of approximate eigensequences is in general not possible. For instance, the bilateral unitary shift operator on $l^2(\mathbb{Z})$ is a decomposable lifting of the backward unilateral shift operator on $l^2(\mathbb{N})$ which does not allow the lifting of approximate eigensequences.

6.3 ESSENTIALLY DECOMPOSABLE SYSTEMS

In an attempt to generalize the notion of essentially normal operators, Albrecht and Mehta (1984) introduced the concept of essentially decomposable operators on Banach spaces. Among other results, they proved that each essentially decomposable semi-Fredholm operator is automatically Fredholm. In the present section we shall indicate that similar constructions are possible in the multidimensional case. We shall only develop the subject as far as we need it for applications.

If E is a Fréchet space, then the linear space E^∞, consisting of all bounded sequences in E, becomes a Fréchet space relative to the semi-norms

$$p_\infty(x) = \sup_k p(x_k) \qquad (x = (x_k) \in l^\infty(E)),$$

where p runs through all continuous semi-norms on E. Using a diagonal process, one can show that the space E^{pc}, consisting of all sequences (x_k) in E such that each subsequence of (x_k) has a convergent subsequence, forms a closed subspace of E^∞. Hence the quotient $E^q = E^\infty/E^{pc}$ becomes a Fréchet space with respect to the quotient topology.

Suppose that $c \in L(E)$ is a continuous linear operator on E. Then the maps $c^\infty: E^\infty \to E^\infty$, $(x_k) \mapsto (cx_k)$, $c^{pc} = c^\infty | E^{pc}$, and $c^q = c^\infty/E^{pc}$ are continuous linear operators. For each tuple $c \in L(E)^n$ (of not necessarily commuting operators), we denote by c^∞, c^{pc}, and c^q the induced n-tuples. We shall say that the system $c \in L(E)^n$ is *essentially commuting* if c^q consists of commuting operators.

Definition 6.3.1 An essentially commuting tuple $a \in L(X)^n$ of continuous linear operators on a complex Banach space is *essentially decomposable* if the induced system $a^q \in L(X^q)^n$ is decomposable.

Of course, each essentially commuting system $T \in L(H)^n$ of *essentially normal operators* on a Hilbert space is essentially decomposable. For, if A denotes the unital C^*-subalgebra of the Calkin algebra $\mathscr{C}(H) = L(H)/K(H)$ generated by $[T_1], \ldots, [T_n]$, then, by composing the Gelfand isomorphism $C(\sigma_A([T_1], \ldots, [T_n])) \to A$ with the contractive algebra homomorphism

$$\mathscr{C}(H) \to L(H^q), \qquad [T] \mapsto T^q,$$

one sees that the tuple T^q is, for instance, generalized scalar.

We next turn to the question of how the sheaf models of a commuting tuple $a \in L(X)^n$ and the induced tuple $a^q \in L(X^q)^n$ are related, if in fact they exist.

Let $\mathscr{F}$ be an analytic Fréchet sheaf on $\mathbb{C}^n$. Then the presheaves defined by

$$F^\infty(U) = \mathscr{F}(U)^\infty, \quad F^{pc}(U) = \mathscr{F}(U)^{pc} \qquad (U \subset \mathbb{C}^n \text{ open}),$$

together with the componentwise restrictions, are really sheaves, i.e they satisfy the sheaf axioms. Componentwise multiplication turns these spaces into Fréchet $\mathcal{O}(U)$-modules. The resulting Fréchet analytic sheaves will be denoted by $\mathcal{F}^{\infty}$ and $\mathcal{F}^{pc}$.

The quotient sheaf $\mathcal{F}^q = \mathcal{F}^{\infty}/\mathcal{F}^{pc}$ is an analytic sheaf on $\mathbb{C}^n$. If $\mathcal{F}^{pc}$ is acyclic on Stein open sets, then the exact sequences

$$0 \to \mathcal{F}^{pc}(U) \to \mathcal{F}^{\infty}(U) \to \mathcal{F}^q(U) \to 0 \qquad (U \subset \mathbb{C}^n \text{ Stein})$$

can be used to turn $\mathcal{F}^q$ into a Fréchet analytic sheaf on $\mathbb{C}^n$.

Theorem 6.3.2 *Each commuting decomposable tuple is essentially decomposable.*

Proof We denote by $\mathcal{F}$ the canonical sheaf model of a, and show that $\mathcal{F}^q$ is a Fréchet soft sheaf model for a^q.

Let $U \subset \mathbb{C}^n$ be open, and let V, W be bounded open sets in $\mathbb{C}^n$ with $\overline{V} \subset W \subset \overline{W} \subset U$. Then we can define a topological epimorphism between Fréchet spaces by

$$\mathcal{F}_{\overline{W}}(U) \oplus \mathcal{F}_{U \setminus V}(U) \to \mathcal{F}(U), \qquad (s, t) \mapsto s + t.$$

Therefore, for each section $t \in \mathcal{F}^{pc}(U)$, there is a section $s \in \mathcal{F}_{\overline{W}}^{pc}(U)$ with $s|_V = t|_V$ (see §22.2(7) in Köthe 1969). Thus we have shown the softness of $\mathcal{F}^{pc}$.

Since a topological epimorphism between Fréchet spaces does not in general allow the lifting of bounded sequences, we need a different argument to show that $\mathcal{F}^{\infty}$ is soft. Using the representation (see Proposition 6.1.12)

$$\mathcal{F}(U) = \varprojlim_{G \Subset U} X/X_a(\mathbb{C}^n \setminus G),$$

we can write a given bounded sequence (t_k) in $\mathcal{F}(U)$ in the form $t_k = ([x_G^k])_{G \Subset U}$, where $(x_G^k)_k$ is a bounded sequence in X for each relatively compact open subset G of U. Because

$$X = X_a(\overline{W}) + X_a(\mathbb{C}^n \setminus V),$$

we may suppose that $x_V^k \in X_a(\overline{W})$ for all k. But then

$$s_k = (x_V^k + X_a(\mathbb{C}^n \setminus G))_{G \Subset U}$$

defines a bounded sequence (s_k) in $\mathcal{F}_{\overline{W}}(U)$ with $s_k|_V = t_k|_V$ for all k. Thus the softness of $\mathcal{F}^{\infty}$ is proved.

Obviously, $\mathcal{F}^{pc}$ and $\mathcal{F}^{\infty}$ are Fréchet soft sheaf models for a^{pc} and a^{∞}. But then the quotient sheaf $\mathcal{F}^q = \mathcal{F}^{\infty}/\mathcal{F}^{pc}$ is soft (see Proposition II.9.7 in Bredon 1967). Moreover, as explained above, $\mathcal{F}^q$ becomes an analytic Fréchet sheaf, and $X^q \cong \mathcal{F}^q(\mathbb{C}^n)$ as Fréchet $\mathcal{O}(\mathbb{C}^n)$-modules. $\square$

Using decomposable resolutions one can show that property (β) implies the essential version of property (β).

Corollary 6.3.3 *Let $a \in L(X)^n$ be a commuting tuple of continuous linear Banach-space operators. Suppose that a satisfies Bishop's property (β) and that $\mathscr{F}$ is the sheaf model of a. Then a^∞, a^{pc}, and a^q satisfy Bishop's property (β) and possess the sheaf models $\mathscr{F}^\infty$, $\mathscr{F}^{pc}$, $\mathscr{F}^q$, respectively.*

Proof Since a satisfies Bishop's property (β), Theorem 6.2.4 implies the existence of a decomposable resolution

$$0 \to (X, a) \to (X_0, a_0) \to (X_1, a_1) \to \cdots \to (X_r, a_r) \to 0.$$

The induced sequence

$$0 \to (X^\infty, a^\infty) \to (X_0^\infty, a_0^\infty) \to (X_1^\infty, a_1^\infty) \to \cdots \to (X_r^\infty, a_r^\infty) \to 0$$

remains exact, and yields a decomposable resolution for a^∞ (see the proof of Theorem 6.3.2). Hence a^∞ satisfies Bishop's property (β), and again by the proof of Theorem 6.3.2, the induced exact sequence of sheaf models is isomorphic to

$$0 \to \mathscr{F}_{a^\infty} \to \mathscr{F}_{a_0}^\infty \to \mathscr{F}_{a_1}^\infty \to \cdots \to \mathscr{F}_{a_r}^\infty \to 0,$$

where $\mathscr{F}_{a^\infty}$ and $\mathscr{F}_{a_i}$ denote the canonical sheaf models of a^∞ and a_i, respectively.

Since on the other hand, for each open set U in $\mathbb{C}^n$, the sequence $0 \to \mathscr{F}(U) \to \mathscr{F}_{a_0}(U) \to \mathscr{F}_{a_1}(U)$ is exact, it follows that

$$0 \to \mathscr{F}^\infty \to \mathscr{F}_{a_0}^\infty \to \mathscr{F}_{a_1}^\infty$$

is also exact. We conclude that $\mathscr{F}^\infty \cong \mathrm{Ker}(\mathscr{F}_{a_0}^\infty \to \mathscr{F}_{a_1}^\infty) \cong \mathscr{F}_{a^\infty}$ as analytic Fréchet sheaves on $\mathbb{C}^n$.

In exactly the same way, it follows that a^{pc} satisfies property (β) and that $\mathscr{F}_{a^{pc}} \cong \mathscr{F}^{pc}$. Since the functor $X \to X^q$ preserves the exactness of complexes of Banach spaces (Lemma 2.6.5), the same reasoning shows that a^q satisfies property (β). The short exact sequence

$$0 \to (X^{pc}, a^{pc}) \to (X^\infty, a^\infty) \to (X^q, a^q) \to 0$$

of commuting tuples with Bishop's property (β) induces a short exact sequence

$$0 \to \mathscr{F}^{pc} \to \mathscr{F}^\infty \to \mathscr{F}_{a^q} \to 0$$

between the sheaf models. Hence $\mathscr{F}_{a^q} \cong \mathscr{F}^\infty / \mathscr{F}^{pc} = \mathscr{F}^q$ as analytic Fréchet sheaves. $\square$

Let $a \in L(X)^n$ be a commuting system with the single-valued extension property, and let $\lambda \in \mathbb{C}^n$ be a point such that $H^p(\lambda - a, X) = 0$ for some p ($0 \le p \le n$). Then $H^i(\lambda - a, X) = 0$ for $i = 0, \ldots, p$. This is a result of Frunză (1975a). Its proof is a direct consequence of Lemma 2.1.5 and the subsequent remark. For tuples with Bishop's property (β) a similar result holds for the essential Koszul complex.

Corollary 6.3.4 *Let $a \in L(X)^n$ be a commuting tuple satisfying Bishop's property (β) (or such that a^q satisfies property (β)), and let $\lambda \in \mathbb{C}^n$. Suppose that $\dim H^p(\lambda - a, X) < \infty$ and that $H^{p+1}(\lambda - a, X)$ is separated for some fixed $p \in \{0, \ldots, n\}$. Then $\dim H^i(\lambda - a, X) < \infty$ for $i = 0, \ldots, p$.*

Proof By Corollary 6.3.3, we know that a^q satisfies property (β). The assumptions made on the cohomology groups of degree p and $p + 1$ ensure that $H^p(\lambda - a^q, X^q) = 0$ (Lemma 2.6.5). Hence by the above remark we conclude that $H^i(\lambda - a^q, X^q) = 0$ for $i = 0, \ldots, p$. Repeated applications of Lemma 2.6.5 yields that $\dim H^i(\lambda - a, X) < \infty$ for $i = 0, \ldots, p$. □

For essentially decomposable systems one obtains a stronger result.

Corollary 6.3.5 *Let $a \in L(X)^n$ be an essentially decomposable commuting n-tuple, and let $\lambda \in \mathbb{C}^n$. If $\dim H^p(\lambda - a, X) < \infty$ and $H^{p+1}(\lambda - a, X)$ is separated for some $p \in \{0, \ldots, n\}$, then a is Fredholm.*

Proof As shown in the last proof, we have $H^i(\lambda - a^q, X^q) = 0$ for $i = 0, \ldots, p$. The dual of the cochain complex $K^\bullet(\lambda - a^q, X^q)$ is isomorphic to the Koszul complex $K_\bullet(\lambda - (a^q)', (X^q)')$ (see Section 2.6). Since $H^{p+1}(\lambda - a, X)$ is separated, the same is true for $H^{p+1}(\lambda - a^q, (X^q))$. This follows easily from Lemma 2.6.5. By Serre's duality lemma (see the remark following Lemma 6.1.7), we obtain that $H_i(\lambda - (a^q)', (X^q)') = 0$ for $i = 0, \ldots, p$. Since $(a^q)'$ is decomposable, the remarks preceding Corollary 6.3.4 imply that the remaining homology groups $H_i(\lambda - (a^q)', (X^q)')$ $(i = p+1, \ldots, n)$ vanish. But then also $K^\bullet(\lambda - a^q, X^q)$ is exact, and a is Fredholm. □

It is well known that, for a normal operator on a Hilbert space, the non-essential spectrum consists precisely of the isolated points in the spectrum that are eigenvalues of finite multiplicity. Our final result in this section shows that there is a much more general principle behind this phenomenon.

Theorem 6.3.6 *Let $a \in L(X)^n$ be a decomposable commuting tuple, and let $\lambda \in \sigma(a) \setminus \sigma_e(a)$. Then $\bigcap_{i=1}^n \operatorname{Ker}(\lambda_i - a_i) \neq \{0\}$, the point λ is isolated in $\sigma(a)$, and the spectral subspace $X_a(\lambda)$ has positive finite dimension.*

Proof Let $a \in L(X)^n$ be decomposable. We first prove, for each closed set F in $\mathbb{C}^n$, the spectral inclusion

$$\sigma(a) \cap \operatorname{Int}(F) \subset \sigma(a, X_a(F)).$$

To do this, let us fix open polydiscs U and V in $\mathbb{C}^n$ with $\overline{V} \subset U \subset \overline{U} \subset \rho(a, X_a(F)) \cap \operatorname{Int}(F)$. Using the local analytic functional calculus (see Lemma 6.1.3), we obtain that $\sigma(a, X_a(\overline{U})) \subset \overline{U}$. As in the proof of Theorem 6.1.13, we conclude that $X_a(\overline{U}) = 0$, and hence that $X = X_a(\mathbb{C}^n \setminus V)$. Therefore,

$H^n(\lambda - a, X) = 0$ for each point λ in V. Since a has the single-valued extension property, it follows that $V \cap \sigma(a) = \varnothing$.

Let $\lambda \in \sigma(a) \setminus \sigma_e(a)$. We choose an open polydisc D with centre λ and $\overline{D} \cap \sigma_e(a) = \varnothing$ and define $Y = X_a(\overline{D})$. The inclusion map $i: Y \to X$ induces a topological embedding $i^q: Y^q \to X^q$ which intertwines $(a \mid Y)^q$ and a^q. From the spectral inclusion

$$\sigma(a^q, \text{Im } i^q) = \sigma((a \mid Y)^q) \subset \sigma(a, Y) \subset \overline{D}$$

we deduce that $Y^q = \{0\}$, or equivalently, that Y is finite dimensional. Since by the first part $\sigma(a) \cap D \subset \sigma(a, Y)$, we conclude that λ is an isolated point in $\sigma(a)$.

It is well known that on a finite-dimensional space each commuting tuple is generalized scalar. Therefore a duality argument, similar to that used in the proof of Corollary 6.3.5, shows that $H^0(\lambda - a, Y) \neq 0$. $\square$

6.4 CONDITION $(\beta)_{\mathscr{E}}$ AND SMOOTH RESOLUTIONS

Bishop's condition (β) characterizes precisely those commuting systems of Banach-space operators that possess a finite resolution to the right by decomposable systems. By definition, property (β) means that

$$\mathscr{O}(U) \perp_{\mathscr{O}(\mathbb{C}^n)} X$$

for each Stein open set U in $\mathbb{C}^n$. In the present section we shall study the question of what happens if, in the above transversality relation, the space $\mathscr{O}(U)$ is replaced by the Fréchet $\mathscr{O}(\mathbb{C}^n)$-module $\mathscr{E}(U)$. The general idea will be to show that if one replaces analytic functions in the definition of property (β) by smooth functions, then everything remains true, provided one also replaces decomposable resolutions by generalized scalar resolutions.

Definition 6.4.1 Let $a \in L(X)^n$ be a commuting tuple of Banach-space operators. The system a is said to satisfy *property* $(\beta)_{\mathscr{E}}$ if X, equipped with the Fréchet $\mathscr{O}(\mathbb{C}^n)$-module structure determined by a, satisfies

$$\mathscr{E}(\mathbb{C}^n) \perp_{\mathscr{O}(\mathbb{C}^n)} X. \tag{6.4.1}$$

Using the Koszul resolution for the Fréchet $\mathscr{O}(\mathbb{C}^n)$-module X, one finds that a commuting tuple $a \in L(X)^n$ possesses property $(\beta)_{\mathscr{E}}$ if and only if the Koszul complex

$$K_{\bullet}(z - a, \mathscr{E}(\mathbb{C}^n) \hat{\otimes} X)$$

is exact in all positive degrees and has separated homology in degree 0. Using locally finite C^{∞}-partitions of unity, one can easily show that, if this condition is satisfied globally on all of $\mathbb{C}^n$, then it is automatically satisfied on all open subsets U of $\mathbb{C}^n$. In other words, the transversality condition (6.4.1) implies that

$$\mathscr{E}(U) \perp_{\mathscr{O}(\mathbb{C}^n)} X$$

holds for all open sets U in $\mathbb{C}^n$.

To prove transversality results of the above type, it is useful to choose a particular system of seminorms generating the topology of the Fréchet space $\mathscr{E}(U, X)$. In exactly the same way as the classical Sobolev embedding lemma is proved, one can establish the inclusion

$$\bigcap_{k \geq 0} W^k(U, X) \subset \mathscr{E}(U, X),$$

where the Banach spaces $W^k(U, X)$ are the vector-valued analogues of the spaces used in Section 6.2 (see Corollary A4.2 in Appendix 4). Hence, for each open set U in $\mathbb{C}^n$, the continuous linear map

$$\Phi: \mathscr{E}(U, X) \to \varprojlim_{\substack{V \Subset U \\ k \geq 0}} W^k(V, X) \tag{6.4.2}$$

given by the canonical restrictions $\mathscr{E}(U, X) \to W^k(V, X)$, $f \mapsto f | V$, is a topological isomorphism between Fréchet spaces. In particular, it follows that the Fréchet space topology of $\mathscr{E}(U, X)$ is generated by the seminorms

$$\|\bar{\partial}^j f\|_{2, V} \quad (j \in \mathbb{N}^n, V \Subset U).$$

If U is an open polydisc in $\mathbb{C}^n$, then the projective limit can be formed with respect to an exhaustion (U_k) of U consisting of open polydiscs. Hence we obtain a representation of $\mathscr{E}(U, X)$ as the limit of a reduced inverse system of Banach spaces (see Lemma A4.4 in Appendix 4).

We start with an analysis of the single-variable case. For $K \subset \mathbb{C}$ arbitrary and $\varepsilon > 0$, we use the notation:

$$K_\varepsilon = \{z \in \mathbb{C}; \text{dist}(z, K) < \varepsilon\}.$$

Proposition 6.4.1 *Let $T \in L(X)$, and let $U \supset \sigma(T)$ be a bounded open set. Then the following conditions are equivalent:*

(i) *T satisfies property $(\beta)_{\mathscr{E}}$;*
(ii) *for every open disc D in $\mathbb{C}$ and every $\varepsilon > 0$, there are $C > 0$ and $k \in \mathbb{N}$ with*

$$\|f\|_{2, D} \leq C \sum_{j=0}^{k} \|(z - T)\bar{\partial}^j f\|_{2, D_\varepsilon} \quad (f \in \mathscr{E}(\mathbb{C}, X));$$

(iii) *there are $C > 0$ and $k \in \mathbb{N}$ such that*

$$\|f\|_{2, U} \leq C \|(z - T)f\|_{W^k(U, X)} \quad (f \in \mathscr{D}(U, X)).$$

Proof If (i) holds, then for D and ε as in (ii), the map

$$z - T: \mathscr{E}(D_\varepsilon, X) \to \mathscr{E}(D_\varepsilon, X)$$

is a topological monomorphism, and thus condition (ii) holds.

To see that (ii) implies (iii), it suffices to apply condition (ii) to an open polydisc containing U with $\varepsilon = 1$.

To prove that (iii) implies (i), let (f_n) be a sequence in $\mathscr{E}(\mathbb{C}, X)$ with

$\lim_{n \to \infty} (z - T)f_n = 0$. Because of $\sigma(T) \subset U$, it suffices to show that (f_n) tends to zero in $\mathscr{E}(U, X)$. Using cut-off functions and condition (iii), one can show that $(\|f_n\|_{2,V})$ tends to zero for each relatively compact open set V in U. Since for each $j \in \mathbb{N}$, the same argument applies to $(\bar{\partial}^j f_n)$ instead of (f_n), the proof is complete. $\square$

The following elementary observation is the key result for proving property $(\beta)_{\mathscr{E}}$ in all concrete cases.

Lemma 6.4.2 *Let U be a bounded open set in $\mathbb{C}$. There is a constant $C_U > 0$ such that, for each operator $T \in L(X)$ and each function $f \in \mathscr{D}(U, X)$, the following estimate holds*:

$$\|f\|_{2,U} \le C_U(\|(\bar{z} - T)\bar{\partial}f\|_{2,U} + \|(\bar{z} - T)\bar{\partial}^2 f\|_{2,U}).$$

Proof Let $f \in \mathscr{D}(U, X)$ be arbitrary. Denote by $g \in \mathscr{D}(U, X)$ the function

$$\bar{\partial}(f - (\bar{z} - T)\bar{\partial}f) = -(\bar{z} - T)\bar{\partial}^2 f.$$

If $\theta \in \mathscr{D}(\mathbb{C})$ is equal to one on $U - U$, then $h = \theta/\pi z \in L^1(\mathbb{C})$ satisfies

$$f(z) - (\bar{z} - T)\bar{\partial}f(z) = \frac{1}{\pi} \int_U \frac{g(w)}{z - w} \, dw = g * h(z) \qquad (z \in U).$$

Hence the assertion follows from the standard estimate for convolutions. $\square$

By combining the last two results we obtain the following criterion.

Proposition 6.4.3 *Let $T, S \in L(X)$ be such that there is a constant $C > 0$ with*

$$\|(\bar{z} - S)x\| \le C\|(z - T)x\|$$

for all $x \in X$ and $z \in \mathbb{C}$. Then T satisfies property $(\beta)_{\mathscr{E}}$. $\square$

As an application we deduce that each normal operator N on a Hilbert space H has property $(\beta)_{\mathscr{E}}$. More generally, each *M-hyponormal operator T* in H satisfies property $(\beta)_{\mathscr{E}}$, since by definition $\|(\bar{z} - T^*)x\| \le M\|(z - T)x\|$ holds for all $z \in \mathbb{C}$ and $x \in H$.

In the single-variable case property $(\beta)_{\mathscr{E}}$ characterizes precisely the restrictions of generalized scalar operators or equivalently the $\mathscr{O}(\mathbb{C})$-submodules of Banach $\mathscr{E}(\mathbb{C})$-modules. The proof of this statement will be based on a division result for distributions, which is of independent interest. To formulate this result, let us say that a continuous linear map $\alpha: E \to F$ between locally convex spaces *allows the lifting of bounded sets* if each bounded set in F is the image of a bounded set in E.

Theorem 6.4.4 *Suppose that $T \in L(X)$ is generalized scalar. Then*

$$\mathscr{E}'(\mathbb{C}) \hat{\otimes} X \xrightarrow{z - T} \mathscr{E}'(\mathbb{C}) \hat{\otimes} X$$

is onto and allows the lifting of bounded sets.

Proof   The proof of Theorem 6.4.4 will be obtained in several steps, including three intermediate lemmas. Let

$$\Phi: \mathscr{E}(\mathbb{C}) \to L(X)$$

be a continuous algebra homomorphism extending the analytic functional calculus of T. Because of the continuity of Φ, there is an open disc Ω containing $\sigma(T)$, and there is an integer $n \geq 0$ such that Φ has an extension to a continuous linear operator.

$$\Phi: W^n(\Omega) \to L(X),$$

again denoted by Φ (see Lemma A.4.4 in Appendix 4). The unique continuous linear operator $\Psi: W^n(\Omega) \hat{\otimes}_\pi X \to X$, with $\Psi(f \otimes x) = \Phi(f)x$ for f in $W^n(\Omega)$ and x in X, satisfies

$$\Psi(zf) = T\Psi(f) \qquad \left(f \in W^n(\Omega) \hat{\otimes}_\pi X\right).$$

We fix an open disc ω containing $\sigma(T)$ with $\bar{\omega} \subset \Omega$. It suffices to show that the map

$$\mathscr{E}'(\omega) \hat{\otimes} X \xrightarrow{w-T} \mathscr{E}'(\omega) \hat{\otimes} X \qquad (6.4.3)$$

is onto and allows the lifting of bounded sets. Then the original assertion can be proved by using a partition of unity and the fact that

$$\mathscr{E}'(\rho(T)) \hat{\otimes} X \xrightarrow{R(w,T)} \mathscr{E}'(\rho(T)) \hat{\otimes} X$$

is a topological isomorphism with inverse given by $w - T$. Here $\rho(T)$ denotes the resolvent set of T, and $R(w,T)$ refers to the resolvent function of T in the usual sense (see the beginning of Section 2.5).

To prove that the map (6.4.3) has the claimed properties, we use the commutative diagram

$$
\begin{array}{ccc}
\mathscr{E}'(\omega) \hat{\otimes} W^n(\Omega) \hat{\otimes}_\pi X & \xrightarrow{(w-z) \otimes 1} & \mathscr{E}'(\omega) \hat{\otimes} W^n(\Omega) \hat{\otimes}_\pi X \\
{\scriptstyle 1 \otimes \Psi} \downarrow & & \downarrow {\scriptstyle 1 \otimes \Psi} \\
\mathscr{E}'(\omega) \hat{\otimes} X & \xrightarrow[w-T]{} & \mathscr{E}'(\omega) \hat{\otimes} X.
\end{array}
$$

Since the map Ψ has the right inverse $X \to W^n(\Omega) \hat{\otimes} X$, $x \mapsto 1 \otimes x$, it remains to check that $(w - z) \otimes 1$ is onto and allows the lifting of bounded sets.

The following result on tensor products of continuous linear mappings shows that it is sufficient to prove the surjectivity of the map $w - z$.

Lemma 6.4.5   *Let E be a complete nuclear (DF)-space, let Y be a reflexive Banach space, and let X be an arbitrary Banach space. Suppose that*

$$T: E \hat{\otimes} Y \to E \hat{\otimes} Y$$

is a surjective continuous linear operator. Then

$$T \otimes 1: E \hat{\otimes} Y \hat{\otimes}_\pi X \to E \hat{\otimes} Y \hat{\otimes}_\pi X$$

remains surjective and allows the lifting of bounded sets.

Proof As a complete nuclear (DF)-space the space E is reflexive, and hence is the strong dual of a nuclear Fréchet space (Lemma A1.9 in Appendix 1). Therefore $E \hat{\otimes} Y$ is the strong dual of a reflexive Fréchet space (see Theorem A1.12 in Appendix 1), and T is the adjoint of a topological monomorphism between the corresponding predual spaces. Hence, apart from the nuclearity conditions, we are precisely in the situation of Lemma 6.1.7 with $\alpha = 0$ and $\beta' = T$.

Exactly as in the proof of Lemma 6.1.7, it follows that T allows the lifting of bounded sets. Moreover, arguing as in the proof of Lemma 6.1.7, one sees that it suffices to show that each bounded set M in $E \hat{\otimes} Y \hat{\otimes}_\pi X$ is contained in the image of the closed unit ball under the canonical map (see Appendix 1 for the notation)

$$(E \hat{\otimes} Y)_A \hat{\otimes}_\pi X_B \to E \hat{\otimes} Y \hat{\otimes}_\pi X$$

for suitable discs (i.e. closed, bounded, and absolutely convex sets) A and B in $E \hat{\otimes} Y$ and X, respectively. By Corollary A1.11 in Appendix 1, there are discs C in E and D in $Y \hat{\otimes}_\pi X$ such that M is contained in the image of the closed unit ball under

$$E_C \hat{\otimes}_\pi (Y \hat{\otimes}_\pi X)_D \to E \hat{\otimes} Y \hat{\otimes}_\pi X.$$

We may of course suppose that D is the closed unit ball in $Y \hat{\otimes}_\pi X$. But the image of the closed unit ball in $E_C \hat{\otimes}_\pi Y$ under $E_C \hat{\otimes}_\pi Y \to E \hat{\otimes} Y$ is contained in the closed absolutely convex hull A of $C \otimes \{y \in Y; \|y\| \leq 1\}$. Hence M is contained in the image of the closed unit ball under the map

$$(E \hat{\otimes} Y)_A \hat{\otimes}_\pi X \to E \hat{\otimes} Y \hat{\otimes}_\pi X. \qquad \square$$

To conclude the proof of Theorem 6.4.4, we prove the surjectivity of the map $w - z: \mathscr{E}'(\omega) \hat{\otimes} W^n(\Omega) \to \mathscr{E}'(\omega) \hat{\otimes} W^n(\Omega)$ by induction on n.

For $n = 0$, the assertion follows from the fact that the normal operator $M_z: L^2(\Omega)' \to L^2(\Omega)'$ satisfies property $(\beta)_{\mathscr{E}}$. For a proof of the inductive step, we make use of the commutative diagram

$$
\begin{array}{ccc}
0 & & 0 \\
\uparrow & & \uparrow \\
0 \to K_3 \to \mathscr{E}'(\omega) \hat{\otimes} W^{n-1}(\Omega) \xrightarrow{w-z} \mathscr{E}'(\omega) \hat{\otimes} W^{n-1}(\Omega) \to 0 \\
\uparrow \bar{\partial}_z \qquad\qquad\qquad\qquad \uparrow \bar{\partial}_z \\
0 \to K_2 \to \mathscr{E}'(\omega) \overline{\otimes} W^n(\Omega) \xrightarrow{w-z} \mathscr{E}'(\omega) \hat{\otimes} W^n(\Omega) \to C_2 \to 0 \\
\uparrow \qquad\qquad\qquad\qquad \uparrow \\
\mathscr{E}'(\omega) \hat{\otimes} L_a^2(\Omega) \xrightarrow{w-z} \mathscr{E}'(\omega) \hat{\otimes} L_a^2(\Omega) \to C_2 \to 0 \\
\uparrow \qquad\qquad\qquad\qquad \uparrow \\
0 \qquad\qquad\qquad\qquad 0
\end{array}
$$

with exact rows and columns. Here K_2, K_3 and C_1, C_2 just denote the corresponding kernels and cokernels, and the exactness of the columns follows from Lemma 6.2.2.

The diagram induces an exact sequence of the form

$$K_2 \to K_3 \xrightarrow{\ d\ } C_1 \to C_2 \to 0.$$

Hence it suffices to prove the surjectivity of the connecting homomorphism denoted by d.

To do this, we first compute the cokernel C_1.

Lemma 6.4.6 *The sequence*

$$0 \to \mathscr{E}'(\omega) \,\hat{\otimes}\, L_a^2(\Omega) \xrightarrow{w-z} \mathscr{E}'(\omega) \,\hat{\otimes}\, L_a^2(\Omega) \xrightarrow{\ m\ } \mathscr{E}'(\omega) \to 0, \quad (6.4.4)$$

where m is the multiplication map, is exact.

Proof Suppose that E is a nuclear locally convex space, which is represented as the limit of a reduced countable inverse system

$$E = \varprojlim_k H_k$$

of Hilbert spaces. For each Hilbert space K, there are unique topological isomorphisms (see §41.6.(3) and §44.5.(5) in Köthe 1979)

$$E \,\hat{\otimes}\, K \to \varprojlim_k H_k \,\hat{\otimes}_\pi\, K \quad \text{and} \quad E \,\hat{\otimes}\, K \to \varprojlim_k H_k \,\hat{\otimes}_\varepsilon\, K$$

acting as $(x_k) \otimes y \mapsto (x_k \otimes y)_k$ on elementary tensors. Since the Hilbertian tensor product (denoted by $\tilde{\otimes}$) is faithful and since the composition

$$E \,\hat{\otimes}\, K \to \varprojlim_k H_k \,\hat{\otimes}_\pi\, K \to \varprojlim_k H_k \,\tilde{\otimes}\, K \to \varprojlim_k H_k \,\hat{\otimes}_\varepsilon\, K$$

is a topological isomorphism, the canonical map $E \,\hat{\otimes}\, K \to \varprojlim_k H_k \,\tilde{\otimes}\, K$ is also a topological isomorphism. Since the inverse system on the right is reduced and defines a reflexive space, we obtain the topological isomorphism (see Theorem IV.4.4 and Theorem IV.9.9 in Schaefer 1966)

$$E' \,\hat{\otimes}\, K' \cong \varinjlim_k H_k' \,\tilde{\otimes}\, K'.$$

Applying these abstract results to the spaces

$$E = \mathscr{E}(\omega) = \varprojlim_k W^k(U_k) \quad \text{and} \quad K = L_a^2(\Omega)',$$

where $(U_k)_{k \geq 1}$ is a suitable exhaustion of ω, we obtain the representation

$$\mathscr{E}'(\omega) \,\hat{\otimes}\, L_a^2(\Omega) = \varinjlim_k W^k(U_k)' \,\tilde{\otimes}\, L_a^2(\Omega).$$

Therefore (6.4.4) can be identified with the direct limit of the sequences

$$0 \to W^k(U_k)' \tilde{\otimes} L^2_a(\Omega) \xrightarrow{w-z} W^k(U_k)' \tilde{\otimes} L^2_a(\Omega) \xrightarrow{m} W^k(U_k)' \to 0.$$

These sequences, and hence also their direct limit, are exact by Proposition 6.2.3.   □

According to the last result, we may identify the cokernel C_1 with $\mathscr{E}'(\omega)$. To check the surjectivity of the map $d: K_3 \to C_1$, we make use of a standard representation of distributions with compact support.

Lemma 6.4.7 *Let $U \subset W$ be bounded open sets in $\mathbb{C}$, and let $m \in \mathbb{Z}^+$. Then each distribution $u \in \mathscr{E}'(U)$ is a finite sum of elements of the form $\Theta \bar{\partial}^j(f \mid U)$, where $\Theta \in \mathscr{D}(U)$, $f \in W^m(W)$, and $j \geq 0$ is an integer.*

Proof Let $u \in \mathscr{E}'(U)$. Using the representation (6.4.2) and the theorem of Hahn–Banach, one obtains an extension $\hat{u} \in W^k(V)'$ of u, where $k \in \mathbb{N}$ and $V \Subset U$ are suitable. Considering $W^k(V)$ as a subspace of $\oplus_{0 \leq j \leq k} L^2(V)$ under the isometric embedding $f \mapsto (\bar{\partial}^j f)_j$, another application of Hahn–Banach gives a family of functions $(f_j)_{0 \leq j \leq k}$ in $L^2(V)$ with

$$u(f) = \sum_{0 \leq j \leq k} \int_V (\bar{\partial}^j f) f_j \, \mathrm{d}z \qquad (f \in \mathscr{E}(U)).$$

Thus we have shown that u is a finite sum of elements of the form $\Theta \bar{\partial}^j(f \mid U)$, where $\Theta \in \mathscr{D}(U)$, $f \in L^2(W)$, and $j \in \mathbb{N}$. In view of Lemma 6.2.2 the assertion follows.   □

To prove the surjectivity of $d: K_3 \to C_1$, we fix an element in $\mathscr{E}'(\omega)$ of the form $u = \Theta \bar{\partial}^j(f \mid \omega)$, where $\Theta \in \mathscr{E}(\omega)$, $f \in W^{n-1}(\Omega)$, and $j \in \mathbb{N}$. Then the continuous linear operator

$$\alpha: \mathscr{E}(\omega) \to W^{n-1}(\Omega), \qquad \varphi \mapsto \bar{\partial}^j(\Theta \varphi) f$$

belongs to K_3. Using the continuous linear right inverse

$$r: W^{n-1}(\Omega) \to W^n(\Omega)$$

for $\bar{\partial}_z$, constructed in the proof of Lemma 6.2.2, we obtain a solution $\beta \in \mathscr{E}'(\omega) \hat{\otimes} W^n(\Omega)$ of $\bar{\partial}_z(\beta) = \alpha$ by setting

$$\beta(\varphi)(z) = \frac{1}{\pi} \int_\Omega \frac{\bar{\partial}^j(\Theta \varphi)(\xi) f(\xi)}{z - \xi} \, \mathrm{d}\xi$$

for $\varphi \in \mathscr{E}(\omega)$ and almost all $z \in \Omega$. The image of α under the connecting homomorphism is by definition the image of $(w - z)\beta$ in C_1. But for almost all $z \in \Omega$, we obtain

$$[((w - z)\beta)(\varphi)](z) = -\frac{1}{\pi} \int_\Omega \bar{\partial}^j(\Theta \varphi)(\xi) f(\xi) \, \mathrm{d}\xi$$

$$= (-1)^{j+1} \frac{1}{\pi} (\Theta \bar{\partial}^j(f \mid \omega))(\varphi).$$

Using tensor product notations the last identity means that

$$(w - z)\beta = (-1)^{j+1} \frac{1}{\pi} (\Theta \bar{\partial}^j (f \mid \omega)) \otimes 1 \in \mathscr{E}'(\omega) \hat{\otimes} W^n(\Omega),$$

and consequently the distribution α is a solution of

$$d\alpha = (-1)^{j+1} \frac{1}{\pi} \Theta \bar{\partial}^j (f \mid \omega).$$

By Lemma 6.4.7 the surjectivity of d is shown, and the proof of Theorem 6.4.4 is complete. □

As an application we obtain the announced characterizations of *subscalar operators*, i.e. of operators occurring as restrictions of generalized scalar operators.

To formulate these results, the following general observation concerning inverse limits will be useful. Suppose that $u: (E_\alpha) \to (F_\alpha)$ is a morphism of countable inverse systems of Fréchet spaces such that (E_α) is reduced, and that E and F are the inverse limits. Then, using the exactness of

$$0 \to \varprojlim \overline{u(E_\alpha)} \xrightarrow{i} F \to \varprojlim F_\alpha / \overline{u(E_\alpha)} \to 0$$

and the fact that $u: E \to \varprojlim \overline{u(E_\alpha)}$ has dense range (Theorem 3.2.4 and Remark 3.2.5), one obtains the topological identifications

$$\overline{\text{Im } u} = \varprojlim \overline{u(E_\alpha)} \quad \text{and} \quad F / \overline{\text{Im } u} = \varprojlim F_\alpha / \overline{u(E_\alpha)}. \quad (6.4.5)$$

If the inverse system (E_α) is not reduced, then the system $(\check{E}_\alpha) = (\overline{\pi_\alpha E_\alpha})$, where the maps $\pi_\alpha: E \to E_\alpha$ are the canonical projections, becomes a reduced inverse system of Fréchet spaces with respect to the induced structural maps. Since $E = \varprojlim \check{E}_\alpha$, we obtain the topological identifications

$$\overline{\text{Im } u} = \varprojlim \overline{u(\check{E}_\alpha)} \quad \text{and} \quad F / \overline{\text{Im } u} = \varprojlim F_\alpha / \overline{u(\check{E}_\alpha)}.$$

Corollary 6.4.8 *A continuous linear Banach space operator $T \in L(X)$ is subscalar if and only if it satisfies property* $(\beta)_{\mathscr{E}}$.

Proof By dualizing Theorem 6.4.4 we see that each generalized scalar operator, and hence also each subscalar operator, possesses property $(\beta)_{\mathscr{E}}$.

Conversely, if $T \in L(X)$ satisfies property $(\beta)_{\mathscr{E}}$, then the map

$$X \to \mathscr{E}(\mathbb{C}, X) / (z - T) \mathscr{E}(\mathbb{C}, X), \quad x \mapsto [x]$$

is a topological embedding. For, if $(x_k - (z - T) f_k)$ tends to zero in $\mathscr{E}(\mathbb{C}, X)$, then applying the $\bar{\partial}$-operator and using property $(\beta)_{\mathscr{E}}$ as well as the exactness of the $\bar{\partial}$-sequence, one sees that (f_k) may be replaced by a sequence in $\mathscr{O}(\mathbb{C}, X)$. But then $\lim_{k \to \infty} x_k = 0$.

Applying (6.4.5) we obtain a representation of the form

$$\mathscr{E}(\mathbb{C}, X)/(z - T)\mathscr{E}(\mathbb{C}, X) = \varprojlim_{\substack{k \geq 0 \\ U \Subset \mathbb{C}}} W^k(U, X)/\overline{(z - T)W^k(U, X)}.$$

Hence for suitable k and U, the map

$$J: X \to \hat{X} = W^k(U, X)/\overline{(z - T)W^k(U, X)}, \qquad x \mapsto [x]$$

defines a topological embedding intertwining $T \in L(X)$ and the generalized scalar operator induced by the multiplication by the argument on $\hat{X}$. □

The following result summarizes some useful characterizations of subscalar operators.

Corollary 6.4.9   *For* $T \in L(X)$, *the following conditions are equivalent*:

(a) *T is subscalar*;
(b) $J: \ X \to \mathscr{E}(\mathbb{C}, X)/\overline{(z - T)\mathscr{E}(\mathbb{C}, X)}, \ x \mapsto [x]$, *is a topological monomorphism*;
(c) *for each* $x' \in X'$, *there is a distribution* $u \in \mathscr{E}'(\mathbb{C}) \mathbin{\hat{\otimes}} X'$ *with* $(z - T')u = 0$ *and* $u(1) = x'$;
(d) *for each bounded open neighbourhood* U *of* $\sigma(T)$, *there are* $k \in \mathbb{N}$ *and* $C > 0$ *such that for all* $f \in \mathscr{D}(U, X)$

$$\|f\|_{2, U} \leq C \, \|(z - T)f\|_{W^k(U, X)};$$

(e) *T satisfies property* $(\beta)_{\mathscr{E}}$;
(f) $z - T: \mathscr{E}(\mathbb{C}, X) \to \mathscr{E}(\mathbb{C}, X)$ *is a topological monomorphism*. □

By dualizing the property described in Corollary 6.4.9(f), one obtains a characterization of quotients of generalized scalar operators.

Theorem 6.4.10   *A continuous linear operator* $T \in L(X)$ *on a complex Banach space* X *is a quotient of a generalized scalar operator if and only if the map*

$$z - T: \mathscr{E}'(\mathbb{C}) \mathbin{\hat{\otimes}} X \to \mathscr{E}'(\mathbb{C}) \mathbin{\hat{\otimes}} X$$

is surjective and allows the lifting of bounded sets.

Proof   Let $S \in L(Y)$ be a generalized scalar lifting for T, i.e suppose that S is generalized scalar, and that there is a surjective continuous linear map $q: Y \twoheadrightarrow X$ intertwining S and T. Then the diagram

$$
\begin{array}{ccc}
\mathscr{E}'(\mathbb{C}) \mathbin{\hat{\otimes}} X & \xrightarrow{\ z - T\ } & \mathscr{E}'(\mathbb{C}) \mathbin{\hat{\otimes}} X \\[4pt]
{\scriptstyle 1 \otimes q} \big\uparrow & & \big\uparrow {\scriptstyle 1 \otimes q} \\[4pt]
\mathscr{E}'(\mathbb{C}) \mathbin{\hat{\otimes}} Y & \xrightarrow[\ z - S\ ]{} & \mathscr{E}'(\mathbb{C}) \mathbin{\hat{\otimes}} Y
\end{array}
$$

is commutative, and the vertical maps as well as the lower horizontal map allow the lifting of bounded sets. Hence the same is true for the upper horizontal map.

Conversely, if $z - T: \mathscr{E}'(\mathbb{C}) \hat{\otimes} X \to \mathscr{E}'(\mathbb{C}) \hat{\otimes} X$ allows the lifting of bounded sets, then the rows and columns in the commutative diagram

$$
\begin{array}{ccccccccc}
& & 0 & & & & 0 & & \\
& & \uparrow & & & & \uparrow & & \\
0 \longrightarrow & L \longrightarrow & \mathscr{O}'(\mathbb{C}) \hat{\otimes} X & \xrightarrow{z-T} & \mathscr{O}'(\mathbb{C}) \hat{\otimes} X & \longrightarrow & 0 & & \\
& & \uparrow{\scriptstyle i^* \otimes 1} & & \uparrow{\scriptstyle i^* \otimes 1} & & & & \\
0 \longrightarrow & K \longrightarrow & \mathscr{E}'(\mathbb{C}) \hat{\otimes} X & \xrightarrow{z-T} & \mathscr{E}'(\mathbb{C}) \hat{\otimes} X & \longrightarrow & 0, & & \\
& & \uparrow{\scriptstyle \bar{\partial}^* \otimes 1} & & \uparrow{\scriptstyle \bar{\partial}^* \otimes 1} & & & & \\
& & \mathscr{E}'(\mathbb{C}) \hat{\otimes} X & \xrightarrow{z-T} & \mathscr{E}'(\mathbb{C}) \hat{\otimes} X & \longrightarrow & 0 & & \\
& & \uparrow & & \uparrow & & & & \\
& & 0 & & 0 & & & &
\end{array}
$$

where L and K are just the corresponding kernels, are exact, and diagram-chasing allows us to choose, for each bounded set C in L, a bounded set B in K with $i^* \otimes 1(B) = C$. To see this, the reader should use Lemma 6.1.7, Lemma 6.4.5, and the fact that $\bar{\partial}^* \otimes 1$ is a topological monomorphism (see §33.6.(a) and §44.4.(6) in Köthe 1979).

Using the identification $\mathscr{O}'(\mathbb{C}) \hat{\otimes} X = L_b(\mathscr{O}(\mathbb{C}), X)$, it is elementary to check that the mapping $L \to K$, $u \mapsto u(1)$, is a topological isomorphism. Hence there is a bounded, and thus equicontinuous, set B in $K \subset L_b(\mathscr{E}(\mathbb{C}), X)$ such that the image of B under $K \to X$, $u \mapsto u(1)$, contains the closed unit ball in X.

Representing $\mathscr{E}(\mathbb{C})$ as the limit of a reduced inverse system of the form (6.4.2), one sees that, for suitable $k \in \mathbb{N}$ and $V \Subset \mathbb{C}$, all elements in the equicontinuous set B can be extended to an element in

$$
\hat{X} = \mathrm{Ker}(L(W^k(V), X) \xrightarrow{z-T} L(W^k(V), X)).
$$

Therefore, for suitable k and V, the operator $\hat{T}$ induced by M_z on $\hat{X}$ is a generalized scalar lifting for T. $\qquad\square$

Using the nuclearity of $\mathscr{E}(\mathbb{C})$, one can replace the space $\hat{X}$ constructed in the last proof by a space of the form

$$
\hat{X} = \mathrm{Ker}(\mathscr{N}(W^k(V), X) \xrightarrow{z-T} \mathscr{N}(W^k(V), X)),
$$

where $\mathscr{N}(W^k(V), X)$ denotes the Banach space of all nuclear operators from $W^k(V)$ into X. Since there are natural maps

$$
\mathscr{N}(W^k(V), X) \cong W^k(V)' \hat{\otimes}_\pi X \to W^k(V)' \hat{\otimes}_\varepsilon X,
$$

one could also use $W^k(V)' \hat{\otimes}_\varepsilon X$ or $W^k(V)' \tilde{\otimes} X$, if X is a Hilbert space.

It is easy to see that, for a given operator $T \in L(X)$, the generalized scalar extension and lifting $\hat{T} \in L(\hat{X})$, constructed in the proof of Corollary 6.4.8 and Theorem 6.4.10, respectively, satisfy

$$\sigma(\hat{T}) \subseteq \sigma(T).$$

By applying Corollary 3.1.16 to the $\bar{\partial}$-resolution, one obtains that, for an arbitrary commuting tuple $a \in L(X)^n$ of Banach-space operators, property $(\beta)_{\mathscr{E}}$ implies property (β). Elementary examples show that the converse is not true in general. For instance, by the proof of Theorem 6.1.13, each quasi-nilpotent operator T is decomposable, hence satisfies Bishop's property (β). If T satisfies property $(\beta)_{\mathscr{E}}$, then T has a quasi-nilpotent generalized scalar extension $\hat{T}$. But each quasi-nilpotent generalized scalar operator is nilpotent, hence T would have to be nilpotent. Thus, the Volterra operator satisfies property (β), but not property $(\beta)_{\mathscr{E}}$.

For a generalized scalar operator $T \in L(X)$, the map

$$\mathscr{E}(\mathbb{C}, X) \xrightarrow{z-T} \mathscr{E}(\mathbb{C}, X)$$

is a topological monomorphism, and the map

$$\mathscr{E}'(\mathbb{C}) \hat{\otimes} X \xrightarrow{z-T} \mathscr{E}'(\mathbb{C}) \hat{\otimes} X$$

is onto and allows the lifting of bounded sets. The first property characterizes the restrictions, and the second property characterizes the quotients of generalized scalar operators. Before we consider the multidimensional situation, we indicate what happens if the complex plane is replaced by the real line.

We shall need two results from the theory of generalized scalar operators that will not be proved here. For the general theory we refer to the monograph of Colojoară and Foiaş (1968). First, we shall exploit the fact that, for a generalized scalar operator $T \in L(X)$ with real spectrum, there is always a C^{∞}-functional calculus over the real line, i.e a continuous algebra homomorphism $\Phi \colon \mathscr{E}(\mathbb{R}) \to L(X)$ with $\Phi(1) = I$ and $\Phi(z) = T$. Secondly, we shall use that each C^{∞}-functional calculus $\Phi \colon \mathscr{E}(\mathbb{C}) \to L(X)$ of a generalized scalar operator $T \in L(X)$ satisfies a spectral mapping theorem of the form

$$\sigma(\Phi(f)) = f(\sigma(T)) \qquad (f \in \mathscr{E}(\mathbb{C})).$$

Both results can be found in Colojoară and Foiaş (1968) (see Theorem V.4.5 and Theorem III.2.1).

Theorem 6.4.11 *Let $T \in L(X)$ be a generalized scalar operator. Then the multiplication operator $t - T \colon \mathscr{E}(\mathbb{R}, X) \to \mathscr{E}(\mathbb{R}, X)$ is a topological monomorphism if and only if $\sigma(T) \cap \mathbb{R}$ is an open subset of $\sigma(T)$.*

Proof Throughout the proof we shall suppose with no loss of generality that $|\mathrm{Re}\, z| < \pi/2$ for all $z \in \sigma(T)$. We choose a continuous functional calculus $\Phi \colon \mathscr{E}(\mathbb{C}) \to L(X)$ for T, and we define $\mathrm{Re}\, T = \Phi(\mathrm{Re}\, z)$, $\mathrm{Im}\, T = \Phi(\mathrm{Im}\, z)$. The proof is divided into several steps.

(1) For $a, b \in \mathbb{R}$, we denote by $\mathscr{D}([a, b], X)$ the space of all $f \in \mathscr{E}(\mathbb{R}, X)$ with $\mathrm{supp}(f) \subset [a, b]$. Let F be the analytic function defined on the open set $\{(z, w) \in \mathbb{C}^2; |\mathrm{Re}\, z| < (3/2)\pi$ and $|\mathrm{Re}\, w| < \pi/2\}$ by

$$F(z, w) = \frac{e^{iz} - e^{iw}}{z - w} \qquad (z \neq w).$$

Since F has no zeros, the induced multiplication operator

$$\mathscr{D}([-\pi/2, \pi/2], X) \xrightarrow{F(t, T)} \mathscr{D}([-\pi/2, \pi/2], X)$$

is a topological isomorphism. We write $\mathscr{E}_0(\mathbb{T}, X)$ for the space of all X-valued C^∞-functions $f \in \mathscr{E}(\mathbb{T}, X)$ on the torus with $f(e^{it}) = 0$ for $\pi/2 \leq |t| \leq \pi$, and we define $A = e^{iT}$. The multiplication operators with $z - A$ on $\mathscr{E}_0(\mathbb{T}, X)$ and $t - T$ on $\mathscr{D}([-\pi/2, \pi/2], X)$ are intertwined by the topological isomorphisms $\mathscr{E}_0(\mathbb{T}, X) \to \mathscr{D}([-\pi/2, \pi/2], X)$ defined by $f \mapsto F(t, T)f(e^{it})$ and $f \mapsto f(e^{it})$, respectively.

By our assumption on the spectrum of T, the map $t - T: \mathscr{E}(\mathbb{R}, X) \to \mathscr{E}(\mathbb{R}, X)$ is a topological monomorphism if and only if $z - A: \mathscr{E}(\mathbb{T}, X) \to \mathscr{E}(\mathbb{T}, X)$ is a topological monomorphism.

(2) The topology of $\mathscr{E}(\mathbb{T}, X)$ is generated by the semi-norms

$$\|f\|_p = \sum_{k=-\infty}^{+\infty} |k|^p \|a_k\| \qquad (p \in \mathbb{N}),$$

where $(a_k)_{k \in \mathbb{Z}}$ is the sequence of Fourier coefficients of f. Note that

$$f = \sum_{k=-\infty}^{+\infty} a_k z^k,$$

where the series converges in $\mathscr{E}(\mathbb{T}, X)$. For f in this form, we have

$$(z - A)f = \sum_{k=-\infty}^{+\infty} (a_{k-1} - Aa_k)z^k.$$

The operator $\mathrm{Re}\, T = \Phi(\mathrm{Re}\, z)$ has an obvious $\mathscr{E}(\mathbb{R})$-functional calculus. Hence there is a natural number m such that the operator $U = e^{i \mathrm{Re}\, T}$ satisfies

$$\|U^k\| = O(|k|^m) \qquad \text{for} \quad |k| \to \infty.$$

But then, for each integer N, the map

$$\psi_N: \mathscr{E}(\mathbb{T}, X) \to \mathscr{E}(\mathbb{T}, X), \qquad f \mapsto f_N,$$

with $f_N = \sum_{k=-\infty}^{+\infty} U^{k-N}a_k z^k$ if f is written as above, is a topological isomorphism. With $|A| = e^{-\mathrm{Im}\, T}$ we obtain

$$(z - |A|)\psi_0 f = \sum_{k=-\infty}^{-\infty} (U^{k-1}a_{k-1} - |A|U^k a_k)z^k = \psi_1(z - A)f,$$

where the a_k $(k \in \mathbb{Z})$ are the Fourier coefficients of f. By applying the result proved in the first part for a second time, one deduces that the multiplication operator $t - T: \mathscr{E}(\mathbb{R}, X) \to \mathscr{E}(\mathbb{R}, X)$ is a topological monomorphism if and only if the same is true for $t - \operatorname{i} \operatorname{Im} T: \mathscr{E}(\mathbb{R}, X) \to \mathscr{E}(\mathbb{R}, X)$.

(3) Suppose that $\sigma = \sigma(T) \cap (\mathbb{C} \setminus \mathbb{R})$ is a closed subset of $\sigma(T)$. Then T and its functional calculus are reduced by the decomposition

$$X = X_T(\mathbb{R}) \oplus X_T(\sigma).$$

Hence we may suppose that T has real spectrum. By the above cited results, T has a continuous functional calculus $\Phi: \mathscr{E}(\mathbb{C}) \to L(X)$ with $\Phi(\operatorname{Im} z) = 0$. By part (2), it suffices to show that the multiplication by $z - I$ induces a topological monomorphism on $\mathscr{E}(\mathbb{T}, X)$. This follows, for instance, from the estimates

$$\sum_{k=-\infty}^{+\infty} |k|^p \|a_{k-1} - a_k\|$$

$$\geq \sum_{k=1}^{\infty} k^p(\|a_{k-1}\| - \|a_k\|) + \sum_{k=1}^{\infty} k^p(\|a_{-k}\| - \|a_{-k-1}\|)$$

$$= \|a_0\| + \sum_{k=1}^{\infty} ((k+1)^p - k^p) \|a_k\| + \|a_{-1}\| + \sum_{k=1}^{\infty} ((k+1)^p - k^p) \|a_{-k-1}\|$$

$$\geq \|a_0\| + \sum_{k=1}^{\infty} k^{p-1} \|a_k\| + \sum_{k=1}^{\infty} k^{p-1} \|a_{-k-1}\|,$$

valid whenever $p \geq 1$ is an integer and $(a_k)_{k \in \mathbb{Z}}$ is a sequence of Fourier coefficients of a function in $\mathscr{E}(\mathbb{T}, X)$.

(4) Conversely, suppose that $t - T: \mathscr{E}(\mathbb{R}, X) \to \mathscr{E}(\mathbb{R}, X)$ is a topological monomorphism. Because of part (2) and $\sigma(\operatorname{i} \operatorname{Im} T) = \operatorname{i} \operatorname{Im} \sigma(T)$, we may confine ourselves to the case where $T = \operatorname{i} \operatorname{Im} T$. The main step will be to show that the restriction S of T onto the space $Y = X_T(\mathbb{C} \setminus \{0\})$ is invertible.

By Theorem 6.1.11, we know that the adjoint of S can be identified with the quotient $T'/X_{T'}(0)$. The last operator is injective. Indeed, if $u \in X'$ is a vector and $f: \mathbb{C} \setminus \{0\} \to X'$ is analytic with $(z - T')f(z) = T'u$ for all z, then

$$g(z) = \frac{f(z) + u}{z} \qquad (z \in \mathbb{C} \setminus \{0\})$$

defines an analytic function with $(z - T')g(z) = u$ for $z \in \mathbb{C} \setminus \{0\}$. Thus we have shown that S has dense range.

Since $t - S: \mathscr{E}(\mathbb{R}, Y) \to \mathscr{E}(\mathbb{R}, Y)$ is a topological monomorphism, there is a closed interval J in $\mathbb{R}$ with $0 \in \operatorname{Int}(J)$ such that, with a suitable integer $n \geq 0$ and a suitable constant $c > 0$, the estimate

$$c \|f(0)\| \leq \|(t - S)f\|_{C^n(J, Y)}$$

holds for all $f \in \mathscr{E}(\mathbb{R}, Y)$. Since $\mathscr{E}(\mathbb{R}, Y) \subset C^n(J, Y)$ is dense, this estimate holds for all $f \in C^n(J, Y)$. Because S is a generalized scalar operator with spectrum contained in the imaginary axis, there is an integer $k \geq 1$ with

$$\|R(t, S)\| = O(1/|t|^k) \quad \text{for} \quad t \in J \setminus \{0\}.$$

The function f defined on J by

$$f(t) = S^{(n+1)k+1} R(t, S) \quad (t \neq 0) \quad \text{and} \quad f(0) = -S^{(n+1)k}$$

belongs to $C^n(J, L(Y))$. To verify this assertion, the reader should observe that

$$t^{(n+1)k+1} R(t, S) - f(t) = \sum_{j=0}^{(n+1)k} t^j S^{(n+1)k-j}$$

holds for $t \in J \setminus \{0\}$. The estimate

$$c \|S^{(n+1)k} x\| \leq \|(t - S) f x\|_{C^n(J, Y)} = \|S^{(n+1)k+1} x\|,$$

valid for all $x \in Y$, shows that S is bounded from below.

To complete the proof, we choose a real number $\varepsilon > 0$ such that

$$\sigma(S) \cap \{z \in \mathbb{C}; |z| < \varepsilon\} = \varnothing,$$

and we show that $\sigma(T)$ contains no point λ in $\mathbb{C}$ with $0 < |\lambda| < \varepsilon$. Suppose that λ is such a point. Then it suffices to choose a closed set F in $\mathbb{C}$ with $\lambda \in \text{Int}(F) \subset F \subset \mathbb{C} \setminus \{0\}$, and to observe that $\sigma(T \mid X_T(F)) \subset \sigma(S)$ and that $\lambda \notin \sigma(T/X_T(F))$. To verify the second of these assertions, note that, for each open disc D around λ with $\overline{D} \subset \text{Int}(F)$, we have

$$X/X_T(F) \cong X_T(\mathbb{C} \setminus D)/X_T(F \cap (\mathbb{C} \setminus D)). \qquad \square$$

To demand that a generalized scalar operator allows the division of real distributions means to impose the same severe restriction on its spectrum.

Theorem 6.4.12 *Let $T \in L(X)$ be a generalized scalar operator. Then*

$$\mathscr{E}'(\mathbb{R}) \hat{\otimes} X \xrightarrow{t-T} \mathscr{E}'(\mathbb{R}) \hat{\otimes} X$$

is onto if and only if $\sigma(T) \cap \mathbb{R}$ is an open subset of $\sigma(T)$.

Proof If T is the adjoint of a generalized scalar operator acting on a suitable predual of X, then the result follows from Theorem 6.4.11 via elementary duality theory (§33.4.(2) in Köthe 1979).

Let us consider the general case. First, let us suppose that the operator $t - T: \mathscr{E}'(\mathbb{R}) \hat{\otimes} X \to \mathscr{E}'(\mathbb{R}) \hat{\otimes} X$ is onto. Then the adjoint

$$t - T': \mathscr{E}(\mathbb{R}, X') \to \mathscr{E}(\mathbb{R}, X')$$

is a weak topological monomorphism (§32.3.(4) in Köthe 1979) relative to the duality with $\mathscr{E}'(\mathbb{R}) \hat{\otimes} X$. As the strong dual of a barrelled space, the space $\mathscr{E}(\mathbb{R}, X')$ is weakly sequentially complete (§41.4.(8) in Köthe 1979). But then the multiplication operator with $t - T'$ is not only a weak, but also a strong, topological monomorphism. Hence, by Theorem 6.4.11, the set $\sigma(T') \cap \mathbb{R}$ is open in $\sigma(T')$.

Conversely, let us suppose that $\sigma(T) \cap \mathbb{R}$ is open in $\sigma(T)$. Arguing as in part (3) of the proof of the previous theorem, we may suppose that T has real spectrum. By the above cited result from Colojoară and Foiaş (1968), there is a continuous algebra homomorphism $\Phi \colon \mathscr{E}(\mathbb{R}) \to L(X)$ with $\Phi(1) = I$ and $\Phi(z) = T$.

For $U \subset \mathbb{R}$ open and $n \in \mathbb{N}$, we denote by $H^n(U)$ the usual L^2-Sobolev space of order n on U, i.e.

$$H^n(U) = \{ f \in L^2(U); f^{(j)} \in L^2(U) \text{ for } 0 \leq j \leq n \}.$$

The classical Sobolev lemma gives rise to the representation

$$\mathscr{E}(\mathbb{R}) = \varprojlim_{n,U} H^n(U),$$

where the projective limit is formed over all $n \in \mathbb{N}$ and all bounded open sets U in $\mathbb{R}$. Hence there is a bounded open set U in $\mathbb{R}$ containing $\sigma(T)$ and an integer $n \geq 0$ such that there is a continuous linear operator $\psi \colon H^n(U) \hat{\otimes}_\pi X \to X$ with

$$\psi(1 \otimes x) = x \quad \text{and} \quad \psi(tf) = T\psi(f)$$

for all $x \in X$ and $f \in H^n(U) \hat{\otimes}_\pi X$.

The upper horizontal map in the commutative diagram.

$$
\begin{array}{ccc}
\mathscr{E}'(\mathbb{R}) \hat{\otimes} H^n(U) \hat{\otimes}_\pi X & \xrightarrow{\;(t-s)\otimes 1\;} & \mathscr{E}'(\mathbb{R}) \hat{\otimes} H^n(U) \hat{\otimes}_\pi X \\
{\scriptstyle 1 \otimes \psi} \downarrow & & \downarrow {\scriptstyle 1 \otimes \psi} \\
\mathscr{E}'(\mathbb{R}) \hat{\otimes} X & \xrightarrow{\;\;\;t-T\;\;\;} & \mathscr{E}'(\mathbb{R}) \hat{\otimes} X
\end{array}
$$

is surjective by the remarks at the beginning of this proof and by Lemma 6.4.5. Hence also the lower horizontal map is surjective.   □

A concrete application of this abstract division result will be given in Chapter 7.

Our next and final aim in this section is to characterize commuting tuples with property $(\beta)_{\mathscr{E}}$.

Proposition 6.4.13   *If X is a Banach $\mathscr{E}(\mathbb{C}^n)$-module, then*

$$\mathscr{E}(\mathbb{C}^n) \perp_{\mathscr{E}(\mathbb{C}^n)} X.$$

Proof   We prove inductively that $K^\bullet(z - w, \mathscr{E}(\mathbb{C}^n) \hat{\otimes} X)$ is exact in degree $p \neq n$ and has separated cohomology in degree $p = n$.

For $n = 1$, this follows from Theorem 6.4.4. We fix an integer $n > 1$ and suppose that the result has been proved for $n - 1$. Writing the elements of $\mathbb{C}^n$

in the form $z = (z_1, z')$ with $z_1 \in \mathbb{C}$ and $z' \in \mathbb{C}^{n-1}$, we can regard the complex $K^{\bullet}(z - w, \mathscr{E}(\mathbb{C}^n) \hat{\otimes} X)$ as the total complex of the double complex $\mathscr{K} = K(z_1 - w_1, z' - w', \mathscr{E}(\mathbb{C}^n) \hat{\otimes} X)$ with rows

$$\bigoplus_{\binom{n-1}{q}} K^{\bullet}\left(z_1 - w_1, \mathscr{E}(\mathbb{C}) \hat{\otimes} X\right) \otimes 1_{\mathscr{E}(\mathbb{C}^{n-1})} \qquad (q = 0, \ldots, n-1)$$

and columns

$$(-1)^p 1_{\mathscr{E}(\mathbb{C})} \otimes K^{\bullet}(z' - w', \mathscr{E}(\mathbb{C}^{n-1}) \hat{\otimes} X) \qquad (p = 0, 1).$$

By induction hypothesis and the exactness results for tensor products of continuous linear maps proved in Appendix 1, it follows that the columns consist of homomorphisms and are exact except in degree $q = n - 1$. Hence $H^p(\mathrm{Tot}\,\mathscr{K}) = 0$ for $p = 1, \ldots, n - 2$, while $H^{n-1}(\mathrm{Tot}\,\mathscr{K})$ and $H^n(\mathrm{Tot}\,\mathscr{K})$ can topologically be identified with the kernel and cokernel, respectively, of

$$(\mathscr{E}(\mathbb{C}) \hat{\otimes} \mathscr{E}(\mathbb{C}^{n-1}, X))/\mathrm{Im}(1 \otimes u) \xrightarrow{z_1 - w_1} (\mathscr{E}(\mathbb{C}) \hat{\otimes} \mathscr{E}(\mathbb{C}^{n-1}, X))/\mathrm{Im}(1 \otimes u),$$

where u is the last map in $K^{\bullet}(z' - w', \mathscr{E}(\mathbb{C}^{n-1}, X))$. The quotient space occurring here can canonically be identified with

$$\mathscr{E}(\mathbb{C}) \hat{\otimes} (\mathscr{E}(\mathbb{C}^{n-1}, X)/\mathrm{Im}\, u).$$

The $\mathscr{E}(\mathbb{C})$-module structure of X obtained via the algebra homomorphism $\mathscr{E}(\mathbb{C}) \to \mathscr{E}(\mathbb{C}^n)$, $f \mapsto f \otimes 1$, turns $\mathscr{E}(\mathbb{C}^{n-1}, X)$ and hence $C = \mathscr{E}(\mathbb{C}^{n-1}, X)/\mathrm{Im}\, u$ into a Fréchet $\mathscr{E}(\mathbb{C})$-module. To conclude the proof, it suffices to check that $\mathscr{E}(\mathbb{C}) \perp_{\mathscr{E}(\mathbb{C})} C$.

In view of Proposition 3.2.11, it suffices to represent C as the limit of a Mittag-Leffler inverse system of Banach $\mathscr{E}(\mathbb{C})$-modules. To do this, we write

$$\mathscr{E}(\mathbb{C}^{n-1}, X) = \varprojlim_{k \geq 0} W^k(D_k, X)$$

as the inverse limit of a reduced inverse system (see Lemma A4.4 in Appendix 4), and we note that by the remarks preceding Corollary 6.4.8 we obtain the desired representation of C:

$$C = \varprojlim_k \left(W^k(D_k, X)/\overline{(\mathrm{Im}\, u)} \right). \qquad \square$$

As a direct application of Proposition 3.2.11, one obtains the following stronger result.

Corollary 6.4.14 *If X is the limit of a countable inverse system of Banach $\mathscr{E}(\mathbb{C}^n)$-modules, then*

$$\mathscr{E}(\mathbb{C}^n) \perp_{\mathscr{E}(\mathbb{C}^n)} X.$$

Proof In view of the last theorem, it suffices to observe that X can be written as the limit of a reduced inverse system of Banach $\mathscr{E}(\mathbb{C}^n)$-modules (see the remarks preceding Corollary 6.4.8), and then to apply Proposition 3.2.11. $\qquad\square$

Let X be a Fréchet space, and let $a \in L(X)^n$ be a commuting tuple of continuous linear operators on X. We shall say that a is a *Mittag-Leffler inverse limit of generalized scalar tuples* if X has a representation

$$X = \varprojlim_{\alpha \in \mathbb{N}} X_\alpha$$

as a limit of a reduced inverse system of Banach spaces, and if a is induced by a family of generalized scalar n-tuples $a_\alpha \in L(X_\alpha)^n$ which is compatible with the inverse system.

Note that, if each X_α is regarded as a Banach $\mathscr{O}(\mathbb{C}^n)$-module via

$$\mathscr{O}(\mathbb{C}^n) \times X_\alpha \to X_\alpha, \qquad (f, x) \mapsto f(a_\alpha)x,$$

then X becomes the limit of an inverse system of Banach $\mathscr{O}(\mathbb{C}^n)$-modules. When speaking of X as a Fréchet $\mathscr{O}(\mathbb{C}^n)$-module, we shall always refer to the module structure of X determined in this way. We should perhaps explicitly mention that the $\mathscr{E}(\mathbb{C}^n)$-module structures of the spaces X_α are not required to be compatible with the inverse system (X_α).

Theorem 6.4.15 *Let $a \in L(X)^n$ be a commuting tuple of continuous linear operators on a complex Banach space X. Then a satisfies property $(\beta)_{\mathscr{E}}$ if and only if there is a finite exact sequence*

$$0 \to X \xrightarrow{i} X^0 \xrightarrow{\delta} X^1 \xrightarrow{\delta} \cdots \xrightarrow{\delta} X^r \to 0 \qquad (6.4.6)$$

consisting of continuous linear operators and Fréchet spaces $X^0, \ldots, X^r$ together with a family $(a_i)_{i=0}^r$ of commuting tuples $a_i \in L(X^i)^n$ such that:

(i) *the sequence (6.4.6) intertwines the tuples $a, a_0, \ldots, a_r$;*
(ii) *each a_i is a Mittag-Leffler inverse limit of generalized scalar tuples.*

Proof Let us suppose that a satisfies property $(\beta)_{\mathscr{E}}$. We fix an open polydisc $D \supset \sigma(a)$ and denote by

$$0 \to \mathscr{O}(D) \to \mathscr{E}(D) \xrightarrow{\bar{\partial}} \mathscr{E}^1(D) \xrightarrow{\bar{\partial}} \cdots \xrightarrow{\bar{\partial}} \mathscr{E}^n(D) \to 0$$

the standard $\bar{\partial}$-resolution. By assumption, $\mathscr{E}(D) \perp_{\mathscr{O}(D)} X$ and hence also $\mathscr{E}^q(D) \perp_{\mathscr{O}(D)} X$ for each $q = 0, \ldots, n$. By applying the functor $* \hat{\otimes}_{\mathscr{O}(D)} X$ to the above sequence, we obtain the exact sequence

$$0 \to X \to \mathscr{E}(D) \hat{\otimes}_{\mathscr{O}(D)} X \to \mathscr{E}^1(D) \hat{\otimes}_{\mathscr{O}(D)} X \to \cdots \to \mathscr{E}^n(D) \hat{\otimes}_{\mathscr{O}(D)} X \to 0$$
$$(6.4.7)$$

of Fréchet $\mathscr{O}(\mathbb{C}^n)$-modules. This sequence intertwines the given tuple a on X and the commuting systems $M_z \otimes 1$ on

$$\mathscr{E}^q(D) \hat{\otimes}_{\mathscr{O}(D)} X \cong \mathscr{E}^q(D, X)/(z - a)\mathscr{E}^q(D, X)^n.$$

Let $D = \bigcup_{j \geq 1} D_j$ be an exhaustion of D by a sequence of open polydiscs $D_j \Subset D_{j+1}$. Then

$$\mathscr{E}(D) \, \hat{\otimes}_{\mathscr{O}(D)} \, X = \varprojlim_{j,k \geq 1} \, W^k(D_j, X)/\overline{(z-a)W^k(D_j, X)}^n$$

is the limit of a Mittag-Leffler inverse system of Banach $\mathscr{E}(\mathbb{C}^n)$-modules. On each of the Banach spaces forming this inverse system the tuple M_z has a continuous $\mathscr{E}(\mathbb{C}^n)$-functional calculus given by the Banach $\mathscr{E}(\mathbb{C}^n)$-module structure. Hence (6.4.7) is a resolution of a by a sequence of Mittag-Leffler inverse limits of generalized scalar tuples.

Conversely, let us suppose that a possesses a resolution of this type. Then (6.4.6) is automatically a resolution of X by Fréchet $\mathscr{O}(\mathbb{C}^n)$-modules. The assumption that a_i is a Mittag-Leffler inverse limit of generalized scalar tuples implies that $\mathscr{E}(\mathbb{C}^n) \perp_{\mathscr{O}(\mathbb{C}^n)} X^i$ (see Proposition 3.2.11 and Proposition 6.4.13). Hence by Corollary 3.1.16 we conclude that $\mathscr{E}(\mathbb{C}^n) \perp_{\mathscr{O}(\mathbb{C}^n)} X$. $\qquad\square$

In Chapter 7 we shall apply the methods developed above to solve concrete division problems in various spaces of distributions.

6.5 INVARIANT SUBSPACES

There are only a few general methods to construct invariant subspaces for concrete classes of operators. Among them are the *Lomonosov technique*, which yields non-trivial invariant subspaces for operators that are not too far removed from compact operators, and the *Scott Brown technique*, which produces invariant subspaces for operators that occur as restrictions or compressions of operators admitting sufficiently rich spectral decompositions.

In his original paper, Scott Brown (1978) proved the existence of invariant subspaces for subnormal operators. His techniques were immediately applied to contractions on Hilbert spaces, using the existence of unitary dilations and the Nagy–Foiaş functional calculus. Many beautiful results concerning the structure of contractions have since been obtained, among them some very general reflexivity results (see for instance Brown and Chevreau 1988).

It was again Scott Brown who proved in 1987 that each hyponormal operator with sufficiently rich spectrum possesses non-trivial invariant subspaces. He used a result of Putinar (1984c), which asserts that each hyponormal operator is subscalar (see Section 6.4). Up to this time, there were only a few applications of the Scott Brown technique to Banach space operators (see for instance Apostol 1981). In an effort to improve Scott Brown's result from 1987, Albrecht and Chevreau (1987) proved that each subdecomposable operator on a quotient of closed subspaces of l^p ($1 < p < \infty$) with sufficiently rich spectrum possesses a non-trivial invariant subspace. Using a result from the theory of finite-dimensional normed spaces due to Zenger (1968), and the canonical models described in Section 6.2, Eschmeier (1989) and Eschmeier and Prunaru (1990) proved the result of Albrecht and Chevreau on arbitrary Banach spaces and under a weaker richness condition on the spectrum. In the following we prove the multidimensional version of this result following Eschmeier (1992b).

To formulate the main result, we introduce some notation. Let V be a bounded open set in $\mathbb{C}^n$. We denote by $H^\infty(V)$ the Banach algebra of all bounded analytic functions on V equipped with the norm $\|f\| = \sup_{z \in V} |f(z)|$.
It is well known that $H^\infty(V)$ is a w^*-closed subspace of $L^\infty(V)$ with respect to the duality $\langle L^1(V), L^\infty(V) \rangle$. Since $L^1(V)$ is separable, this follows easily as an application of the theorem of Krein–Šmulian. In the same way we can show that the point evaluations

$$\varepsilon_\lambda : H^\infty(V) \to \mathbb{C}, \qquad f \mapsto f(\lambda)$$

are w^*-continuous and that a sequence (f_n) in $H^\infty(V)$ is w^*-convergent if and only if it is norm-bounded and uniformly convergent on all compact subsets of V. We shall regard $H^\infty(V)$ as the norm dual of the Banach space $Q = L^1(V)/ {}^\perp H^\infty(V)$.

Following the terminology of Rubel and Shields (1966), we shall say that a given subset σ of $\mathbb{C}^n$ is *dominating* in V if

$$\|f\| = \sup_{z \in \sigma \cap V} |f(z)|$$

holds for each function $f \in H^\infty(V)$. By the separation theorem, this is equivalent to the fact that the closed absolutely convex hull of the set of all point evaluations ε_λ ($\lambda \in \sigma \cap V$) is precisely the closed unit ball of Q. A subset σ of $\mathbb{C}^n$ is *thick* if there is a bounded open set V in $\mathbb{C}^n$ such that σ is dominating in V. Note that of course each subset of $\mathbb{C}^n$ with non-empty interior is thick in this sense.

For a commuting tuple $a \in L(X)^n$, we denote by $\mathrm{Lat}(a)$ the lattice of all closed invariant subspaces for a. We shall say that $\mathrm{Lat}(a)$ is *rich* if there is an infinite-dimensional Banach space Z such that $\mathrm{Lat}(a)$ contains a sublattice which is order isomorphic to the lattice $\mathrm{Lat}(Z)$ of all closed linear subspaces of Z.

Our main invariant-subspace result in this section is the following.

Theorem 6.5.1 *Let* $a \in L(X)^n$ *be a commuting tuple of Banach space operators with Bishop's property* (β). *Then we have*:

(a) *if* $\sigma(a)$ *is thick, then* $\mathrm{Lat}(a)$ *is non-trivial*;
(b) *is* $\sigma_e(a)$ *is thick, then* $\mathrm{Lat}(a)$ *is rich*.

Before beginning the proof of Theorem 6.5.1, we require two more general observations. The first is of a very elementary nature. Let $a \in L(X)^n$ be a commuting tuple on a Banach space X. With the notation of Section 2.6, we call the sets $\sigma_{le}(a) = \sigma_e^{\pi,0}(a)$ and $\sigma_{re}(a) = \sigma_e^{\delta,0}(a)$ the *left* and *right essential spectra* of a.

Lemma 6.5.2 *Let $a \in L(X)^n$ be a commuting system on a Banach space X. A point $z \in \mathbb{C}^n$ belongs to $\sigma_{le}(a)$ if and only if each closed finite-codimensional subspace M of X contains a sequence of unit vectors (x_k) with*

$$\lim_{k \to \infty} (z - a)x_k = 0.$$

Proof Note that $z - a$ stands for the map

$$\Lambda^0(\sigma, X) \xrightarrow{z-a} \Lambda^1(\sigma, X).$$

By definition, $z \in \sigma_{le}(a)$ if this map has infinite-dimensional kernel or non-closed range.

Clearly, we may suppose that $\dim \operatorname{Ker}(z - a) < \infty$. Then, for $z \in \sigma_{le}(a)$ and each closed finite-codimensional subspace M of X with $M \cap \operatorname{Ker}(z - a) = \{0\}$, the space $(z - a)M$ is not closed, and hence M contains a sequence (x_k) with the properties described.

Conversely, suppose that each closed finite-codimensional subspace of X contains such a sequence. Then, in particular, this is true for each direct complement of $\operatorname{Ker}(z - a)$ in X. Hence $\operatorname{Im}(z - a)$ is not closed.   □

The second observation uses the duality theory for decomposable systems developed in the first section of this chapter and plays an essential role in our constructions.

Proposition 6.5.3 *Let $a \in L(X)^n$ be decomposable, and let $z \in \mathbb{C}^n$. Suppose that (x_k) is a sequence in X with $\lim_{k \to \infty}(z - a)x_k = 0$. Then, for each closed set F in $\mathbb{C}^n$ with $z \in \operatorname{Int}(F)$, we have*

$$\lim_{k \to \infty} x_k / X_T(F) = 0.$$

Proof We may suppose that F is the closure of an open polydisc $D = D_1 \times \cdots \times D_n$ with centre z. We choose open polydiscs U and V with centre z and $U \Subset V \Subset D$. If $n = 1$, then the assertion follows from

$$\sigma\big(a, X/X_a(\overline{D})\big) = \sigma\big(a, X_a(\mathbb{C} \setminus V)/X_a(\overline{D} \setminus V)\big) \subset \mathbb{C} \setminus V.$$

To prove the general case, we note that the decomposability of a' implies that

$$X_{a'}(\mathbb{C}^n \setminus D) \subset \sum_{i=1}^n X_{a_i'}(\mathbb{C} \setminus V_i).$$

Using the orthogonality relations proved for the spectral subspaces of a and a' in Theorem 6.1.11, we obtain that

$$X_a(\overline{D})^{\perp} \subset \sum_{i=1}^n X_a(\overline{U}_i)^{\perp}.$$

By the open mapping principle, there is a constant $C > 0$ such that each functional u in the space on the left can be decomposed as $u = u_1 + \cdots + u_n$ with $u_i \in X_{a_i}(\overline{U}_i)^\perp$ and $\max_{1 \le i \le n} \|u_i\| \le C \|u\|$. Since, for each closed subspace Y of X, one can identify $(X/Y)' \cong Y^\perp$ isometrically, we obtain the estimate

$$\|x/X_a(\overline{D})\| \le C\big(\|x/X_{a_1}(\overline{U}_1)\| + \cdots + \|x/X_{a_n}(\overline{U}_n)\|\big) \quad (x \in X),$$

which reduces the assertion to the one-dimensional case.   □

To prove Theorem 6.5.1 it suffices to prove part (b). For, if there is a point $z \in \sigma(a) \setminus \sigma_e(a)$, then $\sum_{i=1}^n (z_i - a_i)X \in \text{Lat}(a)$ is strictly contained in X, because a satisfies the single-valued extension property. Let us fix an open set V in $\mathbb{C}^n$ such that $\sigma_e(a)$ is dominating in V. Set $Y = X'$ and $b = a'$. We choose a canonical decomposable lifting $(Y, b) \xleftarrow{q} (Z, T)$ as explained in Section 6.2.

The general idea of the Scott Brown technique is the following. Suppose that there is a point $\mu \in V$ such that, for suitable sequences (x_j) in X and (y_k) in Y, the relation

$$\delta_{jk} p(\mu) = \langle p(a)x_j, y_k \rangle \tag{6.5.1}$$

holds for all polynomials $p \in \mathbb{C}[z]$ in n variables and all natural numbers j, k. Then

$$M = \bigvee \{p(a)x_j; \, p \in \mathbb{C}[z], j \in \mathbb{N}\},$$

$$N = \bigvee \{p(a)x_j; \, p \in \mathbb{C}[z] \text{ with } p(\mu) = 0, j \in \mathbb{N}\}$$

are spaces in $\text{Lat}(a)$ with $N \subset M$ and $(\mu_i - a_i)M \subset N$ for $i = 1, \ldots, n$. The orthogonality relation (6.5.1) shows that the functionals induced by y_k on M/N together with the family (x_j) form a biorthogonal system on M/N. Since $\dim(M/N) = \infty$ and since the quotient map $\pi: M \to M/N$ induces a lattice embedding

$$\text{Lat}(M/N) \to \text{Lat}(a), \qquad L \mapsto \pi^{-1}(L),$$

the proof of Theorem 6.5.1 would be complete.

To solve the factorization problem (6.5.1), we regard both sides as linear forms in p and try to extend these linear forms to w^*-continuous linear functionals on the Banach algebra $H^\infty(V)$. More precisely, if $x \in X$ and $z \in Z_T(V)$, then using the natural $\mathcal{O}(V)$-module structure of $Z_T(V)$ (see Lemma 6.1.3), we define a w^*-continuous linear functional by

$$x \otimes z: H^\infty(V) \to \mathbb{C}, \qquad f \mapsto \langle x, q(f \cdot z) \rangle.$$

Here $f \cdot z$ is defined using the local analytic functional calculus of T.

We shall show that each infinite matrix $(L_{jk})_{j,k \ge 1}$ with coefficients in Q admits a certain factorization in terms of the above functionals. Applying this result to the matrix $(\delta_{jk}\varepsilon_\mu)_{j,k \ge 1}$, we shall obtain solutions of (6.5.1). The following factorization technique originated in the work of Scott Brown (1978), and was further developed by Bercovici, Chevreau, Foiaş, Pearcy, and others (see Bercovici et al. 1985).

To simplify the notation, we define $F = Z_T(V)$.

Suppose that $x = (x_i)_{1 \le i \le r} \in X^r$ and $z = (z_i)_{1 \le i \le s} \in F^s$ are given vectors. Then we denote by $x \otimes z$ the $r \times s$-matrix with coefficients $x_j \otimes z_k$ ($1 \le j \le r$, $1 \le k \le s$), and we write qz for the vector $(qz_i)_{1 \le i \le s} \in Y^s$. All direct sums or matrix spaces, such as $M(N, Q)$ or $M(\mathbb{N}, Q)$, will be equipped with their maximum norm.

For each integer $N \ge 1$, we write $\mathscr{L}_N$ for the set of those matrices $L = (L_{jk})_{1 \le j, k \le N}$ in $M(N, Q)$ with the property:

Given a natural number s, vectors $a_1, \ldots, a_s \in X^N$, $b_1, \ldots, b_s \in F^N$, and a number $\varepsilon > 0$, there are vectors $x \in X^N$, $z \in F^N$ with $\max(\|x\|, \|qz\|) \le 1$ and $\|L - x \otimes z\| < \varepsilon$, $\max_{1 \le i \le s} \|x \otimes b_i\| < \varepsilon$, $\max_{1 \le i \le s} \|a_i \otimes z\| < \varepsilon$.

Instead of $\mathscr{L}_1$, we shall write $\mathscr{L}$. The spaces $\mathscr{L}_N$ defined in this way are norm-closed subspaces of $M(N, Q)$.

The factorization scheme we outline next is the essential part of each application of the Scott Brown technique. An abstract version can be found in Bercovici *et al.* (1988).

In the first step, we reduce higher-dimensional factorization problems to the corresponding one-dimensional problems.

Lemma 6.5.4   *For each integer $N \ge 1$, we have the inclusion*

$$M(N, \mathscr{L}) \subset N^2 \mathscr{L}_N.$$

Proof   Consider a matrix $L = (L_{jk})_{1 \le j, k \le N}$ in $M(N, \mathscr{L})$. Let $a_1, \ldots, a_s \in X^N$, $b_1, \ldots, b_s \in F^N$, and $\varepsilon > 0$ be given.

The above definitions ensure that we find pairs of vectors $x_\mu \in X$, $z_\mu \in F$ for $\mu \in \{1, \ldots, N\}^2$ with $\max(\|x_\mu\|, \|qz_\mu\|) \le 1$ and

$$\|L_\mu - x_\mu \otimes z_\mu\| < \varepsilon, \qquad \|x_\mu \otimes z_\nu\| < \varepsilon,$$

$$\max_{1 \le i \le s} \|x_\mu \otimes b_i\| < \varepsilon, \qquad \max_{1 \le i \le s} \|a_i \otimes z_\mu\| < \varepsilon$$

for all index pairs μ, ν with $\mu \ne \nu$. Then

$$x = \left(\sum_{i=1}^{N} (x_{ji}/N) \right)_{1 \le j \le N}, \qquad z = \left(\sum_{i=1}^{N} (z_{ik}/N) \right)_{1 \le k \le N}$$

satisfy all relations showing that $L \in N^2 \mathscr{L}_N$.   □

Let us make the additional assumption that there is a positive constant $c > 0$ such that

$$\{L \in Q; \|L\| \le c\} \subset \mathscr{L}. \tag{6.5.2}$$

Before proving the existence of c (see Corollary 6.5.10 below), we show that under this assumption our factorization problem can be solved. The following result is an immediate consequence of Lemma 6.5.4.

Corollary 6.5.5   *For each integer $N \geq 1$, we have*

$$\{L \in M(N,Q); \|L\| \leq c/N^2\} \subset \mathscr{L}_N. \qquad \square$$

We use Corollary 6.5.5 to obtain approximate factorizations of finite matrices such that the constructed factors remain in a prescribed neighbourhood of the given initial data.

Proposition 6.5.6   *Let $N \geq 1$ be an integer. Then for $L \in M(N,Q)$, $x_0 \in X^N$, $z_0 \in F^N$, and $\varepsilon > 0$, there are $x \in X^N$ and $z \in F^N$ with*

$$\|L - x \otimes z\| < \varepsilon,$$

$$\max(\|x - x_0\|, \|q(z - z_0)\|) \leq (N/c^{1/2})\|L - x_0 \otimes z_0\|^{1/2}.$$

Proof   Let us define $d = \|L - x_0 \otimes z_0\|$. Then Corollary 6.5.5 allows us to choose vectors $a \in X^N$ and $b \in F^N$ with norm of a and qb less than or equal to $N(d/c)^{1/2}$, and

$$\|(L - x_0 \otimes z_0) - a \otimes b\| < \delta, \qquad \|a \otimes z_0\| < \delta, \qquad \|x_0 \otimes b\| < \delta,$$

where δ can be chosen arbitrarily small. If $\delta > 0$ is small enough, then $x = x_0 + a$ and $z = z_0 + b$ have the desired properties. $\qquad \square$

In the second step, one defines converging sequences of vectors which asymptotically solve the given factorization problem.

Proposition 6.5.7   *Let $N \geq 1$ be an integer. Suppose that $L \in M(N,Q)$, $x_0 \in X^N$, $z_0 \in F^N$, and $\varepsilon > 0$ are given. Then there are sequences (x_k) in X^N and (z_k) in F^N such that, with $d_0 = \|L - x_0 \otimes z_0\|$:*

(i) $x = \lim_{k \to \infty} x_k, y = \lim_{k \to \infty} qz_k$ *exist and satisfy*

$$\max(\|x - x_0\|, \|y - qz_0\|) < N(d_0/c)^{1/2} + \varepsilon;$$

(ii) $L = \lim_{k \to \infty} x_k \otimes z_k$ *holds in $M(N,Q)$.*

Proof   It suffices to fix a sequence of positive real numbers d_k ($k \geq 1$) such that $(N/c^{1/2})\sum_{k=1}^{\infty} d_k^{1/2} < \varepsilon$, and to define inductively sequences (x_k), (z_k) such that, for all $k \geq 0$,

$$\|L - x_k \otimes z_k\| \leq d_k,$$

$$\max(\|x_{k+1} - x_k\|, \|q(z_{k+1} - z_k)\|) \leq (N/c^{1/2})d_k^{1/2}. \qquad \square$$

The next result describes how to factor infinite matrices.

Proposition 6.5.8   *For each matrix $L = (L_{jk})_{j,k \geq 1}$ in $M(\mathbb{N},Q)$, there are sequences $(x_N)_{N \geq 1}$ and $(z_N)_{N \geq 1}$ satisfying:*

(i) $x_N \in X^N$, $z_N \in F^N$;
(ii) *the limits* $x(j) = \lim_{N \to \infty} x_N(j) \in X$ *and* $y(j) = \lim_{N \to \infty} q z_N(j) \in Y$ *exist for each* $j \geq 1$ *(here* $x_N(j), z_N(j)$ *denote the jth components of* x_N, z_N*);*
(iii) $L_{jk} = \lim_{N \to \infty} x_N(j) \otimes z_N(k)$ *for all* $j, k \geq 1$.

Proof We choose a sequence $(d_N)_{N \geq 0}$ of positive real numbers with

$$\sum_{N=0}^{\infty} (N+1) d_N^{1/2} < \infty.$$

In addition, let us fix a sequence $(c_N)_{N \geq 1}$ of positive real numbers. We set $x_0 = 0 = z_0$ and $M = (c_j c_k L_{jk})_{j,k \geq 1} \in M(\mathbb{N}, Q)$. If the numbers c_N are small enough, then using Proposition 6.5.6, one can define inductively sequences $(x_N)_{N \geq 1}, (z_N)_{N \geq 1}$ with $x_N \in X^N$, $z_N \in F^N$ and

$$\|(M_{jk})_{1 \leq j, k \leq N} - x_N \otimes z_N\| < d_N,$$

$$\max(\|x_N - x_{N-1}\|, \|q(z_N - z_{N-1})\|) \leq N(d_{N-1}/c)^{1/2}$$

for all $N \geq 1$. But then the limits $x = \lim_{N \to \infty} x_N$, $y = \lim_{N \to \infty} q z_N$ exist in $l^\infty(\mathbb{N}, X)$ and $l^\infty(\mathbb{N}, Y)$, respectively. To complete the proof, it suffices to replace x_N and z_N by

$$(x_N(j)/c_j)_{1 \leq j \leq N}, \qquad (z_N(j)/c_j)_{1 \leq j \leq N}. \qquad \square$$

To see how Proposition 6.5.8 can be used to complete the proof of Theorem 6.5.1, we fix a point $\mu \in V$ and choose sequences $(x_N)_{N \geq 1}, (z_N)_{N \geq 1}$ with respect to the matrix $L = (\delta_{jk}, \varepsilon_\mu)_{j, k \geq 1}$, as explained above. Then, for each polynomial $p \in \mathbb{C}[z]$, the identity

$$\delta_{jk} p(\mu) = \lim_{N \to \infty} \langle x_N(j), q p(T) z_N(k) \rangle = \langle p(a) x(j), y(k) \rangle$$

holds for all integers $j, k \geq 1$, which is precisely (6.5.1).

Thus, to conclude the proof of Theorem 6.5.1, it suffices to prove the existence of a constant $c > 0$ satisfying (6.5.2). As a first step we show how to factor convex combinations of point evaluations.

Proposition 6.5.9 *The convex hull of all point evaluations*

$$\mathscr{E} = \{\varepsilon_\lambda; \lambda \in V \cap \sigma_e(a)\}$$

is contained in $2\mathscr{L}$.

Proof Let us fix a convex combination $L = \sum_{k=1}^r c_k \varepsilon_{\lambda_k}$ of elements in $\mathscr{E}$ as well as $a_1, \ldots, a_s \in X$, $b_1, \ldots, b_s \in F$, and $\varepsilon > 0$.

By Corollary 6.3.4, we know that $\sigma_e(a) = \sigma_{re}(a) = \sigma_{le}(b)$. Let $\delta > 0$ be arbitrary. We need the following general result from the theory of normed spaces (see Lemma III.1.1 in Singer 1981). Suppose that $M \in \mathrm{Lat}(Y)$ is finite dimensional. Then there is a closed finite-codimensional subspace N of Y with

$$\delta(M, N) = \inf\{\|x - y\|; \ x \in M \text{ with } \|x\| = 1 \text{ and } y \in N\} > 1 - \delta.$$

For each index $i = 1, \ldots, r$, we choose open polydiscs $W_i \Subset V_i \subset V$ with centre λ_i, and we set $F_i = \overline{W_i}$. We define $Y_0 = \{a_1, \ldots, a_s\}^\perp \subset Y$. Since

$$C = \{q(f \cdot b_i); \ f \in H^\infty(V) \text{ with } \|f\| \le 1, i = 1, \ldots, s\} \subset Y$$

is compact, there are vectors $y_{-t}, \ldots, y_0 \in Y$ with $\min_{0 \le k \le t} \|y - y_{-k}\| < \delta$ for all $y \in C$. We choose a closed finite-codimensional subspace Y_1 of Y with

$$\delta(\vee\{y_{-t}, \ldots, y_0\}, Y_1) > 1 - \delta.$$

Then there are vectors $y_1 \in Y_0 \cap Y_1$, $z_1 \in Z$, and $\tilde{z}_1 \in Z_T(F_1)$ with

$$y_1 = q z_1, \quad \|y_1\| = 1, \quad \|(\lambda_1 - T)z_1\| < \delta, \quad \|z_1 - \tilde{z}_1\| < \delta$$

(see Lemma 6.5.2, Proposition 6.2.5, and Proposition 6.5.3). As a second step one chooses a closed finite-codimensional subspace Y_2 of Y with

$$\delta(\vee\{y_{-t}, \ldots, y_1\}, Y_2) > 1 - \delta$$

and vectors $y_2 \in Y_0 \cap Y_1 \cap Y_2$, $z_2 \in Z$, $\tilde{z}_2 \in Z_T(F_2)$ satisfying the corresponding set of conditions.

Continuing in this way, one obtains unit vectors $y_1, \ldots, y_r \in Y_0$ as well as elements $z_1, \ldots, z_r, \tilde{z}_1, \ldots, \tilde{z}_r \in Z$ with $\tilde{z}_k \in Z_T(F_k)$ and

$$y_k = q z_k, \|(\lambda_k - T)z_k\| < \delta, \qquad \|z_k - \tilde{z}_k\| < \delta,$$

$$\delta(\vee\{y_{-t}, \ldots, y_{k-1}\}, \vee\{y_k, \ldots, y_r\}) > 1 - \delta$$

for $k = 1, \ldots, r$. It easily follows that, for $\alpha_1, \ldots, \alpha_r \in \mathbb{C}$,

$$\max_{1 \le k \le r} |\alpha_k| \le (2/1 - \delta) \left\| \sum_{k=1}^r \alpha_k y_k \right\|$$

and that the canonical projection from $M = \vee_{k=-t}^r y_k$ onto $\vee_{k=1}^r y_k$ has norm less than $2/(1 - \delta)$. By a result due to Zenger (1968), there is a linear form l on M with $\|l\| < 2/(1 - \delta)$, and there are complex numbers $\mu_1, \ldots, \mu_r$ with

$$\left\| \sum_{k=1}^r \mu_k y_k \right\| \le 1, \quad l(\mu_k y_k) = c_k \quad (1 \le k \le r), \qquad l(y_k) = 0 \quad (-t \le k \le 0).$$

Using the isometric identification $M' \cong X/^\perp M$, one can choose an element $x \in X$ of norm less than $2/(1 - \delta)$ representing l on M.

Set $z = \sum_{k=1}^r \mu_k \tilde{z}_k$. Then, for δ small enough, we obtain the estimate $\|qz\| < 1 + \varepsilon$. Observe that there is a constant $\rho > 0$ such that, for each fixed index $k = 1, \ldots, r$ and for each f in the closed unit ball of $H^\infty(V)$, there are functions $f_1, \ldots, f_n \in H^\infty(V_k)$ with $\|f_i\|_{\infty, V_k} \leq \rho$ and

$$f(z) - f(\lambda_k) = \sum_{i=1}^{n} (z_i - \lambda_{k,i}) f_i(z) \qquad (z \in V_k).$$

Therefore, for δ sufficiently small, the estimates

$$|x \otimes z(f) - L(f)| \leq \left| \sum_{k=1}^{r} \mu_k \langle x, q((f|V_k) - f(\lambda_k)) \tilde{z}_k \rangle \right|$$

$$+ \left| \sum_{k=1}^{r} \mu_k \langle x, qf(\lambda_k)(\tilde{z}_k - z_k) \rangle \right| < \varepsilon,$$

$$|x \otimes b_i(f)| = \min_{0 \leq k \leq t} |\langle x, q(fb_i) - y_{-k} \rangle| < \varepsilon,$$

$$|a_i \otimes z(f)| \leq \left| \sum_{k=1}^{r} \mu_k \langle a_1, q((f|V_k) - f(\lambda_k)) \tilde{z}_k \rangle \right|$$

$$+ \left| \sum_{k=1}^{r} \mu_k \langle a_i, qf(\lambda_k)(\tilde{z}_k - z_k) \rangle \right| < \varepsilon$$

hold uniformly for all f in the closed unit ball of $H^\infty(V)$. $\square$

An elementary decomposition into real and imaginary, positive and negative parts, allows one to deduce the complex version of the last result.

Corollary 6.5.10 *The absolutely convex hull of all point evaluations*

$$\mathscr{E} = \{\varepsilon_\lambda; \lambda \in V \cap \sigma_e(a)\}$$

is contained in $32\mathscr{L}$. *In particular,* (6.5.2) *holds with* $c = \frac{1}{32}$.

Proof Consider an absolutely convex combination L of elements in $\mathscr{E}$, and fix vectors $a_1, \ldots, a_s \in X$, $b_1, \ldots, b_s \in F$, as well as a real number $\varepsilon > 0$.

Obviously, L can be written as a combination $L = \sum_{k=1}^{4} \varepsilon_k L_k$ of elements L_k in the convex hull of $\mathscr{E}$ with complex coefficients ε_k of modulus less than or equal to 1. Now for $\delta > 0$ arbitrary, one can choose vectors $x_k \in X$ and $z_k \in F$ ($k = 1, \ldots, 4$) with $\max(\|x_k\|, \|qz_k\|) \leq 1$ and $\|(L_k/2) - x_k \otimes z_k\| < \delta$,

$$\max_{1 \leq i \leq s} \|x_k \otimes b_i\| < \delta, \qquad \max_{1 \leq i \leq s} \|a_i \otimes z_k\| < \delta, \qquad \|x_j \otimes z_k\| < \delta$$

for all $j, k = 1, \ldots, 4$ with $j \neq k$. But then the vectors $x = (\sum_{k=1}^{4} \varepsilon_k x_k)/4$ and $z = (\sum_{k=1}^{4} z_k)/4$ factor $L/32$ in the right sense for δ small enough.

Since $\mathscr{L} \subset Q$ is norm-closed, and since $\sigma_e(a)$ is by assumption dominating in V, we conclude that formula (6.5.2) holds with $c = \frac{1}{32}$. $\square$

Thus the proof of Theorem 6.5.1 is complete. We conclude with a few elementary applications. Using methods from rational approximation theory, one can show that $R(K) = C(K)$ for each compact set K in $\mathbb{C}$ which is not thick (see S. Brown 1987). Moreover, recall from Section 6.2 that in the single-variable case Bishop's property (β) characterizes precisely the class of subdecomposable operators, i.e those operators that are similar to the restriction of a decomposable operator to one of its closed invariant subspaces.

Corollary 6.5.11 *Let $T \in L(X)$ be a subdecomposable operator on a complex Banach space X. If $\sigma(T)$ is thick, then $\mathrm{Lat}(T)$ is non-trivial. If $\sigma_e(T)$ is thick, then $\mathrm{Lat}(T)$ is rich.*   □

Since hyponormal operators are subdecomposable (see Proposition 6.4.3 and Corollary 6.4.8), the last corollary applies in particular to hyponormal operators on Hilbert spaces. For subnormal operators, one can relax the richness condition on the spectrum.

Corollary 6.5.12 *Each subnormal operator T on a Hilbert space H possesses a non-trivial invariant subspace.*   □

To prove this corollary, it suffices to observe that each subnormal operator T with non-thick spectrum possesses a $C(\sigma(T))$-functional calculus, and hence is even normal (see, for instance, Conway 1991).

6.6 REFERENCES AND COMMENTS

In an attempt to develop a duality theory for spectral decompositions of arbitrary Banach space operators, Bishop (1959) introduced condition (β), and observed that the adjoint of an operator with this property admits certain spectral decompositions. In his theory of spectral operators and their functional calculi Dunford (1954) formulated the concepts of local spectrum and single-valued extension property. The definition of decomposability for single operators is due to Foiaş (1963). The theory of decomposable operators, operators with non-analytic functional calculi, and generalized scalar operators was developed systematically in Colojoară and Foiaş (1968). Important contributions to the early theory of decomposable operators are due to Albrecht, Apostol, Foiaş, Frunzá, Radjabalipour, and Vasilescu (see Vasilescu 1982 and the references therein).

Following the fundamental work of J. L. Taylor described before, a multivariable theory of spectral decompositions was initiated in the early seventies by Albrecht and Frunzá. In Albrecht (1972, 1974) decomposable n-tuples, commuting systems with non-analytic functional calculi, and in particular, generalized scalar systems are studied. Frunzá (1975a, 1977) defined the single-valued extension property in the multidimensional case, and proved that commuting systems with a spectral capacity satisfy Bishop's conditions (β). Non-analytic local spectral properties for commuting systems were studied by Albrecht and Vasilescu (1974).

The theory of decomposable operators and functional calculi, as well as many related questions and applications not considered here, are described in detail in the monographs of Colojoară and Foiaş (1968) and Vasilescu (1982), which present the one- and several-variable case, respectively.

The notion of sheaf model was introduced by Putinar (1983a, 1986), who proved, among other results, that each decomposable operator possesses a Fréchet soft sheaf model and that, moreover, sheaves of this type are uniquely determined by their global section spaces. The duality theory for decomposable systems contained in Section 6.1 was developed by Eschmeier and Putinar (1984). The result that the adjoint of a single decomposable operator is decomposable is due to Frunză (1976), while the converse was proved independently by Wang and Liu (1984) and Eschmeier (1984a). The latter paper contains a first multidimensional version of this predual theorem.

The observation that Bishop's condition (β) characterizes precisely the restrictions of decomposable operators is due to Albrecht and Eschmeier (1987). The same paper contains a corresponding characterization of quotients of decomposable operators, and a complete duality theory for restrictions and quotients of decomposable operators. Putinar (1984c) proved that hyponormal operators are subscalar. The fact that condition (β)$_{\mathscr{E}}$ and its dual property characterize restrictions and quotients of generalized scalar operators was proved by Eschmeier and Putinar (1988, 1989). The second paper contains applications to division problems for distributions. The corresponding multidimensional results stating that condition (β) characterizes those tuples which possess a decomposable resolution, and that property (β)$_{\mathscr{E}}$ characterizes the class of tuples admitting a generalized scalar resolution, were obtained by Putinar (1990a). The concrete construction of a canonical decomposable resolution given in Section 6.2 follows Eschmeier (1992b).

The results on essentially decomposable systems described in Section 6.3 extend one-variable results of Albrecht and Mehta (1984), and are contained in Eschmeier (1992b).

The technique used to prove invariant subspace results in Section 6.5 originates in the work of Scott Brown. In S. Brown (1978) the existence of invariant subspaces for subnormal operators is proved, while in S. Brown (1987) hyponormal operators with thick spectrum are shown to possess invariant subspaces. First applications of the Scott Brown technique to the Banach space case were given by Apostol. In Apostol (1981) it is shown that restrictions (with thick spectrum) of unconditionally decomposable Banach space operators possess invariant subspaces. Albrecht and Chevreau (1987) proved invariant subspace results for restrictions and quotients of decomposable operators on quotients of closed subspaces of l^p ($1 < p < \infty$). Corresponding results for arbitrary Banach spaces, and under optimal richness conditions on the spectra, were obtained by Eschmeier (1989), and Eschmeier and Prunaru (1990). The construction of joint invariant subspaces for tuples with Bishop's condition (β) given in Section 6.5 represents a slight improvement of the main result from Eschmeier (1992b).

From the very beginning, the Scott Brown technique has been used to prove invariant subspace and structure results for contractions. The interested reader is referred to the work of Bercovici, Chevreau, Foiaş, and Pearcy (see Bercovici *et al.* 1985). Reflexivity results extending classical results of Sarason (1966) can, for instance, be found in Olin and Thomson (1980), Brown and Chevreau (1988), and Conway and Dudziak (1990).

There are some applications of local spectral theory that have not been, and will not be, touched upon in this book. As examples, let us mention connections to the theory of regular Banach algebras, to harmonic analysis, or problems of automatic continuity. Some sample works treating these topics are Albrecht (1982a), Eschmeier (1982a, 1985), and Laursen and Neumann (1986, 1992). The reader particularly interested in automatic continuity problems should follow the recent work of Laursen and Neumann.

7
Applications to function theory

The abstract notions of quasi-coherence, sub-scalarity, analytic transversality and so on have been related to certain concrete problems which we met in the preceding chapters. These examples offered both a motivation and a validation of the whole theory. The present chapter is intended to bring forward some further examples and applications related mainly to function theory of real or complex variables. This terrain abounds in localization methods and transversality phenomena. Their homological interpretation, based on the preceding chapters, provides a new insight into some old and new results, usually obtained by laborious methods.

We focus below only on two subjects: the division of functions and distributions by analytic functions, and the localization of spaces of analytic functions defined by global conditions. Some other topics which are suitable for a similar treatment are discussed in the next chapters or are left to be discovered in detail by the reader.

7.1 DIVISION LEMMAS

The very definition of a quasi-coherent module over a Stein algebra provides some non-trivial division results, as for instance the next lemma.

Lemma 7.1.1 *Let $U \subset \mathbb{C}^n$ be a Stein open subset, and let $\mathscr{F}$ be a quasi-coherent (Fréchet) $\mathscr{O}_{\mathbb{C}^n}$-module. If $f \in \mathscr{O}(U) \hat{\otimes} \mathscr{F}(\mathbb{C}^n)$ satisfies $f(z, z) = 0$ for $z \in U$, then there are $f_j \in \mathscr{O}(U) \hat{\otimes} \mathscr{F}(\mathbb{C}^n)$ $(1 \leq j \leq n)$ with the property that*

$$f = \sum_{j=1}^{n} (z_j - w_j) f_j.$$

The statement requires some explanation. We have formally denoted by $f(z, z)$ the image of the 'function' f under the restriction and multiplication map $\mathscr{O}(U) \hat{\otimes} \mathscr{F}(\mathbb{C}^n) \to \mathscr{F}(U)$. In the case where $\mathscr{F}$ is a sheaf of functions, this notation is consistent with the usual restriction to the diagonal $\Delta = \{(z, z); z \in U\} \subset U \times \mathbb{C}^n$.

Proof The proposition is a direct corollary of the existence of the Koszul resolution:

$$\cdots [\mathscr{O}(U) \,\hat{\otimes}\, \mathscr{F}(\mathbb{C}^n)]^{n} \xrightarrow{\delta} \mathscr{O}(U) \,\hat{\otimes}\, \mathscr{F}(\mathbb{C}^n) \to \mathscr{F}(U) \to 0,$$

where

$$\delta(f_1,\ldots,f_n) = \sum_{j=1}^{n} (z_j - w_j)f_j$$

(see the end of Section 4.3).   □

Particularly relevant for applications is the case where $\mathscr{F}$ is a soft analytic sheaf. Since such a sheaf is quasi-coherent by Theorem 4.4.1, the conclusion of the previous lemma is true. This observation covers most of the division results required in the construction of appropriate kernels for integral representation formulae in several complex variables (cf. Ramirez' Lemma, Ramirez de Arellano 1970, or the Oka–Hefer Theorem, Henkin and Leiterer 1983).

The coordinate-free analogue of Lemma 7.1.1 can be stated as follows. Here we keep the same notational convention for the restrictions to the diagonal. We recall that an analytic space $(X, \mathscr{O}_X)$ is called *reduced* if the local structural rings $\mathscr{O}_{X,x}$ are integral for each point $x \in X$. Only on reduced analytic spaces can the sections of $\mathscr{O}(X)$ be interpreted as functions.

Lemma 7.1.2 *Let $\mathscr{F}$ be a quasi-coherent $\mathscr{O}_X$-module defined on a reduced Stein space X, and let U denote a Stein open subset of X.*

If $f \in \mathscr{O}_X(U) \,\hat{\otimes}\, \mathscr{F}(X)$ satisfies $f(z,z) = 0$ for every $z \in U$, then there exists an element $F \in \mathscr{O}_X(U) \,\hat{\otimes}\, \mathscr{O}_X(X) \,\hat{\otimes}\, \mathscr{F}(X)$ with the property:

$$f(z,w) = F(z,z,w) - F(z,w,w) \qquad ((z,w) \in U \times X).$$

Proof In analogy with the proof of Lemma 7.1.1, the Bar resolution

$$\cdots \to \mathscr{O}_X(U) \,\hat{\otimes}\, \mathscr{O}_X(X) \,\hat{\otimes}\, \mathscr{F}(X) \to \mathscr{O}_X(U) \,\hat{\otimes}\, \mathscr{F}(X) \to \mathscr{F}(U) \to 0$$

yields the desired conclusion (see Chapter 3 for details concerning this resolution).   □

In fact both the exact sequences which appeared in the previous proofs offer more information. Quite specifically, they describe all solutions of the two division problems, the relations between these solutions, and so on.

A second series of division lemmas is derived from the characterization of subscalar operators obtained in Section 6.4.

Let X be a Banach space, and let $T \in L(X)$ be a bounded linear operator on X. According to Corollary 6.4.8, the operator T is similar to the restriction of a generalized scalar operator to an invariant subspace (we say in short that T is *subscalar*) if and only if the map

$$z - T : \mathscr{E}(\mathbb{C}) \,\hat{\otimes}\, X \to \mathscr{E}(\mathbb{C}) \,\hat{\otimes}\, X$$

is injective with closed range. Analogously, the operator T is similar to the quotient of a generalized scalar operator if and only if the map

$$z - T: \mathscr{E}'(\mathbb{C}) \,\hat{\otimes}\, X \to \mathscr{E}'(\mathbb{C}) \,\hat{\otimes}\, X$$

is surjective and allows the lifting of bounded sets (Theorem 6.4.10).

The last criterion has a series of applications to division problems for distributions which are related to the problem of finding fundamental solutions for partial differential operators with constant coefficients (see Hörmander 1983 for a detailed discussion of these topics).

Lemma 7.1.3 *Let Ω be an open set in $\mathbb{R}^n (n \geq 1)$, let $1 \leq p \leq \infty$, and let $H^{p,m}(\Omega)$ be the usual L^p-Sobolev space of order $m \in \mathbb{Z}$. Then for each function $f: \Omega \to \mathbb{C}$ with the property that f and all its derivatives of order less than or equal to $|m|$ are bounded on Ω, the map*

$$z - f: \mathscr{D}'(U) \,\hat{\otimes}\, H^{p,m}(\Omega) \to \mathscr{D}'(U) \,\hat{\otimes}\, H^{p,m}(\Omega)$$

is surjective for any open subset U of $\mathbb{C}$.

Proof The multiplication operator

$$M_f: H^{p,m}(\Omega) \to H^{p,m}(\Omega), \qquad \varphi \mapsto \varphi f$$

is generalized scalar with the C^∞-functional calculus

$$\Phi: \mathscr{E}(\mathbb{C}) \to L(H^{p,m}(\Omega)), \qquad \Phi(g) = M_{g \circ f}.$$

This completes the proof of Lemma 7.1.3. $\qquad\qquad\qquad\qquad\square$

It is worth mentioning that the space $\mathscr{D}'(U)$ cannot in general be replaced by a smaller space of distributions. For instance, if U and Ω are open subsets of $\mathbb{R}^n$ with $U \Subset \Omega \Subset \mathbb{R}^n$ and $r \geq 1$ is an integer, then for a sufficiently large $s \in \mathbb{N}$, the image of the map

$$z\text{-}w: H^{-r}(U) \hat{\otimes}_\pi H^s(\Omega) \to H^{-r}(U) \hat{\otimes}_\pi H^s(\Omega) \quad (H^m(\Omega) = H^{2,m}(\Omega), m \in \mathbb{Z})$$

is contained in the kernel of the non-trivial map

$$H^{-r}(U) \hat{\otimes}_\pi H^s(\Omega) \to H^{-r}(U), \qquad u \otimes f \mapsto (f|_U) u.$$

The definition of the Sobolev spaces used here can be found in Appendix 4.

Next we discuss a few results on the division of distributions, which depend in a measurable way on an additional parameter, by some distinguished analytic functions.

Lemma 7.1.4 *Let $(\Omega, \mathscr{M}, \mu)$ be a measure space, and let $a_1, \ldots, a_m$ be bounded measurable functions on Ω. Denote by*

$$P(z, w) = z^m + a_1(w) z^{m-1} + \cdots + a_m(w) \qquad (z \in \mathbb{C}, w \in \Omega)$$

the monic polynomial with coefficients $a_1, \ldots, a_m$. Then the map

$$P: \mathscr{D}'(U) \,\hat{\otimes}\, L^p(\Omega) \to \mathscr{D}'(U) \,\hat{\otimes}\, L^p(\Omega)$$

is surjective for each open set U in $\mathbb{C}$ and for each real number p with $1 \leq p \leq \infty$.

Proof Notice that P can be regarded as an element in $\mathscr{E}(\mathbb{C}, L(L^p(\Omega)))$, whence the product Pu makes perfect sense for

$$u \in \mathscr{D}'(U) \,\hat{\otimes}\, L^p(\Omega) \ (1 \leq p \leq \infty).$$

If the order of the polynomial P is one, then the conclusion follows as in Lemma 7.1.3.

In the general case, the polynomial P decomposes into linear factors:

$$P(z, w) = (z - f_1(w)) \cdots (z - f_m(w)),$$

with bounded measurable functions $f_1, \ldots, f_m \colon \Omega \to \mathbb{C}$. Thus this case reduces to the situation considered above.

Let us explain the decomposition of P into measurable linear factors. The mapping

$$\sigma : \mathbb{C}^m \to \mathbb{C}^m, \qquad z \mapsto (\sigma_1(z), \ldots, \sigma_m(z)),$$

where $\sigma_1, \ldots, \sigma_m$ are the fundamental symmetric polynomials in $z_1, \ldots, z_m$, is continuous, surjective and open (see Whitney 1972). According to a theorem of Dixmier concerning much more general situations, there is a Borel measurable right inverse $g \colon \mathbb{C}^m \to \mathbb{C}^m$ for σ (see for instance Theorem 3.4.1 in Arveson 1976). In this case, the functions

$$f_i \colon \Omega \to \mathbb{C}, \qquad f_i(w) = g_i(-a_1(w), a_2(w), \ldots, (-1)^m a_m(w))$$

are bounded and measurable. They yield the above decomposition of the polynomial P.   □

An immediate application of Lemma 7.1.4 is the following classical result.

Theorem 7.1.5 (L. Schwartz). *Let Ω be a domain in $\mathbb{C}^n$, and let $f \in \mathscr{O}(\Omega)$ be an analytic function which is not identically zero. Then the multiplication map*

$$\mathscr{D}'(\Omega) \xrightarrow{\ f\ } \mathscr{D}'(\Omega)$$

is surjective.

Proof By a partition of unity argument, one remarks that it suffices to show that, for each fixed point a in Ω with $f(a) = 0$, there is an open neighbourhood U of a in Ω such that the map $\mathscr{E}'(U) \xrightarrow{\ f\ } \mathscr{E}'(U)$ is surjective.

By the Weierstraß Preparation Theorem (see Whitney 1972), there is a biholomorphic map φ from a suitable open neighbourhood of a onto a polydisc $D = V \times W \subset \mathbb{C} \times \mathbb{C}^{n-1}$ centred at zero with the property that

$$(f \circ \varphi^{-1})(z, w) = g(z, w) P(z, w) \qquad (z \in V, w \in W),$$

where $g \in \mathscr{O}(D)$ has no zeros in D and P is a monic polynomial in z

$$P(z, w) = z^m + a_1(w) z^{m-1} + \cdots + a_m(w)$$

with bounded analytic functions $a_1, \ldots, a_m$ as coefficients.

Since our problem is invariant with respect to biholomorphic maps, we may assume that $\Omega = D$ and $f = P$. For each compact subset K of W, the induced map $\mathscr{E}'(V) \hat{\otimes} L^2(K) \xrightarrow{P} \mathscr{E}'(V) \hat{\otimes} L^2(K)$ is onto by Lemma 7.1.4.

It is well known that each distribution $u \in \mathscr{E}'(D)$ can be written as a finite sum of derivatives (cf. Lemma 6.4.7 and its proof) of suitable functions.

Now fix $u \in \mathscr{E}'(D)$ and a representation of u as a finite sum of elements $\bar{\partial}^\alpha h$, where $h \in \mathscr{C}_c^0(D)$. Suppose that h vanishes outside $V \times K$, where K is a compact subset of W. Then there is a solution g of the equation $Pg = h$ in $\mathscr{E}'(V) \hat{\otimes} L^2(K)$. Hence $\bar{\partial}^\alpha h = P\bar{\partial}^\alpha F$ has a solution F in $\mathscr{E}'(D)$, and consequently u belongs to $P\mathscr{E}'(D)$. □

Theorem 7.1.5 first appeared in Schwartz (1955). In the same paper the author raised a series of problems related to generalizations of this result. Answering in the affirmative one of these questions, Łojasiewicz later proved that Theorem 7.1.5 remains valid for real analytic functions (see Hörmander 1983 for details and bibliography). A second problem, originating in the same paper of Schwartz, concerns the closure of ideals of (real) analytic functions in the space of smooth functions. Its solution was independently settled by Malgrange (1955-56) and Palamodov (1961). The next section will be entirely devoted to a discussion of their result in the complex analytic case.

When trying to extend Lemma 7.1.3 to the real case, one faces some additional difficulties, which go back to Theorem 6.4.12. A typical example is contained in the next result.

Lemma 7.1.6 *Let Ω be a connected open set in $\mathbb{R}^n$, let $1 \leq p \leq \infty$, and let $H^{p,m}(\Omega)$ denote the L^p-Sobolev space of order $m \in \mathbb{Z}$. Consider a function $f \in C^{|m|}(\Omega)$ with the property that f and all its partial derivatives of order less than or equal to $|m|$ are bounded. Then the map*

$$t - f: \mathscr{D}'(\mathbb{R}) \hat{\otimes} H^{p,m}(\Omega) \to \mathscr{D}'(\mathbb{R}) \hat{\otimes} H^{p,m}(\Omega)$$

is surjective if and only if either f is real-valued or $\inf_{\lambda \in \Omega} |\operatorname{Im} f(\lambda)| > 0$.

Proof This is a direct application of Theorem 6.4.12. □

In order to state the real analogue of Lemma 7.1.4 we need the following notation. For a complex n-tuple $a = (a_1, \ldots, a_n) \in \mathbb{C}^n$, consider the polynomial

$$P_a(z) = z^n + a_1 z^{n-1} + \cdots + a_{n-1} z + a_n.$$

If $z_1, \ldots, z_n$ denote the roots of P_a counted with possible multiplicities, then we define

$$\theta(a) = \sum_{\substack{j=1 \\ \operatorname{Im} z_j \neq 0}}^{n} |\operatorname{Im} z_j|^{-1}.$$

If all the roots are real, then we set $\theta(a) = 0$.

Lemma 7.1.7 *Let $(\Omega, \mathcal{M}, \mu)$ be a measure space with a positive measure μ, and let $a_1, \ldots, a_n: \Omega \to \mathbb{C}$ be bounded measurable functions. Let us consider the monic polynomial*

$$P(t, \lambda) = t^n + a_1(\lambda)t^{n-1} + \cdots + a_n(\lambda) \qquad (t \in \mathbb{R}, \lambda \in \Omega)$$

with coefficients $a_1, \ldots, a_n$. Then for each p $(1 \le p \le \infty)$, the multiplication operator

$$\mathscr{D}'(\mathbb{R}) \,\hat{\otimes}\, L^p(\Omega) \overset{P}{\to} \mathscr{D}'(\mathbb{R}) \,\hat{\otimes}\, L^p(\Omega)$$

is surjective if and only if ess-sup$_{\lambda \in \Omega}\ \theta(a_1(\lambda), \ldots, a_n(\lambda)) < \infty$.

Proof As explained in the proof of Lemma 7.1.4, there are bounded measurable functions $f_1, \ldots, f_n: \Omega \to \mathbb{C}$ with

$$P(t, \lambda) = (t - f_1(\lambda)) \cdots (t - f_n(\lambda))$$

for all $t \in \mathbb{R}$ and $\lambda \in \Omega$. Therefore the operator of multiplication by P on $\mathscr{D}'(R) \,\hat{\otimes}\, L^p(\Omega)$ is surjective if and only if, for each $i = 1, \ldots, n$, the operator

$$\mathscr{D}'(\mathbb{R}) \,\hat{\otimes}\, L^p(\Omega) \xrightarrow{t - f_i(\lambda)} \mathscr{D}'(\mathbb{R}) \,\hat{\otimes}\, L^p(\Omega)$$

is surjective, which in turn is equivalent to the corresponding statement with $\mathscr{D}'(\mathbb{R})$ replaced by $\mathscr{E}'(\mathbb{R})$. (The latter reduction is possible because the statement of Lemma 7.1.7 is local in the variable t.)

But for each bounded measurable function $g: \Omega \to \mathbb{C}$, the operator of multiplication by g on $L^p(\Omega)$ is generalized scalar and possesses the spectrum

$$\sigma(M_g) = \{z \in \mathbb{C}; \mu(g^{-1}(U)) > 0 \text{ for each open neighbourhood } U \text{ of } z\}.$$

It is an easy exercise to verify that $\sigma(M_g) \cap \mathbb{R}$ is an open subset of $\sigma(M_g)$ if and only if there is an $\varepsilon > 0$ and a subset N of Ω of μ-measure zero such that either Im $g(\lambda) = 0$ or $|\text{Im } g(\lambda)| > \varepsilon$ holds for all $\lambda \in \Omega \setminus N$. Now it is obvious that Lemma 7.1.7 is just an application of Theorem 6.4.11. $\square$

7.2 IDEALS OF DIFFERENTIABLE FUNCTIONS

The result of L. Schwartz stated in Theorem 7.1.5 can be reformulated, using a simple duality argument, as follows. Let Ω be a domain in $\mathbb{C}^n$, and let $f \in \mathcal{O}(\Omega)$ be a complex analytic function. Then the principal ideal $f\mathscr{E}(\Omega)$ is closed in the Fréchet space topology of the space of smooth functions $\mathscr{E}(\Omega)$. Thus a natural question is whether every finitely generated ideal $(f_1, \ldots, f_r)\mathscr{E}(\Omega) = f_1\mathscr{E}(\Omega) + \cdots + f_r\mathscr{E}(\Omega)$, where $f_1, \ldots, f_r$ are analytic functions, is closed in $\mathscr{E}(\Omega)$. Schwartz (1955) answered this question in the affirmative for complete intersection ideals. As mentioned before, Malgrange and Palamodov gave a solution of this problem, even in the real-analytic case.

It is the aim of the present section to relate these types of problems, at least in the complex analytic case, to the general quasi-coherence criteria contained in Chapter 4 and Chapter 6. We do not enter here into any of the numerous applications of the theorem of Malgrange. They appear for instance in Malgrange (1966), Palamodov (1968), Tougeron (1972), Bănică and Stănăşilă (1977), and Golovin (1986).

First we relate the original question to a transversality problem. Let $\mathscr{F}$ be a coherent $\mathscr{O}$-module on an open set Ω in $\mathbb{C}^n$. We shall use the results of Malgrange to show that $\mathscr{F} \perp_{\mathscr{O}} \mathscr{E}$, that is,

$$\mathscr{F}(U) \perp_{\mathscr{O}(U)} \mathscr{E}(U) \tag{7.2.1}$$

for each Stein open set U in Ω. It is elementary to check that, conversely, this transversality relation implies the closedness of any ideal $f\mathscr{E}(\Omega)$ algebraically generated by a finite system $f = (f_1, \ldots, f_r) \in \mathscr{O}(\Omega)^r$ of analytic functions.

Indeed, by applying the functor $* \hat{\otimes}_{\mathscr{O}(U)} \mathscr{E}(U)$ to the exact sequence

$$\mathscr{O}(U)^n \xrightarrow{f} \mathscr{O}(U) \xrightarrow{q} (\mathscr{O}/f\mathscr{O})(U) \to 0,$$

one obtains the exact sequence

$$\mathscr{E}(U)^n \xrightarrow{f} \mathscr{E}(U) \to (\mathscr{O}/f\mathscr{O})(U)\hat{\otimes}_{\mathscr{O}(U)} \mathscr{E}(U) \to 0.$$

So for each Stein open set U in Ω, the above transversality relation, applied with $\mathscr{F} = \mathscr{O}/f\mathscr{O}$, yields that the ideal $f\mathscr{E}(U)$ is closed in $\mathscr{E}(U)$. Using a partition of unity, one can show that in this case also $f\mathscr{E}(\Omega)$ is closed in $\mathscr{E}(\Omega)$. The next result is proved in Magrange (1966).

Theorem 7.2.1   *Let $\Omega \subset \mathbb{C}^n$ be open. For each point $x \in \Omega$, the ring $\mathscr{E}_x$ of germs of smooth functions at x is flat over $\mathscr{O}_x$. For each system $f = (f_1, \ldots, f_r)$ in $\mathscr{O}(\Omega)^r$, the ideal $f\mathscr{E}(\Omega)$ is closed in $\mathscr{E}(\Omega)$.*   □

Let us indicate how the transversality relation (7.2.1) follows from this result. At the same time, we gather some other consequences of Theorem 7.2.1 which are of independent interest. The $\bar{\partial}$-resolutions occurring in the next result have been used in Chapter 4 (Theorem 4.4.4) to construct resolutions of quasi-coherent sheaves on finite-dimensional Stein spaces by soft analytic Fréchet sheaves.

Corollary 7.2.2   *Let Ω be an open set in $\mathbb{C}^n$, let $\mathscr{F}$ be a coherent analytic sheaf on Ω, and let $f = (f_1, \ldots, f_r) \in \mathscr{O}(\Omega)^r$ be a system of analytic functions. Define the ideal sheaves $\mathscr{I} = f\mathscr{O}_\Omega$ and $\mathscr{J} = f\mathscr{E}_\Omega$.*

(a)   *For each Stein open subset U of Ω, we have*

$$\mathscr{F}(U) \perp_{\mathscr{O}(U)} \mathscr{E}(U).$$

(b) *The complex of sheaves*

$$0 \to \mathscr{O}_\Omega/\mathscr{I} \to \Lambda^0(\mathscr{E}_\Omega/\mathscr{I}) \xrightarrow{\bar{\partial}} \cdots \xrightarrow{\bar{\partial}} \Lambda^n(\mathscr{E}_\Omega/\mathscr{I}) \to 0 \qquad (7.2.2)$$

is exact.

(c) *The complex of sheaves*

$$0 \to \mathscr{I} \to \Lambda^0(\mathscr{I}) \xrightarrow{\bar{\partial}} \cdots \xrightarrow{\bar{\partial}} \Lambda^n(\mathscr{I}) \to 0$$

is exact.

(d) *The sheaf $\mathscr{F} \otimes_\mathscr{O} \mathscr{E}$ is a sheaf of Fréchet spaces, and*

$$\mathrm{Tor}_p^{\mathscr{O}_x}(\mathscr{F}_x, \mathscr{E}_x) = 0 \quad \text{for } p > 0 \text{ and } x \in \Omega.$$

Proof Let us fix a point x in Ω. On a suitable open neighbourhood V of x, the coherent sheaf $\mathscr{F}$ has a resolution of the form

$$0 \leftarrow \mathscr{F}|_V \leftarrow \mathscr{O}_V^{p_0} \leftarrow \mathscr{O}_V^{p_1} \leftarrow \cdots.$$

On each Stein open subset U of V, one obtains an induced topologically free resolution

$$0 \leftarrow \mathscr{F}(U) \leftarrow \mathscr{O}(U)^{p_0} \leftarrow \mathscr{O}(U)^{p_1} \leftarrow \cdots$$

of the $\mathscr{O}(U)$-module $\mathscr{F}(U)$. By Corollary 3.1.13 and its algebraic counterpart (Appendix 2), we have the identifications

$$\hat{\mathrm{Tor}}_p^{\mathscr{O}(U)}(\mathscr{F}(U), \mathscr{E}(U)) \cong H_p\left(\mathscr{O}(U)^{p_\bullet} \hat{\otimes}_{\mathscr{O}(U)} \mathscr{E}(U)\right)$$

$$= H_p\left(\mathscr{O}(U)^{p_\bullet} \otimes_{\mathscr{O}(U)} \mathscr{E}(U)\right) \cong \mathrm{Tor}_p^{\mathscr{O}(U)}(\mathscr{F}(U), \mathscr{E}(U))$$

for all $p \geq 0$. In particular, for $p = 0$, this yields

$$\mathscr{F}(U) \hat{\otimes}_{\mathscr{O}(U)} \mathscr{E}(U) \cong \mathscr{F}(U) \otimes_{\mathscr{O}(U)} \mathscr{E}(U).$$

The flatness part of Theorem 7.2.1 shows that the sequences

$$0 \leftarrow \mathscr{F}_y \otimes_{\mathscr{O}_y} \mathscr{E}_y \leftarrow \mathscr{O}_y^{p_0} \otimes_{\mathscr{O}_y} \mathscr{E}_y \leftarrow \cdots \qquad (y \in V)$$

are exact.

In the particular case where $\mathscr{F} = \mathscr{O}/\mathscr{I}$, there are canonical identifications $\mathscr{F}_y \otimes_{\mathscr{O}_y} \mathscr{E}_y \cong (\mathscr{E}/\mathscr{I})_y$, $(y \in V)$. In this case, we obtain the transversality relation (7.2.1) for each Stein open subset U of V, and moreover,

$$\mathscr{F}(U) \hat{\otimes}_{\mathscr{O}(U)} \mathscr{E}(U) \cong (\mathscr{E}/\mathscr{I})(U)$$

for each such set U. By applying the functor $(\mathscr{O}/\mathscr{I})(U) \hat{\otimes}_{\mathscr{O}(U)} *$ to the ordinary $\bar{\partial}$-sequence

$$0 \to \mathscr{O}(U) \to (\mathscr{E}^\bullet(U), \bar{\partial}),$$

one obtains the exactness of the complex (7.2.2) at the given point x in Ω.

Since the $\bar{\partial}$-sequence $0 \to \mathscr{O}_\Omega \to (\mathscr{E}^\bullet, \bar{\partial})$, regarded as a sequence of sheaves on Ω, is exact, parts (b) and (c) are equivalent.

The sheaves occurring in the complex (7.2.2) are quasi-coherent on each Stein open set U in Ω (Theorem 4.4.1). Hence (7.2.2) is exact if and only if the induced sequence of section spaces is exact on each Stein open set U in Ω. The spaces $\mathscr{E}(U)$ possess canonical representations

$$\mathscr{E}(U) = \varprojlim W^k(U_k) \qquad (U \subset \Omega \text{ open})$$

as limits of countable inverse systems of Banach $\mathscr{E}(\mathbb{C}^n)$-modules, and by the remarks preceding Corollary 6.4.8, the same is true for the Fréchet spaces $(\mathscr{E}/\mathscr{I})(U) = \mathscr{E}(U)/\mathscr{I}(U)$. Hence (Corollary 6.4.14)

$$(\mathscr{E}/\mathscr{I})(U) \perp_{\mathscr{E}(\mathbb{C}^n)} \mathscr{E}(\mathbb{C}^n) \qquad (U \subset \Omega \text{ open}),$$

and Corollary 3.1.16 implies that

$$(\mathscr{O}/\mathscr{I})(U) \perp_{\mathscr{O}(\mathbb{C}^n)} \mathscr{E}(\mathbb{C}^n)$$

for each Stein open set U in Ω.

For each point $x \in \Omega$, there is an open neighbourhood V of x and a sequence

$$0 = \mathscr{F}_0 \subset \mathscr{F}_1 \subset \cdots \subset \mathscr{F}_d = \mathscr{F}|_V$$

of coherent $\mathscr{O}_V$-modules such that $\mathscr{F}_j/\mathscr{F}_{j-1} \cong \mathscr{O}_V/\mathscr{I}_j$ $(j = 1, \ldots, d)$ with suitable finitely generated ideal sheaves $\mathscr{I}_j$ in $\mathscr{O}_V$. The existence of such a composition series follows directly from the corresponding algebraic result for finite modules over Noetherian rings (Chapter VI, §4 in Lang 1984). An ascending induction, based on Proposition 3.1.15, shows that

$$\mathscr{F}(U) \perp_{\mathscr{O}(\mathbb{C}^n)} \mathscr{E}(\mathbb{C}^n) \tag{7.2.3}$$

for each Stein open subset U of V.

By using Čech resolutions with respect to Stein open covers with finite-dimensional nerve, and as an application of Corollary 3.1.16, one obtains that relation (7.2.3) holds for each Stein open set U in Ω, or equivalently, that the Koszul complex $K_\bullet(z - w, \mathscr{E}(\mathbb{C}^n) \hat{\otimes} \mathscr{F}(U))$ is exact in positive degrees and has separated homology in degree zero for each such set U. Since the latter condition is preserved if one replaces $\mathbb{C}^n$ by U, it follows that the transversality relation (7.2.1) holds on each Stein open set U in Ω. This observation completes the proof. $\qquad \square$

To prove the exactness of the complex (7.2.2) (which implies all the remaining parts of Corollary 7.2.2), it suffices, according to the above proof, to show that, for each point x in Ω, the system of all Stein open neighbourhoods U of x with

$$(\mathscr{O}/\mathscr{I})(U) \perp_{\mathscr{O}(U)} \mathscr{E}(U)$$

forms a neighbourhood base at x. For $n = 1$, this transversality relation is easily seen to be true. In this case, the spaces $(\mathscr{O}/\mathscr{I})(U)$ are finite-dimensional complex vector spaces. Hence the $\mathscr{O}(U)$-module structure of these spaces extends to a topological $\mathscr{E}(U)$-module structure, and therefore the above transverality relation follows as an application of Proposition 6.4.13.

For a complete presentation of the techniques involved in the proof of Theorem 7.2.1, the reader may consult Malgrange (1966) and Tougeron (1972).

A few variations of Theorem 7.2.1 are now at hand. For instance, we can state the following result.

Corollary 7.2.3 *Let $U \subset \mathbb{C}^n$ be an open subset, and let $\mathscr{O}'$ denote the sheaf of smooth functions on $\mathbb{C}^n$ which are complex analytic in the first m coordinates $(m \leq n)$. Then for an arbitrary coherent analytic sheaf $\mathscr{F}$ on U, we have:*

$$\mathrm{Tor}_p^{\mathscr{O}_x}(\mathscr{F}_x, \mathscr{O}'_x) = 0 \ (p > 0),$$

and moreover, $\mathscr{F} \otimes_{\mathscr{O}} \mathscr{O}'$ is a sheaf of Fréchet spaces.

Proof It suffices to remark that there exists the following resolution on U:

$$0 \to \mathscr{O}' \to \Lambda^0 \mathscr{E} \xrightarrow{\bar{\partial}'} \Lambda^1 \mathscr{E} \xrightarrow{\bar{\partial}'} \cdots \to \Lambda^m \mathscr{E} \to 0,$$

where $\bar{\partial}'$ is the $\bar{\partial}$-operator with respect to the first m complex coordinates of $\mathbb{C}^n$. Then Theorem 7.2.1 and a descending induction yield the conclusion of the corollary.　　　　　　　　　　　　　　　　　　　　　　　　　□

The above proof of Theorem 7.2.1 fits well with the case when all data depend on parameters. To be more precise, let $\mathscr{O}'$ be the sheaf of analytic functions in $\mathbb{C}^n$, and let $\mathscr{E}''$ be the sheaf of smooth functions in $\mathbb{R}^m$. Let U be an open subset of $\mathbb{C}^n \times \mathbb{R}^m$, and let $\mathscr{A} = \mathscr{O}' \hat{\otimes} \mathscr{E}''$ denote the sheaf of smooth functions on U which are complex analytic in the first n complex variables. Let $\mathscr{J}$ denote a closed, finitely generated ideal of $\mathscr{A}$. The argument given in Lemma 7.2.2 applies with minor changes, and it shows that $\mathscr{J} \cdot \mathscr{E}_U$ is a closed ideal of $\mathscr{E}_U$. Consequently, one can state the next result.

Proposition 7.2.4 *Every Fréchet $\mathscr{A}$-module $\mathscr{F}$ locally of finite presentation of length one on $U \subset \mathbb{C}^n \times \mathbb{R}^m$ satisfies $\mathscr{F} \perp_{\mathscr{A}} \mathscr{E}$. In particular, $\mathscr{F} \otimes_{\mathscr{A}} \mathscr{E}$ is a sheaf of Fréchet spaces.*

Proof One may assume that there exists a finite presentation of $\mathscr{F}$ on U: $\mathscr{A}^p \to \mathscr{A}^q \to \mathscr{F} \to 0$. Then the $\mathscr{A}$-module $\mathscr{F}$ inherits from the canonical filtration of $\mathscr{A}^q$ a finite filtration: $\mathscr{F} = \mathscr{F}^0 \supset \mathscr{F}^1 \supset \cdots \supset \mathscr{F}^q = 0$ such that $\mathscr{F}^j / \mathscr{F}^{j+1} \cong \mathscr{A} / \mathscr{J}_j$, where the $\mathscr{J}_j$ are finitely generated closed ideals of $\mathscr{A}$ $(0 \leq j \leq q - 1)$.

Thus the problem is reduced to considering a sheaf of the form $\mathscr{F} = \mathscr{A} / \mathscr{J}$, where $\mathscr{J}$ is a finitely generated ideal of $\mathscr{A}$. According to the remark preceding the statement of our proposition, there exists a resolution of topological $\mathscr{A}$-modules

$$0 \to \mathscr{A}/\mathscr{J} \to \mathscr{L}^0 \to \mathscr{L}^1 \to \cdots \to \mathscr{L}^n \to 0,$$

where the $\mathscr{L}^k$ are Fréchet $\mathscr{E}$-modules on U $(0 \le k \le n)$. Since there are natural isomorphisms

$$* \hat{\otimes}_{\mathscr{A}} \mathscr{E} \cong * \hat{\otimes}_{\mathscr{E}' \hat{\otimes} \mathscr{E}''} (\mathscr{E}' \hat{\otimes} \mathscr{E}'') \cong * \hat{\otimes}_{\mathscr{E}'} \mathscr{E}',$$

Theorem 7.2.1 shows that $\mathscr{F} \perp_{\mathscr{A}} \mathscr{E}$.

The last assertion follows from the observation that $\mathscr{F} \otimes_{\mathscr{A}} \mathscr{E} \cong \mathscr{F} \hat{\otimes}_{\mathscr{A}} \mathscr{E}$, which in turn holds because $\mathscr{F}$ is a finite-type $\mathscr{A}$-module. $\qquad\square$

A version of Proposition 7.2.4 with a weaker dependence on parameters, for instance of Lipschitz or Sobolev type, is still available by the same method.

7.3   THE MAXIMAL IDEAL SPACE OF CERTAIN ALGEBRAS OF ANALYTIC FUNCTIONS

The computation of the character space of a concrete Banach algebra of analytic functions is in general a difficult problem. The *corona theorem* is perhaps the most famous and important example of this kind. In the present section we exploit the quasi-coherence property of certain topological algebras of analytic functions defined by global conditions in order to describe their character space. This point of view unifies various known results in one or several complex variables. Though most of the facts presented in this section are not original, this new approach opens up some (possibly new) applications, a few of which are included here.

To begin with, we consider a typical situation originally treated by Hakim and Sibony (1980).

Let Ω be a domain in $\mathbb{C}^n$ $(n \ge 1)$, and let $A^{\infty}(\overline{\Omega})$ denote the Fréchet algebra $A^{\infty}(\overline{\Omega}) = \mathscr{O}(\Omega) \cap \mathscr{E}(\overline{\Omega})$. Similarly one defines $A^r(\overline{\Omega}) = \mathscr{O}(\Omega) \cap C^r(\overline{\Omega})$ for any $r \in \mathbb{N}$ and $A^{\alpha}(\overline{\Omega}) = \mathscr{O}(\Omega) \cap \mathrm{Lip}_{\alpha}(\overline{\Omega})$ for a non-integer positive real α.

Our first aim is to compute the character space of these commutative topological algebras, under some geometrical conditions on the boundary $\partial\Omega$ of Ω. Here, by a character of an algebra A, we mean a non-trivial multiplicative linear form on A.

Theorem 7.3.1   *Let Ω be a bounded pseudoconvex domain in $\mathbb{C}^n$ with smooth boundary. Then each character on $A^{\infty}(\overline{\Omega})$ is a point evaluation (at a suitable point of $\overline{\Omega}$).*

Proof Since Ω is a bounded pseudoconvex domain with smooth boundary, the Dolbeault complex

$$0 \to A^\infty(\overline{\Omega}) \to \mathscr{E}^{(0,0)}(\overline{\Omega}) \xrightarrow{\bar{\partial}} \mathscr{E}^{(0,1)}(\overline{\Omega}) \xrightarrow{\bar{\partial}} \cdots \xrightarrow{\partial} \mathscr{E}^{(0,n)}(\overline{\Omega}) \to 0$$

is exact (see Kohn 1973). As an application of Corollary 6.4.14, we obtain the transversality relation $\mathscr{E}(\mathbb{C}^n) \perp_{\mathscr{O}(\mathbb{C}^n)} \mathscr{E}(\overline{\Omega})$. Hence, by Corollary 3.1.16, we know that $\mathscr{E}(\mathbb{C}^n) \perp_{\mathscr{O}(\mathbb{C}^n)} \mathscr{A}^\infty(\overline{\Omega})$. In particular, the Fréchet $\mathscr{O}(\mathbb{C}^n)$-module $A^\infty(\overline{\Omega})$ is quasi-coherent. Since the Koszul complex $K_\bullet(z - w, \mathscr{O}(U) \,\hat{\otimes}\, \mathscr{E}(\overline{\Omega}))$ is exact for each relatively compact Stein open subset U of $\mathbb{C}^n \setminus \overline{\Omega}$, it follows that the quasi-coherent Fréchet $\mathscr{O}_{\mathbb{C}^n}$-sheaf $\mathscr{F}$ corresponding to $A^\infty(\overline{\Omega})$ vanishes on $\mathbb{C}^n \setminus \overline{\Omega}$.

Let χ be a character of $A^\infty(\overline{\Omega})$. Assume that χ is not a point evaluation. Then for each point $z \in \overline{\Omega}$, there is a function $f_z \in A^\infty(\overline{\Omega})$ with $\chi(f_z) = 0$, but $f_z(z) \neq 0$. By compactness there are finitely many functions $f_1, \ldots, f_m$ in $\mathrm{Ker}(\chi)$ such that

$$\inf_{z \in \Omega} |f_1(z)| + \cdots + |f_m(z)| > 0.$$

To obtain a contradiction, it suffices to show that there are functions $g_1, \ldots, g_m \in A^\infty(\overline{\Omega})$ such that

$$f_1 g_1 + \cdots + f_m g_m = 1,$$

or equivalently, such that the Koszul complex $K_\bullet(f, A^\infty(\overline{\Omega}))$ given by $f = (f_1, \ldots, f_m)$ is exact. But the exactness of this complex follows from the exactness of the corresponding complex $K_\bullet(f, \mathscr{F})$ of sheaves, which is obvious because at any point of $\overline{\Omega}$ at least one of the functions $f_1, \ldots, f_m$ is non-zero. $\qquad\square$

An approximation argument can be used to prove, under the same assumptions on Ω, the corresponding result for the Banach algebra $A^0(\overline{\Omega})$ (see Hakim and Sibony 1980). Similarly, with the help of suitable Dolbeault resolutions, one obtains the next result.

Theorem 7.3.2 *Let Ω be a strictly pseudoconvex domain in $\mathbb{C}^n$ with C^2-boundary, and let α be a positive real number. Then each character on $A^\alpha(\overline{\Omega})$ is a point evaluation.*

Proof The Hölder estimates for the $\bar{\partial}$-equation on Ω provide an exact complex of closed operators between Banach spaces:

$$0 \to A^\alpha(\overline{\Omega}) \to K_\bullet\big(\bar{\partial}, \mathrm{Lip}_\alpha(\overline{\Omega})\big)$$

(see Henkin and Leiterer 1983).

Since the domains of the $\bar{\partial}$-operators occurring in this resolution are in a natural way Banach $\mathscr{E}(\mathbb{C}^n)$-modules, we infer, exactly as in the preceding proof, that $A^\alpha(\overline{\Omega})$ is a quasi-coherent $\mathcal{O}(\mathbb{C}^n)$-module such that the corresponding sheaf is supported by $\overline{\Omega}$. In addition, $A^\alpha(\overline{\Omega})$ is a commutative Banach algebra.

The argument given in the last part of the proof of Theorem 7.3.1 also completes the present proof. $\square$

The strict pseudoconvexity of Ω or the smoothness of its boundary can be weakened to conditions which still ensure the existence of Hölder estimates for the $\bar{\partial}$-equations (see for details Henkin and Leiterer 1983).

Let us also give the following equivalent formulation of the last result.

Corollary 7.3.3   *Let Ω be a strictly pseudoconvex domain in $\mathbb{C}^n$ with C^2-boundary. Let $\alpha > 0$ be fixed, and let $f_1,\dots,f_m \in A^\alpha(\overline{\Omega})$ be such that*

$$\inf_{z \in \Omega} (|f_1(z)| + \cdots + |f_m(z)|) > 0.$$

Then the Koszul complex $K_\bullet(f_1,\dots,f_m, A^\alpha(\overline{\Omega}))$ is exact. $\square$

The idea of relating the exactness of a Koszul complex to the computation of the character space of a Banach algebra is explicitly formulated in Hörmander (1967). The main result of this paper can easily be formulated in the framework developed in this section.

Let Ω be a bounded pseudoconvex domain in $\mathbb{C}^n$, and let $p \colon \Omega \to \mathbb{R}$ be a *plurisubharmonic function* which is bounded from below. We denote by $A_p(\Omega)$ the set of all analytic functions f on Ω for which there are constants $C_1, C_2 > 0$ with

$$|f(z)| \le C_1 \exp(C_2 p(z)) \qquad (z \in \Omega).$$

Since one can always replace the function p by $p + k$ ($k \in \mathbb{R}$ arbitrary) without changing the set A_p, we may and shall suppose that p is non-negative. Obviously, $A_p(\Omega)$ is an algebra.

Let us suppose that there are constants $K_1,\dots,K_4 > 0$ with the following property. Whenever $z \in \Omega$, then the closed Euclidean ball B around z with radius $\exp(-K_1 p(z) - K_2)$ is contained in Ω and

$$p(w) \le K_3 p(z) + K_4 \qquad (w \in B).$$

A natural example of a function satisfying this condition is given by $p(z) = \log(1/d(z))$ ($z \in \Omega$), where $d(z)$ is the distance of $z \in \Omega$ from $\mathbb{C}^n \setminus \Omega$.

Under these conditions, one can easily show (Hörmander 1967) that a given analytic function f on Ω belongs to $A_p(\Omega)$ if and only if, for some $K > 0$, the function $f e^{-Kp}$ belongs to $L^2(\Omega)$ (formed with respect to the $2n$-dimensional Lebesgue measure), and that in this case all partial derivatives $\partial f/\partial z_j$ ($1 \le j \le n$) belong to $A_p(\Omega)$. Define

$$L = \{f \in L^1_{\mathrm{loc}}(\Omega); \|f e^{-Kp}\|_{2,\Omega} < \infty \text{ for some constant } K > 0\},$$

and denote by L^q ($0 \le q \le n$) the vector space of all differential forms

$u \in \Lambda^q(d\bar{z}, L)$ for which $\bar{\partial}u$, formed in the sense of distributions, is an element of $\Lambda^{q+1}(d\bar{z}, L)$ (cf. Section 8.1). The fundamental L^2-estimates for the $\bar{\partial}$-operator obtained by Hörmander (1965) yield the exactness of the $\bar{\partial}$-sequence

$$0 \to A_p(\Omega) \to L^0 \xrightarrow{\bar{\partial}} L^1 \xrightarrow{\bar{\partial}} \cdots \xrightarrow{\bar{\partial}} L^n \to 0.$$

The spaces L^q $(0 \le q \le n)$ are $A_p(\Omega)$-modules with respect to component-wise multiplication. Fix an m-tuple $f = (f_1, \ldots, f_m) \in A_p(\Omega)^m$. The Koszul complex $K^\bullet(f, A_p(\Omega))$ is exact if and only if there are functions $g_1, \ldots, g_m$ in $A_p(\Omega)$ with $\sum_{i=1}^m f_i g_i = 1$. An obvious necessary condition for this relation to hold is that there are constants $c_1, c_2 > 0$ with

$$|f_1(z)| + \cdots + |f_m(z)| \ge c_1 \exp(-c_2 p(z)) \qquad (z \in \Omega). \qquad (7.3.1)$$

We want to show that this condition is also sufficient.

Consider the double complex $\mathcal{K} = (\Lambda^p(\sigma, L^q))_{p,q \ge 0}$ with qth row $K^\bullet(f, L^q)$ and pth column given by

$$(-1)^p \bigoplus_{\binom{n}{p}} (L^\bullet, \bar{\partial}).$$

If the estimate (7.3.1) holds, then the spaces L^q $(0 \le q \le n)$ are invariant under multiplication with the functions $g_j = \bar{f}_j / \sum_k |f_k|^2$ $(1 \le j \le m)$. To prove this, note that

$$\bar{\partial}(g_j u) = \left(\bar{\partial}g_j\right) \wedge u + g_j(\bar{\partial}u) \qquad (1 \le j \le m, u \in L^q).$$

In this case, all rows of $\mathcal{K}$ are exact (Lemma 2.2.4), while all columns are exact in all degrees except in degree 0. According to Lemma A2.6 in Appendix 2, the Koszul complex $K^\bullet(f, A_p(\Omega))$ is exact.

Summing up, we have proved the following theorem of Hörmander (1967).

Theorem 7.3.4 *Let $\Omega \subset \mathbb{C}^n$ be a bounded pseudoconvex domain, and let p be a plurisubharmonic function on Ω as above. Then $f_1, \ldots, f_m \in A_p(\Omega)$ generate $A_p(\Omega)$, that is*

$$\sum_{j=1}^m f_j A_p(\Omega) = A_p(\Omega),$$

if and only if (7.3.1) holds with suitable constants $c_1, c_2 > 0$.   □

The idea of using Koszul complexes and generalized $\bar{\partial}$-resolutions has also simplified the original proof of *Carleson's corona theorem* (Hörmander 1967). Without aiming at completeness, we sketch below the basic principle.

The starting point is the resolution

$$0 \to H^\infty(D) \to L^0(D) \xrightarrow{\bar{\partial}} L^1(D) \to 0,$$

where $H^\infty(D)$ is the Banach algebra of all bounded analytic functions on the unit disc D in $\mathbb{C}$, $L^1(D)$ is the Banach space of all Carleson measures on D, and $L^0(D)$ is the space of those distributions $u \in \mathscr{E}'(\mathbb{C})$ which satisfy supp$(u) \subset \overline{D}$, $\bar{\partial}u \in L^1(D)$, and which possess boundary values in $L^\infty(\partial D)$ (see Hörmander 1967 for details).

Let $f = (f_1, \ldots, f_m) \in H^\infty(D)^m$ be an m-tuple of functions with

$$\inf_{z \in D} \sum_{i=1}^m |f_i(z)| > 0.$$

Let $\mathscr{K}$ be the double complex constructed as above. Then the exactness of the first row $K^\bullet(f, L^1(D))$ follows exactly as before, but the exactness proof for $K^\bullet(f, L^0(D))$ fails, since the functions $\partial f_j / \partial z$ need not be bounded. By using results of Carleson and by a modification of the above scheme, one can nevertheless prove the corona theorem along these lines (see Hörmander 1967).

Theorem 7.3.5 (Corona Theorem)   *A system of functions* $f_1, \ldots, f_m \in H^\infty(D)$ *generates* $H^\infty(D)$ *(that is,* $\Sigma_i f_i H^\infty(D) = H^\infty(D)$*) if and only if*

$$\inf_{z \in D} (|f_1(z)| + \cdots + |f_m(z)|) > 0. \qquad \square$$

Our next aim is to deduce various generalizations of this result by using sheaf-theoretic methods and the quasi-coherence of $H^\infty(D)$ as an $\mathscr{O}(\mathbb{C})$-module.

Let Ω be a bounded open set in $\mathbb{C}$, and let $H^\infty(\Omega)$ be the Banach algebra of all bounded analytic functions on Ω equipped with the supremum norm. We say that the *corona theorem holds on* Ω if the set of all point evaluations

$$\varepsilon_\lambda : H^\infty(\Omega) \to \mathbb{C}, \qquad f \mapsto f(\lambda) \qquad (\lambda \in \Omega)$$

is w^*-dense in the character space of $H^\infty(\Omega)$, or equivalently, if for each finite tuple $f \in H^\infty(\Omega)^m$ with $0 \notin \overline{f(\Omega)}$ the Koszul complex $K^\bullet(f, H^\infty(\Omega))$ is exact.

Exactly as in the proof of Lemma 6.2.2 (with L^∞-norms instead of L^2-norms), one obtains an exact $\bar{\partial}$-sequence of the form

$$0 \to H^\infty(\Omega) \to W(\Omega) \xrightarrow{\bar{\partial}} L^\infty(\Omega) \to 0,$$

where, for $U \subset \mathbb{C}$ open, we use the notation

$$W(U) = \{f \in L^\infty(U); \bar{\partial}f \in L^\infty(U)\}.$$

Since $W = W(\Omega)$ and $L = L^\infty(\Omega)$ are Banach $\mathscr{E}(\mathbb{C})$-modules, it follows as before (Proposition 6.4.13 and Proposition 3.1.15) that $H^\infty(\Omega)$ is a quasi-coherent $\mathscr{O}(\mathbb{C})$-module. On an arbitrary open set U in $\mathbb{C}$, the section spaces of the sheaves corresponding to the quasi-coherent $\mathscr{O}(\mathbb{C})$-modules L and W are given by

$$\tilde{L}(U) = \{f \in L^1_{\text{loc}}(U \cap \Omega); f \,|\, V \cap \Omega \in L^\infty(V \cap \Omega) \quad \text{for each } V \Subset U\}$$

and

$$\tilde{W}(U) = \{f \in L^1_{\text{loc}}(U \cap \Omega); f \,|\, V \cap \Omega \in W(V \cap \Omega) \quad \text{for each } V \Subset U\}.$$

To see that this is true, observe that the spaces on the right define soft analytic Fréchet sheaves, and use Corollary 4.4.3. Consequently the sheaf $\mathscr{F} = \mathscr{F}_\Omega = \widehat{H^\infty(\Omega)}$ possesses the section spaces (on each open set U in $\mathbb{C}$):

$$\mathscr{F}(U) = \{f \in \mathscr{O}(U \cap \Omega); \|f\|_{\infty, V \cap \Omega} < \infty \text{ for each } V \Subset U\}.$$

The following result makes precise the fact that the corona problem is a local boundary problem. Its proof is based on the elementary observations that $\mathscr{F}_\Omega | \Omega = \mathscr{O}_\Omega$ and that $\mathscr{F}_\Omega | D = \mathscr{F}_{\Omega \cap D} | D$ for each open set D in $\mathbb{C}$.

Proposition 7.3.6   *Let Ω be a bounded open set in $\mathbb{C}$. The corona theorem holds on Ω if and only if each point $w \in \partial\Omega$ has an open neighbourhood D in $\mathbb{C}$ such that the corona theorem holds on $\Omega \cap D$.*

Proof   For $f \in H^\infty(\Omega)^m$ with $0 \notin \overline{f(\Omega)}$, the exactness of the Koszul complex $K^\bullet(f, H^\infty(\Omega))$ is equivalent to the exactness of the complex of sheaves $K^\bullet(f, \mathscr{F}_\Omega)$ (see Proposition 4.3.7 and the remark following Corollary 3.1.16). Since $\mathscr{F}_\Omega | \Omega = \mathscr{O}_\Omega$, it suffices to prove the exactness of the second complex in each fixed boundary point $w \in \partial\Omega$. By assumption, the corona theorem holds on $\Omega \cap D$ for a suitable neighbourhood D of w. Hence $K^\bullet(f, \mathscr{F}_{\Omega, w}) = K^\bullet(f, \mathscr{F}_{\Omega \cap D, w})$ is exact. $\square$

By the classical solution of the corona problem on the open unit disc D (see Gamelin 1980) there is, for each $\varepsilon > 0$ and each integer $m \geq 1$, a constant $C(\varepsilon, m)$ such that, for any family of functions $f_1, \ldots, f_m \in H^\infty(D)$ with

$$\varepsilon \leq |f_1(z)| + \cdots + |f_m(z)| \quad \text{and} \quad |f_i(z)| \leq 1 \qquad (z \in D, i = 1, \ldots, m),$$

there are functions $g_1, \ldots, g_m \in H^\infty(D)$ with

$$1 = f_1 g_1 + \cdots + f_m g_m \quad \text{and} \quad \|g_j\|_{\infty, D} \leq C(\varepsilon, m) \qquad (1 \leq j \leq m).$$

The Riemann mapping theorem shows that the same result holds, with the same constants, on each bounded open set Ω in $\mathbb{C}$ with the property that all components of Ω are simply connected.

Let us call an open set Ω in $\mathbb{C}$ *locally simply connected at $\partial\Omega$* if each point $w \in \partial\Omega$ possesses an open neighbourhood D such that all components of $\Omega \cap D$ are simply connected.

Corollary 7.3.7   *The corona theorem holds on each bounded open set Ω in $\mathbb{C}$ that is locally simply connected at $\partial\Omega$.* $\square$

The last result applies to each bounded open set Ω in $\mathbb{C}$ with the property that, for a suitable real number $\delta > 0$, all bounded components of $\mathbb{C} \setminus \Omega$ have diameter at least δ. If Ω has this property and if D is an open, simply connected set in $\mathbb{C}$ with diameter less than δ, then all components of $\Omega \cap D$ are simply connected.

Corollary 7.3.8   *The corona theorem holds on each bounded open set Ω in $\mathbb{C}$ with the property that the diameter of all bounded components of $\mathbb{C} \setminus \Omega$ is bounded from below by a fixed positive constant.*   □

Corollary 7.3.9   *The corona theorem holds on each bounded open set Ω in $\mathbb{C}$ with the property that $\mathbb{C} \setminus \Omega$ has only finitely many components.*   □

Different proofs of these and related results are to be found in Gamelin (1970) and Garnett (1981). However, whether the corona theorem is valid or not on an arbitrary domain in the complex plane is still an open question (see the same references).

Let us call the infimum of all constants $C(\varepsilon, m)$, occurring in the section following Proposition 7.3.6, the best possible *corona constant* for ε and m on D. We end this section by giving upper estimates for the best possible corona constants on domains of the type described in Corollary 7.3.8.

Let Ω be a bounded open set in $\mathbb{C}$ such that each bounded component of $\mathbb{C} \setminus \Omega$ has diameter larger than 2δ, where $\delta > 0$ is a fixed constant. It is an elementary exercise to show that there is a finite open cover $\mathcal{U} = (U_i)_{i \in I}$ of $\overline{\Omega}$ and a family $(\alpha_i)_{i \in I}$ of C^∞-functions $\alpha_i \in \mathscr{E}(\mathbb{C})$ with the following properties:

(a) each U_i is a disc of radius not larger than δ;
(b) the intersection of any four sets (with different indices) in $\mathcal{U}$ is empty;
(c) the family $(\alpha_i)_{i \in I}$ is a partition of unity relative to $\mathcal{U}$, that is,

$$\sum_{j \in I} \alpha_j = 1 \text{ near } \overline{\Omega}, \quad 0 \le \alpha_i \le 1, \quad \mathrm{supp}(\alpha_i) \subset U_i \quad (i \in I);$$

(d) $\max_{i \in I} \|\bar{\partial}\alpha_i\|_{\infty, \mathbb{C}} \le C/\delta$ with some universal constant C independent of the particular value of δ.

The fact that condition (d) can be realized follows from standard constructions of suitable cut-off functions (see, for instance, Lemma 1.4.2 in Malgrange 1966). The reader should observe that all components of any set of the form

$$U_0 \cap U_1 \cap \cdots \cap U_k \cap \Omega \quad (U_0, \ldots, U_k \in \mathcal{U}, \ k = 0, 1, 2, \ldots)$$

are automatically simply connected. This follows from the remarks preceding Corollary 7.3.8.

Let $f = (f_1, \ldots, f_m) \in H^\infty(\Omega)^m$ be an m-tuple of functions with

$$\varepsilon \le |f_1(z)| + \cdots + |f_m(z)| \quad \text{and} \quad |f_i(z)| \le 1 \quad (z \in \Omega, i = 1, \ldots, m).$$

Denote by $\mathscr{F}$ the quasi-coherent sheaf associated with $H^\infty(\Omega)$, and denote by $\mathscr{C}^\bullet(\mathcal{U}, \mathscr{F})$ the alternating Čech complex with respect to the open cover $\mathcal{U}$. Let us consider the double complex $\mathscr{K} = K_\bullet(f, \mathscr{C}^\bullet(\mathcal{U}, \mathscr{F}))$ with qth row given by $(-1)^q K_\bullet(f, \mathscr{C}^q(\mathcal{U}, \mathscr{F}))$ and $\oplus_{\binom{m}{p}} \mathscr{C}^\bullet(\mathcal{U}, \mathscr{F})$ as the pth column. We know that all rows of $\mathscr{K}$ are exact, and that all columns are exact except in degree 0. We shall estimate the norm of a solution $g = (g_1, \ldots, g_m) \in H^\infty(\Omega)^m$ of the equation $f_1 g_1 + \cdots + f_m g_m = 1$.

To do this, we work with the following portion of the double complex $\mathcal{K}$:

$$
\begin{array}{cccc}
0 & 0 & 0 & 0 \\
\uparrow & \uparrow & \uparrow & \uparrow \\
K_3(\mathscr{C}^2) \xrightarrow{\delta} K_2(\mathscr{C}^2) \xrightarrow{\delta} K_1(\mathscr{C}^2) \xrightarrow{\delta} K_0(\mathscr{C}^2) \longrightarrow 0 \\
d\uparrow \qquad d\uparrow \qquad d\uparrow \qquad d\uparrow \\
K_3(\mathscr{C}^1) \xrightarrow{\delta} K_2(\mathscr{C}^1) \xrightarrow{\delta} K_1(\mathscr{C}^1) \xrightarrow{\delta} K_0(\mathscr{C}^1) \longrightarrow 0 \\
d\uparrow \qquad d\uparrow \qquad d\uparrow \\
K_2(\mathscr{C}^0) \xrightarrow{\delta} K_1(\mathscr{C}^0) \xrightarrow{\delta} K_0(\mathscr{C}^0) \longrightarrow 0 \\
i\uparrow \qquad i\uparrow \\
H^\infty(\Omega)^m \xrightarrow{\delta} H^\infty(\Omega)
\end{array}
$$

The lifting $g \in H^\infty(\Omega)^m$ of the function identically equal to 1, through the boundary map $\delta \colon H^\infty(\Omega)^m \to H^\infty(\Omega)$, will be constructed by the familiar zigzag method. More precisely, consider $i(1) = x_0^0 \in K_0(\mathscr{C}^0)$, lift this element through δ to $x_1^0 \in K_1(\mathscr{C}^0)$, take $x_1^1 = d(x_1^0)$, and so on, up to $x_3^2 \in K_3(\mathscr{C}^2)$. Then consider a lifting x_3^1 of x_3^2 through d, take $x_2^1 + \delta(x_3^1)$ and lift this element to x_2^0. Finally observe that the required m-tuple g is given by $i(g) = x_1^0 + \delta(x_2^0)$.

The elements x_p^q ($p, q \geq 0$) are tuples with coefficients in the spaces

$$
\mathscr{F}\big(U_{i_0} \cap \cdots \cap U_{i_q} \cap \Omega\big) \qquad \big((i_0, \ldots, i_q) \in I^{q+1}\big).
$$

A careful analysis of the above procedure will allow us to choose the elements x_p^q in such a way that even the supremum norm of these coefficients remains bounded on all of $U_{i_0} \cap \cdots \cap U_{i_q} \cap \Omega$. Let us denote by $\|x_p^q\|$ the maximum of these supremum norms formed over all coefficients of x_p^q. The complex $K_\bullet(\mathscr{C}^0)$ is a direct sum of Koszul complexes of the form $K_\bullet(f, \mathscr{F}(U))(U \in \mathscr{U})$. But all components of $U \cap \Omega$ are simply connected, so that the solution of the corona problem on the unit disc yields a splitting of the complex $K_\bullet(f, H^\infty(U \cap \Omega))$. If $C(\varepsilon, m)$ denotes the best possible corona constant on the unit disc, then one can choose x_1^0 in such a way that

$$
\|x_1^0\| \leq C(\varepsilon, m).
$$

By the very definition of the coboundary operator in the Čech complex we have

$$
\|x_1^1\| \leq 2C(\varepsilon, m).
$$

Therefore there are constants $C_{\varepsilon, m} > 0$, only depending on ε and m, such that x_3^2 can be chosen as a tuple with

$$
\|x_3^2\| \leq C_{\varepsilon, m}.
$$

The estimation of the norm of possible liftings x_3^1 (satisfying the condition $dx_3^1 = x_3^2$) is a different, but still a standard problem. Let us write $x_3^2 = (\varphi_{ijk})$, where $\varphi_{ijk} \in K_3(\mathscr{F}(U_i \cap U_j \cap U_k))$ and $i, j, k \in I$. One possible solution $x_3^1 = (\psi_{ij})$ of the equation $dx_3^1 = x_3^2$ is

$$\psi_{ij}(z) = \psi_{ij}'(z) - \psi_i''(z) + \psi_j''(z) \qquad (z \in U_i \cap U_j \cap \Omega),$$

where

$$\psi_{ij}' = \sum_{k \in I} \varphi_{ijk} \alpha_k,$$

and the functions ψ_i'' are chosen so that $\bar{\partial}\psi_{ij} = 0$. A possible choice is

$$\psi_i'' = \frac{1}{\pi} \sum_{l \in I} \int_{U_i \cap \Omega} \left((\bar{\partial}\psi_{il}') \alpha_l \right)(\xi)(z - \xi)^{-1} \, d\xi \qquad (z \in U_i \cap \Omega).$$

This choice leads to an estimate of the form

$$\|\psi_i''\| \le (3CC_{\varepsilon,m}/\pi\delta) \int_{D_{2\delta}(0)} |\xi|^{-1} \, d\xi.$$

Hence one can choose x_3^1 in such a way that

$$\|x_2^1 + \delta x_3^1\| \le C_{\varepsilon,m}', \tag{7.3.2}$$

where $C_{\varepsilon,m}'$ is a new constant only depending on ε and m.

Finally, the same procedure gives a lifting x_2^0 of $x_2^1 + \delta x_3^1$. Let us write $x_2^1 + \delta x_3^1 = (\varphi_{ij})$. Then $x_2^0 = (\psi_i)$ can be chosen as

$$\psi_i = \sum_{j \in I} \varphi_{ij}\alpha_j - h,$$

where $h \in K_2(\mathscr{E}(\Omega))$ is a (componentwise) solution of

$$\bar{\partial}h \,|\, U_i \cap \Omega = \sum_{j \in I} \varphi_{ij}\bar{\partial}\alpha_j \qquad (i \in I).$$

Note that the sums on the right-hand side are the restrictions of a suitable function $g \in K_2(\mathscr{E}(\Omega))$ onto $U_i \cap \Omega$ ($i \in I$). Again, a possible choice for h is the Cauchy transform

$$h(z) = \frac{1}{\pi} \int_\Omega g(\xi)(z - \xi)^{-1} \, d\xi \qquad (z \in \Omega).$$

This choice gives the estimate

$$\|h\|_{\infty,\Omega} \le (3C_{\varepsilon,m}'C/\pi\delta) \int_\Omega \frac{1}{|\xi - z|} \, d\xi.$$

The remaining integral can be estimated roughly from above by $2\pi \, \mathrm{diam}(\Omega)$. Summarizing these estimates, we obtain the following result.

Theorem 7.3.10   *For $\varepsilon > 0$ and $m \geq 1$, there are constants $C_{\varepsilon,m}$ (depending only on ε and m) with the following property.*

If Ω is a bounded open set in $\mathbb{C}$ with the property that all bounded components of $\mathbb{C} \setminus \Omega$ have diameter larger than δ, then the corona problem with data $f_1, \ldots, f_m \in H^\infty(\Omega)$ such that

$$\varepsilon \leq |f_1(z)| + \cdots + |f_m(z)|, \qquad |f_i(z)| \leq 1 \quad (z \in \Omega, i = 1, \ldots, m)$$

has solutions $g_1, \ldots, g_m \in H^\infty(\Omega)$ (that is, $\sum_i f_i g_i = 1$) satisfying the estimate

$$\|g_i\|_{\infty,\Omega} \leq C_{\varepsilon,m}(1 + \operatorname{diam}(\Omega)\delta^{-1}). \qquad \square \qquad (7.3.3)$$

To solve the corona problem on an arbitrary domain in $\mathbb{C}$, it would be sufficient to show that the best possible corona constants, in the above setting, remain bounded when δ approaches 0 (see Garnett 1981). Although the estimate (7.3.3) is not the best possible, and although it can be obtained by a direct argument based on a partition of unity (see Garnett 1981), the above Čech resolution method deserves attention because of the universal estimate (7.3.2). Whether it is possible to push down this uniform estimate to a solution of the corona problem which does not involve δ remains to be decided in the future.

7.4   A PSEUDOCONVEXITY CRITERION

Let Ω be an open set in $\mathbb{C}^n$. A classical theorem in complex analysis says that Ω is pseudoconvex (that is, possesses a plurisubharmonic exhaustion function) if and only if it is a domain of holomorphy. One possible definition of the latter property, which is suitable for us, is the following. An open set Ω in $\mathbb{C}^n$ is a *domain of holomorphy* if there are no non-empty open sets Ω_1, Ω_2 in $\mathbb{C}^n$ with Ω_2 connected and $\Omega_1 \subset \Omega_2 \cap \Omega \neq \Omega_2$ such that, for each function $f \in \mathcal{O}(\Omega)$, there is a function $F \in \mathcal{O}(\Omega_2)$ with $F|_{\Omega_1} = f|_{\Omega_1}$ (see Section 2.5 in Hörmander 1966).

It is an elementary exercise to show that an open set in $\mathbb{C}^n$ is a domain of holomorphy if, for each point $\lambda \in \mathbb{C}^n \setminus \Omega$, there are functions $f_1, \ldots, f_n$ in $\mathcal{O}(\Omega)$ with

$$\sum_{i=1}^{n} (\lambda_i - z_i) f_i(z) = 1 \qquad (z \in \Omega). \qquad (7.4.1)$$

Indeed, if Ω_1 and Ω_2 were open sets in $\mathbb{C}^n$ satisfying all the above conditions, then one could choose a point $\lambda \in \Omega_2 \setminus \Omega$, functions $f_1, \ldots, f_n$ in $\mathcal{O}(\Omega)$ satisfying (7.4.1), and extensions $F_i \in \mathcal{O}(\Omega_2)$ of $f_i|_{\Omega_1}$ $(1 \leq i \leq n)$. By the identity theorem, (7.4.1) would also hold for the functions F_i on Ω_2, which is impossible. If $\Omega \subset \mathbb{C}^n$ is an open set with $\Omega = \operatorname{Int}(\overline{\Omega})$, then it suffices to find, for each point $\lambda \in \mathbb{C}^n \setminus \overline{\Omega}$, functions $f_1, \ldots, f_n \in \mathcal{O}(\Omega)$ satisfying (7.4.1) to conclude that Ω is a domain of holomorphy.

It is the purpose of this section to develop pseudoconvexity criteria which are formulated in terms of analytic functions that satisfy certain growth conditions near the boundary of their domain. This amounts, via the usual $\bar{\partial}$-techniques, to replacing the analytic cohomology groups $H^q(\Omega, \mathcal{O})$ by the cohomology groups of certain quasi-coherent sheaves, similar to the localized Bergman space. Our approach is motivated by a recent remark due to Ohsawa (1989), which relates the holomorphy of a domain to the vanishing of its analytic L^2-cohomology groups.

Throughout this section, Ω will denote an open set in $\mathbb{C}^n$ with $\Omega = \mathrm{Int}(\bar{\Omega})$. Let $H(\Omega)$ be a Fréchet $\mathcal{O}(\mathbb{C}^n)$-module of analytic functions on Ω which contains the function identically equal to one. Some standard examples of such function spaces are given by $\mathcal{O}(\Omega)$, $A^\alpha(\bar{\Omega})$ ($\alpha > 0$), or $L^p(\Omega) \cap \mathcal{O}(\Omega)$ ($1 \le p \le \infty$), with the notation of the preceding sections.

Theorem 7.4.1 *Let Ω be an open set in $\mathbb{C}^n$ with $\Omega = \mathrm{Int}(\bar{\Omega})$. Each of the following conditions implies that Ω is a domain of holomorphy:*

(a) *$H(\Omega)$ is a quasi-coherent $\mathcal{O}(\mathbb{C}^n)$-module such that the associated sheaf is supported by $\bar{\Omega}$;*
(b) *there exists a finite exact sequence of Fréchet $\mathcal{O}(\mathbb{C}^n)$-modules*

$$0 \to H(\Omega) \to \mathcal{S}^0(\mathbb{C}^n) \to \cdots \to \mathcal{S}^N(\mathbb{C}^n) \to 0$$

with Fréchet soft analytic sheaves $\mathcal{S}^0, \ldots, \mathcal{S}^N$ supported by $\bar{\Omega}$;
(c) *$H(\Omega) \perp_{\mathcal{O}(\mathbb{C}^n)} \mathcal{E}(\mathbb{C}^n)$, and the Fréchet sheaf $H(\Omega) \hat{\otimes}_{\mathcal{O}(\mathbb{C}^n)} \mathcal{E}$ is supported by the set $\bar{\Omega}$.*

Proof Suppose that condition (c) holds. Then the $\bar{\partial}$-complex with coefficients in $H(\Omega) \hat{\otimes}_{\mathcal{O}(\mathbb{C}^n)} \mathcal{E}(\mathbb{C}^n)$ provides a resolution of the space $H(\Omega)$ as in (b) (see the proof of Theorem 6.4.15).

Hence condition (c) implies condition (b). It follows from Theorem 4.4.1 and Corollary 3.1.16 that condition (b) implies the validity of condition (a).

Suppose that condition (a) holds, and consider an arbitrary point $\lambda \in \mathbb{C}^n \setminus \bar{\Omega}$. Let $\mathcal{F}$ be the analytic quasi-coherent sheaf associated with $H(\Omega)$. By assumption $\mathcal{F}$ is supported by $\bar{\Omega}$. The very definitions of the quasi-coherence property and of the associated quasi-coherent sheaf yield the relations

$$H(\Omega) \perp_{\mathcal{O}(\mathbb{C}^n)} \mathcal{O}(V) \quad \text{and} \quad H(\Omega) \hat{\otimes}_{\mathcal{O}(\mathbb{C}^n)} \mathcal{O}(V) = 0$$

for each Stein open neighbourhood V of λ in $\mathbb{C}^n \setminus \bar{\Omega}$ (Chapter 4). Therefore the Koszul complex $K_\bullet(z - w, H(\Omega) \hat{\otimes} \mathcal{O}(V))$ is exact, and similarly, its localization at $w = \lambda$ is exact. In particular, the map

$$\bigoplus_{i=1}^{n} H(\Omega) \xrightarrow{z - \lambda} H(\Omega)$$

is onto. Therefore there are elements $f_1, \ldots, f_n \in H(\Omega)$ satisfying condition (7.4.1). Hence Ω is a domain of holomorphy. $\square$

Theorem 4.4.4 actually implies that conditions (a) and (b) are equivalent. The second condition is stated separately because it is easier to check it in concrete cases. We also remark that, in certain cases, condition (a) is equivalent to the pseudoconvexity of Ω.

Such an example is the Bergman space $L_a^2(\Omega) = L^2(\Omega) \cap \mathcal{O}(\Omega)$. According to Hörmander (1965), this space possesses on each bounded pseudoconvex domain Ω a resolution like that described in condition (b). The next chapter will be entirely devoted to Bergman spaces and the spectral theory of certain natural operators on these spaces. We confine ourselves here to noticing the following consequence of Theorem 7.4.1.

Corollary 7.4.2 *Let Ω be a bounded domain in $\mathbb{C}^n$ with $\Omega = \text{Int}(\overline{\Omega})$. Then Ω is pseudoconvex if and only if the Bergman space $L_a^2(\Omega)$ is a quasi-coherent $\mathcal{O}(\mathbb{C}^n)$-module supported by $\overline{\Omega}$.* $\square$

A similar result can be obtained by using Kohn's estimates for the $\bar{\partial}$-operator on $\mathcal{E}(\overline{\Omega})$, in the case when the boundary of Ω is smooth (see Theorem 7.3.1 above).

Corollary 7.4.3 *Let Ω be a bounded domain in $\mathbb{C}^n$ with smooth boundary. Then Ω is pseudoconvex if and only if $H^q(\bar{\partial}, \mathcal{E}(\overline{\Omega})) = 0$ for $q \geq 1$.* $\square$

Above we have denoted by $H^\bullet(\bar{\partial}, \mathcal{E}(\overline{\Omega}))$ the cohomology of the $\bar{\partial}$-complex with coefficients in the algebra of smooth functions on $\overline{\Omega}$.

By using Henkin's integral representation formula for analytic functions defined on pseudoconvex domains with piecewise smooth boundary, one can state an analogue of Corollary 7.4.3 for the algebra $C(\overline{\Omega})$ instead of $\mathcal{E}(\overline{\Omega})$. Along the same lines, vanishing conditions like $H^q(\bar{\partial}, \text{Lip}_\alpha(\overline{\Omega})) = 0$ provide pseudoconvexity criteria for domains with a sufficiently smooth boundary $\partial\Omega$ (for further details see Henkin and Leiterer 1983).

Finally, it should be mentioned that Sibony (1987) exhibited an example of a pseudoconvex domain Ω in $\mathbb{C}^n$ with smooth boundary which is not $H^\infty(\Omega)$-pseudoconvex in the above sense.

7.5 REFERENCES AND COMMENTS

The division of vector-valued distributions by analytic functions is still, generally speaking, an open subject. An immediate consequence of any result in this direction is the existence of fundamental solutions of linear partial differential operators depending in a prescribed way on parameters (see Mantlik 1990 for such a recent result). The interplay between subscalar operators and the division of distributions by analytic functions is relatively new (see Eschmeier and Putinar 1988, 1989). It is our feeling that this relationship has not yet been fully exploited.

One can show using Corollary 6.4.14 that to prove Corollary 7.2.2 it suffices to prove the exactness of the following Dolbeault complex:

$$0 \to I \to \Lambda^0(J) \xrightarrow{\bar{\partial}} \Lambda^1(J) \xrightarrow{\bar{\partial}} \cdots \xrightarrow{\bar{\partial}} \Lambda^n(J) \to 0,$$

where $I = (f_1, \ldots, f_r)\mathscr{O}(U)$, J denotes the closure of $(f_1, \ldots, f_r)\mathscr{E}(U)$ in the Fréchet space $\mathscr{E}(U)$, and U is a Stein open subset of the given domain Ω. An idea of the proof of the exactness of the above complex, also based on the Whitney spectral theorem and its algebraic corollaries, is discussed in Putinar (1991).

The corona problem still represents an active area of research. First because it is not yet settled for arbitrary domains of the complex plane, and second, because it is still open on any classical domain in $\mathbb{C}^n$. The relation between the corona problem and joint spectra is not new (see, for instance, Li 1992). It is interesting to remark that Hörmander (1967) has used the Koszul complex in an approach to the classical corona problem, prior to Taylor's definition of the joint spectrum (Taylor 1970a).

The zig-zag analysis which appears in the proof of Theorem 7.3.10 is well known to experts in the corona problem (see again Li 1992).

8
Spectral analysis on Bergman spaces

Among the examples on which the techniques developed throughout our book can be tested, there is a privileged one: the class of tuples of analytic Toeplitz operators acting on *Bergman spaces* over bounded pseudoconvex domains in $\mathbb{C}^n$. This is for two reasons: first because of the maturity of the function theory on Bergman spaces, and second because of the recent advances in the spectral theory of single Toeplitz operators on one-variable Bergman spaces. To be more specific, the fundamental L^2-estimates for the $\bar{\partial}$-operator proved by Hörmander (1965) permit the solution of basic division and interpolation problems in spaces of analytic functions with control on the L^2-growth at the boundary (see for instance Mityagin and Henkin 1971; Skoda 1972; Ohsawa 1988). On the other hand, the spectral picture of Toeplitz operators acting on one-dimensional Bergman spaces is well developed (see Axler 1988).

The important idea of Stephan Bergman of studying analytic functions by means of Hilbert space techniques is still successfully exploited in many ways. This chapter represents only a very small, and rather atypical, part of this relationship. However, it fits in with the unspoken desire of any operator theorist to describe his favourite class of operators in terms of analytic models.

For us, the Bergman space over a bounded pseudoconvex domain in $\mathbb{C}^n$ will be the prototype of a quasi-coherent $\mathscr{O}(\mathbb{C}^n)$-module. Although this point of view cannot replace the use of L^2-estimates for the $\bar{\partial}$-operator, it has the advantage of unifying and extending many known results on Bergman spaces and their operators. We focus below on two themes: the spectral picture of analytic Toeplitz tuples, and the study of analytically invariant subspaces of the Bergman space. Questions as to the trace-class estimates of commutators of Toeplitz operators, the (unitary) classification of analytically invariant subspaces, C^*-algebra techniques, or quantization in Bergman spaces will not be treated. They may form the subject of a separate monograph, if not of many. The notes at the end of this chapter contain some bibliographical indications of these actual and active directions of research.

8.1 THE GLEASON PROBLEM

The computations of structure spaces contained in the preceding chapter were all based on the same principle. The problem was localized by using resolutions in terms of Fréchet soft analytic sheaves. The same procedure, modulo some refined L^2-estimates, allows us to compute joint spectra of n-tuples of Toeplitz operators.

This first section is intended to develop a general scheme for carrying the spectral computations. The next section will exploit this construction in the particular case of Toeplitz operators with H^∞-symbol acting on Bergman spaces.

The Gleason problem, to which the title of the present section refers, is a division problem for holomorphic functions in the spirit of Chapter 7. More specifically, Gleason's problem is related to the existence of analytic structures in parts of the structure space of a commutative Banach algebra, which in turn is equivalent, by a fundamental result of Gleason, to the finiteness of the respective maximal ideals (in the natural algebraic sense). A generic problem in this area is the following: let B denote the unit ball in $\mathbb{C}^n$, and let $A^0(B) = \mathscr{O}(B) \cap C^0(\overline{B})$ be the analogue of the disc algebra. Can each element $f \in A^0(B)$ which vanishes at the origin be written in the form $f(z) = z_1 g_1(z) + \cdots + z_n g_n(z)$ $(z \in B)$ with $g_1, \ldots, g_n \in A^0(B)$? For a more general version of this problem and further references see Range (1986).

Our next aim is to develop a general framework in which problems of this type can be studied.

Let U be an open subset of $\mathbb{C}^n$. Suppose that $B(U)$ is a linear subspace of $L^1_{\text{loc}}(U)$ equipped with a norm $\|\cdot\|_B$ that turns $B(U)$ into a Banach space continuously embedded in $L^1_{\text{loc}}(U)$. For $0 \le q \le n$, the differential forms of type $(0, q)$ with coefficients in $B(U)$ form a Banach space $\Lambda^q(d\bar{z}, B)$ relative to the norm

$$\|u\|_q = \max_{|i|=q} \|u_i\|_B,$$

where $u = \sum_{|i|=q} u_i \, d\bar{z}_i$.

For $u \in \Lambda^q(d\bar{z}, L^1_{\text{loc}}(U))$ and $v \in \Lambda^{q+1}(d\bar{z}, L^1_{\text{loc}}(U))$, we write $v = \bar{\partial} u$ if this relation holds in the sense of distributions, that is, if for every test function $\varphi \in \mathscr{D}(U)$ and every strictly increasing sequence $1 \le i_1 < \cdots < i_{q+1} \le n$,

$$\sum_{k=1}^{q+1} (-1)^k \int_U u_{i_1 - \hat{i}_k - i_{q+1}} \left(\bar{\partial}_{i_k} \varphi \right) d\lambda(z) = \int_U v_{i_1 - i_{q+1}} \varphi \, d\lambda(z).$$

Here λ denotes the Lebesgue measure on $\mathbb{C}^n$. The condition that $B(U)$ is a continuously embedded Banach subspace of $L^1_{\text{loc}}(U)$ ensures that, for each q, the linear operator $\bar{\partial} \colon \Lambda^q(d\bar{z}, B(U)) \to \Lambda^{q+1}(d\bar{z}, B(U))$ is closed when taken with domain

$$B^q(U) = \{u \in \Lambda^q(d\bar{z}, B(U)); \bar{\partial} u \in \Lambda^{q+1}(d\bar{z}, B(U))\}.$$

Endowed with the graph norm $\|u\| = \|u\|_q + \|\bar{\partial} u\|_{q+1}$, the space $B^q(U)$ becomes a Banach space. Moreover, the $\bar{\partial}$-sequence

$$0 \to B^0(U) \xrightarrow{\bar{\partial}} B^1(U) \xrightarrow{\bar{\partial}} \cdots \xrightarrow{\bar{\partial}} B^n(U) \to 0 \tag{8.1.1}$$

is a complex of continuous linear operators between Banach spaces.

It is well known that each distributional solution of the equation $\bar{\partial} u = 0$ $(u \in \mathscr{D}'(U))$ is an analytic function in U. Thus the kernel of the first map in the complex (8.1.1) is the space $\mathscr{O}B(U) = \mathscr{O}(U) \cap B(U)$, consisting of those elements in $B(U)$ which have a representative in $\mathscr{O}(U)$.

With the notation of Section 7.3 we list a few examples of natural choices for $B(U)$: $C^r(\bar{U})$; $\mathrm{Lip}_\alpha(\bar{U})$ $(0 < \alpha < 1)$; $L^p(U, \lambda)$ $(1 \leq p \leq \infty)$, and so on.

Let us consider the commuting n-tuple $M_\Omega = (M_{z_1}, \ldots, M_{z_n})$ consisting of the multiplication operators with the coordinate functions on $\mathscr{O}B(\Omega)$, where Ω is a bounded open subset of $\mathbb{C}^n$. The spectral picture of this *generalized Bergman n-tuple* will be computed under certain assumptions on the exactness properties of the sequence (8.1.1), both at the local and global level. We already have a few examples of natural exact $\bar{\partial}$-resolutions from Section 7.4, where it was indicated that the set Ω must satisfy certain geometric conditions, as for instance pseudoconvexity, if the complex (8.1.1) is exact.

To guarantee that the $\bar{\partial}$-sequence (8.1.1) localizes in a natural way to a complex of Fréchet soft sheaves we demand that a rule is given that assigns to each open subset U of Ω a continuously embedded Banach subspace $B(U)$ of $L^1_{\mathrm{loc}}(U)$ such that the following conditions are satisfied:

(i) For $V \subset U \subset \Omega$ open, the restriction maps $B(U) \to B(V)$ are well-defined linear operators;
(ii) via the usual multiplication of functions, each space $B(U)$ is an $\mathscr{E}(\mathbb{C}^n)$-module;
(iii) if $\{W_i\}_{i=1}^r$ is an open covering of an open set W in $\mathbb{C}^n$ and if $f: W \cap \Omega \to \mathbb{C}$ is a measurable function such that $f|_{W_i \cap \Omega} \in B(W_i \cap \Omega)$ for all $i = 1, \ldots, r$, then $f|_{V \cap \Omega} \in B(V \cap \Omega)$ for each open set $V \Subset W$;
(iv) the function identically equal to one belongs to $B(\Omega)$.

It follows immediately that all the above examples satisfy these axioms. By the closed graph theorem, the restriction mappings in (i) are automatically continuous. Moreover, for any open set V in Ω and any open neighbourhood U of $\bar{V}$ in $\mathbb{C}^n$, the space $B(V)$ is a topological $\mathscr{E}(U)$-module via the usual pointwise multiplication of functions.

At this point, the Banach spaces of $(0, q)$ forms $B^q(U)$ can be used to define a Fréchet sheaf of $\mathscr{E}_{\mathbb{C}^n}$-modules in a canonical way. More precisely, we claim that

$$\mathscr{F}(U) = \left\{ u \in \Lambda^q(d\bar{z}, L^1_{\mathrm{loc}}(U \cap \Omega)); u|_{V \cap \Omega} \in B^q(V \cap \Omega) \text{ for } V \Subset U \right\},$$

where U runs through all open subsets U of $\mathbb{C}^n$, is a Fréchet $\mathscr{E}_{\mathbb{C}^n}$-submodule of the sheaf $\Lambda^q(d\bar{z}, \mathscr{L}^1_{\mathrm{loc}} | \Omega)$. We indicate briefly a proof of this assertion. Condition (ii) and the identity

$$\bar{\partial}(\varphi u) = (\bar{\partial}\varphi) \wedge u + \varphi \bar{\partial} u, \tag{8.1.2}$$

valid for any $\varphi \in \mathscr{E}(U)$ and $u \in \Lambda^q(d\bar{z}, L^1_{\mathrm{loc}}(U \cap \Omega))$, show that componentwise multiplication turns $\mathscr{F}(U)$ into an $\mathscr{E}(U)$-module, at least algebraically.

The presheaf $\mathscr{F}(U)$ $(U \in \mathbb{C}^n$ open) satisfies the sheaf axioms by virtue of condition (iii). As another interpretation of the sheaf axioms, we can identify the following spaces

$$\mathscr{F}(U) \cong \lim_{\substack{\longleftarrow \\ V \Subset U}} B^q(V \cap \Omega) \cong \lim_{\substack{\longleftarrow \\ n}} B^q(U_n \cap \Omega)$$

for any relatively compact, open exhaustion $U_n \uparrow U$ and $U \subset\subset \mathbb{C}^n$ open. Via these identifications, $\mathscr{F}(U)$ inherits the structure of a Fréchet $\mathscr{E}(U)$-module. Indeed, the third space is canonically a Fréchet $\mathscr{E}(U)$-module in view of condition (ii) and formula (8.1.2). To check that all the restriction mappings $\mathscr{F}(U) \to \mathscr{F}(V)$ are continuous, the above identifications can again be used.

Let us denote by $\mathscr{B}^q$ the Fréchet soft sheaf described above. It is supported by $\overline{\Omega}$, and its global sections on $\mathbb{C}^n$ form the Banach space $B^q(\Omega) = \mathscr{B}^q(\mathbb{C}^n)$.

In the following, we fix a bounded open subset Ω of C^n, and we suppose that, for every open subset U of Ω, a continuously embedded Banach space $B(U) \subset L^1_{\mathrm{loc}}(U)$ is given such that conditions (i)–(iv) are satisfied, and such that the augmented global $\overline{\partial}$-sequence

$$0 \to \mathscr{O}B(\Omega) \to B^0(\Omega) \overset{\overline{\partial}}{\to} B^1(\Omega) \overset{\overline{\partial}}{\to} \cdots \overset{\overline{\partial}}{\to} B^n(\Omega) \to 0 \qquad (8.1.3)$$

is exact. If in addition, $\Omega = \mathrm{Int}(\overline{\Omega})$, then Theorem 7.4.1 shows that Ω is necessarily pseudoconvex.

Since the spaces $B^q(\Omega)$ $(q = 0, \ldots, n)$ are Banach $\mathscr{E}(\mathbb{C}^n)$-modules, the n-tuple $M_\Omega = (M_{z_1}, \ldots, M_{z_n})$ satisfies condition $(\beta)_{\mathscr{E}}$. On the other hand, the sheaf model $\mathscr{F}$ of M_Ω has the induced $\overline{\partial}$-resolution (see Section 4.4)

$$0 \to \mathscr{F} \to \mathscr{B}^0 \to \mathscr{B}^1 \to \cdots \to \mathscr{B}^n \to 0.$$

It is straightforward to compute the sections of this sheaf on an arbitrary open set $U \subset \mathbb{C}^n$:

$$\mathscr{F}(U) = \{f \in \mathscr{O}(U \cap \Omega); f|_{V \cap \Omega} \in B(V \cap \Omega) \quad \text{for any } V \Subset U\}.$$

This description gives in a natural way the Fréchet topology on $\mathscr{F}$. Secondly, we infer from the same description that $\mathscr{F}|_\Omega = \mathscr{O}|_\Omega$.

To analyze the spectral behaviour of the n-tuple M_Ω on the space of analytic functions $\mathscr{O}B(\Omega)$, we have to study the induced Koszul complex $K_\bullet(\lambda - M_\Omega, \mathscr{O}B(\Omega))$. Since M_Ω satisfies condition (β), and since the associated sheaf is supported by $\overline{\Omega}$, the above complex is exact for $\lambda \notin \overline{\Omega}$.

Let us fix a point $\lambda \in \Omega$, and let $x \in \mathbb{C}^n \setminus \{\lambda\}$. Because the n-tuple of analytic functions $\lambda_1 - z_1, \ldots, \lambda_n - z_n$ has no common zero in a neighbourhood of x, we infer the exactness of the complex $K_\bullet(\lambda - M_\Omega, \mathscr{F}_x)$. On the other hand, because $\mathscr{F}_\lambda \cong \mathscr{O}_\lambda$, the augmented Koszul complex

$$K_\bullet(\lambda - M_\Omega, \mathscr{F}_\lambda) \to \mathscr{O}_\lambda / m_\lambda \to 0$$

is exact (see Chapter 4). Here m_λ stands for the maximal ideal of the local algebra $\mathscr{O}_\lambda$ of germs of analytic functions at λ.

Since $\mathscr{F}$ is acyclic on $\mathbb{C}^n$ (as a quasi-coherent sheaf), one finally obtains the exactness of the complex

$$K_\bullet(\lambda - M_\Omega, \mathscr{O}B(\Omega)) \xrightarrow{\varepsilon_\lambda} \mathbb{C} \to 0,$$

where $\varepsilon_\lambda \colon \mathscr{O}B(\Omega) \to \mathbb{C}$ is the evaluation map at the point $\lambda \in \Omega$. In particular, if a function $f \in \mathscr{O}B(\Omega)$ vanishes at the point λ, then

$$f(z) = (\lambda_1 - z_1)g_1(z) + \cdots + (\lambda_n - z_n)g_n(z) \qquad (z \in \Omega)$$

with suitable functions $g_1, \ldots, g_n \in \mathscr{O}B(\Omega)$. Thus we have solved the Gleason problem for $\mathscr{O}B(\Omega)$.

As a conclusion of the above considerations, we can state the following theorem.

Theorem 8.1.1 *Let M_Ω denote the n-tuple consisting of the multiplication operators with the coordinate functions on a space $\mathscr{O}B(\Omega)$ as above. Then $\sigma(M_\Omega, \mathscr{O}B(\Omega)) = \overline{\Omega}$, and $\Omega \subset \rho_e(M_\Omega, \mathscr{O}B(\Omega))$ with the Fredholm index given by*

$$\mathrm{ind}(\lambda - M_\Omega) = 1 \qquad (\lambda \in \Omega).$$

More precisely, for all $\lambda \in \Omega$, one has

$$\dim H_p(\lambda - M_\Omega, \mathscr{O}B(\Omega)) = \begin{cases} 0, & p \neq 0 \\ 1, & p = 0. \end{cases} \qquad \square$$

As a consequence of the fact that the n-tuple M_Ω possesses property (β), we obtain the following result (see Corollary 6.3.4 and the preceding remarks).

Corollary 8.1.2 *With the same notations as in Theorem 8.1.1, we have*

$$\sigma(M_\Omega, \mathscr{O}B(\Omega)) = \sigma_r(M_\Omega, \mathscr{O}B(\Omega))$$

and

$$\sigma_e(M_\Omega, \mathscr{O}B(\Omega)) = \sigma_{re}(M_\Omega, \mathscr{O}B(\Omega)) \subset \partial\Omega,$$

where equality holds whenever $\Omega = \mathrm{Int}(\overline{\Omega})$. $\square$

We remark that there are examples of domains Ω and spaces $B(\Omega)$ for which $\sigma_e(M_\Omega, \mathscr{O}B(\Omega))$ is different from $\partial\Omega$ (cf. Axler *et al.* 1982).

In the proof of Theorem 8.1.1, we have observed that the n-tuple M_Ω satisfies property $(\beta)_\mathscr{E}$, that is, the Koszul complex

$$K_\bullet\big(w - M_\Omega, \mathscr{E}(\mathbb{C}^n) \,\hat{\otimes}\, \mathscr{O}B(\Omega)\big)$$

is exact in positive degrees and has separated homology in degree zero. By a standard duality argument, one obtains the next result.

Proposition 8.1.3 *Under the previous assumptions on $B(\Omega)$, every distribution $u \in \mathscr{D}'(\mathbb{C}^n) \,\hat{\otimes}\, [\mathscr{O}B(\Omega)]'$ can be decomposed as*

$$u = (w_1 - M'_{z_1})u_1 + \cdots + (w_n - M'_{z_n})u_n,$$

where $u_1, \ldots, u_n \in \mathscr{D}'(\mathbb{C}^n) \,\hat{\otimes}\, [\mathscr{O}B(\Omega)]'$. $\square$

Here [ ]′ denotes the topological duals of the respective Banach spaces or linear operators.

The resolution (8.1.3) provides some significant examples within the theory of decomposable operators. To demonstrate this, let us write

$$Z^q = \mathrm{Ker}(\bar{\partial}: B^q \to B^{q+1}) \qquad (0 \le q \le n),$$

where $B^q = B^q(\Omega)$ and $B^{n+1} = \{0\}$.

The Banach spaces Z^q are quasi-coherent $\mathcal{O}(\mathbb{C}^n)$-modules. The next result describes some properties of the multiplication tuples $M_z \in L(Z^q)^n$ and their sheaf models $\tilde{Z}^q$.

Proposition 8.1.4   *With the preceding assumptions and notations, the following assertions hold*:

(a) $\sigma(M_z, Z^q) = \overline{\Omega}$ $(0 \le q \le n)$;
(b) *for every finite open covering* $(U_i)_{i \in I}$ *of* $\mathbb{C}^n$ *and every* $q \ge 1$, *the sheaves* $\tilde{Z}^q$
   satisfy

$$Z^q = \sum_{i \in I} \Gamma_{\overline{U}_i}(\mathbb{C}^n, \tilde{Z}^q);$$

(c) *the sheaves* $\tilde{Z}^q$ $(0 \le q \le n - 1)$ *are not soft*;
(d) *the systems of operators* $M_z|_{Z^q}$ $(0 \le q \le n)$ *are pairwise non-similar*.

Proof   Let $q \ge 0$ be fixed. The $\mathcal{O}(\mathbb{C}^n)$-module Z^q is quasi-coherent because of the existence of the truncated resolution

$$0 \to Z^q \to B^q \to B^{q+1} \to \cdots \to B^n \to 0.$$

This shows in particular that the sheaf $\tilde{Z}^q$ is supported by $\overline{\Omega}$, whence assertion (a) follows.

For $q \ge 1$, the sheaf $\tilde{Z}^q$ is a quotient of a soft sheaf, and this implies part (b).

Let D be a relatively compact open polydisc contained in Ω. The long exact sequences of cohomology induced by the resolution (8.1.3) yield the following relations for the cohomology spaces with compact support:

$$H_c^p(D, \tilde{Z}^q) \cong \begin{cases} H_c^n(D, \tilde{Z}^0), & \text{if } p + q = n \\ 0, & \text{otherwise.} \end{cases}$$

But $\tilde{Z}^0|_D \cong \mathcal{O}|_D$ and $H_c^n(D, \tilde{Z}^0) \cong \mathcal{O}(D)'$ by Serre duality (see Proposition 6.1.8). This shows that the sheaves $\tilde{Z}^q$ are not soft for $q < n$, and moreover, these sheaves cannot be isomorphic as $\mathcal{O}$-modules, since their cohomology is different.

This completes the proof of Proposition 8.1.4. $\square$

The method used to prove Theorem 8.1.1 admits a direct generalization to the case of tuples of Toeplitz operators with more general analytic symbols. Let us denote by $M(\Omega)$ the set of all analytic functions f on Ω with the property that $B(\Omega)$ is stable under multiplication by each of the functions f, $\bar{\partial}_i f$, and $\varphi \circ f$ $(i = 1, \ldots, n, \ \varphi \in \mathscr{E}(\mathbb{C}))$. Using the identity (8.1.2), the chain rule, and the closed graph theorem, one can show that, for a given function f in $M(\Omega)$, the multiplication operators

$$M_{\varphi \circ f} \colon B^q \to B^q, \quad u \mapsto (\varphi \circ f)u \qquad (0 \le q \le n, \ \varphi \in \mathscr{E}(\mathbb{C}))$$

yield a well-defined continuous C^∞-functional calculus

$$\mathscr{E}(\mathbb{C}) \to L(B^q), \quad \varphi \mapsto M_{\varphi \circ f} \qquad (0 \le q \le n)$$

for the operator $M_f \in L(B^q)$. Here, as before, $B^q = B^q(\Omega)$.

Theorem 8.1.5 *Let $f = (f_1, \ldots, f_m) \in M(\Omega)^m$ be given. Then the m-tuple of multiplication operators $M_f = (M_{f_1}, \ldots, M_{f_m}) \in L(\mathscr{O}B(\Omega))^m$ satisfies property $(\beta)_{\mathscr{E}}$ and*

$$\sigma\big(M_f, \mathscr{O}B(\Omega)\big) = \overline{f(\Omega)}.$$

Proof The unique continuous linear map $\Phi \colon \mathscr{E}(\mathbb{C}^m) \to L(B^q)$ with

$$\Phi(\varphi_1 \otimes \cdots \otimes \varphi_m) = M_{\varphi_1 \circ f_1} \cdot \cdots \cdot M_{\varphi_m \circ f_m} \qquad (\varphi_1, \ldots, \varphi_m \in \mathscr{E}(\mathbb{C}))$$

defines a continuous $\mathscr{E}(\mathbb{C}^m)$-functional calculus for M_f. Since the polynomials in z and $\bar{z}$ are dense in $\mathscr{E}(\mathbb{C}^m)$ (Chapter 15 in Treves 1967), it follows that

$$\Phi(\varphi) = M_{\varphi \circ f} \qquad (\varphi \in \mathscr{E}(\mathbb{C}^m)).$$

Theorem 6.4.15, applied to the exact complex (8.1.3), yields that M_f satisfies property $(\beta)_{\mathscr{E}}$. Since $\Phi(\varphi) = 1$ for each function $\varphi \in \mathscr{E}(\mathbb{C}^m)$ with $\varphi = 1$ on $f(\Omega)$, we have (Lemma 2.2.3 and the proof of Theorem 6.1.13)

$$\sigma\big(M_f, \mathscr{O}B(\Omega)\big) \subset \bigcup_{q=0}^{n} \sigma(M_f, B^q) \subset \overline{f(\Omega)}.$$

To prove the reverse inclusion, observe that the last map in the Koszul complex $K_\bullet(\lambda - M_f, \mathscr{O}B(\Omega))$,

$$\mathscr{O}B(\Omega)^m \xrightarrow{\ \lambda - M_f\ } \mathscr{O}B(\Omega),$$

cannot be onto if $\lambda \in f(\Omega)$. $\square$

So far, this is the most general result of this type known for abstract spaces such as $\mathscr{O}L^p(\Omega)$ $(1 \le p \le \infty)$; $\mathscr{O}\operatorname{Lip}_\alpha(\overline{\Omega})$ $(\alpha \in (0,1))$; $A^r(\overline{\Omega}) = \mathscr{O}C^r(\overline{\Omega})$ $(r \ge 0)$. Of course we have to note that, while conditions (i)–(iv) are obviously satisfied in these cases, the exactness of the complex (8.1.3) is far from being trivial. However, for Ω strictly pseudoconvex with smooth boundary, this condition is also satisfied (cf. Section 7.4).

Next we indicate a possible method for computing essential joint spectra in the same general setting. Although not the most general result of this kind, the next proposition is typical.

Proposition 8.1.6 *Suppose in addition to the previous assumptions that Ω is pseudoconvex. Let $f: \Omega \to \Omega'$ be a proper holomorphic map into an open subset Ω' of $\mathbb{C}^m$ with coordinate functions in $M(\Omega)$. Then $\Omega' \subset \rho_e(M_f, \mathcal{O}B(\Omega))$, and*

$$\mathrm{ind}(\lambda - M_f) = \mathrm{rk}_\lambda(f_* \mathcal{O}_\Omega) \qquad (\lambda \in \Omega').$$

Proof Since M_f has property (β), the complex $K_\bullet(w - M_f, \mathcal{O}_{\Omega'} \,\hat{\otimes}\, \mathcal{O}B(\Omega))$ is exact in all positive degrees. By Proposition 10.1.3, it suffices to show that the homology sheaf of this complex in degree zero is isomorphic on Ω' to the coherent direct image sheaf $f_* \mathcal{O}_\Omega$ (see Section 9.3).

It follows from Proposition 4.3.10 that, for each Stein open set V in $\mathbb{C}^m$, the augmented Koszul complex

$$K_\bullet\!\left(w - M_f, \mathcal{O}(V) \,\hat{\otimes}\, \mathcal{O}(\Omega)\right) \xrightarrow{\Delta} \mathcal{O}(f^{-1}(V)) \to 0$$

is exact, where Δ is the unique continuous linear map with $\Delta(g \otimes h) = (g \circ f)(h \,|\, f^{-1}(V))$ for $g \in \mathcal{O}(V)$ and $h \in \mathcal{O}(\Omega)$. Therefore it suffices to show that the restriction map induces a quasi-isomorphism (that is, an isomorphism at the level of the homology sheaves)

$$K_\bullet\!\left(w - M_f, \mathcal{O}_{\Omega'} \,\hat{\otimes}\, \mathcal{O}B(\Omega)\right) \to K_\bullet\!\left(w - M_f, \mathcal{O}_{\Omega'} \,\hat{\otimes}\, \mathcal{O}(\Omega)\right).$$

Let us denote by $\mathscr{F}$ the sheaf model of M_Ω on $\mathcal{O}B(\Omega)$, and as before, by $\mathscr{B}^q$ $(0 \le q \le n)$ the sheaf model of M_z on $B^q(\Omega)$. To prove that the above morphism of complexes is a quasi-isomorphism, it suffices to show that the corresponding restriction morphisms

$$K_\bullet\!\left(w - M_f, \mathcal{O}_{\Omega'} \,\hat{\otimes}\, B^q(\Omega)\right) \to K_\bullet\!\left(w - M_f, \mathcal{O}_{\Omega'} \,\hat{\otimes}\, \mathscr{B}^q(\Omega)\right)$$

are quasi-isomorphisms for $q = 0, \ldots, n$. Note that both complexes are exact in all positive degrees. For the complex on the right, this assertion follows from Proposition 4.3.10 applied to the quasi-coherent $\mathcal{O}_\Omega$-module $\mathscr{B}^q \,|\, _\Omega$. Fix a relatively compact Stein open set D in Ω', and a function $\theta \in \mathscr{D}(\Omega)$ with $\theta = 1$ on $f^{-1}(\overline{D})$. Our axioms for the spaces $B(U)$ $(U \subset \Omega)$ imply that

$$\mathscr{B}^q(\Omega) \to B^q(\Omega), \qquad u \mapsto \theta u$$

is a well-defined continuous linear operator.

To complete the proof, we show that $1 \otimes \theta$ acts as the identity operator on the spaces

$$H_0\!\left(w - M_f, \mathcal{O}(D) \,\hat{\otimes}\, B^q(\Omega)\right) \quad \text{and} \quad H_0\!\left(w - M_f, \mathcal{O}(D) \,\hat{\otimes}\, \mathscr{B}^q(\Omega)\right).$$

For the space on the right, this follows from Proposition 4.3.10, since the composition

$$\mathcal{O}(D) \,\hat{\otimes}\, \mathscr{B}^q(\Omega) \xrightarrow{1 - 1 \otimes \theta} \mathcal{O}(D) \,\hat{\otimes}\, \mathscr{B}^q(\Omega) \to \mathscr{B}^q(f^{-1}(D))$$

acts as the zero operator. To prove the same assertion for the space on the left, one can use the representations

$$H_0\left(w - M_f, \mathscr{O}(D) \,\hat{\otimes}\, B^q(\Omega)\right) = \varprojlim_{V \in D} B^q(\Omega)/B^q(\Omega)_{M_f}(\mathbb{C}^m \setminus V)$$

and

$$B^q(\Omega)_{M_f}(\mathbb{C}^m \setminus V) = \bigcap \left(\operatorname{Ker} M_{\varphi \circ f};\ \varphi \in \mathscr{E}(\mathbb{C}^m) \ \text{ with } \operatorname{supp}(\varphi) \subset V\right)$$

(see Proposition 6.1.12 and the proof of Theorem 6.1.13), together with the obvious fact that $(\varphi \circ f)(1 - \theta)B^q(\Omega) = 0$ for each function $\varphi \in \mathscr{E}(\mathbb{C}^m)$ with $\operatorname{supp}(\varphi) \subset D$. Thus the proof is complete. $\square$

We remark that the rank of the sheaf $f_* \mathscr{O}_\Omega$ is locally constant on Ω', and that it coincides with the number of the sheets of the finite covering $f\colon \Omega \to \Omega'$ at points of Ω' that are not contained in the ramification locus (for more details, see Whitney 1972).

8.2 TOEPLITZ OPERATORS WITH H^∞-SYMBOL

In this section we specialize the abstract construction presented in the previous section to the case of multivariable Bergman spaces. Exactly as above, the main technical difficulty is the exactness of the resolution (8.1.3), which for L^2-spaces is resolved by the L^2-estimates of Hörmander (1965). Moreover, a refinement of these estimates due to Skoda (1972) will allow us to compute joint spectra of tuples of Toeplitz operators with bounded analytic symbols. The computation of essential joint spectra presents a more delicate problem, which we are going to solve only for domains with smooth boundary. Compared with the case of Toeplitz operators on one-variable Bergman spaces, much remains to be done in this multivariable setting.

Let Ω be a bounded pseudoconvex domain in $\mathbb{C}^n$. The choice $B(\Omega) = L^2(\Omega, \lambda)$ for the abstract $\mathscr{E}(\mathbb{C}^n)$-module $B(\Omega)$ in Section 8.1 obviously satisfies conditions (i)–(iv), where λ denotes the Lebesgue measure in $\mathbb{C}^n$. The space $\mathscr{O}L^2(\Omega, \lambda) = \mathscr{O}(\Omega) \cap L^2(\Omega, \lambda)$ will be denoted by $L_a^2(\Omega)$. This is the *Bergman space associated with the domain* Ω, and it carries a reproducing kernel, the *Bergman kernel*, which reflects both the geometry of Ω and the function theory supported on Ω.

For simplicity, we write $L^2(\Omega) = L^2(\Omega, \lambda)$, $L_{(0,q)}^2(\Omega) = \Lambda^q(d\bar{z}, L^2(\Omega, \lambda))$, and $f = M_f$ for a given function f in $H^\infty(\Omega)$.

The main result of this section is the following generalization of the first part of Theorem 8.1.1.

Theorem 8.2.1 *Let Ω be a bounded pseudoconvex domain in $\mathbb{C}^n$, and let $f \in H^\infty(\Omega)^m$. Then $\sigma(f, L_a^2(\Omega)) = \overline{f(\Omega)}$.*

Proof   The proof relies on techniques of Skoda (1972), and it will be divided into several steps. First, we need an abstract functional analytic lemma.

Lemma 8.2.2   *Let H_1, H_2, and H_3 be Hilbert spaces, let $T_1: H_1 \to H_2$ be a bounded linear operator, and let $T_2: H_1 \to H_3$ be a densely defined, closed operator.*

If $G_2 \subset H_2$ is a closed subspace with $T_1(\operatorname{Ker} T_2) \subset G_2$, then the identity

$$T_1(\operatorname{Ker} T_2) = G_2$$

holds if and only if there is a constant $C > 0$ such that

$$\|T_1^* y + T_2^* z\| \geq C\|y\|$$

holds for all $y \in G_2$ and $z \in \operatorname{Dom} T_2^*$. In this case, there is, for each $y \in G_2$, a vector $x \in \operatorname{Ker} T_2$ with $T_1 x = y$ and $\|x\| \leq (1/C)\|y\|$.

Proof   Since T_1 is a bounded operator, the equality $T_1(\operatorname{Ker} T_2) = G_2$ holds if and only if the compressed operator $P_{\operatorname{Ker} T_2} T_1^* \mid G_2: G_2 \to \operatorname{Ker} T_2$ is bounded from below, that is,

$$\|P_{\operatorname{Ker} T_2} T_1^* x\| \geq C\|x\| \qquad (x \in G_2)$$

holds with a suitable constant $C > 0$. But $(\operatorname{Ker} T_2)^\perp$ is the closure of the range of T_2^*. Hence the last condition is equivalent to

$$\|T_1^* x + T_2^* y\| \geq C\|x\| \qquad (x \in G_2, y \in \operatorname{Dom} T_2^*). \qquad \square$$

To explain how this lemma is used in the proof of Theorem 8.2.1, we need to fix some notation. Let G be an open subset of Ω, and let $f = (f_1, \ldots, f_m) \in H^\infty(\Omega)^m$. We consider the Koszul complex $K_\bullet(f, L^2(G))$ with boundary operators denoted by δ_f. The adjoint of the Hilbert space operator

$$\delta_f: K_{p+1}(f, L^2(G)) \to K_p(f, L^2(G))$$

is the operator

$$\delta^{\bar{f}}: K_p(f, L^2(G)) \to K_{p+1}(f, L^2(G))$$

of exterior multiplication by $\bar{f}_1 e_1 + \cdots + \bar{f}_m e_m$. Here $e_1, \ldots, e_m$ are the generators of the alternating algebra $\Lambda(e)$ (see Section 2.2). For simplicity, we set $K_p L^2(G) = K_p(f, L^2(G))$.

If $\varphi: \Omega \to \mathbb{R}$ is a bounded measurable function, then the inner product

$$\langle g, h \rangle_\varphi = \int_G \bar{g} h e^{-\varphi} \, d\lambda \qquad (g, h \in L^2(G))$$

defines an equivalent norm $\|\cdot\|_{\varphi, G}$ on $L^2(G)$. The space $L^2(G)$ equipped with this norm will be denoted by $L^2(G, \varphi)$. If $\varphi_1, \varphi_2: \Omega \to \mathbb{R}$ are two bounded measurable functions, then the adjoint of the continuous linear operator

$$\delta_f \colon K_{p+1} L^2(G, \varphi_1) \to K_p L^2(G, \varphi_2)$$

is the operator

$$\delta^f \colon K_p L^2(G, \varphi_2) \to K_{p+1} L^2(G, \varphi_1),$$

where $\tilde{f} = (e^{\varphi_1 - \varphi_2} \bar{f}_1, \ldots, e^{\varphi_1 - \varphi_2} \bar{f}_m)$.

Let us suppose in addition that $0 \notin \overline{f(\Omega)}$. Define $g_i = \bar{f}_i / \sum_{j=1}^m |f_j|^2$ for $1 \le i \le m$. Then $g = (g_1, \ldots, g_m) \in (\mathscr{E}(\Omega) \cap L^\infty(\Omega))^m$ with $f_1 g_1 + \cdots + f_m g_m = 1$ on Ω. Hence the Koszul complex $K_\bullet(f, L^2(G))$ is split exact. More precisely, we have the identity

$$\delta_f \delta^g + \delta^g \delta_f = I$$

on $\Lambda(e, L^2(G))$. There is a constant L (independent of G) which is larger than the norms of the operators

$$\delta^g \colon K_p L^2(G) \to K_{p+1} L^2(G) \qquad (p = 0, \ldots, m-1).$$

With this constant, the estimate

$$\|\delta^f \eta\| \ge L^{-1} \|\eta\|$$

holds for all $\eta \in \operatorname{Ker} \delta_f$. In particular, for a homogeneous element $\eta = \sum_{|i| = p} \eta_i e_i \in \operatorname{Ker} \delta_f$, one computes

$$\delta^f \eta = \sum_{|j| = p+1} \sum_{\{i\} \cup \{a\} = \{j\}} \left(\varepsilon_{a,i} \bar{f}_a \eta_i \right) e_j \tag{8.2.1}$$

with suitable numbers $\varepsilon_{a,i} = \pm 1$, and consequently, one can write the previous estimate in the form

$$\int_G \sum_{|j| = p+1} \left| \sum_{\{i\} \cup \{a\} = \{j\}} \varepsilon_{a,i} \bar{f}_a \eta_i \right|^2 d\lambda \ge L^{-2} \int_G \sum_{|i| = p} |\eta_i|^2 \, d\lambda. \tag{8.2.2}$$

We have used the notation $e_i = e_{i_1} \wedge \cdots \wedge e_{i_p}$ for every multi-index $i = (i_1, \ldots, i_p)$ with $1 \le i_1 < \cdots < i_p \le m$.

Since (8.2.2) remains true if G is replaced by arbitrary measurable subsets of G, the same estimate holds, at least almost everywhere, between the integrands occurring on both sides. Therefore, if $\varphi_1, \varphi_2 \colon \Omega \to \mathbb{R}$ are bounded measurable functions and if $\varphi = \varphi_2 - \varphi_1$, then

$$\|\delta^f \eta\|^2_{\varphi_1, G} \ge L^{-2} \|\eta\|_{\varphi_1 + 2\varphi, G} \tag{8.2.3}$$

holds for all $\eta \in \operatorname{Ker} \delta_f$.

At this point, we can explain the idea of the proof of Theorem 8.2.1. Let $\varphi_1, \varphi_2 \colon \Omega \to \mathbb{R}$ be bounded measurable functions, let $G \subset \Omega$ be as before, and let us consider the augmented $\bar{\partial}$-complex with coefficients in $L^2(G, \varphi_1)$

$$0 \to L^2_a(G, \varphi_1) \xrightarrow{i} L^2(G, \varphi_1) \xrightarrow{\bar{\partial}} L^2_{(0,1)}(G, \varphi_1) \xrightarrow{\bar{\partial}} \cdots \xrightarrow{\bar{\partial}} L^2_{(0,n)}(G, \varphi_1) \to 0.$$

The $\bar\partial$-operators in this complex are densely defined linear operators, and as usual $L^2_a(G, \varphi_1)$ denotes the corresponding Bergman space $L^2(G, \varphi_1) \cap \mathscr{O}(G) = \mathrm{Ker}(\bar\partial\colon L^2(G, \varphi_1) \to L^2_{(0,1)}(G, \varphi_1))$.

Let us fix an integer p with $0 \le p \le m - 1$. We regard the kernels

$$Z^a_p = \mathrm{Ker}\big(\delta_f\colon \Lambda_p L^2_a(G) \to \Lambda_{p-1} L^2_a(G)\big),$$

$$Z_p = \mathrm{Ker}\big(\delta_f\colon \Lambda_p L^2(G) \to \Lambda_{p-1} L^2(G)\big)$$

as closed subspaces of $\Lambda_p L^2(G, \varphi_2)$, that is, as spaces carrying the equivalent norm $\|\cdot\|_{\varphi_2}$ rather than the original L^2-norm. We consider the diagram

$$
\begin{array}{ccc}
Z^a_p & \xleftarrow{\ \delta_f\ } & \Lambda_{p+1} L^2_a(G) \\
\downarrow i & & \downarrow i \\
Z_p & \xleftarrow{\ \delta_f\ } & \Lambda_{p+1} L^2(G, \varphi_1) \\
& & \downarrow \bar\partial_0 = \oplus\bar\partial \\
& & \Lambda_{p+1} L^2_{(0,1)}(G, \varphi_1) \\
& & \downarrow \bar\partial_1 = \oplus\bar\partial \\
& & \Lambda_{p+1} L^2_{(0,2)}(G, \varphi_1).
\end{array}
$$

To prove that the upper horizontal map is onto, it suffices in view of Lemma 8.2.2 to find suitable weights φ_1 and φ_2 and a constant $C > 0$ such that the estimate

$$\|\delta^*_f \eta + \bar\partial^*_0 \xi\|^2_{\varphi_1, G} \ge C\|\eta\|^2_{\varphi_2, G} \tag{8.2.4}$$

holds for all $\eta \in Z^a_p$ and $\xi \in \mathrm{Dom}\,\bar\partial^*_0$.

Each element $\xi \in \mathrm{Dom}\,\bar\partial^*_0$ can be written as a sum $\xi = \xi' + \xi''$ with $\xi' \in \mathrm{Ker}\,\bar\partial_1$ and $\xi'' \in (\mathrm{Ker}\,\bar\partial_1)^\perp \subset (\mathrm{Im}\,\bar\partial_0)^\perp = \mathrm{Ker}\,\bar\partial^*_0$. Hence it suffices to check (8.2.4) for all $\xi \in \mathrm{Dom}\,\bar\partial^*_0 \cap \mathrm{Ker}\,\bar\partial_1$.

We shall show that there are weight functions $\varphi_1, \varphi_2\colon \Omega \to \mathbb{R}$ and an exhaustion of Ω by open sets G_n ($n \in \mathbb{N}$) such that the above estimates hold on each G_n with a constant C that is independent of n. Then a normal family argument can be used to complete the proof of Theorem 8.2.1.

Proof of Theorem 8.2.1 Let Ω be a bounded pseudoconvex domain in $\mathbb{C}^n$, and let $f \in H^\infty(\Omega)^m$ with $0 \notin \overline{f(\Omega)}$ and $|f(z)| \le 1$ ($z \in \Omega$) with respect to the Euclidean norm on $\mathbb{C}^m$. Consider a strictly pseudoconvex domain $G \Subset \Omega$ with smooth boundary.

An elementary calculation shows that the function $\varphi\colon \Omega \to \mathbb{R}$, $\varphi(z) = \log(|f(z)|^2)$, satisfies

$$\sum_{\alpha, \beta} \big(\partial_\alpha \bar\partial_\beta \varphi\big) t_\alpha \bar t_\beta = |f|^{-4} \sum_{\beta < \beta'} \left| \sum_\alpha (f_\beta \partial_\alpha f_{\beta'} - f_{\beta'} \partial_\alpha f_\beta) t_\alpha \right|^2 \tag{8.2.5}$$

for all $t = (t_1, \ldots, t_n) \in \mathbb{C}^n$. Let K be an arbitrary constant. We define functions φ_1 and φ_2 on Ω by

$$\varphi_1 = K \log(|f|^2), \qquad \varphi_2 = (K+1)\log(|f|^2).$$

With L as in formula (8.2.2), we shall prove that the estimate (8.2.4) holds for $C = L^{-2}/2$ and for a suitable constant K independent of the choice of G.

For this purpose, let us fix elements $\eta \in Z_a^p$ and $\xi \in \mathrm{Dom}(\bar{\partial}_0^*) \cap \mathrm{Ker}\,\bar{\partial}_1$ (formed with respect to G as explained above). Since the function φ_1 is plurisubharmonic, Hörmander's fundamental estimate (see Hörmander 1965) gives the inequality

$$\|\bar{\partial}_0^* \xi\|_{\varphi_1}^2 \geq \sum_{|j|=p+1} \int_\Omega \sum_{\alpha,\beta} \left(\partial_\alpha \bar{\partial}_\beta \varphi_1\right) \xi_{j,\alpha} \bar{\xi}_{j,\beta} \, e^{-\varphi_1} \, d\lambda, \qquad (8.2.6)$$

where $\xi = \sum_{|j|=p+1} \sum_{\alpha=1}^n (\xi_{j,\alpha} \, d\bar{z}_\alpha) e_j$. We note that the right-hand side of this inequality can be transformed by the formula (8.2.5).

Let $M = 2L^2$. The following sequence of equalities and inequalities is obtained by straightforward computations

$$2\,\mathrm{Re}\langle \delta_f^* \eta, \bar{\partial}_0^* \xi \rangle_{\varphi_1} = 2\,\mathrm{Re}\langle \bar{\partial}_0 \delta_f^* \eta, \xi \rangle_{\varphi_1}$$

$$= 2\,\mathrm{Re} \sum_{|j|=p+1} \sum_{\alpha=1}^n \sum_{\{i\}\cup\{a\}=\{j\}} \langle \varepsilon_{a,i} \bar{\partial}_\alpha (e^{-\varphi} \bar{f}_\alpha) \eta_i, \xi_{j,\alpha} \rangle_{\varphi_1}$$

$$= \sum_{|i|=p} \int_G 2\,\mathrm{Re}\left[\bar{\eta}_i \left(\sum_{\alpha=1}^n \sum_{a \notin \{i\}} \varepsilon_{a,i} \, \partial_\alpha (e^{-\varphi} f_a) \xi_{[i,a],\alpha} \right) \right] e^{-\varphi_1} \, d\lambda$$

$$\geq -\frac{1}{M} \sum_{|i|=p} \int_G |\eta_i|^2 \, e^{-2\varphi - \varphi_1} \, d\lambda$$

$$\quad - M \sum_{|i|=p} \int_G \left| \sum_{\alpha=1}^n \sum_{a \notin \{i\}} \varepsilon_{a,i} e^{\varphi} \partial_\alpha (e^{-\varphi} f_a) \xi_{[i,a],\alpha} \right|^2 e^{-\varphi_1} \, d\lambda.$$

Here, for a p-tuple i of strictly increasing integers $1 \leq i_1 < \cdots < i_p \leq m$ and an integer $a \in \{1, \ldots, m\} \setminus \{i\}$, one denotes by $[i, a]$ the unique strictly increasing $(p+1)$-tuple j with $\{j\} = \{i\} \cup \{a\}$.

For the identity

$$e^{\varphi} \partial_\alpha (e^{-\varphi} f_a) = |f|^{-2} \sum_{\beta=1}^m \bar{f}_\beta (f_\beta \partial_\alpha f_a - f_a \partial_\alpha f_\beta)$$

one concludes that the estimate

$$\left| \sum_{\alpha=1}^n \sum_{a \notin \{i\}} \varepsilon_{a,i} e^{\varphi} \partial_\alpha (e^{-\varphi} f_a) \xi_{[i,a],\alpha} \right|^2$$

$$\leq R|f|^{-4} \sum_{a \notin \{i\}} \sum_{\beta=1}^m \left| \sum_{\alpha=1}^n (f_\beta \partial_\alpha f_a - f_a \partial_\alpha f_\beta) \xi_{[i,a],\alpha} \right|^2$$

holds with a constant R that depends only on m (at least if $|f| \le 1$).

Combining (8.2.3), (8.2.5), (8.2.6), and the last estimates, we obtain that

$$\| \delta_f^* \eta + \bar{\partial}_0^* \xi \|_{\varphi_1, G}^2 = \| \delta_f^* \eta \|_{\varphi_1, G}^2 + \| \bar{\partial}_0^* \xi \|_{\varphi_1, G}^2 + 2 \operatorname{Re} \langle \delta_f^* \xi, \bar{\partial}_0^* \xi \rangle_{\varphi_1, G}$$

$$\ge (L^{-2} - 1/M) \| \eta \|_{\varphi_1 + 2\varphi, G}^2 = C \| \eta \|_{\varphi_2, G}^2$$

with $C = L^{-2}/2$ provided $K \ge 2MR = 4L^2 R$.

The proof of Theorem 8.2.1 is completed by a standard exhaustion argument. Namely, let us consider an increasing sequence of strictly pseudoconvex domains with smooth boundary $G_n \Subset G_{n+1} \Subset \Omega$ such that $\Omega = \bigcup_{n \ge 0} G_n$. Let $f \in H^\infty(\Omega)^m$ satisfy $0 \notin \overline{f(\Omega)}$ and $|f(z)| \le 1$ for $z \in \Omega$.

We choose a constant $C = L^{-2}/2$ as above, and define $k = \sqrt{2} L$, $K = 4L^2 R$ and $\varphi_1 = K \log|f|^2$, $\varphi_2 = (K + 1)\log|f|^2$. Let $y \in \Lambda_p L_a^2(\Omega)$ be a form with $\delta_f y = 0$. By Lemma 8.2.2 and the above estimates, for each $n \ge 0$, there is a form $x_n \in \Lambda_{p+1} L_a^2(G_n)$ satisfying $y = \delta_f x_n$ on G_n and $\|x_n\|_{\varphi_1, G_n} \le k\|y\|_{\varphi_2, \Omega}$.

For each $N \ge 1$, the sequence $(x_n |_{G_N})_{n \ge N}$ is bounded in the Fréchet–Montel topology of the space $\Lambda_{p+1} \mathcal{O}(G_N)$. By a standard diagonal selection process, we may suppose that, for each $N \ge 1$, the sequence $(x_n|_{G_N})_{n \ge N}$ converges in $\Lambda_{p+1} \mathcal{O}(G_N)$. But then obviously there exists a form $x \in \Lambda_{p+1} L_a^2(\Omega)$ with $y = \delta_f x$ and $\|x\|_{\varphi_1, \Omega} \le k\|y\|_{\varphi_2, \Omega}$.

This finishes the proof of Theorem 8.2.1. $\square$

Exactly as in the preceding section, the simple observation that $f(\Omega)$ is always contained in $\sigma_r(f, L_a^2(\Omega))$ proves the following result.

Corollary 8.2.3 *Let Ω be a bounded pseudoconvex domain in $\mathbb{C}^n$, and let $f \in H^\infty(\Omega)^m$. Then*

$$\sigma(f, L_a^2(\Omega)) = \sigma_r(f, L_a^2(\Omega)) = \overline{f(\Omega)}.$$ $\square$

The remaining part of this section is devoted to the computation of the essential spectrum of f in a slightly more particular case. For this purpose, we shall use the sheaf model of the Bergman n-tuple M_Ω and the characterization of the essential spectrum as the joint spectrum relative to the functor $E^q = E^\infty / E^{pc}$ (see Section 2.6).

Theorem 8.2.4 *Let Ω be a bounded pseudoconvex domain in $\mathbb{C}^n$, and let $f \in H^\infty(\Omega)^m$. Then*

$$\sigma_e(f, L_a^2(\Omega)) \subset \bigcap \{ \overline{f(U \cap \Omega)}; \ U \supset \sigma_e(M_\Omega, L_a^2(\Omega)) \text{ open} \}. \quad (8.2.7)$$

Proof We recall the notation $M_\Omega = (M_{z_1}, \dots, M_{z_n})$ from Section 8.1 and the fact that the n-tuple M_Ω has property (β). Let $\mathscr{F}$ denote its sheaf model.

Suppose that $0 \notin \overline{f(U \cap \Omega)}$ for some open neighbourhood U of $\sigma_e(M_\Omega, L_a^2(\Omega))$. According to Section 2.6, we have to prove that the complex

$K_{\bullet}(f, L_a^2(\Omega)^q)$ is exact. By Corollary 6.3.3 this is equivalent to the exactness of the complex $K_{\bullet}(f, \mathcal{F}^q)$. Recall that $\mathcal{F}^q = \mathcal{F}^{\infty}/\mathcal{F}^{pc}$ is also a quasi-coherent sheaf, and that its support is exactly the essential spectrum of M_{Ω}.

Let $\lambda \in \sigma_e(M_{\Omega}, L_a^2(\Omega))$ be arbitrary, and let $D \Subset U$ be an open polydisc centred at λ. It is evident, from the abstract construction of the spaces $\mathcal{O}B(\Omega)$ and their associated sheaves, that starting with the pseudoconvex open set $\Omega \cap D$ instead of Ω one obtains a sheaf $\mathcal{G}$ with the property $\mathcal{G}|_D \cong \mathcal{F}|_D$. Thus we are led to prove that the complex $K_{\bullet}(f, \mathcal{G}^q)$ is exact in a neighbourhood of λ. But according to Theorem 8.2.1, even the complex $K_{\bullet}(f, L_a^2(\Omega \cap D))$ is exact (since $0 \notin \overline{f(\Omega \cap D)}$. Consequently the complex $K_{\bullet}(f, L_a^2(\Omega \cap D)^q)$ is exact, and hence also $K_{\bullet}(f, \mathcal{G}^q)$ is an exact complex. □

We should remark at this point that the inclusion in Theorem 8.2.4 is an equality for $m = n = 1$ (see Axler 1982). This phenomenon occurs in all known examples, and it is natural to conjecture that (8.2.7) is an equality in general. Next we discuss two results which support this conjecture.

Proposition 8.2.5 *Let Ω be a bounded pseudoconvex domain in $\mathbb{C}^n$, and let $f \in A^0(\overline{\Omega})^m$. Then*

$$\sigma_e(f, L_a^2(\Omega)) = \bigcap \{ \overline{f(U \cap \Omega)}; \, U \supset \sigma_e(M_{\Omega}, L_a^2(\Omega)) \text{ open} \}$$
$$= f(\sigma_e(M_{\Omega}, L_a^2(\Omega))).$$

Proof The set in the middle can be characterized as the set of all points $w \in \mathbb{C}^m$ with the property that there is a point $z \in \sigma_e(M_{\Omega}, L_a^2(\Omega))$ and a sequence (z_k) in Ω converging to z such that $w = \lim_k f(z_k)$. In particular, this proves the second equality in the statement.

The first equality is a consequence of the general spectral mapping principles explained in Section 2.6. Let A be the Banach algebra $A^0(\overline{\Omega})$, and let M be its character space. Note that A can be identified isometrically with a closed commutative subalgebra of $L(L_a^2(\Omega))$. For any $f \in A^m$ ($m \geq 1$), the Taylor joint and essential joint spectra can be computed as

$$\sigma(f, L_a^2(\Omega)) = \hat{f}(\Delta), \qquad \sigma_e(f, L_a^2(\Omega)) = \hat{f}(\Delta_e),$$

where $\Delta_e \subset \Delta$ are closed subsets of M (see Proposition 2.6.1).

Theorem 8.2.1 shows that $\hat{f}(\Delta) = f(\overline{\Omega})$ for any $m \geq 1$ and $f \in A^m$. This implies that $\Delta = \overline{\Omega}$. In particular,

$$\sigma(M_{\Omega}, L_a^2(\Omega)) = \Delta \qquad \text{and} \qquad \sigma_e(M_{\Omega}, L_a^2(\Omega)) = \Delta_e,$$

again via the embedding $\overline{\Omega} \subset M$.

In conclusion, $\sigma_e(f, L_a^2(\Omega)) = \hat{f}(\Delta_e) = f(\sigma_e(M_{\Omega}, L_a^2(\Omega)))$, and the proof of Proposition 8.2.5 is complete.   □

The same proof shows for instance that

$$\sigma_e(f, L_a^2(\Omega)) = \sigma_{re}(f, L_a^2(\Omega)) \qquad \left(f \in A^0(\overline{\Omega})^m\right),$$

where the latter is the right essential joint spectrum

$$\sigma_{re}(f, L_a^2(\Omega)) = \left\{ \lambda \in \mathbb{C}^m; \dim L_a^2(\Omega) \middle/ \left(\sum_{j=1}^m (\lambda_j - f_j) L_a^2(\Omega) \right) = \infty \right\}.$$

Another case where equality holds in (8.2.7) is described in the next result.

Theorem 8.2.6 *Let $\Omega \subset \mathbb{C}^n$ be a strictly pseudoconvex domain with C^2-boundary. Then $\sigma_e(M_\Omega, L_a^2(\Omega)) = \partial\Omega$, and for each tuple $f \in H^\infty(\Omega)^m$, we have*

$$\sigma_e(f, L_a^2(\Omega)) = \cap \left\{ \overline{f(U \cap \Omega)}; U \supset \partial\Omega \text{ open} \right\}.$$

Proof   Let (z_k) be a sequence in Ω that converges to a point $\lambda \in \partial\Omega$ and such that $\lim_k f(z_k) = 0$. We have to prove that $0 \in \sigma_e(f, L_a^2(\Omega))$.

Since the domain Ω is strictly pseudoconvex with $\mathscr{C}^2$-boundary, one can choose a function $h \in \mathscr{O}(\Omega)$ with $h(\Omega) \subset \mathbb{D}$, the unit disc, and $\lim_{z \to \lambda} h(z) = 1$ (see for instance Corollary VI.1.14 in Range 1986). By passing to a suitable subsequence, we may suppose that the map

$$H^\infty(\mathbb{D}) \to l^\infty, \qquad g \mapsto [(g \circ h)(z_k)]_k$$

is onto (see Hoffman 1962, p. 204). But then also the map

$$H^\infty(\Omega) \to l^\infty, \qquad g \mapsto (g(z_k))_k$$

is onto. By the open mapping principle there is a sequence (f_N) in $H^\infty(\Omega)^m$ such that (f_N) converges componentwise to zero in $H^\infty(\Omega)^m$ and

$$f_N(z_k) = \begin{cases} 0, & k < N \\ f(z_k), & k \geq N. \end{cases}$$

Let us define $g_N = f - f_N$, and let us denote by $\varepsilon_k: L_a^2(\Omega) \to \mathbb{C}$ the evaluation map at the point $z_k \in \Omega$ $(k \geq 1)$. Since, for each $N \geq 1$, the composition of the two maps

$$K_1(g_N, L_a^2(\Omega)) \xrightarrow{\delta_{g_N}} K_0(g_N, L_a^2(\Omega)) \xrightarrow{\varepsilon_k} \mathbb{C}$$

is zero for $k \geq N$, the homology groups $H_0(g_N, L_a^2(\Omega))$ are infinite dimensional. Therefore $0 \in \sigma_{re}(g_N, L_a^2(\Omega)) \subset \sigma_e(g_N, L_a^2(\Omega))$. But $\lim_N(g_N - f) = 0$ in the operator norm. Hence $0 \in \sigma_{re}(f, L_a^2(\Omega)) \subset \sigma_e(f, L_a^2(\Omega))$.

The same argument applied to the n-tuple $M_\Omega = (M_{z_1}, \ldots, M_{z_n})$ yields the identity $\sigma_e(M_\Omega, L_a^2(\Omega)) = \partial\Omega$.

This completes the proof of Theorem 8.2.6. $\qquad\qquad\qquad\qquad\qquad\square$

The idea used in the last part of the preceding proof shows that, under the assumptions of Corollary 8.2.3, the fibres $f^{-1}(w)$ are finite for each point $w \in \mathbb{C}^m \setminus \sigma_e(f, L_a^2(\Omega))$. Since for $m < n$ these fibres cannot possess any isolated points, it follows that in this case

$$\sigma(f, L_a^2(\Omega)) = \sigma_e(f, L_a^2(\Omega)) = \overline{f(\Omega)}$$

$$= \bigcap \big(\overline{f(U \cap \Omega)}; \ U \supset \sigma_e(M_\Omega, L_a^2(\Omega)) \text{ open}\big).$$

The proof of Theorem 8.2.6 immediately implies the next result.

Corollary 8.2.7 *Under the conditions of Theorem 8.2.6, one has*

$$\sigma_{re}(f, L_a^2(\Omega)) = \sigma_e(f, L_a^2(\Omega)).$$ □

When compared with the results obtained in Section 8.1 (cf. Theorem 8.1.5 and Proposition 8.1.6), what is missing in the case of arbitrary H^∞-symbols is property (β). Although there is some evidence for the validity of condition (β) for tuples with H^∞-symbol on Bergman spaces, this question remains open.

Also it would be desirable to have an analogue of Theorem 8.2.1 for spaces such as $\mathscr{O}L^p(\Omega)$ $(1 \leq p \leq \infty)$. But the lack of suitable substitutes for Hörmander's L^2-estimates in this setting turns this into a rather difficult open problem.

8.3 ANALYTICALLY INVARIANT SUBSPACES

This section contains some global results on the structure of the analytically invariant closed subspaces of a multidimensional Bergman space. It must be pointed out from the very beginning that our approach is limited by some serious technical difficulties. For instance, we shall only study Bergman spaces supported by strictly pseudoconvex domains with smooth boundary or by polydomains. On the other hand, it is not expected that a complete classification of the analytically invariant subspaces of Bergman spaces is possible. For instance, in the case of the unit disc $\mathbb{D} \subset \mathbb{C}$, there is a principal structural difference between the lattice of analytically invariant subspaces of the Hardy space $H^2(\partial \mathbb{D})$ (cf. the classical theorem of Beurling) and of the Bergman space $L_a^2(\mathbb{D})$ (see Bercovici *et al.* 1985).

In this section we focus on some qualitative results which will eventually distinguish a class of analytically invariant subspaces which can be investigated by the homological methods developed throughout this book. A series of open problems which reflect the early stage of this investigation are spread throughout the text.

Compared with the remaining parts of the book, the material included in this section is not self-contained. A few auxiliary algebraic and complex analytic lemmas needed in the technical proofs will be stated with a bibliographical reference.

Let Ω be a bounded pseudoconvex domain in $\mathbb{C}^n$. There are two natural classes of analytically invariant subspaces of the Bergman space $L_a^2(\Omega)$. The first class consists of the extended ideals

$$I \cdot L_a^2(\Omega) = \operatorname{Im}\left(I \otimes_{\mathscr{O}(\overline{\Omega})} L_a^2(\Omega) \xrightarrow{m} L_a^2(\Omega) \right),$$

where m is the multiplication map and $I \subset \mathscr{O}(\overline{\Omega})$ is a closed ideal. The second class is given by the subspaces of the form

$$S(V) = \{ f \in L_a^2(\Omega); \; f \,|\, V = 0 \},$$

where V is a closed subset of Ω. It is obvious that a subspace $S \subset L_a^2(\Omega)$ as above satisfies $fS \subset S$ for any function $f \in H^\infty(\Omega)$. This property will be called in short the *analytic invariance* of S.

A subspace such as $I \cdot L_a^2(\Omega)$ above is not necessarily closed in $L_a^2(\Omega)$, while $S(V)$ always is. If, for a given closed subset $V \subset \Omega$, one denotes by $I(V)$ the ideal in $\mathscr{O}(\overline{\Omega})$ consisting of all functions that vanish on V, then $I(V) \cdot L_a^2(\Omega) \subset S(V)$. Whether this inclusion is an equality or not (i.e. a Nullstellensatz holds) is one of the questions to be treated below.

First we prove that the analytically invariant subspaces of finite codimension in $L_a^2(\Omega)$ are necessarily of the first type above.

Theorem 8.3.1 *Let Ω be a strictly pseudoconvex domain in $\mathbb{C}^n$ with smooth boundary. Then each analytically invariant subspace S of finite codimension in $L_a^2(\Omega)$ is of the form*

$$S = \sum_{k=1}^{m} P_k \cdot L_a^2(\Omega),$$

where $P = (P_1, \ldots, P_m)$ is a system of polynomials with common zero set contained in Ω.

Proof Let us consider the quotient $\mathscr{O}(\mathbb{C}^n)$-module $Q = L_a^2(\Omega)/S$. Since $\dim Q < \infty$, the support of this $\mathscr{O}(\mathbb{C}^n)$-module is finite and contained in $\overline{\Omega}$. Moreover, because there are no L^2-bounded point evaluations at points in the boundary of a strictly pseudoconvex domain with C^2-boundary (see Corollary VI.1.14 in Range 1986), the support of Q is contained in Ω.

As Fréchet $\mathscr{O}(\mathbb{C}^n)$-modules, both $L_a^2(\Omega)$ and Q are quasi-coherent. Let $\mathscr{F}$ and $\mathscr{Q}$ denote their associated sheaves. Since $\operatorname{supp}(\mathscr{Q})$ is finite, there exists by Hilbert's Syzygy Theorem a global finite resolution

$$0 \to \mathscr{O}_{\mathbb{C}^n}^{m_r} \xrightarrow{P_r} \cdots \to \mathscr{O}_{\mathbb{C}^n}^{m_2} \xrightarrow{P_2} \mathscr{O}_{\mathbb{C}^n}^{m_1} \xrightarrow{P_1} \mathscr{O}_{\mathbb{C}^n} \to \mathscr{Q} \to 0, \tag{8.3.1}$$

where the boundary operators are induced by polynomials $p_1, \ldots, p_r$ (see for instance Mac Lane 1963).

Since $\operatorname{supp}(\mathscr{Q}) \subset \Omega$ and $\mathscr{F}|_\Omega \simeq \mathscr{O}|_\Omega$, by tensoring the sequence (8.3.1) with $\mathscr{F}$ over $\mathscr{O}_{\mathbb{C}^n}$, one obtains a new exact complex

$$0 \to \mathscr{F}^{m_r} \xrightarrow{P_r} \cdots \to \mathscr{F}^{m_2} \xrightarrow{P_2} \mathscr{F}^{m_1} \xrightarrow{P_1} \mathscr{F} \to \mathscr{Q} \to 0.$$

But both sheaves $\mathscr{F}$ and $\mathscr{E}$ are acyclic on $\mathbb{C}^n$. Hence the exactness is preserved at the level of global sections

$$0 \to L_a^2(\Omega)^{m_r} \xrightarrow{p_r} \cdots \to L_a^2(\Omega)^{m_2} \xrightarrow{p_2} L_a^2(\Omega)^{m_1} \xrightarrow{p_1} L_a^2(\Omega) \to Q \to 0.$$

In particular, this shows that $S = \operatorname{Im}(p_1)$, and the proof of Theorem 8.3.1 is finished.   □

Actually in the preceding proof we have also obtained the following result.

Corollary 8.3.2 *Under the conditions of Theorem 8.3.1, one has*

$$S \perp_{\mathscr{O}(\mathbb{C}^n)} L_a^2(\Omega),$$

and S is a quasi-coherent $\mathscr{O}(\mathbb{C}^n)$-module.   □

Moreover, Theorem 8.3.1 establishes a one-to-one correspondence between the analytically invariant subspaces S of finite codimension in $L_a^2(\Omega)$ and the polynomial ideals $I \subset \mathbb{C}[z_1, \ldots, z_n]$ with the property that the natural quotient map

$$\mathbb{C}[z_1, \ldots, z_n]/I \to L_a^2(\Omega)/S$$

is an isomorphism. This remark, combined with the following lemma, completes the classification of the spaces S appearing in Theorem 8.3.1.

Lemma 8.3.3 (Grothendieck) *Let I_1, I_2 be two ideals of $\mathscr{O}_{\mathbb{C}^n, 0}$ of height greater or equal than 2. Then $I_1 = I_2$ if and only if*

$$I_1 \otimes_{\mathscr{O}_{\mathbb{C}^n, 0}} M \cong I_2 \otimes_{\mathscr{O}_{\mathbb{C}^n, 0}} M \tag{8.3.2}$$

for every finite-dimensional $\mathscr{O}_{\mathbb{C}^n, 0}$-module M.   □

For a proof of Lemma 8.3.3, see Douglas *et al.* (1995).

Theorem 8.3.4 *Let $\Omega \subset \mathbb{C}^n$ ($n > 1$) be a bounded strictly pseudoconvex domain with smooth boundary, and let S_1, S_2 be two analytically invariant subspaces of finite codimension in $L_a^2(\Omega)$. Then S_1 and S_2 are isomorphic as topological $\mathscr{O}(\mathbb{C}^n)$-modules if and only if they coincide.*

Proof According to Theorem 8.3.1 there are two ideals $I_1, I_2 \subset \mathbb{C}[z_1, \ldots, z_n]$ such that $S_j = I_j L_a^2(\Omega)$ ($j = 1, 2$).

Let M be a finite-dimensional $\mathbb{C}[z_1, \ldots, z_n]$-module, and suppose that S_1 and S_2 are isomorphic as Fréchet $\mathscr{O}(\mathbb{C}^n)$-modules. According to Lemma 8.3.3, we have to prove that the isomorphism (8.3.2) holds. Since M is a direct sum of submodules supported by a single point, we may suppose that $\operatorname{supp} M = \{\lambda\}$ for some point $\lambda \in \mathbb{C}^n$. In this case, it suffices by localization to check that the modules $I_1 \cdot M$ and $I_2 \cdot M$ are isomorphic over the local ring $\mathbb{C}[z_1, \ldots, z_n]_\lambda$, or equivalently, over $\mathscr{O}_{\mathbb{C}^n, \lambda}$.

If $\lambda \notin \mathrm{supp}(\mathbb{C}[z_1,\ldots,z_n]/I_1) \cap \mathrm{supp}(\mathbb{C}[z_1,\ldots,z_n]/I_2)$, then the last assertion is true simply because $I_{1,\lambda} = I_{2,\lambda} = \mathbb{C}[z_1,\ldots,z_n]_\lambda$. Otherwise, $\lambda \in \Omega$ and we have the following series of canonical isomorphisms of $\mathscr{O}_{\mathbb{C}^n,\lambda}$-modules:

$$I_1 \cdot M \cong I_{1,\lambda} \cdot M \cong \tilde{S}_{1,\lambda} \otimes_{\mathscr{O}_{\mathbb{C}^n,\lambda}} M \cong \tilde{S}_{2,\lambda} \otimes_{\mathscr{O}_{\mathbb{C}^n,\lambda}} M \cong I_{2,\lambda} \cdot M \cong I_2 \cdot M,$$

where $\tilde{S}_j$ is the analytic quasi-coherent sheaf derived from the subspace S_j ($j = 1, 2$). Here we have implicitly used the isomorphism $\tilde{S}_1 \cong \tilde{S}_2$ derived from $S_1 \cong S_2$.

This finishes the proof of Theorem 8.3.4. □

The behaviour of the analytically invariant subspaces of finite condimension in $L_a^2(\Omega)$ is a model for what is expected to happen in the infinite-codimensional case. However, the results in this direction obtained so far are only partial, and they reveal some new difficulties. The rest of this section is devoted to a study of the infinite-codimensional subspaces. Some complementary results will be discussed in Sections 8.4 and 8.5.

The next result is an extreme case of a Nullstellensatz phenomenon. For an ideal I in $\mathscr{O}(\overline{\Omega})$, we denote by $V(I)$ the common zero set of the elements of I.

Proposition 8.3.5 *Let* $\Omega \subset \mathbb{C}^n$ *be a domain with smooth real-analytic boundary, and let* $I \subset \mathscr{O}(\overline{\Omega})$ *be an ideal. Suppose that* $V(I) \cap \overline{\Omega}$ *is a finite subset of* $\partial\Omega$. *Then* I *is dense in the Bergman space* $L_a^2(\Omega)$.

Proof Since the boundary of Ω is real-analytic, a theorem of Frisch (1967) shows that the ideal I is finitely generated.

Then Theorems 4.1 and 4.8 of Nagel (1974) imply that the space $I \cdot \mathscr{E}(\overline{\Omega})$ is closed in the Fréchet space topology of $\mathscr{E}(\overline{\Omega})$, and $I \cdot \mathscr{E}(\overline{\Omega}) \cap A^\infty(\overline{\Omega}) = I \cdot A^\infty(\overline{\Omega})$. In particular, this shows that the space $I \cdot A^\infty(\overline{\Omega})$ is closed in $A^\infty(\overline{\Omega})$.

Assume that the ideal I is not dense in the Bergman space $L_a^2(\Omega)$. Then there is a non-trivial element $g \in L_a^2(\Omega)$ which is orthogonal to I. The scalar product with g induces a linear functional

$$L: A^\infty(\overline{\Omega})/I \cdot A^\infty(\overline{\Omega}) \to \mathbb{C}, \qquad [u] \mapsto \langle u, g \rangle.$$

In virtue of the above observations and of the Hahn–Banach theorem, L can be extended to a continuous linear functional $\tilde{L}: \mathscr{E}(\overline{\Omega})/I \cdot \mathscr{E}(\overline{\Omega}) \to \mathbb{C}$. In its turn, $\tilde{L}$ can be lifted to a distribution $u \in \mathscr{E}'(\mathbb{C}^n)$ which is supported by the finite set $V(I) \cap \overline{\Omega}$. Therefore u is a finite sum of derivatives of Dirac measures at points of $\overline{\Omega} \cap V(I)$. This means that there exists an integer $k > 0$ such that u vanishes on the ideal $\underline{m}_\lambda^k \mathscr{E}(\overline{\Omega})$ for every $\lambda \in \overline{\Omega} \cap V(I)$, where $\underline{m}_\lambda$ denotes the maximal ideal of $\mathscr{O}_\lambda$. Since $\mathscr{O}(\overline{\Omega})$ is dense in $L_a^2(\Omega)$ (again because $\partial\Omega$ is smooth and strictly pseudoconvex), the original function g is orthogonal to $\underline{m}_\lambda^k L_a^2(\Omega)$ for every $\lambda \in \overline{\Omega} \cap V(I)$.

But $\underline{m}_\lambda L_a^2(\Omega) = \sum_{j=1}^n (z_j - \lambda_j) L_a^2(\Omega)$ is a dense subspace of $L_a^2(\Omega)$, because otherwise λ would be a bounded $L_a^2(\Omega)$-evaluation point. Hence $\underline{m}_\lambda^k L_a^2(\Omega)$ is also a dense subspace of $L_a^2(\Omega)$.

In conclusion, $g = 0$, which contradicts our assumption and proves that the ideal I is dense in $L_a^2(\Omega)$. $\square$

The above proof works with minor modifications for each of the spaces $\mathcal{O}L^p(\Omega)$ ($1 \leq p < \infty$), but not for instance for $\mathcal{O}\,\mathrm{Lip}_\alpha(\overline{\Omega})$, since there are bounded evaluation points on the boundary of Ω in this case.

The interesting subspaces are however those with zeros inside Ω. We now discuss a generic case where for the first time the geometric properties of $\partial\Omega$ and of the zero set play a role.

We shall use the notation from the beginning of this section, and moreover, for a closed subset V of some neighbourhood of $\overline{\Omega}$, we define $I(V)$ to be the radical ideal

$$I(V) = \{f \in \mathcal{O}(\overline{\Omega}); f\,|\,V = 0\}.$$

Theorem 8.3.6   *Let Ω be a bounded strictly pseudoconvex domain with smooth boundary in $\mathbb{C}^n$, and let V be a smooth complex submanifold of a neighbourhood of $\overline{\Omega}$.*

If V and $\partial\Omega$ are transversal, then

$$\mathcal{O}(\overline{\Omega})/I(V) \perp_{\mathcal{O}(\mathbb{C}^n)} L_a^2(\Omega) \qquad and \qquad I(V)\cdot L_a^2(\Omega) = S(V).$$

Proof   Since $\partial\Omega$ is a real hypersurface in this case, transversality simply means that at any intersection point $\lambda \in V \cap \partial\Omega$ the real tangent spaces satisfy $T_\lambda V \not\subset T_\lambda(\partial\Omega)$.

By our assumptions, the compact set $\overline{\Omega}$ has a fundamental system of pseudoconvex open neighbourhoods. Therefore the sheaf of ideals $\mathcal{I} \subset \mathcal{O}_{\overline{\Omega}}$ associated with $I(V)$ has a finite free resolution with finite-type $\mathcal{O}_{\overline{\Omega}}$-modules $\mathcal{L}_\bullet$. We have to prove that $\mathcal{L}_\bullet(\overline{\Omega}) \otimes_{\mathcal{O}(\overline{\Omega})} L_a^2(\Omega)(\cong \mathcal{L}_\bullet(\overline{\Omega}) \hat{\otimes}_{\mathcal{O}(\mathbb{C}^n)} L_a^2(\Omega))$ is exact in positive degrees and has separated homology in degree zero. Denoting by $\mathcal{F}$ the associated quasi-coherent sheaf of $L_a^2(\Omega)$, the last assertion can be localized. More precisely, it suffices to prove that $\mathcal{L}_\bullet \otimes_{\mathcal{O}_{\overline{\Omega}}} \mathcal{F}$ is exact in all positive degrees and that its homology sheaf in degree zero is a Fréchet sheaf.

If $\lambda \in \Omega$, then $\mathcal{F}_\lambda \cong \mathcal{O}_{\Omega,\lambda}$, where $(\mathcal{L}_\bullet \otimes_{\mathcal{O}_{\overline{\Omega}}} \mathcal{F})_\lambda \cong \mathcal{L}_{\bullet,\lambda}$, and the assertion follows. Also, for $\lambda \in \mathbb{C}^n \setminus \overline{\Omega}$, one has $(\mathcal{L}_\bullet \otimes_{\mathcal{O}_{\overline{\Omega}}} \mathcal{F})_\lambda = 0$, while for $\lambda \in \overline{\Omega} \setminus V$, $\mathcal{L}_{\bullet,\lambda}$ is a split exact complex of free $\mathcal{O}_{\overline{\Omega},\lambda}$-modules and the assertion is true.

Let $\lambda \in V \cap \partial\Omega$. By a local change of coordinates we may suppose that $\lambda = 0$, that for a small polydisc D centred at 0,

$$V \cap D = \{z \in D; z_1 = \cdots = z_p = 0\},$$

and that the normal of the real hypersurface $\partial\Omega$ at λ is contained in the linear subspace V. A resolution of the module $\mathcal{O}_D/\mathcal{I}_D$ is given by the Koszul complex $K_\bullet(z', \mathscr{O}_D)$, where $z' = (z_1, \ldots, z_p)$, $z'' = (z_{p+1}, \ldots, z_n)$, and $z = (z', z'') \in D$. Thus we are led to prove that $K_\bullet(z', \mathscr{F}_D)$ is exact in all positive degrees and has separated homology in degree zero.

In other words, we have to prove that, for a fixed q $(0 \le q \le p)$ and a sequence (η_m) in $K_q(z', \mathscr{F}(D))$ satisfying $\lim_{m \to \infty} \delta \eta_m = 0$, there exists a smaller polydisc D' centred at 0 and a sequence (ζ_m) in $K_{q+1}(z', \mathscr{F}(D'))$ such that $\lim_{m \to \infty} (\delta \zeta_m - \eta_m) = 0$ in $K_q(z', \mathscr{F}(D'))$. This assertion is obviously true for the sheaf $\mathscr{O}_\Omega$ instead of $\mathscr{F}$. We shall exploit this fact, remembering that $\mathscr{F}|_\Omega \cong \mathscr{O}_\Omega$ and deforming the domain $D \cap \Omega$.

Let τ be the exterior normal vector of length one to $\partial \Omega$ at the point $0 \in \partial \Omega$. We can choose a smaller polydisc $D' \Subset D$ centred at 0 such that $\partial \Omega \cap D'$ is the graph of a defining function for $\partial \Omega$ depending on the real variables of the orthogonal complement of $\mathbb{R}\tau$ in $\mathbb{C}^n$. Moreover, we can fix a constant $\varepsilon_0 > 0$ with the property that $D' \cap \Omega$ is relatively compact in each of the translated domains $D \cap \Omega + \varepsilon \tau$ $(0 < \varepsilon < \varepsilon_0)$.

Let q and (η_m) be chosen as before and fix an ε $(0 < \varepsilon < \varepsilon_0)$. Then the sequence $\delta(z') \eta_m (z', z'' - \varepsilon \tau)$ converges to zero in $K_{q-1}(z', L_a^2(\Omega \cap D + \varepsilon \tau))$ and *a fortiori* in the weaker topology of the space $K_{q-1}(z', \mathscr{O}(\Omega \cap D + \varepsilon \tau))$. Therefore there exists a sequence (ζ_m^ε) in $K_{q+1}(z', \mathscr{O}(\Omega \cap D + \varepsilon \tau))$ with the property that

$$\lim_{m \to \infty} [\eta_m(z', z'' - \varepsilon \tau) - \delta(z') \zeta_m^\varepsilon(z', z'')] = 0$$

for $(z', z'') \in \Omega \cap D + \varepsilon \tau$ in the topology of $K_q(z', \mathscr{O}(\Omega \cap D + \varepsilon \tau))$.

By taking restrictions to $\Omega \cap D'$ and by a diagonal selection process (for $\varepsilon \to 0$ and $m \to \infty$), we can choose a sequence (ζ_m) in $K_{q+1}(z', \mathscr{O}(\overline{\Omega \cap D'}))$ such that $\lim_{m \to \infty}(\delta \zeta_m - \eta_m) = 0$ in $K_q(z', L_a^2(\Omega \cap D'))$.

The same argument proves the second assertion in the statement. Namely, take a function $f \in \mathscr{F}(D)$ which vanishes on $V \cap D$. Then the translated function $f(z', z'' - \varepsilon \tau)$ is in $\Gamma(\Omega \cap D + \varepsilon \tau, \mathscr{I}) = (z_1, \ldots, z_p) \mathscr{O}(\Omega \cap D + \varepsilon \tau)$ by the classical Nullstellensatz. By passing to the limit with $\varepsilon \to 0$ and by restricting the functions to $\Omega \cap D'$, one finds that $f|_{\Omega \cap D'}$ belongs to the closure of the range of the map $\delta \colon K_1(z', L_a^2(\Omega \cap D')) \to K_0(z', L_a^2(\Omega \cap D'))$. But we already know that this map has closed range. Hence $f|_{\Omega \cap D'}$ belongs to $(z_1, \ldots, z_p) L_a^2(\Omega \cap D')$.

This concludes the proof of Theorem 8.3.6. $\qquad\qquad\square$

Unfortunately the rather restrictive conditions in Theorem 8.3.6 cannot be relaxed very much when using the preceding homological approach. The next result is a typical statement which can be reduced to Theorem 8.3.6.

Corollary 8.3.7 *Let Ω be a bounded strictly pseudoconvex domain in $\mathbb{C}^n$, and let us consider an ascending chain of ideals*

$$\mathscr{I} = \mathscr{I}_0 \subset \mathscr{I}_1 \subset \mathscr{I}_2 \subset \cdots \subset \mathscr{I}_N = \mathscr{O}_\Omega,$$

with consecutive quotients $\mathcal{I}_k/\mathcal{I}_{k-1} \cong \mathcal{O}_{V_k}$ $(1 \le k \le N)$, where the V_k are complex subvarieties of a neighbourhood of $\overline{\Omega}$ transversal to $\partial\Omega$. Then $\mathcal{A}(\overline{\Omega}) \perp_{\mathcal{O}(\mathbb{C}^n)} L_a^2(\Omega)$, and there is an $\mathcal{O}(\mathbb{C}^n)$-filtration

$$\mathcal{I}_0(\overline{\Omega}) \hat{\otimes}_{\mathcal{O}(\mathbb{C}^n)} L_a^2(\Omega) \subset \cdots \subset \mathcal{I}_N(\overline{\Omega}) \hat{\otimes}_{\mathcal{O}(\mathbb{C}^n)} L_a^2(\Omega)$$

with the consecutive quotients equal to $L_a^2(\Omega)/S(V_k)$.   □

The proof consists in a repeated application of Theorem 8.3.6 and of the propagation of analytic transversality through short exact sequences.

It would be interesting to extend Theorem 8.3.6 to more general singular analytic sets. The most comprehensive results in this direction rely on L^2-estimates for $\bar{\partial}$, but they are not as precise as in the case of transversal intersections discussed above (see Skoda 1972).

However, there is a better situation, encountered in the theory of deformations which characterizes the analytic transversality of a subspace of $L_a^2(\Omega)$ and of $L_a^2(\Omega)$ in purely geometric terms. It is our next aim to comment briefly on this result, with originates in A. Douady's thesis (see Douady 1966, Pourcin 1975, and for more recent advances Bingener and Kosarew 1987).

Let $\Omega = \Omega_1 \times \cdots \times \Omega_n$ be a polydomain in $\mathbb{C}^n$. For a point $\lambda \in \overline{\Omega}$, we set $I_\lambda = \{i \in \{1,\ldots,n\}; \; \lambda_i \in \partial\Omega_i\}$, and we consider the corresponding projection

$$I_\lambda : \overline{\Omega} \to \prod_{i \in I_\lambda} \mathbb{C}.$$

We set $\overline{\Omega}^p = \{\lambda \in \overline{\Omega}; \; \mathrm{card}\, I_\lambda \ge p\}$. It is obvious that $\overline{\Omega}^1 = \partial\Omega$ and that $\overline{\Omega}^n$ coincides with the distinguished boundary $\partial\Omega_1 \times \partial\Omega_2 \times \cdots \times \partial\Omega_n$ of $\overline{\Omega}$.

On the other hand, consider a coherent sheaf of ideals $\mathcal{I} \subset \mathcal{O}_U$ and its global sections $I = \mathcal{I}(U) \subset \mathcal{O}(U)$ defined on a Stein open neighbourhood of $\overline{\Omega}$. The support of $\mathcal{O}_U/\mathcal{I}$ has a canonical stratification obtained from closed analytic sets defined by a cohomological invariant called the profounder

$$S_p(\mathcal{O}_U/\mathcal{I}) = \{\lambda \in U; \; \mathrm{prof}_\lambda(\mathcal{O}_U/\mathcal{I}) \le p\} \qquad (p \ge 0)$$

(see for details Serre 1965). We remark that $\dim_{\mathbb{C}} S_p(\mathcal{O}_U/\mathcal{I}) \le p$ for any $p \ge 0$.

Theorem 8.3.8 *Let Ω be a bounded polydomain with $\Omega = \mathrm{Int}\,\overline{\Omega}$ and let $\mathcal{I} \subset \mathcal{O}_U$ be a coherent sheaf of ideals defined on a (Stein) open neighbourhood U of $\overline{\Omega}$. Then the following assertions are equivalent:*

(a) $\mathcal{O}(U)/\mathcal{I} \perp_{\mathcal{O}(U)} L_a^2(\Omega)$, where $I = \mathcal{A}(U)$;

(b) *for any integer p $(0 \le p \le n)$, $S_p(\mathcal{O}_U/\mathcal{I}) \cap \overline{\Omega}^{p+1} = \varnothing$;*

(c) $I \cdot L_a^2(\Omega)$ *is a closed subspace of $L_a^2(\Omega)$;*

(d) $I \cdot L_a^2(\Omega) = \{f \in L_a^2(\Omega); \; f|_\Omega \in \mathcal{A}(\Omega)\}$.   □

A complete proof of this result would go beyond the limits of this chapter, so we refer to Pourcin (1975) for full details. Instead, let us make a few remarks.

First note that assertions (a) and (d) are very similar to the conclusion of Theorem 8.3.6. The novelty and the strength of Theorem 8.3.8 consists in the equivalence between (a) and the apparently much weaker condition (c), and on the other hand, in the equivalence of these conditions with the purely geometric condition (c). A couple of examples will clarify these comments.

Let $P = \mathbb{D} \times \mathbb{D}$ denote the unit polydisc in $\mathbb{C}^2$, and let H be a complex analytic hypersurface (with possible singularities) defined in a neighbourhood U of $\bar{P}$. Let $\mathcal{I}$ be the radical ideal of H. Then one remarks that $S_1(\mathcal{O}_U/\mathcal{I}) = H$ and that $\bar{P}^2 = \partial \mathbb{D} \times \partial \mathbb{D}$. Thus the four assertions of Theorem 8.3.8 are valid in this case if and only if $H \cap (\partial \mathbb{D} \times \partial \mathbb{D}) = \varnothing$. For instance, for $\alpha \in \mathbb{C}$ fixed, the subspace $(z_2 - \alpha z_1) L_a^2(P)$ is closed in $L_a^2(P)$ (and it coincides with $S(H) = \{f \in L_a^2(P); \; f(z, \alpha z) = 0, |z| < 1\}$) if and only if $|\alpha| \neq 1$.

As a second example, let Ω be an arbitrary bounded polydomain in $\mathbb{C}^n$, and let V be a smooth p-dimensional complex submanifold of an open neighbourhood U of $\bar{\Omega}$. Let $\mathcal{I}$ again be the radical ideal of V. Then one easily computes the profounder by using a local Koszul resolution of $\mathcal{O}_U/\mathcal{I}$, and one remarks that $S_r(\mathcal{O}_U/\mathcal{I}) = \varnothing$ unless $r = p$, in which case $S_p(\mathcal{O}_U/\mathcal{I}) = V$. Thus the four assertions of Theorem 8.3.8 are valid if and only if $V \cap \bar{\Omega}^{p+1} = \varnothing$. In particular, $\mathcal{I}(\bar{\Omega}) L_a^2(\Omega) = S(V)$ if and only if $V \cap \bar{\Omega}^{p+1} = \varnothing$. The reader will easily find concrete examples of this type.

At this moment it appears quite natural to ask about the gap between Theorems 8.3.6 and 8.3.8. More specifically, is there any chance of having four equivalent conditions, as in Theorem 8.3.8, in the case treated by Theorem 8.3.6? We close this discussion by two simple examples which point towards a positive solution of this problem.

Proposition 8.3.9 *Let Ω be a bounded convex domain in $\mathbb{C}^n$ with smooth boundary, and let L be a linear variety which intersects $\bar{\Omega}$. Then the following conditions are equivalent*:

(a) $I_L \cdot L_a^2(\Omega)$ *is a closed subspace of* $L_a^2(\Omega)$;
(b) $I_L \cdot L_a^2(\Omega) = \{f \in L_a^2(\Omega); \; f|_{\Omega \cap L} = 0\}$;
(c) $\mathcal{O}(\mathbb{C}^n)/I_L \perp_{\mathcal{O}(\mathbb{C}^n)} L_a^2(\Omega)$;
(d) $L \cap \Omega \neq \varnothing$.

Here $I_L = \{f \in \mathcal{O}(\mathbb{C}^n); \; f|_L = 0\}$.

Proof There are only two possibilities: $L \cap \Omega \neq \varnothing$ or L is tangent to Ω (and intersects $\partial \Omega$). In the second case, Proposition 8.3.5 shows that assertions (a), (b), and (c) are not true. Indeed, if $L \cap \Omega \neq \varnothing$, then the intersection $L \cap \partial \Omega$ is transversal, and Theorem 8.3.6 implies the validity of (a), (b), and (c). $\square$

Another simple example in which the analytic transversality of a Bergman submodule can be tested in geometric terms is contained in the following result.

Proposition 8.3.10 *Let Ω be a bounded strictly pseudoconvex domain with smooth boundary in $\mathbb{C}^n$, and let H be a smooth complex analytic hypersurface defined in a neighbourhood of $\overline{\Omega}$. Assume that for every point $\lambda \in \partial\Omega \cap H$, H is either transversal to $\partial\Omega$ at λ, or λ is an isolated point of $\partial\Omega \cap H$. Then the following assertions are equivalent:*

(a) *$\mathcal{I}_H(\overline{\Omega}) \cdot L_a^2(\Omega)$ is a closed subspace of $L_a^2(\Omega)$;*
(b) *$\mathcal{I}_H(\overline{\Omega}) \cdot L_a^2(\Omega) = \{ f \in L_a^2(\Omega);\ f|_{H \cap \Omega} = 0 \}$;*
(c) *$\mathcal{O}_H(\overline{\Omega}) \perp_{\mathcal{O}(\overline{\Omega})} L_a^2(\Omega)$;*
(d) *H intersects $\partial\Omega$ transversally.*

Here $\mathcal{I}_H$ is the reduced ideal of $\mathcal{O}_{\overline{\Omega}}$ associated with H.

Proof At every point of intersection $\lambda \in H \cap \partial\Omega$, the hypersurface H is either transversal to $\partial\Omega$ or tangential to $\partial\Omega$. In the latter case, λ is an isolated point of $H \cap \overline{\Omega}$. Thus $\mathcal{O}_H = \mathcal{O}_{H_1} \oplus \mathcal{O}_{H_2}$, where H_2 is the collection of the tangential points and H_1 is the complement of H_2 in $H \cap \overline{\Omega}$.

According to Proposition 8.3.5, assertion (a) is valid if and only if $H_2 = \varnothing$, that is, if and only if assertion (d) is true. But a hypersurface is given locally by a single equation, hence on the Stein compact set $\overline{\Omega}$ there is a function $f \in \mathcal{O}(\overline{\Omega})$ which generates the ideal $\mathcal{I}_H$ both locally and globally. In other words, there exists a resolution of $\mathcal{O}_H$ of length one,

$$0 \to \mathcal{O}_{\overline{\Omega}} \overset{f}{\to} \mathcal{O}_{\overline{\Omega}} \to \mathcal{O}_H \to 0.$$

The corresponding complex which computes the invariants

$$\mathrm{T\hat{o}r}_p^{\mathcal{O}(\overline{\Omega})}\big(\mathcal{O}_H(\overline{\Omega}), L_a^2(\Omega) \big)$$

is given by

$$L_a^2(\Omega) \overset{f}{\to} L_a^2(\Omega).$$

Since the multiplication by f is injective on $L_a^2(\Omega)$, we infer that conditions (a) and (c) are equivalent. Finally if the equivalent conditions (a), (c), (d) hold, then by Theorem 8.3.6 one obtains (b). Conversely, if (b) holds, then (a) follows, and the proof is complete. $\square$

A statement similar to Proposition 8.3.10 can be obtained for $n = 2$ and a smooth (not necessarily connected) complex subvariety V of a neighbourhood of $\overline{\Omega}$. In this case, V has connected components of dimension 1, and thus hypersurfaces and isolated points. Theorem 8.3.1 and Proposition 8.3.10 then imply the equivalence of the four conditions above.

8.4 RIGIDITY OF BERGMAN SUBMODULES

Theorem 8.3.4 reveals a rigidity phenomenon for analytically invariant subspaces of finite codimension in the Bergman space. It is the aim of the present section to extend this rigidity result to a larger class of submodules of the Bergman space. We shall prove that two closed analytically invariant subspaces of $L_a^2(\Omega)$ coincide whenever there exists an isometric analytic isomorphism between them, that is, an isometric isomorphism that is compatible with the $\mathcal{O}(\mathbb{C}^n)$-module structures of both spaces. Elementary examples show that the corresponding result fails if one replaces isometric isomorphisms by topological isomorphisms. We describe conditions under which the result nevertheless remains true in the case of topological isomorphisms.

The classification of analytically invariant subspaces of classical function spaces is a principal theme in operator theory. The celebrated theorem of Beurling, which asserts that all closed analytic subspaces of the Hardy space on the torus are unitarily equivalent, is perhaps the most familiar example. In an attempt to extend Beurling's theorem to several complex variables, new rigidity phenomena have been discovered (see Agrawal *et al.* 1986, Douglas and Paulsen 1989, and Douglas and Yan 1990). The case of the Bergman space presented in this section is in many respects more elementary than the Hardy space case treated by these authors. However, even in the simpler case of Bergman spaces, the variety of rigidity results suggests that a common and more general explanation should be possible.

The first theorem proved below also follows as a simple consequence of the theory of subnormal operators.

Theorem 8.4.1 *Let Ω be a bounded pseudoconvex domain in $\mathbb{C}^n$, and let S_1, S_2 be two closed analytically invariant subspaces of the Bergman space $L_a^2(\Omega)$. Suppose that there exists a unitary analytic isomorphism between S_1 and S_2. Then $S_1 = S_2$.*

Proof Let $u: S_1 \to S_2$ be a unitary analytic isomorphism. Let us suppose that $S_1 \neq 0$, and let us fix a non-zero function $h \in S_1$. We denote by $\tilde{S}_i$ the closure of $\mathbb{C}[z, \bar{z}]S_i$ in $L^2(\Omega)$ $(i = 1, 2)$, where $z = (z_1, \ldots, z_n)$. We claim that u admits a $\mathbb{C}[z, \bar{z}]$-linear extension to a unitary operator $\tilde{u}: \tilde{S}_1 \to \tilde{S}_2$.

Indeed, for $x_j \in S_1$, $p_j, q_j \in \mathbb{C}[z]$ $(1 \leq j \leq m)$, we have

$$\left\| \sum_{j=1}^m p_j(z)\overline{q_j(z)}x_j \right\|_2^2 = \sum_{j,k=1}^m \langle p_j(z)q_k(z)x_j, p_k(z)q_j(z)x_k \rangle.$$

Thus the unique continuous linear operator $\tilde{u}: \tilde{S}_1 \to \tilde{S}_2$ with

$$\tilde{u}\left(p(z)\overline{q(z)}x \right) = p(z)\overline{q(z)}u(x) \qquad (p, q \in \mathbb{C}[z], \ x \in S_1)$$

is a unitary operator.

For any function $f \in L_a^2(\Omega)\setminus\{0\}$, the subspace $\mathbb{C}[z, \bar{z}]f$ is dense in $L^2(\Omega)$ (otherwise there would be a function $g \in L^2(\Omega)\setminus\{0\}$ such that the measure $f\bar{g}d\lambda$ would vanish on $\mathbb{C}[z, \bar{z}]$). Hence $\tilde{S}_1 = \tilde{S}_2 = L^2(\Omega)$. Moreover, the $\mathbb{C}[z, \bar{z}]$-linear, unitary operator $\tilde{u}: L^2(\Omega) \to L^2(\Omega)$ acts as the multiplication by a unimodular function $\varphi \in L^\infty(\Omega)$ (see for instance Zhu 1990).

Thus in particular, $uh = \varphi h$ with h as above. Hence, on the complement of the zero set V of h, the function φ is analytic, hence constant. Since V has Lebesgue measure zero and since $\Omega \setminus V$ is connected, the function φ is constant. Therefore $S_2 = S_1$, and the proof is complete.   □

In order to relax the hypothesis of the above theorem, we consider a restricted class of submodules.

Let Ω be a bounded pseudoconvex domain in $\mathbb{C}^n$ with $\Omega = \mathrm{Int}(\overline{\Omega})$ and such that $\overline{\Omega}$ has a fundamental system of Stein open neighbourhoods. Let $\mathcal{I} \subset \mathcal{O}_{\overline{\Omega}}$ be a coherent sheaf of ideals which fulfils the following conditions:

(i) $\mathcal{O}(\overline{\Omega})/\mathcal{I}(\overline{\Omega}) \perp_{\mathcal{O}(\overline{\Omega})} L_a^2(\Omega)$;

(ii) $\mathcal{I}(\overline{\Omega}) \cdot L_a^2(\Omega) = \{ f \in L_a^2(\Omega); \ f|_\Omega \in \mathcal{I}(\Omega) \}$.

Since $\overline{\Omega}$ is Stein compact, the sheaf $\mathcal{I}$ is generated by finitely many functions $f_1, \ldots, f_p \in \mathcal{O}(\overline{\Omega})$, and by the Syzygy theorem it admits a finite resolution with free finite type $\mathcal{O}_{\overline{\Omega}}$-modules

$$0 \to \mathcal{O}_{\overline{\Omega}}^r \to \cdots \to \mathcal{O}_{\overline{\Omega}}^q \to \mathcal{O}_{\overline{\Omega}}^p \to \mathcal{I} \to 0.$$

The two conditions (i) and (ii) are then equivalent to the exactness of the following induced complex

$$0 \to L_a^2(\Omega)^r \to \cdots \to L_a^2(\Omega)^p \to L_a^2(\Omega) \xrightarrow{r} \mathcal{O}(\Omega)/(f_1, \ldots, f_p)\mathcal{O}(\Omega) \to 0,$$

where r is the composite of the inclusion $L_a^2(\Omega) \to \mathcal{O}(\Omega)$ and of the canonical quotient map.

Theorem 8.4.2 *Let Ω be a bounded pseudoconvex domain in $\mathbb{C}^n$ such that $\Omega = \mathrm{Int}(\overline{\Omega})$ and $\overline{\Omega}$ is Stein, and let S_1, S_2 be closed analytically invariant subspaces of $L_a^2(\Omega)$ of the form $S_j = \mathcal{I}_j(\overline{\Omega})L_a^2(\Omega)$ ($j = 1, 2$), where the coherent ideals $\mathcal{I}_1, \mathcal{I}_2$ satisfy conditions (i) and (ii). Suppose that*

$$\max_{j=1,2} \left(\dim_{\mathbb{C}} \mathrm{supp}\, \mathcal{O}_{\overline{\Omega}}/\mathcal{I}_j \right) \le n - 2.$$

Then S_1 and S_2 are isomorphic as topological $\mathcal{O}(\mathbb{C}^n)$-modules if and only if they coincide.

Proof Suppose that S_1 and S_2 are isomorphic as Fréchet $\mathcal{O}(\mathbb{C}^n)$-modules. Since both are quasi-coherent by condition (i), their associated sheaves $\mathcal{S}_1, \mathcal{S}_2$ are isomorphic as $\mathcal{O}_{\mathbb{C}^n}$-modules. Due to the fact that the localization of $L_a^2(\Omega)$ coincides with $\mathcal{O}_\Omega$ on Ω, for any point $\lambda \in \Omega$, one finds that $\mathcal{S}_{j,\lambda} \cong \mathcal{I}_{j,\lambda}$ ($j = 1, 2$).

According to Lemma 8.3.3, one obtains $\mathcal{I}_{1,\lambda} = \mathcal{I}_{2,\lambda}$ for any point $\lambda \in \Omega$. Hence $\mathcal{I}_1(\Omega) = \mathcal{I}_2(\Omega)$ as ideals in $\mathcal{O}(\Omega)$. But by condition (ii) this implies $S_1 = S_2$ as subspaces of $L_a^2(\Omega)$.   □

The case of hypersurfaces is different. For instance, if $\mathcal{I}_1, \mathcal{I}_2$ are coherent ideals as above (satisfying conditions (i) and (ii)) and if their zero sets

$V(\mathcal{I}_1), V(\mathcal{I}_2)$ are purely $n-1$ dimensional, then $\mathcal{I}_i = f_i \mathcal{O}_{\overline{\Omega}}$ for some analytic functions $f_i \in \mathcal{O}(\overline{\Omega})$ $(i = 1, 2)$. In this case,

$$L_a^2(\Omega) \xrightarrow{f_i} \mathcal{I}_i(\overline{\Omega}) L_a^2(\Omega) \qquad (i = 1, 2)$$

are isomorphisms of $\mathcal{O}(\overline{\Omega})$-modules, and consequently the two Bergman submodules $\mathcal{I}_1(\overline{\Omega}) L_a^2(\Omega)$ and $\mathcal{I}_2(\overline{\Omega}) L_a^2(\overline{\Omega})$ are analytically isomorphic without being equal.

8.5 REFERENCES AND COMMENTS

Operator theory on Bergman spaces is built upon the model of Toeplitz operators on the Hardy space of the unit torus in one complex dimension. A comprehensive monograph on the latter subject is Nikolskii (1986). A few fundamental themes in this area are: the classification of Toeplitz-like operators up to unitary equivalence or similarity, the computation of various spectra for these operators, index formulae for concrete classes of Fredholm operators, functional models and dilation theory, the existence and classification of invariant subspaces for Toeplitz operators, the structure theory of the C^*-algebra generated by these operators, and so on. In the case of the Bergman space each of these subjects has more open problems than positive results. Without aiming at completeness, we mention below a few bibliographical references in these areas.

The survey paper by Axler (1988) and Section II.8 of the monograph by Conway (1991) contain excellent introductions to the spectral theory of the Bergman operator $B_\Omega = M_z$ acting on the Bergman space $L_a^2(\Omega)$ on an arbitrary bounded open set Ω in $\mathbb{C}$. For instance, the spectral picture of this operator is completely determined in these works. The observation that the invariant subspace lattice of B_D, where D is a disc, is extremely rich and complicated is due to Bercovici *et al.* (1988). Remarkable progress in understanding a certain class of invariant subspaces of B_D was recently made by several authors (see Duren *et al.* 1992). These authors discuss 'contractive divisors', which are functions in the Bergman space on the disc and which replace the Blaschke products of the Hardy space.

The multivariable spectral theory of Toeplitz operators acting on Bergman spaces associated with domains in $\mathbb{C}^n$ is much less investigated than its one-dimensional counterpart. Besides the two topics developed in Chapter 8, the computation of joint spectra and the classification of some analytically invariant subspaces, there are some other aspects worth mentioning.

First there is the general question of how smooth is the commutator $[T_f, T_g]$ of two Toeplitz operators acting on the Bergman space associated with a bounded pseudoconvex domain in $\mathbb{C}^n$. Significant progress was made in this direction in the case of Cartan domains by several authors, including Arazy, Berger, Coburn, Koranyi, Misra, Upmeier (see for references Zhu 1990; Koranyi 1993).

The case of Toeplitz operators with trace class commutators deserves special attention because exactly this condition is necessary for establishing trace formulae for commutators or multi-commutators. The general theory of the principal function developed by Pincus, Carey, Xia, and other authors produces refined index invariants for these operators or systems of operators. The book Xia (1983) amply refers to this subject. Among many articles we mention only Helton and Howe (1975), Carey and Pincus (1977, 1985). An important abstract criterion for the smoothness of commutators of operators is contained in the Berger–Shaw theorem (see Conway 1991, Chapter IV).

The structure of the C^*-algebra T_Ω generated by all Toeplitz operators with continuous symbol on the Bergman space $L_a^2(\Omega)$ of a bounded pseudo-convex domain in $\mathbb{C}^n$ has been intensively studied in the last decade. The general principle here is that the geometry of the boundary of Ω determines the type of T_Ω, and moreover, a composition series for T_Ω. The pioneering works of Curto and Muhly (1985) and Upmeier (1987) were continued by Salinas *et al.* (1989) and by several other authors. Besides the composition series of the C^*-algebra T_Ω, the geometry of the domain Ω is related via K-theory to some relative index formulae for Toeplitz operators (with respect to the faces of Ω). The recent works of Upmeier incorporate these results in a general quantization theory for classical domains.

As we have already mentioned in the text of Chapter 8, an important tool in studying operators on function spaces is the concept of Hilbert module developed by Douglas and his school. Besides the rigidity phenomenon isolated, for instance, in Douglas and Paulsen (1989), Douglas and Yan (1990), Douglas *et al.* (1995), this theory yields some well-adapted resolutions for Hilbert modules corresponding to the classical dilation theory of Sz.-Nagy and Foiaş (see Chapter 3 in Douglas and Paulsen 1989). Presumably, this approach will produce more unitary invariants for systems of commuting operators and will lead to a better understanding of the operator theory on concrete function spaces.

9

Finiteness theorems in analytic geometry

It is well known that a compact analytic manifold carries, besides the usual topological invariants (its Euler characteristic, the Betti numbers), some numerical invariants like the degree of the canonical divisor, the Hodge numbers, the arithmetic genus, and so on. A common property of all these invariants is that they coincide with the dimension of the cohomology spaces of certain analytic vector bundles or analytic sheaves. In this way, these invariants become accessible to algebraic and homological methods.

The main purpose of the present chapter is to give a proof of Grauert's theorem on the coherence of direct image sheaves (Grauert 1960). Although Grauert's theorem has its origin and principal applications in the deformation theory of complex structures, we include a complete proof, mainly because Kiehl and Verdier (1971) introduced the notion of analytic transversality as a tool to simplify the original proof of Grauert's result. It should be mentioned that the original proof of Grauert (1960) was quite involved. At present this theorem has a variety of simpler proofs (see for example Kiehl and Verdier 1971, Forster and Knorr 1971, Knorr 1971, Grauert and Remmert 1984, Levy 1987b).

9.1 THE CARTAN–SERRE THEOREM

The fundamental result which provides numerical invariants associated with analytic objects is the following theorem of Cartan and Serre (1953).

Theorem 9.1.1 (Cartan–Serre) *Let $\mathscr{F}$ be a coherent analytic sheaf on the compact analytic space X. Then the cohomology spaces $H^q(X, \mathscr{F})$ are finite-dimensional vector spaces for every $q \geq 0$.*   □

The proof of this theorem exploits the observation that the restriction map $\mathscr{F}(U) \to \mathscr{F}(V)$ is nuclear whenever V is a relatively compact subset of U. Quite specifically, let $\mathscr{U} = (U_i)_{i \in I}$ be a finite covering of X by Stein open subsets. Since X is compact, there exists a covering $\mathscr{V} = (V_i)_{i \in I}$ subordinate to $\mathscr{U}$ with relatively compact Stein open subsets $V_i \Subset U_i$ ($i \in I$).

By Leray's theorem, the cohomology spaces of each of the Čech complexes

$\mathscr{C}^{\bullet}(\mathscr{U},\mathscr{F})$ and $\mathscr{C}^{\bullet}(\mathscr{V},\mathscr{F})$ are canonically isomorphic to the cohomology groups $H^p(X,\mathscr{F})$ $(p \geq 0)$. Hence the restriction mapping

$$\rho: \mathscr{C}^{\bullet}(\mathscr{U},\mathscr{F}) \to \mathscr{C}^{\bullet}(\mathscr{V},\mathscr{F})$$

is a quasi-isomorphism.

Thus the proof of Theorem 9.1.1 is completed by the next result.

Proposition 9.1.2 *Let* $\rho: E^{\bullet} \to F^{\bullet}$ *be a compact morphism of complexes of Fréchet spaces. If* ρ *is a quasi-isomorphism, then* $\dim H^q(E^{\bullet})$ $(= \dim H^q(F^{\bullet}))$ *is finite for any integer* q.

The proof of Proposition 9.1.2 is, in its turn, a consequence of the following perturbation lemma.

Lemma 9.1.3 (Schwartz) *Let* $\rho: E \to F$ *be an epimorphism of Fréchet spaces. If* $k: E \to F$ *is a compact operator, then* $\dim \mathrm{Coker}(\rho + k)$ *is finite.* □

A proof of Schwartz's lemma is given in Appendix 1.

Proof of Proposition 9.1.2 Let q be fixed. Let us consider the commutative diagram

$$
\begin{array}{ccc}
E^{q-1} & \xrightarrow{\;d_E\;} & \mathrm{Ker}\, d_E^q \\
\downarrow{\scriptstyle\rho} & & \downarrow{\scriptstyle\rho} \\
F^{q-1} & \xrightarrow{\;d_F\;} & \mathrm{Ker}\, d_F^q.
\end{array}
$$

Since ρ induces an isomorphism between $\mathrm{Coker}\,(d_E)$ and $\mathrm{Coker}\,(d_F)$, the map

$$(d_F, \rho): F^{q-1} \oplus \mathrm{Ker}\, d_E^q \to \mathrm{Ker}\, d_F^q$$

is surjective. As ρ was assumed to be a compact operator, Schwartz's lemma yields the finiteness of the space $H^q(F^{\bullet}) = \mathrm{Ker}\, d_F^q / \mathrm{Im}\, d_F^{q-1}$. □

In principle the above scheme for proving the Cartan–Serre theorem also works in the relative case of proper families of analytic spaces. Our next aim is to gather the results that will be used in the proof of Grauert's direct image theorem.

9.2 RELATIVE NUCLEARITY AND PERTURBATION LEMMAS

In the present section the perturbation lemma of L. Schwartz (Lemma 9.1.3) is generalized to morphisms of topological modules over Fréchet algebras.

This will be the main tool in the proof of Grauert's theorem on direct images (sometimes this theorem is called the coherence theorem). However, some of the results below are of an independent interest. We closely follow the paper of Kiehl and Verdier (1971) and Douady's exposition in Séminaire de Géométrie Analytique (1974).

Throughout this section A will denote a Fréchet algebra which is assumed to be commutative and unital.

Definition 9.2.1 A morphism of Fréchet A-modules $f: M \to N$ is called *A-nuclear* if it is of the form

$$f(x) = \sum_{n=0}^{\infty} \lambda_n u_n(x) y_n \ (x \in M), \tag{9.2.1}$$

where (y_n) is a bounded sequence in N, $(\lambda_n) \in l^1(\mathbb{N}, \mathbb{C})$, and the mappings $u_n: M \to A$ form an equicontinuous sequence of morphisms of Fréchet A-modules.

The morphism $f: M \to N$ is called *A-subnuclear* if there is a nuclear Fréchet A-module Q, an A-nuclear morphism $g: Q \to N$, and an epimorphism $h: Q \to M$ such that $g = f \circ h$:

$$M \xrightarrow{\quad f \quad} N$$

As usual, the A-nuclear morphisms form a two-sided ideal in the following sense. A product $f_3 \circ f_2 \circ f_1$ with $f_1 \in \mathrm{Hom}_A(M_1, M_2)$, $f_2 \in \mathrm{Hom}_A(M_2, M_3)$, and $f_3 \in \mathrm{Hom}_A(M_3, M_4)$ is A-nuclear whenever f_2 has this property. Note also that $1_A \otimes f: A \hat{\otimes} E \to A \hat{\otimes} F$ is an A-nuclear morphism whenever f is a nuclear $\mathbb{C}$-linear operator between Fréchet spaces E and F. This remark essentially yields the proof of the next result.

Proposition 9.2.2 *Let S be an analytic space, and let $V \Subset U \Subset \mathbb{C}^n$ be open sets. Then the restriction map*

$$\rho: \mathscr{O}(S \times U) \to \mathscr{O}(S \times V)$$

is $\mathscr{O}(S)$-nuclear. Moreover, if S and U are Stein and if $\mathscr{F}$ is a coherent analytic sheaf on $S \times U$, which is globally generated by a finite number of sections, then the restriction map

$$\rho: \mathscr{F}(S \times U) \to \mathscr{F}(S \times V)$$

is $\mathscr{O}(S)$-subnuclear.

Proof The first part follows from the fact that there is a natural isomorphism $\mathscr{O}(S \times U) = \mathscr{O}(S) \hat{\otimes} \mathscr{O}(U)$ of Fréchet $\mathscr{O}(S)$-modules which is compatible with restrictions. To prove the second part, we use the fact that under the stated conditions, by Cartan's Theorem B, there is an epimorphism $h: \mathscr{O}(S \times U)^r \to \mathscr{F}(S \times U)$ of Fréchet $\mathscr{O}(S)$-modules, and that moreover the resulting diagram

$$
\begin{array}{ccc}
\mathscr{O}(S \times U)^r & \xrightarrow{\oplus \rho} & \mathscr{O}(S \times V)^r \\
\downarrow h & & \downarrow h \\
\mathscr{F}(S \times U) & \xrightarrow{\rho} & \mathscr{F}(S \times V)
\end{array}
$$

is commutative. Indeed, the sheaf $\mathscr{F}$ is generated by finitely many sections on $S \times U$, and by Cartan's Theorem B, the same sections generate $\mathscr{F}(S \times U)$ as an $\mathscr{O}(S \times U)$-module.   □

The next technical result gives a first idea of the usefulness of the notion of transversality for the theory of A-nuclear operators.

Lemma 9.2.3 *Let $i: A \to B$ be a $\mathbb{C}$-nuclear morphism of nuclear (commutative and unital) Fréchet algebras. Suppose that $f: M \to N$ is an A-subnuclear morphism of Fréchet A-modules with $f(M) \subset N_1$ for some closed A-submodule N_1 of N. If $B \perp_A M$, $B \perp_A N$ and $B \perp_A (N/N_1)$, then the map*

$$
1_B \otimes f: B \,\hat{\otimes}_A\, M \to B \,\hat{\otimes}_A\, N_1
$$

is B-subnuclear.

Proof Let g, h with $g = fh$ be chosen as described in Definition 9.2.1, and let $g(x) = \sum_n \lambda_n u_n(x) y_n$ be a nuclear decomposition of $g: Q \to N$. As a nuclear operator, i can be written as a product $A \to A_1 \to B$ of continuous $\mathbb{C}$-linear operators with a suitable Banach space A_1. The continuous linear map $i \otimes g: Q = A \,\hat{\otimes}_A\, Q \to B \,\hat{\otimes}_A\, N$ admits a continuous linear factorization

$$
Q \xrightarrow{\sigma} l^1(N, A_1) \xrightarrow{\tau} B \,\hat{\otimes}_A\, N.
$$

Since Q is nuclear, the map $i \otimes g$ is $\mathbb{C}$-nuclear. We denote by f_1, g_1 the operators f, g with range space replaced by N_1. According to the transversality assumptions, $B \,\hat{\otimes}_A\, N_1$ can be regarded as a closed subspace of $B \,\hat{\otimes}_A\, N$ (see Proposition 3.1.15). Therefore the nuclearity of Q and $i \otimes g$ imply the nuclearity of $i \otimes g_1$ (see Appendix 1). Since the diagram

$$
\begin{array}{ccc}
B \,\hat{\otimes}\, Q & \xrightarrow{1_B \otimes (i \otimes g_1)} & B \,\hat{\otimes}\, (B \,\hat{\otimes}_A\, N_1) \\
\downarrow 1 \otimes_A h & & \downarrow m \\
B \,\hat{\otimes}_A\, M & \xrightarrow{1_B \otimes f_1} & B \,\hat{\otimes}_A\, N_1
\end{array}
$$

commutes, where m denotes the continuous B-linear map induced by the B-module structure of $B \hat{\otimes}_A N_1$, the operator $1_B \otimes f_1$ is B-subnuclear.   □

The only reason to demand the nuclearity of the space Q in the definition of A-subnuclearity (Definition 9.2.1) was to ensure that the proof of Lemma 9.2.3 works.

For the proof of the Cartan–Serre theorem indicated in Section 9.1, it suffices to have a version of Schwartz's lemma for nuclear perturbations of epimorphisms between Fréchet spaces. The first perturbation result of this section generalizes this version of Schwartz's lemma.

Proposition 9.2.4 *Let $f, u: M \to N$ be morphisms of Fréchet A-modules. Assume that f is onto and u is A-subnuclear. Let $i: A \to B$ be a morphism of Fréchet algebras, which factors through a Banach algebra (all assumed to be commutative and unital).*

Suppose that $B \perp_A M$ and that $B \perp_A N$. Then the cokernel of the morphism $1 \otimes (f - u): B \hat{\otimes}_A M \to B \hat{\otimes}_A N$ is a finitely generated B-module.

Proof We divide the proof into several steps.

(a) Assume that $M = N$, A is a Banach algebra, $f = 1_M$, and that u is A-nuclear. We claim that Coker $(1_M - u)$ is a finitely generated A-module. Each representation of the form described in (9.2.1) yields a canonical A-linear factorization $u = u_2 \circ u_1$

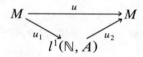

such that u_1 is again A-nuclear. Hence $u_1 \circ u_2: l^1(\mathbb{N}, A) \to l^1(\mathbb{N}, A)$ is an A-nuclear operator acting on a Banach A-module. Since u induces the identity operator on Coker $(1 - u)$, the morphism u_1 yields an isomorphism Coker$(1_M - u_2 \circ u_1) \cong$ Coker$(1_{l^1(\mathbb{N}, A)} - u_1 \circ u_2)$ of A-modules whose inverse is induced by u_2. Hence it suffices to show that the latter space is a finitely generated A-module.

As an A-nuclear operator on a Banach module, $u_1 \circ u_2$ can be written in the form:

$$u_1 \circ u_2 = v + w,$$

where $\|v\| < 1$ and Im(w) is a finite rank A-module. But then $1 - u_1 \circ u_2 = 1 - v - w = (1 - v)(1 - (1 - v)^{-1}w)$ is clearly a morphism with a finitely generated cokernel.

(b) The reader will easily be convinced, using the definition of A-subnuclearity, that one may suppose M to be a free Fréchet A-module of the form $M = A \hat{\otimes} Q$, and that one may suppose u to be A-nuclear.

But in this case u can be factorized as $f \circ v$ with a suitable A-nuclear operator $v: M \to M$. Namely, suppose that

$$u(x) = \sum_n \lambda_n u_n(x) y_n \qquad (x \in M)$$

as in (9.2.1). Since there is always a decomposition $\lambda_n = \lambda'_n \cdot \lambda''_n$ such that $(\lambda'_n) \in l^1(\mathbb{N})$ and $\lim_n \lambda''_n = 0$, we may suppose that $\lim_n y_n = 0$. Using the surjectivity of f, one can choose a sequence (x_n) in M with $f(x_n) = y_n$ and $\lim_n x_n = 0$. Then the operator v can be defined by

$$v(x) = \sum_n \lambda_n' u_n(x) x_n.$$

(c) Next one may suppose that $M = N$ and $f = 1_M$. Indeed, with the notation of part (b), the map $1_B \otimes f: B \hat{\otimes}_A M \to B \hat{\otimes}_A N$ induces a surjection $\operatorname{Coker}(1 \otimes (1 - v)) \to \operatorname{Coker}(1 \otimes (f - u))$.

(d) At this point, we consider A to be a commutative Fréchet algebra. Let $A \to A_1 \to B$ be a factorization of i through a Banach algebra A_1. As explained in (b) and (c), we may suppose that $M = N$, $u: M \to M$ is A-nuclear, $f = 1_M$, and $M = A \hat{\otimes} Q$ for a suitable Fréchet space Q. Then $A_1 \hat{\otimes}_A M = A_1 \hat{\otimes} Q$ as Fréchet A_1-modules, and $1_{A_1} \hat{\otimes} u: A_1 \hat{\otimes}_A M \to A_1 \hat{\otimes}_A M$ is A_1-nuclear. Hence $\operatorname{Coker}(1 - (1_{A_1} \otimes u))$ is a finitely generated A_1-module by part (a).

We choose a morphism $g: A_1^n \to A_1 \hat{\otimes} Q$ of Fréchet A_1-modules such that the map

$$\left(1 - (1_{A_1} \otimes u), g\right): A_1 \hat{\otimes} Q \oplus A_1^n \to A_1 \hat{\otimes} Q$$

is surjective. Applying the functor $B \hat{\otimes}_{A_1} *$ one obtains an epimorphism $B \hat{\otimes} Q \oplus B^n \to B \hat{\otimes} Q$ of Fréchet B-modules, the restriction of which to $B \hat{\otimes} Q$ is equal to $1_B \otimes (1_M - u): B \hat{\otimes}_A M \to B \hat{\otimes}_A M$ modulo the isomorphism $B \hat{\otimes} Q = B \hat{\otimes}_A M$ described in Lemma 3.1.9.   □

The reader will observe that, for the proof of Proposition 9.2.4, the nuclearity condition on Q demanded in the definition of A-subnuclearity is not needed.

Definition 9.2.5 A *nuclear chain of algebras* is an inductive system $\mathbb{A} = (A_t, \rho_s^t)_{0 \le s \le t \le 1}$ of nuclear Fréchet algebras (commutative and unital) and morphisms $\rho_s^t: A_s \to A_t$ of unital Fréchet algebras such that ρ_s^t factors through a Banach algebra whenever $s < t$.

The chain $\mathbb{A}$ is called *transversal* to a Fréchet A_0-module M if $A_t \perp_{A_0} M$ for every $t \in [0, 1]$. In this case, we write $\mathbb{A} \perp_{A_0} M$.

The principal result à la Schwartz in this section is the following theorem.

Theorem 9.2.6 *Let* $\mathbb{A} = (A_t, \rho_s^t)$ *be a nuclear chain of algebras. Suppose that* $f: M^\bullet \to N^\bullet$ *is a quasi-isomorphism of complexes of nuclear Fréchet A_0-modules. Let* $M^\bullet$ *and* $N^\bullet$ *be bounded to the right.*

If the components of f are A_0-subnuclear and $\mathbb{A} \perp_{A_0} M^\bullet, \mathbb{A} \perp_{A_0} N^\bullet$, *then there is a complex $L^\bullet$ of free finite-type A_1-modules and a quasi-isomorphism* $L^\bullet \to A_1 \hat{\otimes}_{A_0} M^\bullet$.

Proof   The complex $L^\bullet$ is constructed by descending induction.

We shall use the abbreviations $M_t^n = A_t \, \hat{\otimes}_{A_0} \, M^n$, $N_t^n = A_t \, \hat{\otimes}_{A_0} \, N^n$ for arbitrary n and $t \in [0, 1]$. We choose an integer r with $M^n = N^n = 0$ for $n > r$, and fix an arbitrary sequence $(t_n)_{n \le r}$ of real numbers with $0 < t_n < t_{n-1} < 1$ for all $n \le r$.

By descending induction, we shall define a sequence $(k_n)_{n \le r}$ of integers $k_n \ge 0$, and for each $n \le r$, a complex

$$L_{(n)}^\bullet \colon 0 \to A_{t_n}^{k_n} \to A_{t_n}^{k_{n+1}} \to \cdots \to A_{t_n}^{k_r} \to 0$$

as well as a morphism $h_{(n)}^\bullet \colon L_{(n)}^\bullet \to M_{t_n}^\bullet$ such that:

(1) the cone of the morphism $h_{(n)}^\bullet$ is exact in degree greater or equal to n;

(2) the effect of applying the functor $A_{t_{n-1}} \, \hat{\otimes}_{A_{t_n}} \, *$ to $h_{(n)}^\bullet \colon L_{(n)}^\bullet \to M_{t_n}^\bullet$ is the same as if one replaces in $h_{(n-1)}^\bullet \colon L_{(n-1)}^\bullet \to M_{t_{n-1}}^\bullet$ the space $A_{t_{n-1}}^{k_{n-1}}$ occurring on the extreme left of $L_{(n-1)}^\bullet$ by the zero space.

To explain the main idea, we suppose for the moment that, in addition to the above hypotheses, $n \le r$ is an integer such that the complexes $M^\bullet$ and $N^\bullet$ are exact in degree greater than n.

For each integer k, we define $Z^k = \mathrm{Ker}(d^k \colon M^k \to M^{k+1})$ and $Z'^k = \mathrm{Ker}(d^k \colon N^k \to N^{k+1})$. In the case where $n < r$, the conditions

$$\mathbb{A} \perp_{A_0} M^{r-1}, \mathbb{A} \perp_{A_0} M^r$$

imply, by Proposition 3.1.15, that $\mathbb{A} \perp_{A_0} Z^{r-1}$. An iteration of this argument yields that $\mathbb{A} \perp_{A_0} Z^n, \mathbb{A} \perp_{A_0} Z'^n$, and that $\mathbb{A} \perp_{A_0} N^n / Z'^n$. The image of the A-subnuclear map $f \colon Z^n \to N^n$ is contained in Z'^n. According to Lemma 9.2.3, the induced map $f \colon A_t \, \hat{\otimes}_{A_0} \, Z^n \to A_t \, \hat{\otimes}_{A_0} \, Z'^n$ is A_t-subnuclear for every $t \in (0, 1]$.

Since f is a quasi-isomorphism, the map

$$(d, f) \colon A_t \, \hat{\otimes}_{A_0} \, N^{n-1} \oplus A_t \, \hat{\otimes}_{A_0} \, Z^n \to A_t \, \hat{\otimes}_{A_0} \, Z'^n$$

is an epimorphism of A_t-modules. For $0 < t < t_n$, Proposition 9.2.4 applied to the morphism $\rho_t^{t_n} \colon A_t \to A_{t_n}$ yields that $\mathrm{Coker}(d \colon A_{t_n} \, \hat{\otimes}_{A_0} \, N^{n-1} \to A_{t_n} \, \hat{\otimes}_{A_0} \, Z'^n)$ is a finite-type A_{t_n}-module.

Our transversality conditions ensure that the functor $A_{t_n} \, \hat{\otimes}_{A_0} \, *$ preserves the exactness of the complex

$$0 \to Z'^n \to N^n \to N^{n+1} \to \cdots \to N^r \to 0.$$

In particular, it follows that $H^n(A_{t_n} \, \hat{\otimes}_{A_0} N^\bullet)$ is an A_{t_n}-module of finite type. According to Lemma 3.1.17, the morphism $f \colon M_{t_n}^\bullet \to N_{t_n}^\bullet$ remains a quasi-isomorphism. Hence also $H^n(A_{t_n} \, \hat{\otimes}_{A_0} \, M^\bullet)$ is a finitely generated A_{t_n}-module. Consequently, with k suitable, there is a morphism $g \colon L = (A_{t_n})^k \to A_{t_n} \hat{\otimes}_{A_0} M^n$ of Fréchet A_{t_n}-modules such that

$$(d, g) \colon A_{t_n} \, \hat{\otimes}_{A_0} \, M^{n-1} \oplus L \to Z^n \big(A_{t_n} \, \hat{\otimes}_{A_0} \, M^\bullet \big)$$

is onto. Using again the fact that $f\colon M_{t_n}^\bullet \to N_{t_n}^\bullet$ is a quasi-isomorphism, we deduce that the corresponding statement with M^{n-1} and $M^\bullet$ replaced by N^{n-1} and $N^\bullet$, and g replaced by fg, also holds.

Since the morphism

$$
\begin{array}{ccccccc}
\cdots \longrightarrow & 0 & \longrightarrow & L & \longrightarrow & 0 & \longrightarrow \cdots \\
& \downarrow & & \downarrow & & \downarrow & \\
\cdots \longrightarrow & M_{t_n}^{n-1} & \longrightarrow & M_{t_n}^{n} & \longrightarrow & M_{t_n}^{n+1} & \longrightarrow \cdots
\end{array}
$$

is an n-quasi-isomorphism, its cone $\overline{M}^\bullet = C^\bullet(g)$ is exact in degrees greater than $n-1$. The same argument applies to the complex $\overline{N}^\bullet = C^\bullet(fg)$. Moreover, the morphism $\overline{M}^\bullet \to \overline{N}^\bullet$ induced by f is an A_{t_n}-subnuclear quasi-isomorphism (cf. Lemma 9.2.3). Therefore the above construction can be repeated with $M^\bullet, N^\bullet, n$ replaced by $\overline{M}^\bullet, \overline{N}^\bullet, n-1$, and $\mathbb{A}$ replaced by $(A_t)_{t \in [t_n, 1]}$.

If one starts the inductive procedure with $n = r$, then one obtains successively a sequence of complexes $\overline{M}_r^\bullet, \overline{M}_{r-1}^\bullet, \ldots$ To complete the inductive construction, it suffices to observe that, for each $n \leq r$, the complex $\overline{M}_n^\bullet$ is the cone of a morphism $h_{(n)}^\bullet\colon L_{(n)}^\bullet \to M_{t_n}^\bullet$ such that $L_{(n)}^\bullet$ and $h_{(n)}^\bullet$ satisfy the conditions described at the beginning.

In particular, the above inductive construction yields a sequence $(k_n)_{n \leq r}$ of natural numbers such that $L_{(n)}^\nu = A_{t_n}^{k_\nu}$ for all $n \leq \nu \leq r$. The proof is completed by the observation that there is a unique complex $L^\bullet = (A_1)_{n \leq r}^{k_n}$, and a uniquely determined morphism $L^\bullet \to A_1 \hat{\otimes}_{A_0} M^\bullet$ with the property that, for each $n \leq r$, the morphism of complexes

$$
\begin{array}{ccccccccc}
\cdots \longrightarrow & 0 & \longrightarrow & L^n & \longrightarrow & L^{n+1} & \longrightarrow \cdots \longrightarrow & L^r & \longrightarrow 0 \\
& \downarrow & & \downarrow & & \downarrow & & \downarrow & \\
\cdots \longrightarrow & M_1^{n-1} & \longrightarrow & M_1^n & \longrightarrow & M_1^{n+1} & \longrightarrow \cdots \longrightarrow & M_1^r & \longrightarrow 0
\end{array}
$$

arises by applying the functor $A_1 \hat{\otimes}_{A_{t_n}} *$ to the morphism $h_{(n)}^\bullet\colon L_{(n)}^\bullet \to M_{t_n}^\bullet$. Clearly, the morphism $L^\bullet \to A_1 \hat{\otimes}_{A_0} M^\bullet$ is a quasi-isomorphism of the desired type. $\square$

Using the remark immediately following the proof of Proposition 9.2.4 one easily verifies that the nuclearity of the Fréchet modules N^n was not needed in the proof of Theorem 9.2.6. The main application that motivates Theorem 9.2.6 is contained in Section 9.3.

9.3 GRAUERT'S DIRECT IMAGE THEOREM

This section contains a proof of Grauert's theorem and a few of its geometric applications.

Theorem 9.3.1 (Grauert)   *Let* $f: X \to Y$ *be a proper morphism of analytic spaces, and let* $\mathscr{F}$ *be a coherent* $\mathscr{O}_X$*-module. Then the sheaf* $f_*\mathscr{F}$ *is* $\mathscr{O}_Y$*-coherent as well as all its derived functors* $\mathscr{R}^q f_*\mathscr{F}$ *(*$q \geq 0$*).*

Recall that a continuous map $f: X \to Y$ is called *proper* if $f^{-1}(K)$ is compact for each compact subset K of Y. A proper map f extends continuously to the one-point compactifications (mapping $\infty \in \hat{X}$ to $\infty \in \hat{Y}$), whence it is closed. In particular, for each $y \in Y$, the sets $f^{-1}(U)$, U running through a neighbourhood basis of y form a neighbourhood basis of $f^{-1}(y)$.

The *higher direct images* $\mathscr{R}^q f_*\mathscr{F}$ of the sheaf $\mathscr{F}$ with respect to the morphism $f: X \to Y$ are the $\mathscr{O}_Y$-modules associated with the presheaves

$$U \mapsto H^q(f^{-1}(U), \mathscr{F}) \qquad (U \subset Y \text{ open}).$$

They are the derived functors of f_* in the following sense. To each short exact sequence

$$0 \to \mathscr{F}' \to \mathscr{F} \to \mathscr{F}'' \to 0$$

of $\mathscr{O}_X$-modules there corresponds a long exact sequence

$$0 \to f_*\mathscr{F}' \to f_*\mathscr{F} \to f_*\mathscr{F}'' \to \mathscr{R}^1 f_*\mathscr{F}' \to \mathscr{R}^1 f_*\mathscr{F} \to \mathscr{R}^1 f_*\mathscr{F}'' \to \cdots .$$

If Y is a one-point space, then the space X is compact, and $\mathscr{R}^q f_*\mathscr{F} = H^q(X, \mathscr{F})$ for any $q \geq 0$.

Proof of Theorem 9.3.1   By the very definition of the direct image functors, for each open set V in Y and each q, there is an isomorphism $\mathscr{R}^q f_*(\mathscr{F}) \mid V \cong \mathscr{R}^q g_*(\mathscr{F} \mid U)$ of $\mathscr{O}_Y \mid V$-modules, where $U = f^{-1}(V)$ and $g: U \to V$ is the morphism induced by f. Thus we may suppose that there is an open polydisc Ω in a suitable numerical space $\mathbb{C}^n$ and a closed holomorphic embedding $\varphi: Y \to \Omega$. Since in this case there are canonical isomorphisms $\mathscr{R}^q(\varphi \circ f)_*\mathscr{F} = \varphi_*(\mathscr{R}^q f_*\mathscr{F})$ $(q \geq 0)$, the assertion is reduced to the case where $Y = \Omega$ is actually an open polydisc in $\mathbb{C}^n$.

Let $y_0 \in f(X)$ be fixed. Using the fact that f is proper one can choose concentric open polydiscs P_0, P_1 centred at y_0 with $P_1 \Subset P_0 \Subset \Omega$ and a finite number of closed holomorphic embeddings $\varphi_i: E_i \to U_i$ $(i = 1, \ldots, r)$ of suitable open subsets E_i of X into open polydiscs $U_i \subset \mathbb{C}^{n_i}$ such that there are open polydiscs $W_i \Subset V_i \subset U_i$ satisfying

$$f^{-1}(P_0) \subset \bigcup_{i=1}^{r} \varphi_i^{-1}(W_i).$$

Let $(P_t)_{t \in [0,1]}$ be a family of open polydiscs with $P_t \Subset P_s$ for $0 \leq s < t \leq 1$. Then the Fréchet algebras $\mathscr{O}(P_t)$ $(t \in [0,1])$ form a nuclear chain in the terminology of the preceding section. For $i = 1, \ldots, r$, we denote by $\Psi_i = (f|_{E_i}) \vee \varphi_i: E_i \to \Omega \times U_i$ the closed embedding with coordinate functions $f|_{E_i}$ and φ_i. For each $t \in [0,1]$, the sets $X'_{i,t} = \Psi_i^{-1}(P_t \times V_i)$ and $X''_{i,t} = \Psi_i^{-1}(P_t \times W_i)$ form Stein open covers X'_t and X''_t of $f^{-1}(P_t)$. The alternating Čech complexes $\mathscr{C}^\bullet(X'_t, \mathscr{F}), \mathscr{C}^\bullet(X''_t, \mathscr{F})$ are complexes of Fréchet

$\Gamma(f^{-1}(P_t), \mathcal{O}_X)$-modules. Via the unital morphism of Fréchet algebras $\mathcal{O}(P_t) \to \Gamma(f^{-1}(P_t), \mathcal{O}_X)$ induced by f they become complexes of Fréchet $\mathcal{O}(P_t)$-modules. We define $M^\bullet = \mathcal{C}^\bullet(X_0', \mathcal{F})$, $N^\bullet = \mathcal{C}^\bullet(X_0'', \mathcal{F})$, and we observe that by Leray's theorem and Cartan's Theorem B the refinement map $M^\bullet \to N^\bullet$ is a quasi-isomorphism of complexes of nuclear Fréchet $\mathcal{O}(P_0)$-modules, which are bounded to the right.

Fix for the moment a $(q+1)$-tuple i of indices $1 \leq i_0 < \cdots < i_q \leq r$, and define $H_t = X_{i_0, t}' \cap \cdots \cap X_{i_q, t}'$, $E = E_{i_0} \cap \cdots \cap E_{i_q}$, $V = V_{i_0} \times \cdots \times V_{i_q}$, $U = U_{i_0} \times \cdots \times U_{i_q}$ for all $t \in [0,1]$. For simplicity, we write H instead of H_0. Then the closed holomorphic embedding $\rho \colon H \to P_0 \times V$ induced by $f_i = f \vee \varphi_{i_0} \vee \cdots \vee \varphi_{iq}$ gives rise to the isomorphism $\mathcal{F}(H) \to \Gamma(P_0 \times V, \mathcal{G})$ of Fréchet $\mathcal{O}(P_0)$-modules, where $\mathcal{G} = \rho_*(\mathcal{F}|_H)$.

Since $\mathcal{G}$ can be regarded as the restriction of the coherent $\mathcal{O}_{\Omega \times U}$-module $f_{i*}(\mathcal{F}|_E)$ onto $P_0 \times V \Subset \Omega \times U$, Cartan's Theorem A implies the existence of an exact sequence

$$\cdots \to \mathcal{O}_{P_0 \times V}^{p_2} \to \mathcal{O}_{P_0 \times V}^{p_1} \to \mathcal{O}_{P_0 \times V}^{p_0} \to \mathcal{G} \to 0$$

of $\mathcal{O}_{P_0 \times V}$-modules. By Cartan's Theorem B the induced sequence of sections $\mathcal{O}(P_0 \times V)^{p_\bullet} \to \mathcal{G}(P_0 \times V) \to 0$ remains exact. If one regards $\mathcal{O}(P_0 \times V)^{p_\bullet}$ as a complex of free Fréchet $\mathcal{O}(P_0)$-modules and if one applies the functor $\mathcal{O}(P_t) \hat{\otimes}_{\mathcal{O}(P_0)} *$ to this complex, then one obtains the complex $\mathcal{O}(P_t \times V)^{p_\bullet}$ where the boundary maps are the section maps induced by the above resolution of $\mathcal{G}$. Again by Cartan's Theorem B, the augmented complex $\mathcal{O}(P_t \times V)^{p_\bullet} \to \mathcal{G}(P_t \times V) \to 0$ is exact. By Corollary 3.1.13, we conclude that $\mathcal{O}(P_t) \perp_{\mathcal{O}(P_0)} \mathcal{G}(P_0 \times V)$ and we obtain an isomorphism of Fréchet $\mathcal{O}(P_t)$-modules

$$\mathcal{O}(P_t) \hat{\otimes}_{\mathcal{O}(P_0)} \mathcal{G}(P_0 \times V) \to \mathcal{G}(P_t \times V).$$

The reader may easily (and should) check that, modulo canonical identifications, this isomorphism becomes just the unique continuous linear map $\Pi_i \colon \mathcal{O}(P_t) \hat{\otimes}_{\mathcal{O}(P_0)} \mathcal{F}(H) \to \mathcal{F}(H_t)$ with

$$\Pi_i(f \otimes \gamma) = f(\gamma|_{H_t}) \qquad (f \in \mathcal{O}(P_t), \gamma \in \mathcal{F}(H)).$$

In the following, we shall write $X_{i,t}'$ for H_t, and we denote by $X_{i,t}''$ the corresponding intersection of sets of the Stein open cover X_t''. In exactly the same way as above, one obtains corresponding relations for $\mathcal{F}(X_{i,0}'')$. Moreover, since there is a canonical commuting diagram of Fréchet $\mathcal{O}(P_0)$-modules

$$
\begin{array}{ccc}
\mathcal{F}(X_{i,0}') & \xrightarrow{\sim} & \Gamma(P_0 \times V, \mathcal{G}) \\
\text{res} \downarrow & & \downarrow \text{res} \\
\mathcal{F}(X_{i,0}'') & \xrightarrow{\sim} & \Gamma(P_0 \times W, \mathcal{G}),
\end{array}
$$

the restriction map on the left is $\mathcal{O}(P_0)$-subnuclear by Proposition 9.2.2.

As a conclusion we obtain that all conditions needed for the application of Theorem 9.2.6 are satisfied for the refinement map $M^{\bullet} \to N^{\bullet}$. Using the fact that the maps Π_i yield isomorphisms $\mathcal{O}(P_1) \hat{\otimes}_{\mathcal{O}(P_0)} M^{\bullet} \to \mathscr{C}^{\bullet}(X'_1, \mathscr{F})$, we therefore obtain a complex $L^{\bullet}$ of free, finite-type $\mathcal{O}(P_1)$-modules together with a quasi-isomorphism

$$h: L^{\bullet} \to \mathscr{C}^{\bullet}(X'_1, \mathscr{F})$$

of complexes of Fréchet $\mathcal{O}(P_1)$-modules. We denote by $\mathscr{L}^{\bullet}$ the sequence of $\mathcal{O}_{P_1}$-modules associated with $L^{\bullet}$.

Let D be a Stein open subset of P_1. Denote by X'_D the Stein open cover of $f^{-1}(D)$ consisting of the sets $f^{-1}(D) \cap \varphi_i^{-1}(V_i)$ $(1 \leq i \leq r)$. Then $\mathcal{O}(D) \perp_{\mathcal{O}(P_1)} L^{\bullet}$ and, exactly as above, it follows that $\mathcal{O}(D) \perp_{\mathcal{O}(P_1)} \mathscr{C}^{\bullet}(X'_1, \mathscr{F})$ and that there is an isomorphism

$$\mathcal{O}(D) \hat{\otimes}_{\mathcal{O}(P_1)} \mathscr{C}^{\bullet}(X'_1, \mathscr{F}) \to \mathscr{C}^{\bullet}(X'_D, \mathscr{F})$$

of complexes of Fréchet $\mathcal{O}(D)$-modules. By Lemma 3.1.17, there is a quasi-isomorphism $\mathscr{L}^{\bullet}(D) \to \mathscr{C}^{\bullet}(X'_D, \mathscr{F})$. It is routine to verify that, for each q, the induced cohomology isomorphisms

$$\Gamma(D, \mathscr{H}^q(\mathscr{L}^{\bullet})) \to H^q(f^{-1}(D), \mathscr{F}),$$

where D is a Stein open subset of P_1, form a family of homomorphisms which is compatible with restrictions.

Thus $(\mathscr{R}^q f_* \mathscr{F})|_{P_1} \cong \mathscr{H}^q(\mathscr{L}^{\bullet})$ as $\mathcal{O}_{P_1}$-modules, and the proof of Theorem 9.3.1 is complete. $\qquad \square$

Now the Cartan–Serre theorem (Theorem 9.1.1) becomes a simple consequence obtained by choosing Y as a reduced one-point space. Another remarkable application of Grauert's theorem is the following result due to Remmert (1956).

Corollary 9.3.2 *The image of an analytic set by a proper holomorphic map is an analytic set.*

Proof If $A \subset X$ is an analytic subset of an analytic space X, then by the Oka–Cartan theorem the ideal sheaf $\mathscr{I}_A$ of A is a coherent $\mathcal{O}_X$-module. Let $i: A \to X$ be the holomorphic inclusion of the closed complex subspace defined by $\mathscr{I}_A$. Then $f \circ i: A \to Y$ is a proper holomorphic map with image $f(A)$ whenever $f: X \to Y$ is a proper holomorphic map.

Hence we may suppose that $A = X$. But then $f(X) = \text{supp}(f_* \mathcal{O}_X)$ is analytic as the support of the coherent $\mathcal{O}_Y$-module $f_*(\mathcal{O}_X)$ (see Grauert and Remmert 1984). $\qquad \square$

This corollary, usually called the *proper mapping theorem* of Remmert, has non-trivial applications to the determination of analytic dependence between variables related by implicit holomorphic conditions (see Remmert 1956, Chirka 1985).

Further consequences of Grauert's theorem are discussed in the book of Grauert and Remmert (1984).

9.4   SEMICONTINUITY THEOREMS

Most of the analytic invariants which have appeared in the present chapter vary semicontinuously, with some control on the jumping sets, relative to analytic deformations of the initial data. This behaviour, which has close connection with Grauert's direct image theorem, lies at the heart of the modern theory of analytic deformations (see Palamodov 1990 for an excellent account of this field).

Some similar semicontinuity phenomena appear in the multidimensional Fredholm theory of abstract operators. Mainly because of the latter subject, which will be resumed in Chapter 10, we devote this section to semicontinuity results.

Throughout this section, $m_x \subset {}_X\mathcal{O}_x$ stands for the maximal ideal of ${}_X\mathcal{O}_x$ (to simplify the notation we write the subscript referring to the space on the left side). For each ${}_X\mathcal{O}$-module $\mathcal{F}$ and each $x \in X$, we regard

$$\mathcal{F}(x) = \mathcal{F}_x \otimes_{{}_X\mathcal{O}_x} ({}_X\mathcal{O}_x/m_x)$$

as a $\mathbb{C}$-vector space. If $d\colon \mathcal{F} \to \mathcal{G}$ is a morphism of ${}_X\mathcal{O}$-modules, then $d(x) = d \otimes 1_{{}_X\mathcal{O}_x/m_x}\colon \mathcal{F}(x) \to \mathcal{G}(x)$ becomes a $\mathbb{C}$-linear map for any $x \in X$.

It is an elementary exercise in linear algebra to check that the rank of a continuous family $A(z)$ of matrices defined on an arbitrary topological space X depends lower semicontinuously on z in X. The next lemma is a direct consequence of this observation.

Lemma 9.4.1   *Let X be a reduced analytic space, and let $d\colon {}_X\mathcal{O}^m \to {}_X\mathcal{O}^n$ be a morphism of ${}_X\mathcal{O}$-modules. Then:*

(1) *the function $x \mapsto \operatorname{rk} d(x)$ is lower semicontinuous;*
(2) *the function $x \mapsto \operatorname{rk} d(x)$ is locally constant on X if and only if* Coker d *is a locally free ${}_X\mathcal{O}$-module.*

Proof   As we have recalled in Chapter 4, the morphism d is given by an analytic function $d \in \mathcal{O}(X, L(\mathbb{C}^m, \mathbb{C}^n))$, which we again denote by d. Indeed, modulo the canonical identifications $\mathbb{C}^m = {}_X\mathcal{O}^m(x), \mathbb{C}^n = {}_X\mathcal{O}^n(x)$, the map $d(x)$ is the $\mathbb{C}$-linear operator

$$\mathbb{C}^m \to \mathbb{C}^n, t \mapsto d(x)t.$$

Hence assertion (1) is clear. In order to prove assertion (2), we define $\mathcal{F} = \operatorname{Coker} d$ and observe that, for each $x \in X$, the sequence

$$\mathbb{C}^m \to \mathbb{C}^n \to \mathcal{F}(x) \to 0$$

is exact. Therefore $\operatorname{rk} d(x)$ is locally constant if and only if $\dim \mathcal{F}(x)$ is locally constant.

Fix a point $x_0 \in X$ and set $p = \dim \mathscr{F}(x_0)$. As an easy consequence of the Nakayama Lemma (see for instance Serre 1965), p is the minimal number of generators of the $_X\mathscr{O}_{x_0}$-module $\mathscr{F}(x_0)$. Hence, in a suitable open neighbourhood U of x_0, there is an exact sequence of the form

$$_X\mathscr{O}^q|_U \overset{g}{\to} {_X\mathscr{O}^p}|_U \overset{h}{\to} \mathscr{F}|_U \to 0.$$

Suppose that rk $d(x)$ is constant on U. Since the last sequence remains exact, if localized to an arbitrary point $x \in X$, it follows that $g(x) = 0$ for $x \in U$. Therefore h is an isomorphism.

The proof of the reverse implication is obvious.   □

The following statement is obtained by applying Lemma 9.4.1 to a complex of finite-type free $_X\mathscr{O}$-modules.

Proposition 9.4.2 *Let $(\mathscr{L}^\bullet, d^\bullet)$ be a complex of finite-type free $_X\mathscr{O}$-modules defined on the reduced analytic space X. Then:*

(1) *the function $x \mapsto \dim H^q(\mathscr{L}^\bullet(x))$ is upper semicontinuous for every $q \in \mathbb{Z}$;*
(2) *the function $x \mapsto \dim H^q(\mathscr{L}^\bullet(x))$ is locally constant if and only if $\operatorname{Coker} d^{q-1}$ and $\operatorname{Coker} d^q$ are locally free $_X\mathscr{O}$-modules;*
(3) *if the complex $\mathscr{L}^\bullet$ is bounded, then the function*

$$x \mapsto \sum_{q \in \mathbb{Z}} (-1)^q \dim H^q(\mathscr{L}^\bullet(x))$$

is constant on X.

Proof If $\mathscr{L}^q = {_X\mathscr{O}^{m_q}}$ for $q \in \mathbb{Z}$, then for $x \in X$,

$$\dim H^q(\mathscr{L}^\bullet(x)) = m_q - \operatorname{rk} d^q(x) - \operatorname{rk} d^{q-1}(x).$$

Thus assertions (1) and (2) follow immediately from Lemma 9.4.1. For the proof of (2), use in addition the fact that X is locally connected (see Grauert and Remmert 1984, p. 178). If the complex $\mathscr{L}^\bullet$ is bounded, then the sum occurring in (3) is equal to $\sum_q (-1)^q m_q$ for arbitrary $x \in X$.   □

Before we come to an infinite-dimensional version of Proposition 9.4.2, we have to discuss some elementary properties of complexes of Hilbert-free $_X\mathscr{O}$-modules. By a *Hilbert-free $_X\mathscr{O}$-module*, we mean a sheaf which is isomorphic as a topological $_X\mathscr{O}$-module to a sheaf of the form $_X\mathscr{O} \,\hat{\otimes}\, H$ for a suitable Hilbert space H.

Proposition 9.4.3 *Let F, G, and H be Hilbert spaces, and let*

$$_X\mathscr{O} \,\hat{\otimes}\, F \overset{d}{\to} {_X\mathscr{O}} \,\hat{\otimes}\, G \overset{d'}{\to} {_X\mathscr{O}} \,\hat{\otimes}\, H$$

be a sequence of topological $_X\mathscr{O}$-modules over the reduced analytic space X.

Suppose that $x_0 \in X$ is a point such that $\operatorname{Ker} d'(x_0) = \operatorname{Im} d(x_0)$ *and* $\operatorname{Im} d'(x_0)$ *is closed. Then there is an open neighbourhood U of x_0 such that*

$$_X\mathscr{O} \hat{\otimes} F|_U \xrightarrow{d} {}_X\mathscr{O} \hat{\otimes} G|_U \xrightarrow{d'} {}_X\mathscr{O} \hat{\otimes} H|_U$$

and

$$F \xrightarrow{d(x)} G \xrightarrow{d'(x)} H \qquad (x \in U)$$

are exact complexes.

Proof We use the same notations for the morphisms d, d', and the induced analytic operator-valued functions $d \in \mathscr{O}(X, L(F, G))$, $d' \in \mathscr{O}(X, L(G, H))$ (see Chapter 4). Our assumptions guarantee the existence of operators $r \in L(G, F), r' \in L(H, G)$ with $d(x_0)r + r'd'(x_0) = 1_G$. We choose an open neighbourhood U of x_0 with the property that, for $x \in U$,

$$\|d(x) - d(x_0)\| < (3\|r\|)^{-1}, \qquad \|d'(x) - d'(x_0)\| < (3\|r'\|)^{-1},$$

and define analytic functions $f \in \mathscr{O}(U, L(G, F))$, $g \in \mathscr{O}(U, L(G))$ by setting

$$f(x) = r(1_G - r'(d'(x_0) - d'(x))),$$
$$g(x) = r'(d'(x_0) - d'(x)) + (d(x_0) - d(x))f(x).$$

Thus we obtain a well-defined analytic function $h: U \to L(G, F)$ if we set $h(x) = f(x)(1_G - g(x))^{-1}$. We claim that

$$d(x)h(x)y = y,$$

whenever $x \in U$ and $y \in \operatorname{Ker} d'(x)$.

To show this, we fix a point $x \in U$. An easy calculation shows that $\operatorname{Ker} d'(x)$ is invariant under $g(x)$, hence also under $(1_G - g(x))^{-1}$. Hence it suffices to verify that

$$d(x)f(x)y = (1_G - g(x))y$$

for all $y \in \operatorname{Ker} d'(x)$. Again the elementary verification of this identity is left to the reader.

If $t \in ({}_X\mathscr{O} \hat{\otimes} G)(V) = \mathscr{O}(V, G)$ satisfies $d't = 0$ for some open subset V of U, then $s = ht \in \mathscr{O}(V, F)$ is a solution of $ds = t$. Thus the assertion is proved.
$\square$

The kernel of a morphism between two Hilbert-free ${}_X\mathscr{O}$-modules will, apart from some trivial situations, almost never be a Hilbert-free ${}_X\mathscr{O}$-module. The next result describes such a special setting.

Lemma 9.4.4 *Let $\mathscr{F} \xrightarrow{\alpha} \mathscr{G} \xrightarrow{\beta} \mathscr{H}$ be a sequence of Hilbert-free ${}_X\mathscr{O}$-modules on a reduced analytic space X satisfying $\operatorname{Im} \beta = \mathscr{H}$ and $\dim(\operatorname{Ker} \beta(x_0)/\operatorname{Im} \alpha(x_0)) < \infty$ for a given fixed point $x_0 \in X$. Then there is an open neighbourhood V of x_0 such that $\operatorname{Ker} \beta$ restricted to V is a Hilbert-free*

$_V \mathcal{O}$-*module, and such that the inclusion map* i: $\operatorname{Ker}\beta \to \mathscr{G}$ *induces an isomorphism of vector spaces* $i(x)$: $(\operatorname{Ker}\beta)(x) \to \operatorname{Ker}(\beta(x))$ *for each* $x \in V$.

Proof We may suppose that $\mathscr{F} = {}_X\mathcal{O}\,\hat{\otimes}\,F$, $\mathscr{G} = {}_X\mathcal{O}\,\hat{\otimes}\,G$, $\mathscr{H} = {}_X\mathcal{O}\,\hat{\otimes}\,H$ for some Hilbert spaces F, G, H. Since β is onto, we obtain the exact sequence

$$(\operatorname{Ker}\beta)(x_0) \xrightarrow{i(x_0)} \mathscr{G}(x_0) \xrightarrow{\beta(x_0)} \mathscr{H}(x_0) \to 0.$$

Let $\{[h_1], \ldots, [h_p]\}$ be a basis of $\operatorname{Ker}\beta(x_0)/\operatorname{Im}\alpha(x_0)$. On a suitable open neighbourhood V of x_0 there are functions $f_1, \ldots, f_p \in \mathcal{O}(V, G)$ with $\beta f_i = 0$ and $f_i(x_0) = h_i$ for all i. We choose a direct complement L of $\operatorname{Ker}\alpha(x_0)$ in F and define $K = L \oplus \mathbb{C}^p$,

$$\gamma(x): \mathbb{C}^p \to G, (\alpha_i) \mapsto \sum_{i=1}^{p} \alpha_i f_i(x) \qquad (x \in V).$$

The complex

$$0 \to K \xrightarrow{(\alpha(x),\gamma(x))} G \xrightarrow{\beta(x)} H \to 0$$

is exact at $x = x_0$, hence for x in a suitable open neighbourhood W of x_0. Since the induced complex of Hilbert-free $\mathcal{O}_W$-modules remains exact on W, we may identify $\mathcal{O}_W \,\hat{\otimes}\, K \cong (\operatorname{Ker}\beta)|_W$ as topological $\mathcal{O}_W$-modules. Obviously, for $x \in W$, the map

$$i(x): (\operatorname{Ker}\beta)(x) \to \mathscr{G}(x)$$

composed with the isomorphism $K \cong (\operatorname{Ker}\beta)(x)$ is just the map $(\alpha(x), \gamma(x))$ from above. Therefore the map $i(x)$ is one-to-one for $x \in W$. $\qquad\square$

Now we have gathered all details required to prove the announced infinite-dimensional version of Proposition 9.4.2.

Proposition 9.4.5 *Let* $(\mathcal{M}^\bullet, d^\bullet)$ *be a complex of Hilbert-free* $\mathcal{O}_X$-*modules on a reduced analytic space* X *with* $\dim H^q(\mathcal{M}^\bullet(x)) < \infty$ *for all* $q \in \mathbb{Z}$ *and* $x \in X$.
   If the complex $\mathcal{M}^\bullet$ *is bounded to the right, then:*

(1) *the function* $x \mapsto \dim H^q(\mathcal{M}^\bullet(x))$ *is upper semicontinuous on* X *for every* $q \in \mathbb{Z}$;
(2) *the discontinuity points of the same function form an analytic subset of* X *for every* $q \in \mathbb{Z}$;
(3) *if* $\mathcal{M}^\bullet$ *is also bounded to the left, then the function* $x \mapsto \sum_q (-1)^q \dim H^q(\mathcal{M}^\bullet(x))$ *is locally constant on* X;
(4) *if for a certain* $q \in \mathbb{Z}$ *the function* $x \mapsto \dim H^q(\mathcal{M}^\bullet(x))$ *is locally constant on* X, *then* $\bigcup_{x \in X} \operatorname{Ker} d^q(x)$ *and* $\bigcup_{x \in X} \operatorname{Im} d^{q-1}(x)$ *are (locally trivial) analytic Hilbert bundles on* X.

Proof We fix a point $x_0 \in X$ and an integer n with $\mathcal{M}^q = 0$ for $q > n$.
   By descending induction, we construct, for each $p \leq n$, on a suitable open

neighbourhood V of x_0 a complex $(\mathscr{L}^\bullet, u^\bullet)$ of finite type-free $\mathscr{O}_V$-modules and a morphism $h: \mathscr{L}^\bullet \to \mathscr{M}^\bullet|_V$ such that

(i) $\mathscr{L}^i = 0$ for $i > n$ or $i \leq p$;
(ii) $h: \mathscr{L}^\bullet \to \mathscr{M}^\bullet|_V$ and $h(x): \mathscr{L}^\bullet(x) \to \mathscr{M}^\bullet(x)$ $(x \in V)$ are $(p+1)$-quasi-isomorphisms;
(iii) $u^{i+1}(x_0) = 0$, $h^{i+1}(x_0)$ is injective, and $\operatorname{Im} d^i(x_0) \oplus \operatorname{Im} h^{i+1}(x_0) = \operatorname{Ker} d^{i+1}(x_0)$ holds for all $i \geq p$.

To reduce the number of indices, we have partly suppressed the dependence on p.

For $p = n$, we define $(\mathscr{L}^\bullet, u^\bullet)$ to be the trivial complex, i.e. $\mathscr{L}^i = 0$ for all i. Assume that, for a fixed number p, a morphism h as before has been defined. To simplify the notation, we shall assume that $V = X$. By assumption there is a sequence $(M^i)_{i \in \mathbb{Z}}$ of Hilbert spaces and a sequence $(L^i)_{i \in \mathbb{Z}}$ of finite-dimensional spaces with $\mathscr{M}^i = \mathscr{O}_X \otimes M_i$, $\mathscr{L}^i = \mathscr{O}_X \hat{\otimes} L^i$. The coboundary maps $d^i: \mathscr{M}^i \to \mathscr{M}^{i+1}$ are induced by continuous linear maps $d^i(x): M^i \to M^{i+1}$ depending analytically on $x \in X$. In the same sense, $\mathscr{L}^\bullet$ and h are given by analytically parametrized operator functions.

The cone $(\mathscr{C}^\bullet, \alpha^\bullet)$ of the morphism h is a complex of the form

$$\mathscr{M}^{p-1} \xrightarrow{d^{p-1}} \mathscr{M}^p \oplus \mathscr{L}^{p+1} \xrightarrow{(d^p, k^{p+1})} \mathscr{M}^{p+1} \oplus \mathscr{L}^{p+2} \to \cdots \to \mathscr{M}^n \to 0,$$

where (d^p, k^{p+1}) is the boundary operator as defined in Appendix 2. By (ii) and (iii), we know that $H^i(\mathscr{C}^\bullet) = 0 = H^i(\mathscr{C}^\bullet(x))$ for $i > p$ and each $x \in V = X$, and that $H^p(\mathscr{C}^\bullet(x_0)) = H^p(\mathscr{M}^\bullet(x_0))$. A repeated application of Lemma 9.4.4 allows us to choose an open neighbourhood of x_0, again denoted by V, such that for each $i \geq p$ the complex

$$\mathscr{C}^{i-1} \xrightarrow{\alpha^{i-1}} \mathscr{C}^i \xrightarrow{\alpha^i} \operatorname{Ker} \alpha^{i+1}$$

restricted to V consists of Hilbert-free $\mathscr{O}_V$-modules and the inclusion map induces vector space isomorphisms

$$(\operatorname{Ker} \alpha^i)(x) \xrightarrow{i(x)} \operatorname{Ker}(\alpha^i(x)) \qquad (x \in V).$$

In particular, if $\{[h_1], \ldots, [h_r]\}$ is a basis of $H^p(\mathscr{M}^\bullet(x_0))$, then on a possibly still smaller open neighbourhood of x_0, again denoted by V, there are analytic functions $f_1, \ldots, f_r \in \mathscr{O}(V, M^p \oplus L^{p+1})$ with $\alpha^p f_i = 0$ and $f_i(x_0) = h_i$ $(1 \leq i \leq r)$.

If X is a complex analytic manifold, then the existence of functions f_i with the above properties follows directly from the remark following Lemma 2.1.5, even in the case when $\mathscr{M}^\bullet$ is a complex of Banach-free $\mathscr{O}_X$-modules.

We define $L^p = \mathbb{C}^r$, $\mathscr{L}^p = \mathscr{O}_V \hat{\otimes} L^p$, and set

$$k^p(x): L^p \to M^p \oplus L^{p+1}, (t_i)_{i=1}^r \mapsto \sum_{i=1}^r t_i f_i(x) \qquad (x \in V).$$

For reasons that will shortly become clear, we write $k^p(x)$ in the form $k^p(x) = ((-1)^p h^p(x), u^p(x))$. For $x = x_0$, the complex

$$M^{p-1} \oplus L^p \xrightarrow{(d^{p-1}, k^p)(x)} M^p \oplus L^{p+1} \xrightarrow{(d^p, k^{p+1})(x)} M^{p+1} \oplus L^{p+2}$$

is exact and the second map has closed range. Hence this complex and the induced complex of Hilbert-free $\mathcal{O}_V$-modules is exact on a possibly smaller open neighbourhood V of x_0. Again, if X is a complex analytic manifold, then the corresponding result is true in the Banach space setting. The inductive construction is completed by the observation that the morphism

$$
\begin{array}{ccccccccc}
\cdots \to & 0 & \longrightarrow & \mathcal{L}^p & \to & \mathcal{L}^{p+1} & \to \cdots \to & \mathcal{L}^n & \to 0 \\
& \downarrow & & \downarrow h^p & & \downarrow h^{p+1} & & \downarrow h^n & \\
\cdots \to & \mathcal{M}^{p-1}|_V & \to & \mathcal{M}^p|_V & \to & \mathcal{M}^{p+1}|_V & \to \cdots \to & \mathcal{M}^n|_V & \to 0
\end{array}
$$

satisfies conditions (i), (ii), and (iii) for $p-1$ instead of p. Note that for $H^p(\mathcal{M}^\bullet(x_0)) = 0$ everything remains true with $L^p = 0$.

Since with the notations established above the estimate

$$\dim H^p(\mathcal{M}^\bullet(x)) \le \dim H^p(\mathcal{L}^\bullet(x)) \le r = \dim H^p(\mathcal{M}^\bullet(x_0))$$

holds for all $x \in V$, assertion (1) has been proved.

If in addition $H^q(\mathcal{M}^\bullet(x)) = 0$ for $q \le p$ and $x \in V$, then on a suitable neighbourhood V of x_0 we have $\mathcal{H}^q(\mathcal{C}^\bullet(x)) = 0$ for all $q \in \mathbb{Z}$. By Proposition 9.4.2, the sum

$$\sum_{q \in \mathbb{Z}} (-1)^q \dim H^q(\mathcal{M}^\bullet(x)) = \sum_{q \in \mathbb{Z}} (-1)^q \dim H^q(\mathcal{L}^\bullet(x))$$

is constant on V. Thus assertion (3) has been proved.

To prove (2), let $q \in \mathbb{Z}$ and $x_0 \in X$ be arbitrary. We choose an open neighbourhood V of x_0 and a morphism $h: \mathcal{L}^\bullet \to \mathcal{M}^\bullet|_V$ as explained above for $p = q - 2$. Since $H^q(\mathcal{L}^\bullet(x)) \cong H^q(\mathcal{M}^\bullet(x))$ for $x \in V$, we may suppose that $V = X$ and $\mathcal{M}^\bullet = \mathcal{L}^\bullet$. By Propositions 9.4.2, the set of discontinuity points of the function $x \to \dim H^q(\mathcal{L}^\bullet(x))$ coincides with the analytic set (cf. Grauert and Remmert 1984, p. 92)

$$\bigcup_{i=q-1,q} \{x \in X; \text{Coker } d^i \text{ is not locally free at } x\}.$$

Let us suppose that $\dim H^q(\mathcal{M}^\bullet(x))$ is constant for $x \in V$, and that V and $h: \mathcal{L}^\bullet \to \mathcal{M}^\bullet|_V$ have been chosen as above for $p = q - 2$. By condition (iii), it follows that $u^{q-1}(x) = 0$, $u^q(x) = 0$ for all $x \in V$. If M is a direct complement of $\text{Ker } d^{q-1}(x_0)$ in M^{q-1}, then on a suitable smaller neighbourhood W of x_0 the complex

$$0 \to M \oplus L^q \xrightarrow{(d^{q-1}, h^q)(x)} M^q \xrightarrow{d^q(x)} M^{q+1}$$

is pointwise exact. The injectivity of the map

$$L^q \to H^q(\mathcal{M}^\bullet(x)), \qquad \xi \mapsto [h^q(x)\xi] \qquad (x \in V)$$

implies that $\text{Im } d^{q-1}(x) = d^{q-1}(x)M$ for $x \in W$. Therefore

$$0 \to M \xrightarrow{d^{q-1}(x)} \text{Im } d^{q-1}(x) \to 0$$

is an exact sequence for all $x \in W$. Thus, if $x \mapsto \dim H^q(\mathcal{M}^\bullet(x))$ is locally constant on X, then $\bigcup_{x \in X} \operatorname{Ker} d^q(x)$ and $\bigcup_{x \in X} \operatorname{Im} d^{q-1}(x)$ become locally trivial analytic Hilbert bundles in a natural way.

This completes the proof of Proposition 9.4.5. $\qquad\square$

For later use we gather some additional observations, which are contained in the proof of Proposition 9.4.5.

Remark 9.4.6 (a) If X is a complex analytic manifold, then the first three assertions of Proposition 9.4.5 remain valid even for complexes of Banach-free $\mathcal{O}_X$-modules.

(b) If $\mathcal{M}^\bullet = (\mathcal{M}^p, d^p)_{p=0}^n$ is a finite complex of Banach (Hilbert)-free $\mathcal{O}_X$-modules on a complex analytic manifold (reduced analytic space), and if $x \in X$ is a point with $\dim H^p(\mathcal{M}^\bullet(x)) < \infty$ for all p, then on a suitable open neighbourhood V of x, there is a complex $\mathcal{L}^\bullet = (\mathcal{L}^p, u^p)_{p=0}^n$ of finite-type free $\mathcal{O}_V$-modules and a morphism $h \colon \mathcal{L}^\bullet \to \mathcal{M}^\bullet|_V$ such that h and all localizations $h(x) \colon \mathcal{L}^\bullet(x) \to \mathcal{M}^\bullet(x)$ $(x \in V)$ are quasi-isomorphisms.

The construction of the quasi-isomorphism $h \colon \mathcal{L}^\bullet \to \mathcal{M}^\bullet$ in the above proof transfers a series of properties of complexes of free, finite-type $\mathcal{O}_X$-modules to complexes whose components are only Hilbert-free $\mathcal{O}_X$-modules. For instance, one can state the next result.

Corollary 9.4.7 *Under the assumptions of Proposition 9.4.5, the discontinuity sets of the functions*

$$x \mapsto \dim H^q(\mathcal{M}^\bullet(x)) \qquad (q \in \mathbb{Z})$$

are analytic subsets of X.

Furthermore, for any $q, k \in \mathbb{Z}$, the closed sets

$$S_k = \{x \in X; \dim H^q(\mathcal{M}^\bullet(x)) \geq k\}$$

are analytic. $\qquad\square$

Returning to proper holomorphic maps, we conclude this chapter by the following important result, due to Grauert (1960).

Theorem 9.4.8 *Let $f \colon X \to Y$ be a proper morphism of reduced analytic spaces, and let $\mathcal{F}$ be a coherent $\mathcal{O}_X$-module which is $\mathcal{O}_Y$-flat. Then we have:*

(1) *the function $y \mapsto \dim H^q(X_y, \mathcal{F}_y)$ is upper semicontinuous;*
(2) *if the function $y \mapsto \dim H^q(X_y, \mathcal{F}_y)$ is locally constant on Y, then $\mathcal{R}^q f_* \mathcal{F}$ is a locally free $\mathcal{O}_Y$-module;*
(3) *the function $y \mapsto \sum_{q \in \mathbb{Z}} (-1)^q \dim H^q(X_y, \mathcal{F}_y)$ is locally constant on Y.*

Proof As usual, the fibres $X_y = f^{-1}(y)$ are endowed with the structural sheaves $\mathcal{O}_{X_y} = \mathcal{O}_X/m_y \mathcal{O}_X$. Similarly, $\mathcal{F}_y = \mathcal{F} \otimes_{\mathcal{O}_X} \mathcal{O}_{X_y}$. The theorem will be reduced to Proposition 9.4.2 using the quasi-isomorphism h constructed in the proof of Theorem 9.3.1.

Since the sheaf $\mathscr{F}$ was assumed to be flat over Y, Theorem 4.2.4 implies the following transversality relation:

$$\mathscr{F}(X_{i,t}) \perp_{\mathscr{O}(P_t)} \mathscr{O}_{Y,y}/m_y \text{ for every } i \in I \text{ and } t \in [0,1].$$

Then one may apply the functor $\bullet \hat{\otimes}_{\mathscr{O}(P_t)} \mathscr{O}_{Y,y}/m_y$ to the quasi-isomorphism h, and one finds the isomorphisms

$$H^q(\mathscr{L}^\bullet(y)) = H^q(X_y, \mathscr{F}_y) \qquad (q \geq 0).$$

It remains to remark that

$$(\mathscr{R}^q f_* \mathscr{F})_y = H^q(f^{-1}(y), \mathscr{F}),$$

because the map f is proper (see the introduction to Section 9.3 above). This completes a sketch of the proof of Theorem 9.4.8. □

For applications of this last result, see Bănică and Stănăşilă (1977), Palamodov (1990), and Chirka (1985).

From the previous results we shall use only the semicontinuity and continuity assertions contained in Proposition 9.4.5.

9.5   REFERENCES AND COMMENTS

Beginning with the theory of compact Riemann surfaces, the finiteness theorems are to be found everywhere in complex analytic geometry. A few references which illustrate various applications of the finiteness theorems are: Bănică and Stănăşilă (1977), Bingener and Kosarew (1987), Grauert and Remmert (1984), Hirzebruch (1965), and Palamodov (1990).

The notion of analytically parametrized Fredholm complexes, which is used in Section 9.4, is closely related to the perturbation theory of linear operators. Namely, $T - z \colon X \to X$ is the simplest analytically parametrized Fredholm complex attached to any Fredholm operator T acting on a Banach space X. However, the results contained in Proposition 9.4.5 are not often used in operator theory. For instance, similar conclusions are derived by power series arguments and, in general, by *ad hoc* computations in the analytic perturbation theory of single operators (see Kato 1966).

10
Multidimensional index theory

By far the most studied spectral invariant of a linear operator is the Fredholm index. This integer, associated with an operator with finite-dimensional kernel and cokernel, has remarkable stability and functorial properties. The computation of the Fredholm index of concrete integro-differential operators turns out to be the key to understanding the nature of the majority of analytic invariants encountered in modern global analysis. The celebrated Atiyah–Singer index theorem and its recent ramifications are the fundamental results in this direction.

At the abstract level of spectral theory developed throughout this book, the Fredholm index of a single operator, or a commutative n-tuple of operators, is simply the algebraic intersection multiplicity of the associated analytic module with the local ring of germs of analytic functions at zero. Although this interpretation is not surprising, its immediate consequences explain the classical Fredholm theory and much more. It is the aim of the present chapter to develop these ideas, with emphasis again on quasi-coherent modules. Besides this rather limited purpose, we shall focus in the second part of the chapter on a proof of the Riemann–Roch theorem for complex analytic spaces with singularities. Apart from some technical, but by now standard, methods, this approach to the proof of the Riemann–Roch theorem is practically a continuation of the Fredholm theory of commutative n-tuples of operators. The second part of the chapter closely follows Levy (1987a). For the reader's convenience, Appendix 3 recalls some results from analytic and topological K-theory and various equivalent statements of the Riemann–Roch theorem.

The index theory of Hilbert space operators is more naturally studied in the framework of the theory of C^*-algebras. For instance the K-homology theory of C^*-algebras, the so-called *Brown–Douglas–Fillmore theory*, was motivated by the index theory of a concrete class of operators (see Brown *et al.* 1973). The multivariable index theory, in the sense of this chapter, has a variety of connections with C^*-algebras and their theory, developed especially in the last two decades. We completely omit these topics, which are sufficient to form the body of a separate monograph. The notes at the end of the chapter indicate a few bibliographical references in this direction. Also, we do not refer in the present chapter to any interpretation of the index data in terms of cyclic cohomology or trace formulas.

Knowledge of the classical Fredholm theory of a single operator, as for instance presented in the first chapters of the monograph by Kato (1966), will guide the reader through the apparently more sophisticated algebraic constructions contained in the present chapter. However, we hope that at the end of this chapter the advantages of this new language will be obvious. We have to mention that, beginning with Proposition 10.3.10, the text becomes more dependent on external references.

10.1 FREDHOLM COMPLEXES

For technical reasons, in the multidimensional setting it is necessary to study complexes of locally convex spaces with finite-dimensional homology spaces rather than single Fredholm operators. These so-called Fredholm complexes form a direct generalization of the concept of a linear operator with finite index. The present section is devoted to a few specific questions related to Fredholm complexes which were not treated in the previous chapters.

The familiar difference construction in K-theory extends canonically to Fredholm complexes which are analytically or continuously parametrized on a base space X. This construction associates with a Fredholm complex, depending continuously on a parameter $x \in X$, a formal index (or Euler characteristic) in $K(X)$. The algebraic and topological properties of this index are presented in the sequel. Except for some technical details, due to the fact that we need a non-smooth base space X and complexes of Fréchet spaces instead of complexes of Banach spaces, the contents of this section are rather standard. Consequently, we shall deliberately simplify the formal proofs.

Let $(X, \mathcal{O}_X)$ be a complex space (as usually supposed to be countable at infinity), and let $\mathscr{C}_X$ denote the sheaf of continuous functions on X. When there is no confusion, we simply denote by $\mathcal{O}$ and $\mathscr{C}$ the respective sheaves on X. Let $(E_\bullet, \alpha_\bullet(z))$ be a bounded, analytically parametrized complex of Banach spaces over X, that is, $\alpha_n \in \mathcal{O}_X(X) \hat{\otimes} L(E_n, E_{n-1})$ for any integer n. The following result was proved in Section 2.1 for the particular case where X is an open subset of $\mathbb{C}^n$.

Proposition 10.1.1 *Let X be a complex manifold, and let $(E_\bullet, \alpha_\bullet(x))$ be a bounded complex of Banach spaces, analytically parametrized on X. The following conditions are equivalent:*

(i) *$\mathcal{O}_X \hat{\otimes} E_\bullet$ is an exact complex of sheaves;*
(ii) *$\mathscr{C}_X \hat{\otimes}_\varepsilon E_\bullet$ is an exact complex of sheaves;*
(iii) *$\mathscr{C}(X) \hat{\otimes}_\varepsilon E_\bullet$ is an exact complex of Fréchet spaces;*
(iv) *for every Stein open subset U of X, the complex $\mathcal{O}_X(U) \hat{\otimes} E_\bullet$ is exact;*
(v) *the complex $(E_\bullet, \alpha_\bullet(x))$ is exact for any $x \in X$.* □

In the following we need a version of this result valid on complex spaces with possible singularities and for complexes $E_\bullet$ of Fréchet spaces instead of Banach spaces. It remains an open question whether Proposition 10.1.1 is valid or not on an arbitrary complex space X. Fortunately this is not an essential question for the rest of the chapter.

Corollary 10.1.2 *Let $(X, \mathcal{O}_X)$ be a complex space, and let $(E_\bullet, \alpha_\bullet(x))$ be an analytically parametrized, bounded complex of Fréchet spaces on X. The following implications between the assertions of Proposition 10.1.1 hold:*

$$\text{(i)} \Leftrightarrow \text{(iv)}; \qquad \text{(ii)} \Leftrightarrow \text{(iii)} \Leftrightarrow \text{(v)}',$$

where the last condition is:

(v)' *the complex $(E_\bullet, \alpha_\bullet(x))$ is locally uniformly exact.*

Moreover, if the spaces E_p ($p \in \mathbb{Z}$) are Banach spaces or nuclear Fréchet spaces, then the implication (iv) $\Rightarrow$ (ii) holds.

Proof The equivalences (i) $\Leftrightarrow$ (iv) and (ii) $\Leftrightarrow$ (iii) follow from the acyclicity of the sheaves $\mathcal{O}_X$ and $\mathscr{C}_X$, respectively. For the equivalence of (iii) and (v)', see Theorem 2.1.7.

In order to prove that (iv) implies (ii), let us suppose that the complex $\mathcal{O}(U) \hat{\otimes} E_\bullet$ is exact for a given Stein open set U. For $\lambda \in U$, an application of the functor $* \hat{\otimes}_{\mathcal{O}(U)} \mathcal{O}_\lambda / m_\lambda$ to the complex $\mathcal{O}(U) \hat{\otimes} E_\bullet$ shows that $(E_\bullet, \alpha_\bullet(\lambda))$ is exact. If the spaces E_p ($p \in \mathbb{Z}$) are Banach spaces, then Lemma 2.1.6 shows that (iv) implies (ii). If (iv) holds and if all spaces E_p ($p \in \mathbb{Z}$) are nuclear, then one can tensor the exact complexes $\mathcal{O}(U) \hat{\otimes} E_\bullet$ ($U \subset X$ Stein) with $* \hat{\otimes}_{\mathcal{O}(U)} \mathscr{C}_U$ to obtain the exactness of the complex under (ii) (see the remark following Corollary 3.1.16). $\qquad\qquad\qquad\qquad\qquad\qquad\qquad\qquad\qquad\qquad\qquad\qquad\quad\square$

We put aside the question of whether the implication (iv) $\Rightarrow$ (ii) holds in general for arbitrary Fréchet spaces E_p.

The most important examples of (locally) uniformly exact complexes of analytic Fréchet-free modules are, at least for our applications, the complexes derived from Koszul resolutions. For instance, for each open polydisc D in $\mathbb{C}^n$, the augmented Koszul complexes:

$$K_\bullet(z - \lambda, \mathcal{O}(D)) \xrightarrow{\delta_\lambda} \mathbb{C} \to 0 \qquad (\delta_\lambda(f) = f(\lambda))$$

form a locally uniformly exact family of complexes with respect to the parameter $\lambda \in D$. Indeed, by the preceding corollary it suffices to recall (end of Section 4.3) that the complex

$$K_\bullet(z - \lambda, \mathcal{O}(D) \hat{\otimes} \mathscr{C}(D)) \to \mathscr{C}(D) \to 0$$

is exact.

For the same reason, for any Fréchet space E, the following complex is exact

$$K_\bullet\left(z - \lambda, \mathcal{O}(D) \hat{\otimes} \mathscr{C}(D) \hat{\otimes}_\varepsilon E\right) \to \mathscr{C}(D) \hat{\otimes}_\varepsilon E \to 0.$$

The transition from exact complexes to complexes which have finite-type homology is more delicate than the above comparison between complexes of

Banach-free and Fréchet-free analytic modules. First we begin by recalling the better understood case of Banach-free modules. The following result was implicitly proved in Proposition 9.4.5.

Proposition 10.1.3 *Let X be a complex manifold, and let $(E_\bullet, \alpha_\bullet)$ be a bounded analytically parametrized complex of Banach spaces on X. Then the following conditions are equivalent:*

  (i) *the complex $(E_\bullet, \alpha_\bullet(x))$ has finite-dimensional homology spaces for each point $x \in X$;*
 (ii) *the homology sheaves $\mathcal{H}_\bullet(\mathcal{O}_X \hat{\otimes} E_\bullet)$ are coherent;*
(iii) *for every Stein compact subset $K \subset X$, there are an open neighbourhood V of K and a bounded complex of locally free, finite-type $\mathcal{O}_V$-modules which is quasi-isomorphic to $(\mathcal{O}_X \hat{\otimes} E_\bullet)|V$.*

Proof In view of Remark 9.4.6, it suffices to show that condition (ii) implies conditions (i) and (iii).

Let us suppose that $E_p = 0$ if $p > n$ or $p < 0$ for a given natural number n. Suppose that all homology sheaves $\mathcal{H}_p = \mathcal{H}_p(\mathcal{O}_X \hat{\otimes} E_\bullet)$ $(p \geq 0)$ are coherent. An elementary descending induction on p shows that all kernel and image sheaves Ker α_p and Im α_p $(p \geq 0)$ are acyclic on each Stein open subset of X.

Fix a point $x \in X$ and a Stein compact set K in X. Inductively, one can show that, for each $p \geq 0$, there are a Stein open neighbourhood V of x (of K, respectively), a complex $(\mathcal{L}_\bullet, u_\bullet)$ of finite-type free $\mathcal{O}_V$-modules with $\mathcal{L}_q = 0$ if $q < 0$ or $q \geq p$, and a $(p-1)$-quasi-isomorphism $h \colon \mathcal{L}_\bullet \to \mathcal{O}_V \hat{\otimes} E_\bullet$.

Suppose that $\mathcal{L}_\bullet$ and h have been defined for some $p \geq 0$. Let δ be the composition of the morphisms

$$\text{Ker } u_{p-1} \xrightarrow{h_{p-1}} (\text{Ker } \alpha_{p-1})|_V \xrightarrow{q} \mathcal{H}_{p-1}|_V,$$

where q is the quotient map. After shrinking V if necessary, we may suppose that there are epimorphisms

$$\mathcal{O}_V^r \xrightarrow{\alpha} \text{Ker } \delta \qquad \text{and} \qquad \mathcal{O}_V^s \xrightarrow{\beta} \mathcal{H}_p|_V$$

such that β factors through the quotient map $q \colon \text{Ker } \alpha_p \to \mathcal{H}_p$, that is, there is a morphism $g \colon \mathcal{O}_V^s \to (\text{Ker } \alpha_p)|_V$ with $\beta = q \circ g$.

Since $\text{Im}(h_{p-1} \circ \alpha)$ is contained in the image of $\alpha_p \colon \mathcal{O}_X \hat{\otimes} E_p \to \mathcal{O}_X \hat{\otimes} E_{p-1}$, one can arrange that in addition

$$\alpha_p \circ f = h_{p-1} \circ \alpha,$$

with a suitable morphism $f \colon \mathcal{O}_V^r \to (\mathcal{O}_X \hat{\otimes} E_p)|_V$. Now h (and $\mathcal{L}_\bullet$) can be extended to a p-quasi-isomorphism by setting $\mathcal{L}_p = \mathcal{O}_V^r \oplus \mathcal{O}_V^s$, $u_p = (\alpha, 0) \colon \mathcal{L}_p \to \mathcal{L}_{p-1}$, and

$$h_p = (f, g) \colon \mathcal{L}_p \to \mathcal{O}_V \hat{\otimes} E_p.$$

To complete the proof, we finally choose an $(n+1)$-quasi-isomorphism $h: \mathscr{L}_\bullet \to \mathscr{O}_V \hat{\otimes} E_\bullet$ on a Stein open neighbourhood V of x (of K, respectively) as explained above. The mapping cone $(\mathscr{C}_\bullet, d_\bullet)$ of h is a complex of the form (see Appendix 2)

$$0 \to \mathscr{L}_{n+1} \xrightarrow{d_{n+2}} \mathscr{L}_n \xrightarrow{d_{n+1}} \mathscr{L}_{n-1} \oplus \mathscr{O}_V \hat{\otimes} E_n \xrightarrow{d_n} \cdots \xrightarrow{d_1} \mathscr{O}_V \hat{\otimes} E_0 \to 0,$$

with $\mathscr{H}_p(\mathscr{C}_\bullet, d_\bullet) = 0$ for $p = 0, \ldots, n+1$. Since all kernel and image sheaves $\operatorname{Ker} d_p$ and $\operatorname{Im} d_p$ occurring in this complex are acyclic on V (see the beginning of the proof), a finite induction starting on the right-hand side of the complex can be used to show that $H_p(\mathscr{C}_\bullet(V), d_\bullet) = 0$ for $p = 0, \ldots, n+1$. Hence also the localized complexes

$$\mathscr{C}_\bullet(V) \hat{\otimes}_{\mathscr{O}(V)} (\mathscr{O}_z/m_z) \qquad (z \in V)$$

are exact in degrees $p = 0, \ldots, n+1$ (see the remark following Corollary 3.1.16). But then all localizations

$$h(z): \mathscr{L}^\bullet(z) \to (E^\bullet, \alpha^\bullet(z)) \qquad (z \in V)$$

are $(n+1)$-quasi-isomorphisms, and condition (i) follows.

By Lemma 9.4.1 and Remark 9.4.6,

$$\dim \operatorname{Ker} d_{n+1}(z) = \dim \operatorname{Im} d_{n+2}(z)$$

is continuous as a function of z in V. Therefore (see Lemma 9.4.1) $\mathscr{L}'_n = \operatorname{Coker}(d_{n+2})$ is a locally free $\mathscr{O}_V$-module. The proof is completed by the observation that the induced complex

$$0 \to \mathscr{L}'_n \to \mathscr{L}_{n-1} \to \cdots \to \mathscr{L}_0 \to 0$$

is quasi-isomorphic to $\mathscr{O}_V \hat{\otimes} E_\bullet$ on V.   □

A result similar to Proposition 10.1.3 holds in the category of modules over the sheaf $\mathscr{C}_X$ of continuous functions, with the only difference being that the smoothness condition on the base space X is not necessary.

Let E be a Banach space, and let T be a bounded linear operator on X. The operator T is said to be *Fredholm* if $\operatorname{Ker} T$ and $\operatorname{Coker} T$ are finite dimensional. In this case, there is an open neighbourhood U of 0 in $\mathbb{C}$ such that the operators $T - z$ are Fredholm for all $z \in U$ (see Kato 1966). On the set U, the operator T induces a complex

$$0 \to \mathscr{O}_U \hat{\otimes} E \xrightarrow{T-z} \mathscr{O}_U \hat{\otimes} E \to 0 \tag{10.1.1}$$

of the type described in Proposition 10.1.3. It is by now classical to call a parametrized complex $(E_\bullet, \alpha_\bullet(x))$ Fredholm if it satisfies one of the equivalent conditions of Proposition 10.1.3 (see Segal 1970; Hörmander 1985). Thus the following definition is justified by the preceding examples and by Proposition 10.1.3.

Definition 10.1.4 Let $(X, \mathscr{O}_X)$ be a complex space, and let $(E_\bullet, \alpha_\bullet(x))$ be a bounded complex of Fréchet spaces analytically parametrized on X. The complex $(E_\bullet, \alpha_\bullet(x))$ is said to be *Fredholm* if, for every point $x \in X$, there exists an open neighbourhood U of x and a bounded complex of free $\mathcal{O}_U$-modules of finite type which is quasi-isomorphic, as a complex of $\mathcal{O}_U$-modules, to $(\mathcal{O}_X \hat{\otimes} E_\bullet) \mid U$.

In view of Corollary 10.1.2, if the complexes

$$\left(\mathcal{O}_X \hat{\otimes} E_\bullet\right)|_U \cong \mathcal{O}_U \hat{\otimes} F_\bullet$$

are quasi-isomorphic in the category of $\mathcal{O}_U$-modules, then the complexes of $\mathscr{C}_U$-modules

$$\left(\mathscr{C}_X \hat{\otimes}_\varepsilon E_\bullet\right)|_U \cong \mathscr{C}_U \hat{\otimes}_\varepsilon F_\bullet$$

are also quasi-isomorphic provided that either $E_\bullet$ and $F_\bullet$ are complexes of Banach spaces or complexes of nuclear Fréchet spaces.

The fundamental abelian invariant of a Fredholm complex $(E_\bullet, \alpha_\bullet(x))$ is its index, which we shall define abstractly at the level of the K-theory groups. Since for the rest of this section we are only concerned with the topological properties of Fredholm complexes and their indices, it is more convenient to work within the category of $\mathscr{C}_X$-modules.

In the remainder of Section 10.1, let X be a locally compact Hausdorff space such that the topology of X possesses a countable basis. Then X is metrizable by Urysohn's metrization theorem. In particular, the space X is paracompact. We adopt the following working definition.

Definition 10.1.4$'$ Let $(\mathscr{C}_X \hat{\otimes}_\varepsilon E_\bullet, \alpha_\bullet)$ be a bounded complex of Fréchet-free $\mathscr{C}_X$-modules on X. The complex $\mathscr{C}_X \hat{\otimes}_\varepsilon E_\bullet$ is said to be *Fredholm* if it is locally quasi-isomorphic to a bounded complex of finite type $\mathscr{C}_X$-modules.

In order to define properly the K-theoretic index of a Fredholm complex in the above sense, we need the following technical result.

Lemma 10.1.5 *Let* $\mathscr{E}_\bullet, \mathscr{E}'_\bullet$, *and* $\mathscr{E}''_\bullet$ *be bounded complexes of Fréchet $\mathscr{C}_X$-modules. Let* $f\colon \mathscr{E}_\bullet \to \mathscr{E}'_\bullet$ *and* $g\colon \mathscr{E}'_\bullet \to \mathscr{E}''_\bullet$ *be continuous $\mathscr{C}_X$-linear morphisms of complexes.*

(a) *If two of the morphisms f, g, or $g \circ f$ are quasi-isomorphisms, then so is the third.*

(b) *If f is a quasi-isomorphism and if $u'\colon \mathscr{F}_\bullet \to \mathscr{E}'_\bullet$ is a continuous $\mathscr{C}_X$-linear morphism of complexes starting on a bounded complex $\mathscr{F}_\bullet$ of finite-type free $\mathscr{C}_X$-modules, then there exists a continuous $\mathscr{C}_X$-linear morphism of complexes $u\colon \mathscr{F}_\bullet \to \mathscr{E}_\bullet$ with the property that u' is homotopic to $f \circ u$ in the category of continuous $\mathscr{C}_X$-linear morphisms of complexes.*

Proof For part (a), one uses the cones of the corresponding morphisms and the algebraic relations between them (see Appendix 2 or Gelfand and Manin 1988).

Part (b) is again purely algebraic, via the cone of the morphism f. We leave the details to the reader. $\square$

The next lemma prepares the definition of the abstract index of a Fredholm complex.

Lemma 10.1.6 *Let as before X be a locally compact Hausdorff space with a countable basis, and let $\mathscr{E}_\bullet$ be a bounded complex of Fréchet-free $\mathscr{C}_X$-modules on X. Suppose that $\mathscr{E}_\bullet$ is Fredholm. Then, on any compact subset F of X, there exists a bounded complex of locally finite $\mathscr{C}_F$-modules $\mathscr{L}_\bullet$ which is quasi-isomorphic to the complex $\mathscr{E}_\bullet^F$ of Fréchet-free $\mathscr{C}_F$-modules induced by $\mathscr{E}_\bullet$ on F.*
The class $[\mathscr{L}_\bullet] = \sum_{n \in \mathbb{Z}} (-1)^n [\mathscr{L}_n] \in K(F)$ depends only on $\mathscr{E}_\bullet$.

Proof The existence of the locally finite complex $\mathscr{L}_\bullet$ with the desired properties follows by a partition of unity argument. Suppose that there are two quasi-isomorphisms $f \colon \mathscr{L}_\bullet \to \mathscr{E}_\bullet^F$ and $f' \colon \mathscr{L}'_\bullet \to \mathscr{E}_\bullet^F$ as described in the lemma. By Lemma 10.1.5(b), modulo homotopy equivalence, there exists a lifting $g \colon \mathscr{L}'_\bullet \to \mathscr{L}_\bullet$ of the morphism f', and by Lemma 10.1.5(a), g itself is a quasi-isomorphism. Therefore we have

$$[\mathscr{L}_\bullet] = [\mathscr{H}_\bullet(\mathscr{L}_\bullet)] = [\mathscr{H}_\bullet(\mathscr{L}'_\bullet)] = [\mathscr{L}'_\bullet]. \qquad \square$$

Definition 10.1.7 The class $[\mathscr{E}_\bullet] \in \mathscr{K}(X)(= \varprojlim \{K(F); \ F \Subset X\})$ of a Fredholm complex $\mathscr{E}_\bullet$ of Fréchet-free $\mathscr{C}_X$-modules on X is the unique element $[\mathscr{E}_\bullet]$ in $\mathscr{K}(X)$ such that, for each compact subset F of X, its component in $K(F)$ is the class $[\mathscr{L}_\bullet]$ introduced in Lemma 10.1.6.

A few invariance properties of the abstract index of a Fredholm complex are listed in the following proposition.

Proposition 10.1.8 *Let $\mathscr{E}_\bullet$, $\mathscr{E}'_\bullet$, and $\mathscr{E}''_\bullet$ be bounded Fredholm complexes of Fréchet-free $\mathscr{C}_X$-modules. Then the following assertions hold:*

(a) *If $f_\bullet \colon \mathscr{E}'_\bullet \to \mathscr{E}''_\bullet$ is a quasi-isomorphism, then $[\mathscr{E}'_\bullet] = [\mathscr{E}''_\bullet]$.*
(b) *If $0 \to \mathscr{E}'_\bullet \to \mathscr{E}_\bullet \to \mathscr{E}''_\bullet \to 0$ is an exact sequence of complexes of Fréchet $\mathscr{C}_X$-modules, then $[\mathscr{E}_\bullet] = [\mathscr{E}'_\bullet] + [\mathscr{E}''_\bullet]$.*
(c) *Let $x \in X$ be a fixed point, and let $\mathrm{rk}_x \colon \mathscr{K}(X) \to \mathbb{Z}$ be the rank evaluation map at x. Then*

$$\mathrm{rk}_x[\mathscr{E}_\bullet] = \sum_{n \in \mathbb{Z}} (-1)^n \dim H_n(\mathscr{E}_\bullet(x)).$$

Proof The assertions (a) and (c) follow easily from Definition 10.1.7. In order to prove assertion (b), let $\mathscr{C}_\bullet$ denote the cone of the morphism of complexes $\mathscr{E}'_\bullet \to \mathscr{E}_\bullet$ occurring in the statement. Then $\mathscr{C}_\bullet$ and $\mathscr{E}''_\bullet$ are quasi-isomorphic. On the other hand, the complexes $\mathscr{C}_\bullet$ and $\mathscr{E}_\bullet \oplus \mathscr{E}'_\bullet [1]$ are homotopically equivalent. Hence

$$[\mathscr{E}''_\bullet] = [\mathscr{C}_\bullet] = [\mathscr{E}_\bullet \oplus \mathscr{E}'_\bullet[1]] = [\mathscr{E}_\bullet] - [\mathscr{E}'_\bullet]. \qquad \square$$

The applications of the above abstract constructions to the index theory of commutative n-tuples will be discussed in Section 10.3.

10.2 ESSENTIAL FREDHOLM COMPLEXES

The natural commutation relation in many concrete index problems is commutativity modulo the ideal of compact operators. Therefore the notion of Fredholm complexes introduced in the previous section is restrictive from the very beginning. However, as we shall see at the end of the present section, this restriction is not of a serious nature. Any essential Fredholm complex of Banach spaces is locally a compact perturbation of a Fredholm complex. As usual, the term 'essential' is used to indicate that we are working modulo the ideal of compact operators.

In view of the applications that we shall present later in this chapter, we confine ourselves in the present section to complexes of Banach spaces, more precisely, to cochain complexes, which are closer to the index conventions in operator theory.

Definition 10.2.1 An *essential complex of Banach spaces* is a bounded sequence

$$0 \to E^0 \xrightarrow{\delta^0} E^1 \xrightarrow{\delta^1} E^2 \to \cdots \to E^n \to 0,$$

where $\delta^q \in L(E^q, E^{q+1})$ and $\delta^q \delta^{q-1} \in K(E^{q-1}, E^{q+1})$ is a compact operator for every $q \in \mathbb{N}$. Here as usual $E^q = 0$ for $q < 0$ or $q > n$. The essential complex $(E^\bullet, \delta^\bullet)$ is *Fredholm* if the induced complexes

$$L(F, E^\bullet)/K(F, E^\bullet)$$

are exact for each Banach space F.

In fact it suffices to demand the last exactness condition for the Banach space $F = E^0 \oplus \cdots \oplus E^n$. This follows immediately from the proof of the next result.

Proposition 10.2.2 *An essential complex* $(E^\bullet, \delta^\bullet)$ *is Fredholm if and only if there are operators* $\varepsilon^q \in L(E^q, E^{q-1})$ $(q \in \mathbb{N})$ *with the property*

$$\delta^{q-1}\varepsilon^q + \varepsilon^{q+1}\delta^q \in I + K(E^q) \qquad (q \in \mathbb{N}). \tag{10.2.1}$$

Proof The sufficiency is clear because, for any Banach space F, the operators (ε^q) yield a trivial homotopy of the complex $L(F, E^\bullet)/K(F, E^\bullet)$.

The necessity will be proved by descending induction on q. Suppose that operators $\varepsilon^{q+1}, \varepsilon^{q+2}, \ldots$ have been constructed with the properties (10.2.1). Then the complex

$$L(E^q, E^{q-1})/K(E^q, E^{q-1}) \xrightarrow{\delta_*^{q-1}} L(E^q)/K(E^q)$$

$$\xrightarrow{\delta_*^q} L(E^q, E^{q+1})/K(E^q, E^{q+1})$$

is exact.

The class $I - \varepsilon^{q+1}\delta^q$ in the middle term satisfies $\delta_*^q[I - \varepsilon^{q+1}\delta^q] = 0$. Hence there exists an operator $\varepsilon^q \in L(E^q, E^{q-1})$ such that

$$I - \varepsilon^{q+1}\delta^q \in \delta^{q-1}\varepsilon^q + K(E^q).$$

This completes the proof of Proposition 10.2.2.   □

Let us remark that any other sequence of operators $(\tilde{\varepsilon}^q)$ which fulfils (10.2.1) is homotopic to (ε^q) within the same class. Indeed, for each $t \in [0, 1]$, the sequence

$$((1 - t)\varepsilon^q + t\tilde{\varepsilon}^q)_{q \in \mathbb{N}}$$

satisfies (10.2.1).

A sequence $(\varepsilon^q)(q \in \mathbb{N})$ as above is called a (*trivial essential*) *homotopy* of the essential complex $(E^\bullet, \delta^\bullet)$.

Let $(E^\bullet, \delta^\bullet)$ be an essential Fredholm complex with the homotopy $\varepsilon^\bullet$. The operator

$$D = \delta^\bullet + \varepsilon^\bullet: \bigoplus_q E^{2q} \to \bigoplus_q E^{2q+1}$$

is Fredholm. Indeed, if one considers the same expression $D' = \delta^\bullet + \varepsilon^\bullet$ defined on $\oplus_q E^{2q+1}$ and with values in $\oplus_q E^{2q}$, then the relations (10.2.1) imply that

$DD' = $ invertible + compact   and $D'D = $ invertible + compact.

Since two homotopies of the essential complex $(E^\bullet, \delta^\bullet)$ can be continuously deformed into each other, the index

$$\mathrm{ind}(E^\bullet, \delta^\bullet) = \mathrm{ind}\, D$$

is well defined.

Proposition 10.2.3   *Let $(E^\bullet, \delta^\bullet)$ be an essential Fredholm complex. Its index is stable under compact or small perturbations of the coboundary operators (δ^q).*

Proof   The assertion concerning compact perturbations is obvious. To prove the second assertion, let us fix a trivial essential homotopy (ε^q) for $(E^\bullet, \delta^\bullet)$, that is, a family of operators satisfying

$$\delta^{q-1}\varepsilon^q + \varepsilon^{q+1}\delta^q \in I + K(E^q) \qquad (q \in \mathbb{N})$$

By assumption the complex $(\mathscr{C}(\oplus_q E^q, E^\bullet), \delta_*^\bullet)$ is exact. Using Lemma 2.1.3 one can choose a number $r > 0$ with the property that, for each essential complex of the form $(E^\bullet, \tilde{\delta}^\bullet)$ with

$$\|\tilde{\delta}^q - \delta^q\| < r \qquad (q \in \mathbb{N}),$$

the resulting complex $(\mathscr{C}(\oplus_q E^q, E^\bullet), \tilde{\delta}_*^\bullet)$ remains exact. By choosing r small enough, one can achieve that at the same time, for each such complex, the operator

$$\tilde{\delta}^\bullet + \varepsilon^\bullet: \bigoplus_q E^{2q} \to \bigoplus_q E^{2q+1}$$

is Fredholm with $\mathrm{ind}(E^\bullet, \delta^\bullet) = \mathrm{ind}(\tilde{\delta}^\bullet + \varepsilon^\bullet)$, and moreover that

$$\sigma_e(\tilde{\delta}^{q-1}\varepsilon^q + \varepsilon^{q+1}\tilde{\delta}^q) \subset \{z \in \mathbb{C}; |z - 1| < 1\}$$

holds for all q.

Let $(E^\bullet, \tilde{\delta}^\bullet)$ be an essential complex close enough to $(E^\bullet, \delta^\bullet)$ in the above sense. If $(\tilde{\varepsilon}^q)$ is a homotopy for $(E^\bullet, \tilde{\delta}^\bullet)$, then the operators

$$D_t = \tilde{\delta}^\bullet + (1 - t)\varepsilon^\bullet + t\tilde{\varepsilon}^\bullet : \bigoplus_q E^{2q} \to \bigoplus_q E^{2q+1} \qquad (t \in [0,1])$$

are Fredholm. Indeed, if one defines

$$D_t' = \tilde{\delta}^\bullet + (1 - t)\varepsilon^\bullet + t\tilde{\varepsilon}^\bullet : \bigoplus_q E^{2q+1} \to \bigoplus_q E^{2q},$$

then one can show that $D_t D_t'$ and $D_t' D_t$ can both be written as the sum of a compact operator and an essentially invertible operator. To check this, the reader should note that a matrix operator, which is given by an upper triangular matrix with essentially invertible operators on the diagonal, is essentially invertible. Since D_t depends continuously on the parameter $t \in [0,1]$, it follows that

$$\mathrm{ind}(E^\bullet, \tilde{\delta}^\bullet) = \mathrm{ind}(\tilde{\delta}^\bullet + \tilde{\varepsilon}^\bullet) = \mathrm{ind}(\tilde{\delta}^\bullet + \varepsilon^\bullet) = \mathrm{ind}(E^\bullet, \delta^\bullet) \qquad \square$$

The above defined index for essential complexes coincides for usual Fredholm complexes with their *Euler characteristic*. More precisely, we have the next result.

Lemma 10.2.4 *Let* $(E^\bullet, \delta^\bullet)$ *be a complex of Banach spaces which is Fredholm. Then* $\dim H^q(E^\bullet) < \infty$ *for* $q \in \mathbb{N}$, *and*

$$\mathrm{ind}(E^\bullet, \delta^\bullet) = \chi(E^\bullet, \delta^\bullet) = \sum_q (-1)^q \dim H^q(E^\bullet, \delta^\bullet). \qquad (10.2.2)$$

Proof The finiteness of the cohomology spaces has been proved in Lemma 2.6.13.

The formula for the index is proved by induction on the length n of $(E^\bullet, \delta^\bullet)$. For $n = 1$, the matrix $D = D(\varepsilon^\bullet)$ is just the operator

$$D = \delta^0 : E^0 \to E^1.$$

Let us assume that the result is true for all complexes $(E^\bullet, \delta^\bullet)$ of length $n - 1$ ($n \geq 2$ fixed), and let $(E^\bullet, \delta^\bullet)$ be a Fredholm complex of length n. By Lemma 2.6.13, for each p, one can choose closed subspaces L^p and N^p of E^p with

$$E^p = \mathrm{Ker}\, \delta^p \oplus L^p \qquad \text{and} \qquad \mathrm{Ker}\, \delta^p = \mathrm{Im}\, \delta^{p-1} \oplus N^p.$$

The operators $\varepsilon^p : E^p \to E^{p-1}$ defined by

$$\varepsilon^p(d^{p-1}x + y) = x \qquad (x \in L^{p-1}, y \in N^p \oplus L^p)$$

form an essential homotopy for $(E^\bullet, \delta^\bullet)$ such that $k^p = I_{E^p} - (\varepsilon^{p+1}\delta^p + \delta^{p-1}\varepsilon^p)$ is the projection of E^p onto N^p along Im $\delta^{p-1} \oplus L^p$. In particular, the space $\tilde{E}_1 = L^1 \oplus N^1$ is a complement of Im δ^0 in E^1 containing Im ε^2. The complex $\tilde{E}^\bullet$ given by

$$0 \to \tilde{E}^1 \xrightarrow{\tilde{\delta}^1} E^2 \xrightarrow{\delta^2} \cdots \xrightarrow{\delta^{n-1}} E^n \to 0 \qquad \left(\tilde{\delta}^1 = \delta^1 \mid \tilde{E}_1\right)$$

is Fredholm and possesses the essential homotopy $(\tilde{\varepsilon}^2, \varepsilon^3, \ldots, \varepsilon^n)$, where we write $\tilde{\varepsilon}^2$ for ε^2 regarded as a map with values in $\tilde{E}^1$. The induction hypothesis implies that

$$\chi(E^\bullet) = \dim \text{Ker}(\delta^0) - \chi(\tilde{E}^\bullet) = \dim(\text{Ker } \delta^0) - \text{ind}(\tilde{E}^\bullet).$$

By definition we have

$$\text{ind}(\tilde{E}^\bullet) = \text{ind} \begin{pmatrix} \tilde{\delta}^1 & \varepsilon^3 & \\ & \delta^3 & \varepsilon^5 & 0 \\ & 0 & \delta^5 & \\ & & & \ddots \end{pmatrix} = -\text{ind} \begin{pmatrix} \tilde{\varepsilon}^2 & & \\ \delta^2 & \varepsilon^4 & 0 \\ & \delta^4 & \varepsilon^6 \\ & 0 & \ddots \end{pmatrix}.$$

The last matrix defines an operator

$$D: E^2 \oplus E^4 \oplus \cdots \to \tilde{E}^1 \oplus E^3 \oplus E^5 \oplus \cdots \;.$$

If one multiplies the operator $I_{\text{Im } \delta^0} \oplus D$ on the right by the operator

$$\delta^0 \oplus I_{E^2 \oplus E^4 \oplus \cdots} : E^0 \oplus E^2 \oplus E^4 \oplus \cdots \to \text{Im } \delta^0 \oplus E^2 \oplus E^4 \oplus \cdots,$$

then one obtains the matrix $D(\varepsilon^\bullet) = \delta^\bullet + \varepsilon^\bullet$, which was used to define $\text{ind}(E^\bullet, \delta^\bullet)$. Hence it follows that

$$\chi(E^\bullet) = \dim(\text{Ker } \delta^0) + \text{ind } D = \text{ind } D(\varepsilon^\bullet) = \text{ind}(E^\bullet, \delta^\bullet). \qquad \square$$

The results proved so far in this section hold in the more general setting of complexes of Banach-free modules continuously parametrized on a suitable base X. In this case the index can be defined as an element in the group $\mathscr{K}(X)$, as in the previous section.

Our next aim is to prove that essential Fredholm complexes of Banach-free $\mathscr{C}_X$-modules are locally equivalent to compact perturbations of Fredholm complexes. In this way essential Fredholm complexes appear as a necessary technical notion, which finally reduces locally to the theory developed in the first section of this chapter.

Theorem 10.2.5   *Let X be a Hausdorff topological space, and let*

$$0 \to E^0 \xrightarrow{\delta^0(x)} E^1 \to \cdots \to E^{n-1} \xrightarrow{\delta^{n-1}(x)} E^n \to 0$$

be a family of essential complexes of Banach spaces that are Fredholm such that the coboundary operators $\delta^q(x)$ depend continuously on $x \in X$.

Then, for each point $x_0 \in X$, there is a neighbourhood U of x_0 such that there are compact perturbations $\tilde{\delta}^q(x)$ of $\delta^q(x)$ $(x \in U)$ depending continuously on $x \in U$ such that $\tilde{\delta}^q(x)\tilde{\delta}^{q-1}(x) = 0$ for $1 \leq q \leq n - 1$.

Proof Fix a point $x_0 \in X$. We prove the assertion by induction on n. For $n = 1$, there is nothing to prove.

Let E_0^n be a complement of Im $\delta^{n-1}(x_0)$, and let P_0^n be the projection of E^n onto E_0^n along Im $\delta^{n-1}(x_0)$. By assumption, $\mathrm{rk}(P_0^n) < \infty$ (see Lemma 2.6.13). Hence $\delta^{n-1}(x) - P_0^n \delta^{n-1}(x)$ is a finite rank perturbation of $\delta^{n-1}(x)$, which, regarded as an operator from E^{n-1} to $(I - P_0^n)E^n$, is onto for $x = x_0$. Thus, without loss of generality, we may suppose that the coboundary operator $\delta^{n-1}(x_0)$ is onto.

Let E_0^{n-1} be a direct complement of $E_1^{n-1} = \mathrm{Ker}(\delta^{n-1}(x_0))$ in E^{n-1} (see Lemma 2.6.13). Let P_0^{n-1} be the projection of E^{n-1} onto E_0^{n-1} along E_1^{n-1}, and let $P_1^{n-1} = I - P_0^{n-1}$. For x close enough to x_0, the operators

$$\delta_0^{n-1}(x) = \delta^{n-1}(x)|_{E_0^{n-1}} : E_0^{n-1} \to E^n$$

are isomorphisms. Let $\delta_1^{n-1}(x)$ denote the restriction of $\delta^{n-1}(x)$ to E_1^{n-1}. Correspondingly, let us define

$$\delta_i^{n-2}(x) = P_i^{n-1}\delta^{n-2}(x) \qquad (i = 0,1).$$

Next we replace the operator $\delta^{n-1} = (\delta_0^{n-1}, \delta_1^{n-1})$ by

$$(\delta_0^{n-1}, \delta_1^{n-1})\begin{pmatrix} 1 & -(\delta_0^{n-1})^{-1}\delta_1^{n-1} \\ 0 & 1 \end{pmatrix},$$

and we replace the operator δ^{n-2} by

$$\begin{pmatrix} 1 & (\delta_0^{n-1})^{-1}\delta_1^{n-1} \\ 0 & 1 \end{pmatrix}\begin{pmatrix} \delta_0^{n-2} \\ \delta_1^{n-2} \end{pmatrix}.$$

The resulting operators are $(\delta_0^{n-1}, 0)$ and $(a, \delta_1^{n-2})^T$, respectively, where

$$a = \delta_0^{n-2} + (\delta_0^{n-1})^{-1}\delta_1^{n-1}\delta_1^{n-2} = (\delta_0^{n-1})^{-1}\delta^{n-1}\delta^{n-2}$$

is a compact operator. Moreover, for x in a suitable neighbourhood of x_0, the operator $\delta_1^{n-2}(x): E^{n-2} \to E_1^{n-1}$ has finite-codimensional range. Thus, by replacing δ^{n-2} by its compact perturbation

$$\delta_1^{n-2} = \delta^{n-2} - \begin{pmatrix} a \\ 0 \end{pmatrix},$$

and by replacing the space E^{n-1} by E_1^{n-1}, we have reduced the statement to a complex of length $n - 1$.

Thus the inductive proof of Theorem 10.2.5 is complete. $\square$

A global version of Theorem 10.2.5 appears in Levy (1989). However, the proof given there contains a gap illustrated by the following example.

Lemma 10.2.6   *There is a Hausdorff space X such that there are families of Banach space operators having closed range*

$$a(x): E \to F \quad \text{and} \quad b(x): F \to G \quad (x \in X)$$

with the following properties. Each product $b(x)a(x)$ is a finite-rank operator, but there is no continuous family of finite-rank operators $r(x): E \to F$ $(x \in X)$ with the property that $b(x)(a(x) + r(x)) = 0$ for $x \in X$.

Proof   We construct the example as follows. Let H be a complex infinite-dimensional Hilbert space. Define $X = E = F = H$, $G = \mathbb{C}$, and $a(x) = I$ (identity operator on H), $b(x) = \langle \cdot, x \rangle$.

Assume that there is a continuous family $r(x)$ of finite rank operators on H with the property that $\langle \cdot, x \rangle (I + r(x)) = 0$ for $x \in H$. Then $r(x)^*x = -x$ for each $x \in H$, and in particular, for $t \in [0, 1]$, we obtain

$$r(tx)^*tx = -tx.$$

Equivalently,

$$r(tx)^*x = -x.$$

By passing to the limit $t \to 0$, one finds that $r(0) = -I$, which contradicts the fact that $r(0)$ was supposed to be a finite-rank operator. $\square$

In the rest of this chapter we shall not need a global version of Theorem 10.2.5.

10.3   INDEX THEORY OF COMMUTATIVE TUPLES OF OPERATORS

The principal motivation for the abstract constructions presented in the previous two sections comes from the spectral theory of commuting operators. At this level, the new notions become more transparent and comparable to the situation for single linear operators. We develop the contents of the present section without aiming at completeness; instead we try to clarify various concepts of indices for commuting tuples of operators and the relations between them. In the final part some concrete applications are given.

For a commuting tuple of operators, the classical Fredholm index associates with each point of its essential resolvent set an integer. In the following we shall show that there are two natural K-theoretic invariants, the analytic and the topological index. The latter two indices were introduced by Levy (1989), and they turn out to be essential for a deeper understanding of the spectral picture of a commuting system of operators.

For simplicity, we shall only treat the case of Hilbert space operators. Some minor changes will in general suffice to extend the results to the setting of Banach space operators.

Let H be a (separable) complex Hilbert space. We know from Chapter 2 that an n-tuple of commuting bounded operators $T = (T_1, \ldots, T_n)$ on H should be considered together with its associated Koszul complex $K_\bullet(T, H)$. In a similar way to the case of a single operator, one introduces the following terminology.

Definition 10.3.1 The n-tuple T is said to be *Fredholm* if $K_\bullet(T, H)$ is a Fredholm complex. The *index* of a Fredholm tuple T is the integer

$$\text{ind } T = \chi(K_\bullet(T, H)) = \sum_{k=0}^{n} (-1)^k \dim H_k(T, H).$$

For $n = 1$, the preceding index is equal to minus the Fredholm index of T. If H denotes the Hilbert $\mathscr{O}(\mathbb{C}^n)$-module attached to T (see Chapter 5), then T is Fredholm if and only if

$$\dim \text{Tor}_q^{\mathscr{O}(\mathbb{C}^n)}(H, \mathscr{O}(\mathbb{C}^n)/\underline{m}_0) < \infty \qquad (q \geq 0),$$

where $\underline{m}_0$ denotes the maximal ideal in $\mathscr{O}(\mathbb{C}^n)$ consisting of all functions that vanish at zero.

According to Proposition 10.2.3 this index has the expected stability properties.

Theorem 10.3.2 *The Fredholm property and the index of n-tuples of commuting operators are stable under small or compact perturbations.*

Exactly as in Lemma 10.2.4, one proves that the index of an n-tuple T (even for n-tuples of essentially commuting operators) is given by

$$\text{ind}(T) = \text{ind}(\delta + \delta^*),$$

where δ is the boundary map of the Koszul complex of T,

$$\delta + \delta^*: \bigoplus_q K_{2q}(T, H) \to \bigoplus_q K_{2q+1}(T, H),$$

and $\text{ind}(\delta + \delta^*)$ denotes the classical Fredholm index.

For $n = 2$ and $n = 3$, the operator $\delta + \delta^*$ has the matrix representations

$$\delta + \delta^* = \begin{pmatrix} T_1 & T_2^* \\ T_2 & -T_1^* \end{pmatrix} \qquad (n = 2)$$

and

$$\delta + \delta^* = \begin{pmatrix} T_1 & T_2^* & T_3^* & 0 \\ T_2 & -T_1^* & 0 & T_3^* \\ T_3 & 0 & -T_1^* & -T_2^* \\ 0 & T_3 & -T_2 & T_1 \end{pmatrix} \qquad (n = 3).$$

In order to study the functorial behaviour of the index of a system of operators, as well as for investigating the structure of the essential resolvent set, it will be useful to pass from the pointwise-defined index to a K-theoretic version of the index. The Koszul complex $K_{\bullet}(z - T, \mathscr{O}^{H})$, regarded as a complex of Hilbert-free analytic sheaves on $\mathbb{C}^{n}$, will be the main tool with respect to this aim.

Before we give the next result, let us recall the notion of the rank for coherent analytic sheaves. Let U be a connected open set in $\mathbb{C}^{n}$ (or a normal complex space), and let $\mathscr{F}$ be a coherent $\mathscr{O}_{U}$-module. Then the set S consisting of all points z in U for which $\mathscr{F}$ is not locally free at z is a nowhere dense analytic subset of U, and the complement of S in U is connected (Grauert and Remmert 1984, p. 92 and p. 145). The *rank* $\mathrm{rk}_{U}\mathscr{F}$ of $\mathscr{F}$ on U is the constant value of $\mathrm{rk}_{\lambda}\mathscr{F}$ for $\lambda \in U \backslash S$. For convenience we also define $\mathrm{rk}_{\lambda}(\mathscr{F}) = \mathrm{rk}_{U}(\mathscr{F})$ for $\lambda \in S$.

By applying Proposition 10.1.3 to our particular context, one obtains the next result.

Proposition 10.3.3 *Let T be a commutative n-tuple of operators on the Hilbert space H. Then $\lambda - T$ is Fredholm if and only if the sheaves $\mathrm{T\hat{o}r}_{q}^{\mathscr{O}(\mathbb{C}^{n})}(H, \mathscr{O}) = \mathscr{H}_{q}(z - T, \mathscr{O}^{H})$ are coherent in a neighbourhood of λ for each $q \geq 0$. In this case,*

$$\mathrm{ind}(\lambda - T) = \sum_{q=0}^{\infty} (-1)^{q} \, \mathrm{rk}_{\lambda} \mathrm{T\hat{o}r}_{q}^{\mathscr{O}(\mathbb{C}^{n})}(H, \mathscr{O}). \qquad (10.3.1)$$

Proof Only formula (10.3.1) requires an argument. Let $\lambda - T$ be Fredholm. Then there are a connected open neighbourhood U of λ and a proper analytic subset $S \subset U$ such that the sheaves $\mathrm{T\hat{o}r}_{q}^{\mathscr{O}(\mathbb{C}^{n})}(H, \mathscr{O})_{U \backslash S}$ are locally free of rank h_{q}, respectively. In view of Theorem 10.3.2, we have

$$\mathrm{ind}(\lambda - T) = \mathrm{ind}(\mu - T) = \sum_{q=0}^{n} (-1)^{q} \dim \mathrm{Tor}_{q}^{\mathscr{O}(\mathbb{C}^{n})}\left(H, \mathscr{O}(\mathbb{C}^{n})/\underline{m}_{\mu}\right)$$

$$= \sum_{q=0}^{n} (-1)^{q} \dim\left[\mathrm{T\hat{o}r}_{q}^{\mathscr{O}(\mathbb{C}^{n})}(H, \mathscr{O})_{\mu} \otimes_{\mathscr{O}_{\mu}} (\mathscr{O}_{\mu}/m_{\mu})\right]$$

$$= \sum_{q=0}^{n} (-1)^{q} h_{q}$$

for $\mu \in U \backslash S$ arbitrary.

The equality of the two dimensions occurring above is a consequence of the fact that the homology sheaves of the complex $K_{\bullet}(z - T, \mathscr{O}^{H})$ are locally free in a neighbourhood of μ. Indeed, according to Remark 9.4.6(b), on a suitable open neighbourhood V of μ, there is a complex $\mathscr{L}_{\bullet} = (\mathscr{L}_{p})_{p=0}^{n}$ of finite type free $\mathscr{O}_{V}$-modules and a morphism $h: \mathscr{L}_{\bullet} \to K_{\bullet}(z - T, \mathscr{O}_{V}^{H})$ such that h and all localizations $h(x): \mathscr{L}_{\bullet}(x) \to K_{\bullet}(x - T, X)$ $(x \in V)$ are quasi-isomorphisms. After shrinking V we may suppose that all homology sheaves of $\mathscr{L}_{\bullet}$ are finite-type free $\mathscr{O}_{V}$-modules.

An inductive application of Lemma 9.4.4 shows that all kernel and image sheaves occurring in $\mathscr{L}_\bullet$ are finite-type free analytic modules near μ, and that there are canonical identifications

$$H_q(\mathscr{L}_\bullet(\mu)) \to \mathscr{H}_q(\mathscr{L}_\bullet)(\mu) \qquad (q \geq 0).$$

Hence

$$\dim H_q(\mu - T, X) = \mathrm{rk}_\mu(\mathscr{H}_q(\mathscr{L}_\bullet)) = h_q \qquad (q \geq 0). \qquad \square$$

Corollary 10.3.4 *Let T be a commuting n-tuple with property (β). Then $\lambda - T$ is Fredholm if and only if the sheaf model $\mathscr{F}_T = H \hat{\otimes}_{\mathscr{O}(\mathbb{C}^n)} \mathscr{O}$ of T is coherent in a neighbourhood of λ. In this case we have*

$$\mathrm{ind}(\lambda - T) = \mathrm{rk}_\lambda(\mathscr{F}_T).$$

Proof It suffices to recall that property (β) implies that $\mathrm{T\hat{o}r}_q^{\mathscr{O}(\mathbb{C}^n)}(H, \mathscr{O}) = 0$ for $q > 0$. $\qquad \square$

This corollary shows in particular that a commuting tuple with property (β) cannot have negative index with our sign convention.

Let us recall from Section 2.6 that the *essential spectrum* of a commuting n-tuple T of Hilbert space operators is given by

$$\sigma_e(T) = \{\lambda \in \mathbb{C}^n; \lambda - T \text{ is not Fredholm}\}.$$

Its complement $\rho_e(T) = \mathbb{C}^n \setminus \sigma_e(T)$ is called the *essential resolvent set* of T. The description of the essential spectrum obtained in Corollary 10.3.4 enables us to show that, for tuples with Bishop's property (β), the essential spectrum is invariant with repect to quasi-similarities. More precisely, one obtains the following result.

Theorem 10.3.5 *Let $T \in L(H)^n$ and $S \in L(K)^n$ be commuting tuples with Bishop's property (β). Suppose that there are continuous linear operators $A: H \to K$ and $B: K \to H$ with dense range such that*

$$AT_j = S_j A \quad \text{and} \quad BS_j = T_j B \quad (1 \leq j \leq n).$$

Then $\sigma(T) = \sigma(S)$, $\sigma_e(T) = \sigma_e(S)$, and $\mathrm{ind}(\lambda - T) = \mathrm{ind}(\lambda - S)$ for each point $\lambda \in \sigma(T) \setminus \sigma_e(S)$.

Proof Let $\mathscr{F}_T$ and $\mathscr{F}_S$ denote the sheaf models of T and S. Then the operators A and B induce continuous morphisms of analytic Fréchet sheaves $a: \mathscr{F}_T \to \mathscr{F}_S$ and $b: \mathscr{F}_S \to \mathscr{F}_T$. The representations

$$\mathscr{F}_T(U) = \mathscr{O}(U, H) \Big/ \sum_{j=1}^n (z_j - T_j)\mathscr{O}(U, H) \quad (U \subset \mathbb{C}^n \text{ Stein open set}),$$

and the analogous representations of the sheaf model of S, show that the induced section maps $a: \mathcal{F}_T(U) \to \mathcal{F}_S(U)$ and $b: \mathcal{F}_S(U) \to \mathcal{F}_T(U)$ have dense range for each Stein open set U in $\mathbb{C}^n$.

Let $\lambda \in \mathbb{C}^n \setminus \sigma(T)$. Then, for each sufficiently small neighbourhood U of λ the space $\mathcal{F}_T(U)$, and hence also the space $\mathcal{F}_S(U)$, is zero. Hence $\lambda \notin \sigma(S)$ (see Theorem 2.1.8).

Next, let us suppose that $\lambda \in \mathbb{C}^n \setminus \sigma_e(T)$. Then there exists an open neighbourhood U of λ such that $\mathcal{F}_T \mid U$ is a coherent $\mathcal{O}_U$-module. We set $\mathcal{F} = \mathcal{F}_T \mid U$ and $\mathcal{G} = \mathcal{F}_S \mid U$. For each Stein open set $V \subset U$, the composition $u = b \circ a: \mathcal{F}(V) \to \mathcal{F}(V)$ is an operator with dense range. On the other hand, the image $u\mathcal{F} \subset \mathcal{F}$ is a coherent subsheaf, and therefore

$$u\mathcal{F}(V) = (u\mathcal{F})(V) \subset \mathcal{F}(V)$$

is a closed subspace of $\mathcal{F}(V)$ for each Stein open subset V of U (Abgeschlossenheitssatz in Chapter V, §6 in Grauert and Remmert 1979). Hence $u\mathcal{F} = \mathcal{F}$. For $z \in U$ arbitrary, let $u_z: \mathcal{F}_z \to \mathcal{F}_z$ be the $\mathcal{O}_z$-module homomorphism induced by u. Since $\mathcal{F}_z$ is a noetherian $\mathcal{O}_z$-module, the sequence of submodules $\mathrm{Ker}(u_z^k)$ ($k \geq 1$) becomes stable. Hence, for sufficiently large k, the restriction of u_z to $\mathrm{Im}(u_z^k)$ is injective. Therefore u_z is not only surjective, but also injective. Thus $u: \mathcal{F} \to \mathcal{F}$ is an isomorphism. But then, for each Stein open subset V of U, the map $a: \mathcal{F}_T(V) \to \mathcal{F}_S(V)$ is a topological monomorphism with dense range. We conclude that $a: \mathcal{F} \to \mathcal{G}$ is an isomorphism of $\mathcal{O}_{\mathbb{C}^n}$-modules. Hence $\mathcal{G} = \mathcal{F}_S \mid U$ is coherent, and $\lambda \notin \sigma_e(S)$ by Corollary 10.3.4.

Since the hypothesis is symmetric in S and T, it follows that $\sigma(S) = \sigma(T)$ and that $\sigma_e(S) = \sigma_e(T)$. Because the sheaf models of S and T are locally isomorphic on $\rho_e(S) = \rho_e(T)$, the description of the index given in Corollary 10.3.4 implies that S and T have the same index at all points of the essential resolvent set. $\square$

The descriptions of the essential spectrum and of the index obtained in Proposition 10.3.3 and Corollary 10.3.4 remain true for commuting tuples T of Banach space operators if the essential spectrum $\sigma_e(T)$ of T is defined as in Section 2.6 (see the definition following Corollary 2.6.3). With this definition of $\sigma_e(T)$ Theorem 10.3.5 also remains true (with exactly the same proof) in the Banach space setting.

Lemma 10.3.6 *Let T be a commuting n-tuple of operators. Then $\sigma_e(T) \subset \sigma(T)$ is a non-empty set, and $\sigma(T) \cap \rho_e(T)$ is a closed analytic subset of $\rho_e(T)$.*

Proof According to Definition 10.2.1, we have

$$\sigma_e(T) = \sigma(T, L(H)/K(H)).$$

Hence $\sigma_e(T) \neq \varnothing$ (see also Section 2.6). On the other hand, Proposition 10.3.3 shows that $\sigma(T) \cap \rho_e(T)$ is the support of the coherent $\mathcal{O}$-module $\bigoplus_{q=0}^{n} \mathrm{T\hat{o}r}_q^{\mathcal{O}(\mathbb{C}^n)}(H, \mathcal{O})\mid_{\rho_e(T)}$. $\square$

We shall see later that the analytic set $\sigma(T) \cap \rho_e(T)$ is subject to a few additional restrictions.

Let T be a commuting n-tuple of operators on H. Since the complex of Hilbert-free $\mathcal{O}$-modules $K_\bullet(z - T, \mathcal{O}^H)$ has coherent homology on the open set $\rho_e(T) \subset \mathbb{C}^n$, the abstract K-theoretic constructions described in Section 9.1 apply and yield a K-theory class

$$i_t(T) = \left[K_\bullet\left(z - T, \mathcal{C}_{\rho_e(T)} \hat{\otimes}_\varepsilon H\right)\right] \in \mathcal{K}^0(\rho_e(T)),$$

where $\mathcal{K}^0(\rho_e(T)) = \mathcal{K}(\rho_e(T))$ is as in Definition 10.1.7, called the *topological index* of T. Analogously one can introduce, following Levy (1989), the *analytic index* of T as

$$i_a(T) = \sum_{q=0}^{n} (-1)^q \left[\mathrm{T\hat{o}r}_q^{\mathcal{O}(\mathbb{C}^n)}(H, \mathcal{O})|_{\rho_e(T)}\right] \in K_0^a(\rho_e(T)).$$

Here K_0^a stands for the *Grothendieck ring* of classes of analytic coherent sheaves (see Appendix 3).

The topological index $i_t(T)$ makes perfect sense for essentially commuting tuples of operators. In particular, $i_t(T)$ is invariant under arbitrary compact perturbations of T. More precisely, we set by definition $i_t(T) = i_t(\delta + \delta^*)$, where $\delta + \delta^*$ denotes the Fredholm complex of Hilbert-free $\mathcal{C}_{\rho_e(T)}$-modules given by the function

$$\rho_e(T) \to L\left(\bigoplus_q K_{2q}(T, H), \bigoplus_q K_{2q+1}(T, H)\right), \qquad z \mapsto \delta_{z-T} + \delta^*_{z-T}.$$

The relationship between the above indices is immediate. Namely, for each point $\lambda \in \rho_e(T)$, we have

$$\mathrm{ind}(\lambda - T) = \mathrm{rk}_\lambda i_t(T) = \mathrm{rk}_\lambda i_a(T) \tag{10.3.2}$$

(see Proposition 10.1.8).

However, the K-theoretic indices i_a and i_t contain much more information than the classical numerical index.

These abstract indices are related by a natural map

$$\alpha \colon K_0^a(\rho_e(T)) \to \mathcal{K}^0(\rho_e(T)),$$

which is explained in Appendix 3. Roughly speaking, $\alpha[\mathcal{F}]$ is the class of a locally free resolution of the sheaf $\mathcal{F}$ on compact subsets of the underlying space $\rho_e(T)$. Because $i_t(T) = \alpha i_a(T)$, the class $i_t(T)$ is 'algebraizable'. This fact has some cohomological consequences when passing to Chern classes (see Levy 1989).

On the other hand, if the homology sheaves $\mathcal{H}_q = \mathcal{H}_q(z - T, \mathcal{O}^H)$ $(q \geq 0)$ have ranks $r_{q,i} = \mathrm{rk}\,\mathcal{H}_q|_{\rho_i}$ on the highest-dimensional irreducible components ρ_i of the analytic set $\rho_e(T) \cap \sigma(T)$, then a result due to Grothendieck asserts that

$$\mathrm{ch}\, i_t(T) = \sum_i \sum_q (-1)^q r_{q,i}[\rho_i] + \text{higher classes}. \tag{10.3.3}$$

In the above equation, $[\rho_i] \in H^{2n-2i}(\rho_e(T), \mathbb{Q})$ denotes the fundamental class of the analytic subset $\rho_i \subset \rho_e(T)$ (see Appendix 3 and the references cited there for details).

Before continuing, it is worth computing the preceding abstract indices for some very simple particular cases.

Let $S \in L(H)$ be a single operator. Then the essential resolvent set $\rho_e(S)$ consists of the resolvent set $\rho(S)$, some bounded connected components of the complement of $\sigma_e(S)$, and the non-essential isolated points in $\sigma(S)$. Let U be a connected component of $\rho_e(S)$ with $U \subset \sigma(S)$. The Fredholm index $\text{ind}(\lambda - S) = k_U$ is constant for $\lambda \in U$, and we have

$$i_a(S)|_U = [\text{Coker}(z - S)] - [\text{Ker}(z - S)] \in K_0^a(U).$$

Similarly,

$$i_t(S)|_U = \left[\mathscr{C}_U \hat{\otimes}_\varepsilon H \xrightarrow{z-S} \mathscr{C}_U \hat{\otimes}_\varepsilon H \right] \in \mathscr{K}^0(U).$$

As a second simple example, we consider an n-tuple T with Bishop's property (β). Let $\mathscr{F} = H \hat{\otimes}_{\mathscr{O}(\mathbb{C}^n)} \mathscr{O}$ denote its sheaf model. Then we obtain

$$i_a(T) = [\mathscr{F}|_{\rho_e(T)}] \in K_0^a(\rho_e(T)),$$

and

$$i_t(T) = \alpha[\mathscr{F}|_{\rho_e(T)}].$$

For instance, let B denote the unit ball in $\mathbb{C}^2$, let a, b be positive integers, and let us consider the pair of analytic Toeplitz operators $T = (z_1^a, z_2^b)$, acting on the Bergman space $L_a^2(B)$. According to Proposition 8.2.5, the essential spectrum of T is

$$\sigma_e(T) = \{(\lambda, \mu) \in \mathbb{C}^2; |\lambda|^{2/a} + |\mu|^{2/b} = 1\}.$$

The interior of this ellipsoid is the set

$$E = \sigma(T) \setminus \sigma_e(T) = \{(\lambda, \mu) \in \mathbb{C}^2; |\lambda|^{2/a} + |\mu|^{2/b} < 1\}.$$

The pair T satisfies Bishop's property (β), and its sheaf model $\mathscr{F}_T$ is supported by $\bar{E}$ (see Section 8.1). Thus $i_a(T) = [\mathscr{F}_T | E]$. On the other hand, according to the proof of Proposition 8.1.6, we have

$$\mathscr{F}_T | E = f_* \mathscr{O}_B,$$

where $f: B \to E$ is the finite analytic map $f(z_1, z_2) = (z_1^a, z_2^b)$. The multiplicity of this map, or equivalently, the rank of the coherent sheaf $\mathscr{F}_T | E$, is ab. Hence

$$i_t(T) | E = ab[1] \in \mathscr{K}^0(E),$$

where [1] denotes the element of $\mathscr{K}^0(E)$ with all components represented by the trivial rank-one bundle. At the same time, the sheaf

$$(f_* \mathscr{O}_B)_{(\lambda, \mu)} = \bigoplus_{\substack{z_1^a = \lambda \\ z_2^b = \mu}} \mathscr{O}_{(z_1, z_2)} \qquad ((\lambda, \mu) \in E)$$

contains much more analytic information than its rank ab.

One of the fundamental properties of the indices i_a and i_t is their functorial behaviour with respect to the analytic functional calculus.

Theorem 10.3.7 *Let $f\colon U \to \mathbb{C}^m$ be an analytic map defined on an open neighbourhood U of the joint spectrum of the commuting n-tuple T. Let $D = f^{-1}(\rho_e(f(T)))$. If $D = \varnothing$, then $\rho_e(f(T)) = \rho(f(T))$. If $D \neq \varnothing$, then*

$$\mathcal{H}_q\big(w - f(T), \mathcal{O}^H_{\rho_e(f(T))}\big) = f_*\big(\mathcal{H}_q(z - T, \mathcal{O}^H_D)\big) \qquad (q \geq 0).$$

In this case, $f_*(i_a(T)|_D) = i_a(f(T))$.

Proof Suppose that $D \neq \varnothing$. We define $E = \rho_e(f(T))$, and we write the function $f\colon D \to E$ as a composition

$$D \xrightarrow{i} D \times E \xrightarrow{j} \mathbb{C}^n \times E \xrightarrow{p} E,$$

where i is the holomorphic embedding defined by $i(z) = (z, f(z))$ ($z \in D$), j is the inclusion map, and p denotes the projection onto the last m coordinates.

For $W \subset D \times E$ open, we define a morphism of complexes

$$\rho_W\colon K_\bullet((z - T, w - f(z)), \mathcal{O}(W, H)) \to K_\bullet(z - T, \mathcal{O}(i^{-1}(W), H))$$

by setting $\rho_W(g s_{i_1} \wedge \cdots \wedge s_{i_q}) = g \circ i s_{i_1} \wedge \cdots \wedge s_{i_q}$ if $i_q \leq n$ and $\rho_W(g s_{i_1} \wedge \cdots \wedge s_{i_q}) = 0$ otherwise. The family $(\rho_W)_W$ induces a quasi-isomorphism

$$\rho\colon K_\bullet((z - T, w - f(z)), \mathcal{O}^H_{D \times E}) \to i_* K_\bullet(z - T, \mathcal{O}^H_D).$$

It suffices to check that ρ is stalkwise a quasi-isomorphism. This is non-trivial only at points $(z, f(z))$ ($z \in D$). For each $z \in D$ and each open set $W = G \times P$, where $z \in G \subset D$ and P is an open polydisc containing $f(G)$, the map ρ_W is a homotopy equivalence with homotopy inverse

$$\sigma_W\colon K_\bullet(z - T, \mathcal{O}(G, H)) \to K_\bullet((z - T, w - f(z)), \mathcal{O}(W, H))$$

defined by $\sigma_W(g s_{i_1} \wedge \cdots \wedge s_{i_q}) = g \otimes 1 s_{i_1} \wedge \cdots \wedge s_{i_q}$.

Since i_* is an exact functor (see p. 20 in Grauert and Remmert 1984), we have

$$\mathcal{H}_q\big(i_* K_\bullet(z - T, \mathcal{O}^H_D)\big) \cong i_* \mathcal{H}_q(z - T, \mathcal{O}^H_D)$$

for arbitrary $q \geq 0$. Hence

$$i_* \mathcal{H}_q(z - T, \mathcal{O}^H_D) \cong \mathcal{H}_q((z - T, w - f(z)), \mathcal{O}^H_{D \times E})$$

as $\mathcal{O}_{D \times E}$-modules. On the other hand, there are isomorphisms

$$\mathcal{H}_q((z - T, w - f(z)), \mathcal{O}^H_{D \times E}) \cong \mathcal{H}_q((z - T, w - f(T)), \mathcal{O}^H_{D \times E}). \qquad (10.3.4)$$

Indeed, using induction on m and the homological version of Proposition

2.5.9 (see also the remark following Proposition 2.5.9), one can construct a family $(\varphi_{U,V})$ $(U \subset D, V \subset E$ Stein open sets) of isomorphisms

$$\varphi_{U,V} \colon H_q((z-T,w-f(z)),\mathscr{O}(U \times V,H))$$

$$\to H_q((z-T,w-f(T)),\mathscr{O}(U \times V,H))$$

which is compatible with restrictions. This isomorphism of presheaves induces the claimed isomorphism (10.3.4) between the associated sheaves. Thus we have shown that

$$i_*\mathscr{H}_q(z-T,\mathscr{O}_D^H) \cong \mathscr{H}_q((z-T,w-f(T)),\mathscr{O}_{D \times E}^H) \qquad (q \geq 0).$$

Using the fact that $f^{-1}(E) \subset D$, one easily obtains that

$$j_*\mathscr{H}_q((z-T,w-f(T)),\mathscr{O}_{D \times E}^H) \cong \mathscr{H}_q((z-T,w-f(T)),\mathscr{O}_{\mathbb{C}^n \times E}^H) \qquad (q \geq 0)$$

as $\mathscr{O}_{\mathbb{C}^n \times E}$-modules.

Thus it remains to prove the assertion of the theorem for a projection of the form $p \colon \mathbb{C}^{n+m} \to \mathbb{C}^m$, $(z,w) \to w$, instead of f, and for a commuting $(n+m)$-tuple $(T,S) = (T_1,\ldots,T_n,S_1,\ldots,S_m)$ in place of T. By writing p as the composition of n suitable projections, each lowering the dimension by one, the assertion is reduced to the case $n = 1$. Therefore let us suppose that $T \in L(H)$ is a single operator. To complete the proof, we show that

$$p_*\mathscr{H}_q((z-T,w-S),\mathscr{O}_{\mathbb{C} \times E}^H) \cong \mathscr{H}^q(w-S,\mathscr{O}_E^H) \qquad (q \geq 0),$$

where $E = \rho_e(S)$.

The sheaves $\mathscr{H}_q((z-T,w-S),\mathscr{O}_{\mathbb{C} \times E}^H)$ and $\mathscr{H}_q(w-S,\mathscr{O}_E^H)$ are coherent by Proposition 10.1.3. As remarked in the proof of Proposition 10.1.3, all kernel and image sheaves occurring in the complexes $K_\bullet((z-T,w-S),\mathscr{O}_{\mathbb{C} \times E}^H)$ and $K_\bullet(w-S,\mathscr{O}_E^H)$ are acyclic on all Stein open subsets of $\mathbb{C} \times E$ and E, respectively. It follows that, for each Stein open subset V of E, the canonical maps

$$H_q((z-T,w-S),\mathscr{O}(\mathbb{C} \times V,H)) \to \mathscr{H}_q((z-T,w-S),\mathscr{O}_{\mathbb{C} \times E}^H)(\mathbb{C} \times V),$$

$$H_q(w-S,\mathscr{O}(V,H)) \to \mathscr{H}_q(w-S,\mathscr{O}_E^H)(V)$$

are isomorphisms. Hence to conclude the proof, it suffices to construct a family of isomorphisms

$$H_q(w-S,\mathscr{O}(V,H)) \to H_q((z-T,w-S),\mathscr{O}(\mathbb{C} \times V,H)) \qquad (V \subset E \text{ Stein})$$

of $\mathscr{O}(V)$-modules which is compatible with restrictions. This can be done as follows. Because of the exactness of the augmented Koszul complex

$$K_\bullet(z-T,\mathscr{O}(\mathbb{C} \times V,H)) \to \mathscr{O}(V,H) \to 0$$

the maps $\pi_V \colon K_\bullet(w-S,\mathscr{O}(V,H)) \to K_\bullet((z-T,w-S),\mathscr{O}(\mathbb{C} \times V,H))$ defined by

$$\pi_V\left(gs_{i_1} \wedge \cdots \wedge s_{i_q}\right) = 1 \otimes gs_{i_1} \wedge \cdots \wedge s_{i_q} \qquad (1 \leq i_1 < \cdots < i_q \leq m)$$

form a family π_V ($V \subset E$ Stein open subset) of quasi-isomorphisms (see Lemma A2.6 in Appendix 2) that is compatible with restrictions (following the proof of Lemma 1.1 from Eschmeier 1985, one can also directly prove that the maps π_V are in fact homotopy equivalences). Thus the proof is complete.   □

Again the last result remains true, with exactly the same proof, in the Banach space setting, if the essential spectrum $\sigma_e(T)$ of a commuting tuple of Banach space operators is defined as in Section 2.6. The assumption that the set E, occurring in the above proof, is the essential resolvent set of $f(T)$ was only used to guarantee that the homology sheaves of the Koszul complexes $K_\bullet(w - S, \mathcal{O}_E^H)$ and $K_\bullet((z - T, w - S), \mathcal{O}_{\mathbb{C} \times E}^H)$ are acyclic on all Stein open subsets of their domains. This is certainly true if both S and (T, S) satisfy Bishop's property (β). A careful analysis of the preceding proof therefore leads to the next result.

Corollary 10.3.8   *Let* $T \in L(X)^n$ *be a commuting tuple of continuous linear operators on a complex Banach space* X. *Let* $f \in \mathcal{O}(U)^m$ *be an analytic map on an open neighbourhood* U *of* $\sigma(T)$. *Suppose that* T *satisfies Bishop's property* (β). *Then* $f(T)$ *satisfies Bishop's property* (β), *and there is an isomorphism of Fréchet* $\mathcal{O}(\mathbb{C}^m)$*-modules*

$$f_*(\mathcal{F}_T \,|\, U) \cong \mathcal{F}_{f(T)}$$

relating the sheaf models of T *and* $f(T)$.   □

Corollary 10.3.9   *Under the assumptions of Theorem* 10.3.7, $f_* i_t(T) = i_t(f(T))$.

Proof   This is a consequence of the Riemann–Roch theorem, which asserts that $f_* \alpha = \alpha f_*$ for the natural morphism $\alpha \colon K_0^a \to \mathcal{K}^0$ (see Theorem 10.4.1 below).   □

Levy (1989) proves Corollary 10.3.9 directly for more general maps f defined only in neighbourhoods of $\sigma_e(T)$. In the same paper a few interesting applications of Theorem 10.3.7 are discussed.

We have already remarked that the topological index $i_t(T)$ has a meaning for any essentially commuting n-tuple T. On the other hand, $i_a(T)$ is defined only for commuting n-tuples, and in this case $\operatorname{supp} i_a(T) \subset \sigma(T) \setminus \sigma_e(T)$ is an analytic set. This observation can be exploited for finding obstructions for an essentially commuting n-tuple to be a compact perturbation of a commuting n-tuple. For instance the next result holds.

Proposition 10.3.10   *If* T *is a compact perturbation of a commutative* n-tuple of operators and $i_t(T) \neq 0$, *then* $\sigma_e(T)$ *contains the boundary of a bounded analytic subset of* $\mathbb{C}^n \setminus \sigma_e(T)$.

Proof   Because $i_t(T) \neq 0$, the set $\rho_e(T) \cap \sigma(T)$ necessarily has dimension greater than or equal to one.   □

In fact the relation between $i_t(T)$ and the dimension of the analytic set $\rho_e(T) \cap \sigma(T)$ can be made more precise (see Levy 1989). A simple application of this comparison principle between the analytic and the topological index of an n-tuple of operators is sketched in what follows.

Let $T_f = PM_f$ denote the Toeplitz operator with continuous symbol f in $C(\mathbb{T})$, acting on the Hardy space $H^2(\mathbb{T})$. As usual P denotes the orthogonal projection onto the Hardy subspace $H^2(\mathbb{T}) \subset L^2(\mathbb{T})$, and M_f stands for the operator of multiplication by f. If $f_1, \ldots, f_n \in A(\mathbb{D})$ is an n-tuple of functions belonging to the disc algebra $A(\mathbb{D}) = \mathcal{O}(\mathbb{D}) \cap C(\overline{\mathbb{D}})$, then obviously the n-tuple of Toeplitz operators $(T_{f_1}, \ldots, T_{f_n})$ is commutative. Our next aim is to prove a partial converse of this statement.

Proposition 10.3.11   *Let $f = (f_1, \ldots, f_n): \mathbb{T} \to \mathbb{C}^n$ be an injective map of class C^2. The essentially commutative n-tuple of Toeplitz operators $(T_{f_1}, \ldots, T_{f_n})$ is a compact perturbation of a commutative n-tuple if and only if there is a homeomorphism $h: \mathbb{T} \to \mathbb{T}$ such that $f_1 \circ h, \ldots, f_n \circ h \in A(\mathbb{D})$.*

Proof   Suppose that $h: \mathbb{T} \to \mathbb{T}$ is a homeomorphism such that $f_1 \circ h, \ldots, f_n \circ h$ are restrictions of functions $g_i \in A(\mathbb{D})$. By comparing the $A(\mathbb{D})$-functional calculus of the contraction $T_{h^{-1}}$ with the C^0-functional calculus of the essentially normal operator T_z, we obtain that the operators $g_i(T_{h^{-1}}) - T_{g_i \circ h^{-1}} = g_i(T_{h^{-1}}) - T_{f_i}$ $(1 \le i \le n)$ are compact.

Suppose that the tuple $T_f = (T_{f_1}, \ldots, T_{f_n})$ is a compact perturbation of a commuting n-tuple. The essential spectrum of T_f is $\sigma_e(T_f) = f(\mathbb{T})$, for instance by a Gelfand transform argument. Under the above assumptions the set $f(\mathbb{T})$ is the boundary of a bounded analytic subset S of $\mathbb{C}^n \setminus f(\mathbb{T})$. But $f(\mathbb{T})$ is a connected one-dimensional real manifold. Therefore, dim $S = 1$. A theorem due to Harvey and Lawson (1975) shows that, modulo a subset of $f(\mathbb{T})$ of linear measure zero, the set $S \cup f(\mathbb{T})$ is a 1-smooth manifold with boundary $f(\mathbb{T})$. Since $f(\mathbb{T})$ is connected, the manifold S must be irreducible.

Therefore there exists a proper holomorphic map $g: \mathbb{D} \to S$ with Im $g = S$. By standard arguments (see for instance Chirka 1982), the map g extends to a C^1-function $g: \overline{\mathbb{D}} \to \mathbb{C}^n$. Furthermore, the map g induces a homeomorphism $g: \mathbb{T} \to f(\mathbb{T})$. The map $h = f^{-1} \circ (g \mid \mathbb{T}): \mathbb{T} \to \mathbb{T}$ is a homeomorphism such that the functions $f_i \circ h$ are restrictions of functions in $A(\mathbb{D})$. $\square$

Some further applications of Proposition 10.3.10 are contained in Levy (1989). The same paper contains an ample discussion of the relationship between the topological index $i_t(T)$ and the Brown–Douglas–Fillmore extension theory.

Besides Theorem 10.3.7 there are a few other useful methods of operating with the analytic index i_a. The next result is an example.

Proposition 10.3.12   *Let T be a commuting n-tuple, and let $\mathcal{L}_\bullet$ be a finite* *complex of finite-type free $\mathcal{O}_U$-modules defined in an open neighbourhood U of the joint spectrum $\sigma(T)$.*

If $\mathcal{L}_\bullet \mid \sigma_e(T)$ is exact, then the complex $L_\bullet = \mathcal{L}_\bullet(U) \hat{\otimes}_{\mathcal{O}(U)} H$ is Fredholm and

$$\text{ind } L_\bullet = \langle i_a(T), [\mathcal{L}_\bullet] \rangle,$$

where $\langle , \rangle$ denotes the Tor-intersection number.

Proof The definition of the intersection number will be given below.

For simplicity, we shall suppose that U is a Stein open set in $\mathbb{C}^n$. The argument that can be used to treat the general case will be described in detail in the proof of Theorem 10.3.13. For U Stein, the augmented Koszul complex

$$K_\bullet(z - T, \mathcal{O}(U) \hat{\otimes} H) \to H \to 0$$

is exact (see the end of Section 4.3). Therefore the complex

$$L_\bullet = \mathcal{L}_\bullet(U) \hat{\otimes}_{\mathcal{O}(U)} H$$

is quasi-isomorphic to the total complex $C_\bullet$ of the double complex

$$K_\bullet(z - T, \mathcal{O}(U) \hat{\otimes} H) \hat{\otimes}_{\mathcal{O}(U)} \mathcal{L}_\bullet(U) = K_\bullet\left(z - T, \mathcal{L}_\bullet(U) \hat{\otimes} H\right).$$

By assumption the homology sheaves $\mathcal{H}_\bullet(\mathcal{L})$ are supported by an analytic subset of $U \setminus \sigma_e(T)$. Choose an open neighbourhood W of $\sigma_e(T)$ such that $\mathcal{L}_\bullet \mid W$ is exact. The simple complex $\text{Tot}(K_\bullet(z - T, \mathcal{O}^H) \otimes_\mathcal{O} \mathcal{L}_\bullet)$ has coherent homology (use Lemma 3.1.17 and Remark 9.4.6) supported by an analytic subset $F = F \cap \sigma(T) \cap (\mathbb{C}^n \setminus W)$ of $U \cap \rho_e(T)$. Being also compact, the set F is necessarily finite.

On the other hand, we have by definition

$$C_\bullet = \Gamma\left(U, \text{Tot}\left(K_\bullet(z - T, \mathcal{O}^H) \otimes_\mathcal{O} \mathcal{L}_\bullet\right)\right).$$

Therefore $C_\bullet$ is quasi-isomorphic to the complex

$$\Gamma\left(V, \text{Tot}\left(K_\bullet(z - T, \mathcal{O}^H) \otimes_\mathcal{O} \mathcal{L}_\bullet\right)\right),$$

where $V \subset U \cap \rho_e(T)$ is a Stein open neighbourhood of the finite set F.

We can choose V as a finite disjoint union $V = \bigcup_{k=1}^r V_k$ of Stein open sets V_k each containing at most one point of F. Moreover, by Proposition 10.1.3, we may suppose that there is a finite complex $\mathcal{M}_\bullet$ of finite-type free $\mathcal{O}_V$-modules which is quasi-isomorphic to $K_\bullet(z - T, \mathcal{O}^H)$ on V. Then (see Lemma 3.1.17) the complex $C_\bullet$ is quasi-isomorphic to the complex $\Gamma(V, \text{Tot}(\mathcal{M}_\bullet \otimes_\mathcal{O} \mathcal{L}_\bullet))$.

An elementary application of Rückert's Nullstellensatz (Chapter 3, §2 in Grauert and Remmert 1984) shows that the homology groups of the last complex are finite-dimensional complex vector spaces. Hence $L_\bullet$ is Fredholm, and

$$\text{ind } L_\bullet = \chi(C_\bullet) = \chi\Gamma(V, \text{Tot}(\mathcal{M}_\bullet \otimes_\mathcal{O} \mathcal{L}_\bullet)) =: \langle \mathcal{M}_\bullet, \mathcal{L}_\bullet \rangle, \quad (10.3.5)$$

where the last equality is the definition of the intersection number. Again by definition we have

$$\langle i_a(T), [\mathscr{L}_\bullet] \rangle = \langle \mathscr{M}_\bullet, \mathscr{L}_\bullet \rangle,$$

and the proof of Proposition 10.3.12 is complete.   □

Next we focus on a particular case of Proposition 10.3.12 in which a more concrete index compuation is possible and which has applications to the index theory for systems of Toeplitz operators.

Let T be a continuous linear operator on a complex Banach space X. For each point $w \in \sigma(f(T)) \setminus \sigma_e(f(T))$ and each analytic function $f \in \mathscr{O}(\sigma(T))$, the following formula

$$\mathrm{ind}(w - f(T)) = \sum_{z \in f^{-1}(w)} \nu_z(f) \mathrm{ind}(z - T) \qquad (10.3.6)$$

holds (see Gramsch 1967). Here the sum is formed over all points $z \in f^{-1}(w)$ with $\mathrm{ind}(z - T) \neq 0$, and $\nu_{z_0}(f)$ denotes the order of the zero of $f(z) - w$ at z_0.

Let $T = (T_1, \ldots, T_n)$ be a commuting tuple of continuous linear operators on the Banach space X, and let $f = (f_1, \ldots, f_m)$ be an m-tuple of analytic functions defined in a neighbourhood of Taylor's joint spectrum $\sigma(T)$. For $\nu = 0, \ldots, n$, the restriction of the cohomology sheaf $\mathscr{H}^\nu$ of the complex $K^\bullet(z - T, \mathscr{O}_{\mathbb{C}^n}^X)$ onto $\rho_e(T) = \mathbb{C}^n \setminus \sigma_e(T)$ is coherent (see Proposition 10.1.3). Hence

$$\sigma(T) \setminus \sigma_e(T) = \bigcup_{\nu=0}^{n} \mathrm{supp}(\mathscr{H}^\nu | \rho_e(T))$$

is an analytic subset of $\rho_e(T)$ (see Chapter 4, §1 in Grauert and Remmert 1984). As a consequence one obtains that, for each point w in $\sigma(f(T)) \setminus \sigma_e(f(T))$, the fibre $f^{-1}(w) \cap \sigma(T)$ is a compact analytic, hence finite, subset of $\mathbb{C}^n$.

The only interesting case for which a formula like (10.3.6) might be expected is the case $m = n$. Indeed, if $m > n$, then the set $\sigma(f(T)) = f(\sigma(T))$ has no interior points by Sard's lemma, and hence $\mathrm{ind}(w - f(T))$ is zero for each $w \notin \sigma_e(f(T))$. If $m < n$, then elementary dimension theory shows that the set $f^{-1}(w) \cap C$ is empty for each component C of $\rho_e(T)$ contained in $\sigma(T)$. Hence the index set of the sum occurring in formula (10.3.6) is empty in this case. But (for instance, by using the analytic index formula for tensor product systems proved in Eschmeier 1988b), one can construct examples of commuting tuples $T \in L(X)^n$ and analytic functions $f \in \mathscr{O}(\sigma(T))^m$ $(m < n)$ for which the left-hand side of (10.3.6) does not vanish.

In order to state the analogue of formula (10.3.6) in the multivariable setting, we recall some basic facts concerning the multiplicity of a finite analytic map. Let U and V be open sets in $\mathbb{C}^n$, and let $f: U \to V$ be an analytic map with finite fibres (that is, the set $f^{-1}(y)$ is finite for each point $y \in V$). Then the *multiplicity* $\nu_x(f)$ of the map f at a point $x \in f^{-1}(y)$ can be defined as the number of sheets of the ramified covering f passing through the point x.

A more convenient topological descriptioin of this multiplicity is

$$\nu_x(f) = \deg(f, \Omega, y),$$

where Ω is an open, relatively compact neighbourhood of x in U such that $\overline{\Omega} \cap f^{-1}(y) = \{x\}$. Here $\deg(f, \Omega, y)$ denotes the *Brouwer mapping degree* (see §8 in Deimling 1974).

To formulate the index formula, we denote by $\mathscr{C}$ the family of all bounded components of $\mathbb{C}^n \setminus \sigma_e(T)$ and, for $C \in \mathscr{C}$, by $\mathrm{ind}_C(T)$ the constant value of $\mathrm{ind}(z - T)$ for $z \in C$.

Theorem 10.3.13 *Let $T \in L(X)^n$ be a commuting tuple of Banach space operators, and let $f \in \mathcal{O}(U)^n$ be an n-tuple of analytic functions defined on an open neighbourhood U of $\sigma(T)$. For each point $w \in \sigma(f(T)) \setminus \sigma_e(f(T))$, we have*

$$\mathrm{ind}(w - f(T)) = \sum_{C \in \mathscr{C}} \deg(f, C, w) \mathrm{ind}_C(T)$$

$$= \sum_{z \in f^{-1}(w) \cap \sigma(T)} \nu_z(f) \mathrm{ind}(z - T). \tag{10.3.7}$$

Proof A standard compactness argument shows that there are at most finitely many components $C \in \mathscr{C}$ with $\deg(f, C, w) \neq 0$ (see the beginning of the proof of Satz 11.1 in Deimling 1974). Using the excision property and the additivity of the mapping degree (Satz 8.1 in Deimling 1971), one easily verifies that the two sums occurring in (10.3.7) define the same integer.

Both sides of the index formula are invariant under small changes in w. Hence Sard's lemma allows one to reduce the assertion to the case where w is a regular value for f, the whole fibre $f^{-1}(w)$ is finite in U and is contained in $\sigma(T)$. Moreover, we may of course suppose that $w = 0$.

Let $z_1, \ldots, z_r$ be the distinct zeros of the map f in $\sigma(T)$. We choose pairwise disjoint Stein open neighbourhoods V_k of z_k in U and an open Euclidean ball B in $\mathbb{C}^n$ such that f induces biholomorphic maps $f \colon V_k \to B$ $(k = 1, \ldots, r)$. We set $V = V_1 \cup \cdots \cup V_r$, and we choose a Stein open cover $\mathscr{U} = (U_i)_{i \in \mathbb{N}}$ of U with $U_0 = V$ and $U_i \cap \{z_1, \ldots, z_r\} = \varnothing$ for $i > 0$.

We denote by $\mathscr{C}^{\bullet}(U)$ the alternating Čech complex with coefficients in $\mathcal{O}_U$ relative to the open cover $\mathscr{U}$. We regard $\mathcal{O}(V)$ as the trivial complex with $\mathcal{O}(V)$ as the space in degree 0 and zero elsewhere. The kernel $K^{\bullet}$ of the canonical epimorphism $r \colon \mathscr{C}^{\bullet}(\mathscr{U}) \to \mathcal{O}(V)$ becomes a complex of Fréchet $\mathcal{O}(U)$-modules. We denote by

$$0 \to K_1 \to K_2 \to K_3 \to 0$$

the induced short exact sequence between the double complexes

$$K_1 = K^\bullet \hat{\otimes}_{\mathscr{O}(U)} K^\bullet(z - T, \mathscr{O}(U, X)),$$

$$K_2 = \mathscr{C}^\bullet(\mathscr{U}) \hat{\otimes}_{\mathscr{O}(U)} K^\bullet(z - T, \mathscr{O}(U, X)),$$

$$K_3 = \mathscr{O}(V) \hat{\otimes}_{\mathscr{O}(U)} K^\bullet(z - T, \mathscr{O}(U, X)) = K^\bullet(z - T, \mathscr{O}(V, X)).$$

We know that

$$H^p(\mathrm{Tot}(K_2)) \cong \begin{cases} 0 & (p \neq n) \\ X & (p = n) \end{cases}$$

as topological $\mathscr{O}(U)$-modules (see Theorem 5.1.5 and the uniqueness result contained in Theorem 5.2.4). We obtain induced short exact sequences

$$0 \to \mathrm{Tot}(K_1) \to \mathrm{Tot}(K_2) \to \mathrm{Tot}(K_3) \to 0$$

between the corresponding total complexes and between double complexes

$$0 \to \tilde{K}_1 \xrightarrow{i} \tilde{K}_2 \xrightarrow{r} \tilde{K}_3 \to 0,$$

where

$$\tilde{K}_i = \mathrm{Tot}(K_i) \hat{\otimes}_{\mathscr{O}(U)} K^\bullet(f, \mathscr{O}(U)) \qquad (i = 1, 2, 3).$$

By definition the pth column of the double complex $\tilde{K}_1$ is, up to the factor $(-1)^p$, a direct sum of complexes of the form $K^\bullet(f, \mathscr{O}(W, X))$, where W is a Stein open subset of $U \backslash f^{-1}(0)$. Hence the total complex associated with $\tilde{K}_1$ is exact, and $r: \mathrm{Tot}(\tilde{K}_2) \to \mathrm{Tot}(\tilde{K}_3)$ is a quasi-isomorphism. Note that, for $p = 0, \ldots, n$, we have the vector space isomorphisms (Lemma A2.6 in Appendix 2)

$$H^p(f(T), X) \cong H^{p+n}\big(\mathrm{Tot}(\tilde{K}_2)\big) \cong H^{p+n}\big(\mathrm{Tot}(\tilde{K}_3)\big).$$

But the double complex $\tilde{K}_3$ has the form

$$K^\bullet(z - T, \mathscr{O}(V, X)) \hat{\otimes}_{\mathscr{O}(U)} K^\bullet(f, \mathscr{O}(U))$$

$$\cong K^\bullet(z - T, \mathscr{O}(V, X)) \hat{\otimes}_{\mathscr{O}(V)} K^\bullet(f, \mathscr{O}(V)).$$

Note that

$$K^\bullet(f, \mathscr{O}(V)) = \bigoplus_{k=1}^r K^\bullet(f, \mathscr{O}(V_k)) \cong \bigoplus_{k=1}^r K^\bullet(z, \mathscr{O}(B)),$$

where the last isomorphism of complexes is induced by the biholomorphic maps $f: V_k \to B$. Therefore,

$$H^p(f, \mathscr{O}(V)) = \begin{cases} 0, & p \neq n \\ \mathscr{O}(V)/(f), & p = n, \end{cases}$$

where $\mathscr{O}(V)/(f) = \oplus_{k=1}^r \mathscr{O}(V_k)/(f)$. Using the identifications

$$\mathscr{O}(V_k)/(f) \to \mathbb{C}, \quad [g] \mapsto g(z_k) \qquad (1 \le k \le r),$$

one finally obtains the chain of vector space isomorphisms

$$H^{p+n}\big(\mathrm{Tot}(\tilde{K}_3)\big) \cong H^p(z-T, H^n(f, \mathcal{O}(V, X)))$$

$$\cong \bigoplus_{k=1}^{r} H^p(z-T, H^n(f, \mathcal{O}(V_k, X))) \cong \bigoplus_{k=1}^{r} H^p(z_k - T, X).$$

The proof is completed by the observation that

$$\mathrm{ind}(f(T)) = \sum_{k=1}^{r} \sum_{p=0}^{n} (-1)^{n-p} \dim H^p(z_k - T, X)$$

$$= \sum_{k=1}^{r} \mathrm{ind}(z_k - T). \qquad \square$$

As an application of the previous result one can prove an index formula for n-tuples of Toeplitz operators with H^∞-symbol acting on the Bergman space of a pseudoconvex domain in $\mathbb{C}^n$. The lack of a definitive result on the essential spectrum of Toeplitz tuples with H^∞-symbol (see Theorem 8.2.4) is responsible for the particular form of the next result. The notations below are those of Section 8.2.

Corollary 10.3.14   *Let Ω be a bounded pseudoconvex domain in $\mathbb{C}^n$, and let $f \in H^\infty(\Omega)^n$. For each point $w \notin \cap(\overline{f(U \cap \Omega)}; \ U \supset \partial\Omega$ open), we have the formula*

$$\mathrm{ind}(f - w, L_a^2(\Omega)) = \sum_{z \in f^{-1}(w)} \nu_z(f). \tag{10.3.8}$$

Proof   By assumption there exists an open neighbourhood U of $\partial\Omega$ with $w \notin \overline{f(U \cap \Omega)}$. In particular, the set $f^{-1}(w)$ if finite.

Let $\mathcal{F}$ denote the sheaf model of the quasi-coherent $\mathcal{O}(\mathbb{C}^n)$-module $L_a^2(\Omega)$. Choose a relatively compact, Stein open subset Ω_0 of Ω with $\Omega \cap (\mathbb{C}^n \setminus U) \subset \Omega_0$. Note that the complex $K^\bullet(f - w, \mathcal{F}(W))$ is exact for each Stein open set W in $\mathbb{C}^n$ with $w \notin \overline{f(W \cap \Omega)}$. This follows from Theorem 8.2.1 and the fact that the sheaf model of the space $L_a^2(W \cap \Omega)$ coincides on W with $\mathcal{F}$.

Let V be a Stein open neighbourhood of $f^{-1}(w)$ in Ω which is disjoint from U. By using a Čech complex with coefficients in $\mathcal{F}$ one can show that the morphism of complexes

$$K_\bullet(f - w, L_a^2(\Omega)) \to K_\bullet(f - w, \mathcal{O}(V))$$

induced by the restriction map $\mathcal{F}(\mathbb{C}^n) = L_a^2(\Omega) \to \mathcal{F}(V) = \mathcal{O}(V)$ is a quasi-isomorphism. Since we may replace Ω in the above considerations by Ω_0, and since $\sigma(M_{\Omega_0}, L_a^2(\Omega_0)) = \overline{\Omega}_0 \subset \Omega$, Theorem 10.3.13 implies the assertion

$$\mathrm{ind}(f - w, L_a^2(\Omega)) = \mathrm{ind}\big(f(M_{\Omega_0}) - w, L_a^2(\Omega_0)\big) = \sum_{z \in f^{-1}(w)} \nu_z(f).$$

Here we have used the fact that $\mathrm{ind}(z - M_{\Omega_0}) = 1$ for each point $z \in \Omega_0$ (see Theorem 8.1.1). □

The index formula (10.3.7) admits several extensions. To conclude this section we briefly indicate one of these formulas.

Let H be a separable complex Hilbert space, and let $T = (T_1, \ldots, T_n)$ denote an *essentially normal n-tuple* of continuous linear operators on H (that is, all commutators $[T_i, T_j]$, $[T_i, T_j^*]$ $(1 \leq i, j \leq n)$ are compact). The essential spectrum $\sigma_e(T)$ of T can be defined as the joint spectrum of the multiplication tuple induced by T on the Calkin algebra $L(H)/K(H)$. For a point $w \in \mathbb{C}^n \setminus \sigma_e(T)$, the index of $w - T$ is by definition the index of the essential Fredholm complex $K^\bullet(w - T, H)$ (see Section 10.2). As above we denote by $\mathscr{C}$ the collection of all bounded connected components of $\mathbb{C}^n \setminus \sigma_e(T)$. It is clear that a continuous functional calculus modulo the compact operators exists, that is, $f(T)$ (mod $K(H)$) makes sense for each continuous function f on $\sigma_e(T)$. In fact, the C^*-algebra generated in $L(H)/K(H)$ by $[T_1], \ldots, [T_n]$ is isomorphic to $C(\sigma_e(T))$. Thus with each m-tuple of continuous functions $g \in C(\sigma_e(T))^m$ we associate (non-uniquely) an essentially normal m-tuple $g(T)$. The essential spectrum of $g(T)$ satisfies

$$\sigma_e(g(T)) = g(\sigma_e(T)),$$

and the index of $g(T)$, whenever it exists, is independent of the chosen representatives (for details see Douglas 1980).

Theorem 10.3.15 *Let T be an essentially normal n-tuple of Hilbert space operators, and let $g: \sigma_e(T) \to \mathbb{C}^n$ be a continuous map. Then for each point $w \notin g(\sigma_e(T))$, we have the formula*

$$\mathrm{ind}(w - g(T)) = \sum_{C \in \mathscr{C}} \deg(g, C, w)\mathrm{ind}_C(T). \tag{10.3.9}$$

Proof For simplicity, we write $\sigma = \sigma_e(T)$. The essentially normal n-tuple T determines a class in the extension group $\mathrm{Ext}_1(\sigma)$ (see Douglas 1980) and, via the Chern character ch: $\mathrm{Ext}_1(\sigma) \to H_{\mathrm{odd}}(\sigma, \mathbb{Q})$, a class $\tau \in H_{\mathrm{odd}}(\sigma, \mathbb{Q})$. For a matrix of continuous functions $A \in GL(k, C(\sigma))$, we have the following index formula:

$$\mathrm{ind}\, A(T) = \langle \mathrm{ch}[A], \tau \rangle$$

(see Douglas 1980). Above $[A]$ denotes the class of A in $K^1(\sigma)$, and $\langle \ , \ \rangle$ is the pairing between cohomology and homology.

In the case of the n-tuple $w - g(T)$, we have to compute the index of the Fredholm essential complex $K^\bullet(w - g(T), H)$, or equivalently, of the matrix $d(w - g(T)) + d(w - g(T))^*$, where

$$d(w - g(T)): K^{\mathrm{even}}(w - g(T), H) \to K^{\mathrm{odd}}(w - g(T), H)$$

is the coboundary operator.

On the other hand, the corresponding matrix $d(z) + d(z)^*$ associated with the Koszul complex $K^\bullet(z, \mathbb{C})$ $(z \in S^{2n-1})$ generates the cyclic group $K^1(S^{2n-1})$. Here S^{2n-1} is the $(2n-1)$-sphere centred at $0 \in \mathbb{C}^n$. Since $H^{\mathrm{odd}}(S^{2n-1}, \mathbb{Q}) \cong H^{2n-1}(S^{2n-1}, \mathbb{Q}) \cong \mathbb{Q}$, we obtain

$$\mathrm{ch}[d(w - g(T)) + d(w - g(T))^*] = \mathrm{ch}^{2n-1}[d(w - g(T)) + d(w - g(T))^*]$$

$$\in H^{2n-1}(\sigma, \mathbb{Q}).$$

By Alexander duality $H^{2n-1}(\sigma, \mathbb{Q}) \overset{D}{\cong} \tilde{H}_0(\mathbb{C}^n \setminus \sigma, \mathbb{Q}) = \oplus_{C \in \mathscr{C}} \mathbb{Q}[C]$, and via this identification we have

$$D\,\mathrm{ch}^{2n-1}[d(h) + d(h)^*] = \gamma \sum_{C \in \mathscr{C}} \oplus \deg(h, C, 0)[C],$$

where $h \colon \sigma \to S^{2n-1}$ is a continuous map and $\gamma \in \mathbb{Q}$ is a universal constant (see for instance Atiyah and Singer 1968).

Summing up we have proved the formula

$$\mathrm{ind}(w - g(T)) = \sum_{C \in \mathscr{C}} \deg(g, C, w) a_C(T), \qquad (10.3.10)$$

where $a_C(T) \in \mathbb{Q}$ are some coefficients still to be determined.

Let $C \in \mathscr{C}$ be a fixed bounded component of $\mathbb{C}^n \setminus \sigma$, and let $w \in C$ be a point. For the identity function $g(z) = z$, we have

$$\deg(g, C', w) = \begin{cases} 0, & C' \neq C \\ 1, & C' = C. \end{cases}$$

According to formula (10.3.10) we obtain

$$\mathrm{ind}_C(T) = \mathrm{ind}(w - T) = a_C(T).$$

Thus formula (10.3.9) is proved. $\qquad\qquad\qquad\qquad\qquad\qquad\qquad\qquad\square$

A more elementary proof of Theorem 10.3.15 is possible, via a few homotopy properties of the topological degree and some general properties of the index of essentially commuting systems.

10.4 RIEMANN–ROCH THEOREM ON COMPLEX SPACES

The purpose of this section is to give an outline of a proof of the following theorem.

Theorem 10.4.1 *Let $(X, \mathscr{O}_X)$ be a complex space that is countable at infinity. Then there is a $\mathbb{Z}$-linear morphism*

$$\alpha_X \colon K_0^a(X) \to \mathscr{K}_0^t(X),$$

with the following properties:

(a) *If X is a smooth compact manifold and E is a locally free $\mathscr{O}_X$-module of finite rank, then $\alpha_X([E]) = [E]$ via the canonical isomorphism $\mathscr{K}_0^t(X) \cong K_t^0(X)$.*

(b) *For a proper morphism of complex spaces* $f: (X, \mathscr{O}_X) \to (Y, \mathscr{O}_Y)$, *the diagram*

$$\begin{array}{ccc} K_0^a(X) & \xrightarrow{f_*} & K_0^a(Y) \\ \alpha_X \downarrow & & \downarrow \alpha_Y \\ \mathscr{K}_0^t(X) & \xrightarrow{f_*} & \mathscr{K}_0^t(Y) \end{array}$$

is commutative. □

This theorem was proved in Levy (1987a), and it contains as particular cases the *Hirzebruch–Riemann–Roch theorem* for vector bundles on compact manifolds, the *Grothendieck–Riemann–Roch theorem* for algebraic coherent sheaves, and some more recent Riemann–Roch theorems for analytic coherent sheaves. In Appendix 3 we recall the precise statements of these theorems, the necessary terminology, and the relation between them.

For us the proof of Theorem 10.4.1 constitutes the principal motivation for writing this section. We do not discuss any of the numerous applications of this result (for some of them see Fulton and Lang 1985 and the references there). We shall constantly make use in what follows of the notion of Fredholm complex and its index. Along with some standard techniques of K-theory and complex analytic geometry this will be the main tool in proving Theorem 10.4.1.

The proof of Theorem 10.4.1 is divided into several steps, corresponding to different degrees of generality for the base space X.

Step 1 X is a Stein space of finite embedding dimension.

In this case we embed X into a numerical space $\mathbb{C}^n$, and we pass via Alexander duality to the usual K-theory groups

$$D: \mathscr{K}_0^t(X) \xrightarrow{\sim} \mathscr{K}_t^0(\mathbb{C}^n, \mathbb{C}^n \setminus X)$$

(see Appendix 3). The action of α_X, modulo the duality operator D, can be described as follows.

Let $\mathscr{F}$ be a coherent $\mathscr{O}_X$-module. By extending $\mathscr{F}$ trivially to the whole space $\mathbb{C}^n$, we obtain a coherent sheaf, still denoted by $\mathscr{F}$. By Corollory 4.2.5, the augmented Koszul complex

$$K_\bullet\big(z - w, \mathscr{O}_{\mathbb{C}^n} \hat{\otimes} \mathscr{F}(X)\big) \to \mathscr{F} \to 0 \tag{10.4.1}$$

is exact in the category of $\mathscr{O}_{\mathbb{C}^n}$-modules. Moreover, the proof of Proposition 10.1.3 shows that $K_\bullet(z - w, \mathscr{O}_{\mathbb{C}^n} \hat{\otimes} \mathscr{F}(X))$ is a Fredholm complex of Fréchet-free $\mathscr{O}_{\mathbb{C}^n}$-modules. We define

$$\alpha_X(\mathscr{F}) = \big[K_\bullet\big(z - w, \mathscr{O}_{\mathbb{C}^n} \hat{\otimes} \mathscr{F}(X)\big)\big] \in \mathscr{K}_t^0(\mathbb{C}^n, \mathbb{C}^n \setminus X).$$

The additivity of the map α_X follows from Proposition 10.1.8.

Part (b) of Theorem 10.4.1 will be proved by writing the morphism $f: X \to Y$ as the composite of the graph embedding $i: X \to \mathbb{C}^n \times Y$ and the projection $p: \mathbb{C}^n \times Y \to Y$.

At the level of analytic K-theory groups we have the relation

$$i_*[\mathcal{F}] = \sum_{q=0}^{\infty} (-1)^q [R^q i_*(\mathcal{F})] = [i_*\mathcal{F}],$$

and analogously the Gysin map $i_*\colon \mathcal{K}_0^t(X) \to \mathcal{K}_0^t(\mathbb{C}^n \times Y)$ is simply the extension by zero. Thus one deduces property (b) for the map i.

For the case of the projection p one can embed $Y \subset \mathbb{C}^m$ as above and consider the extended map $p\colon \mathbb{C}^n \times \mathbb{C}^m \to \mathbb{C}^m$. Let $i_*\mathcal{F}$ be the coherent $\mathcal{O}_{\mathbb{C}^n \times \mathbb{C}^m}$-module obtained from the initial sheaf $\mathcal{F}$. Denoting by z', w' the coordinates on $\mathbb{C}^m$, the associated class of $i_*\mathcal{F}$ is

$$\alpha_{\mathbb{C}^n \times \mathbb{C}^m}(i_*\mathcal{F}) = \left[K_\bullet\big((z-w, z'-w'), \mathcal{O}_{\mathbb{C}^n \times \mathbb{C}^m} \,\hat{\otimes}\, \mathcal{F}(X)\big) \right]$$
$$= \left[K_\bullet\big(z-w, \mathcal{O}_{\mathbb{C}^n} \,\hat{\otimes}\, K_\bullet(z'-w', \mathcal{O}_{\mathbb{C}^m} \,\hat{\otimes}\, \mathcal{F}(X))\big) \right],$$

while on Y we have

$$\alpha_Y(f_*\mathcal{F}) = \left[K_\bullet\big(z'-w', \mathcal{O}_{\mathbb{C}^m} \,\hat{\otimes}\, \mathcal{F}(X)\big) \right].$$

But at the topological level, the Gysin morphism p_* deletes the Thom class represented in our case by the Koszul complex $K_\bullet(z-w, \mathcal{O}_{\mathbb{C}^n} \,\hat{\otimes}\, \cdot)$ (see Appendix 3).

Notice that $\mathcal{F}(X)$ must be interpreted as $(f_*\mathcal{F})(Y)$ in the second equality and that $f_*\mathcal{F}$ is a coherent $\mathcal{O}_Y$-module in virtue of Grauert's coherence theorem. This observation finishes the first step of the proof of Theorem 10.4.1.

Step 2 X is a locally closed subspace of $\mathbb{C}^n$.

This means that X is a closed analytic subspace of an open subset $U \subset \mathbb{C}^n$. This case is similar to the Stein space case. Namely, let $\mathcal{U}$ be a locally finite Stein open covering of U. Then the double complex $K_\bullet(z-w, \mathcal{O}_U \,\hat{\otimes}\, \mathscr{C}^\bullet(\mathcal{U}, \mathcal{F}))$ associated with a coherent $\mathcal{O}_x$-module $\mathcal{F}$ is quasi-isomorphic on U to the trivial extension of $\mathcal{F}$ by zero. According to Proposition 10.1.3, applied to a local quasi-ismorphism of the later complex to a finite-type free resolution of $\mathcal{F}$, we infer that the simple complex $\mathrm{Tot}\, K_\bullet(z-w, \mathcal{O}_U \,\hat{\otimes}\, \mathscr{C}^\bullet(\mathcal{U}, \mathcal{F}))$ is a Fredholm complex of Fréchet-free $\mathcal{O}_X$-modules. As before we define

$$\alpha_X(\mathcal{F}) = \left[\mathrm{Tot}\, K_\bullet\big(z-w, \mathcal{O}_U \,\hat{\otimes}\, \mathscr{C}^\bullet(\mathcal{U}, \mathcal{F})\big) \right] \in \mathcal{K}_t^0(U, U \backslash X).$$

In view of Proposition 10.1.8(a) this definition does not depend on the covering $\mathcal{U}$.

The proof of the functoriality of α_X under direct image morphisms is analogous to the proof given in Step 1.

The remaining part of the proof is a repetition of the arguments given in Steps 1 and 2. However, in the case of an arbitrary complex space X, two additional difficulties arise. First, X is not necessarily embeddable in a complex manifold, and secondly, if one embeds X locally into Euclidean spaces, then it is necessary to compare and to glue together the complexes which locally represent the images of the map α_X. The first technical obstacle will be overcome by using an almost complex embedding of X into a suitable differentiable manifold (this is the main original contribution of Levy 1987a). The second patching procedure is already standard, namely it uses the notion of simplicial systems of sheaves introduced in complex analytic geometry by Forster and Knorr (1971, 1972).

Step 3   Almost complex embeddings of complex spaces.

A complex space $(X, \mathscr{O}_X)$ has local models of the form

$$(\mathrm{supp}(\mathscr{O}_U/\mathscr{I}), \mathscr{O}_U/\mathscr{I}),$$

where $U \subset \mathbb{C}^n$ is an open subset and $\mathscr{I} \subset \mathscr{O}_U$ is a coherent sheaf of ideals. Of course, the local chart U is determined only up to an analytic automorphism $h: U \to U$ which leaves invariant the ideal $\mathscr{I}: h^*\mathscr{I} = \mathscr{I}$. Thus one can unambiguously speak of an *atlas* of X, corresponding to a locally finite covering (by local models).

Henceforth we shall suppose that the space X can be covered by finitely many local models. If not, then one can use an exhaustion by such spaces and pass to the limit in the corresponding K-groups. Let $\mathscr{W} = (W_i)_{i \in I}$ be such a finite atlas with $W_i = \mathrm{supp}(\mathscr{O}_{U_i}/\mathscr{I}_i)$ and $U_i \subset \mathbb{C}^n$ for $i \in I$.

Definition 10.4.2   An *almost complex embedding* of $(X, \mathscr{O}_X)$ into the differentiable manifold M is a closed embedding $X \overset{\rho}{\hookrightarrow} M$ at the level of topological spaces together with a set of diffeomorphisms $\varphi_i: U_i \to U_i' (i \in I)$, where U_i' are open subsets of M, with the properties:

(i) $\varphi_i | W_i = \rho | W_i$ $(i \in I)$;

(ii) the connecting maps $\varphi_{ij} = \varphi_i^{-1} \circ \varphi_j: \varphi_j^{-1}(U_i' \cap U_j') \to \varphi_i^{-1}(U_i' \cap U_j')$ are *admissible*, that is,

$$\varphi_{ij}^*\left(\mathscr{O}_{U_j} + \mathscr{I}_j\mathscr{E}_{U_j}\right)_{ij} = \left(\mathscr{O}_{U_i} + \mathscr{I}_i\mathscr{E}_{U_i}\right)_{ij}, \tag{10.4.2}$$

$$\varphi_{ij}^*\left(\mathscr{I}_j\mathscr{E}_{U_j}\right)_{ij} = \left(\mathscr{I}_i\mathscr{E}_{U_i}\right)_{ij}, \tag{10.4.3}$$

and the diagram

$$\left(\frac{\mathscr{O}_{U_j} + \mathscr{I}_j\mathscr{E}_{U_j}}{\mathscr{I}_j\mathscr{E}_{U_j}}\right)_{ij} \xrightarrow{\quad \varphi_{ij}^* \quad} \left(\frac{\mathscr{O}_{U_i} + \mathscr{I}_i\mathscr{E}_{U_i}}{\mathscr{I}_i\mathscr{E}_{U_i}}\right)_{ij}$$

$$\Big\uparrow{\scriptstyle\sim} \qquad\qquad\qquad\qquad \Big\uparrow{\scriptstyle\sim} \qquad\qquad (10.4.4)$$

$$(\mathscr{O}_{U_j}/\mathscr{I}_j)_{ij} \qquad\qquad\qquad\qquad (\mathscr{O}_{U_i}/\mathscr{I}_i)_{ij}$$

$$\underset{\sim}{\searrow} \qquad\qquad \underset{\sim}{\swarrow}$$

$$(\mathscr{O}_X)_{ij}$$

commutes for any pair of indices $i, j \in I$.

The index ij means that restrictions to $\varphi_j^{-1}(U_i' \cap U_j')$ or $\varphi_i^{-1}(U_i \cap U_j)$ have to be considered. The vertical maps are isomorphisms by the Theorem of Malgrange (Theorem 7.2.1).

In fact, in an almost complex embedding as above the differentiable manifold M carries globally defined sheaves $\mathscr{A}$ and $\mathscr{T}_X$ with the properties

$$\mathscr{A}|_{U_i'} = (\varphi_i)_* (\mathscr{O}_{U_i} + \mathscr{T}_i \mathscr{E}_{U_i}) \subset \mathscr{E}_{U_i'},$$

and

$$\mathscr{T}_X|_{U_i'} = (\varphi_i)_* (\mathscr{T}_i \mathscr{E}_{U_i}) \subset \mathscr{E}_{U_i'}.$$

In particular,

$$\mathscr{O}_X = \mathscr{A}/\mathscr{T}_X|_X.$$

Moreover, at a point $x \in X$, the connecting map φ_{ij} satisfies $\bar{\partial}_{\varphi_{ij}}(x) = 0$ as a consequence of the commutativity of the diagram (10.4.4). Thus the complex structure of the tangent spaces $T_x U_i$ is canonically identified with a complex structure on $T_{\varphi_i(x)} U_i'$. Thus the tangent bundle $TM|_X$ becomes a complex vector bundle. Since the complex structure on TM extends to a neighbourhood of X in M, we may suppose that M is indeed an almost complex manifold.

As usual, two almost complex embeddings of X with the same underlying topological support, but with different atlases (W, φ) and (W', φ'), are said to be *equivalent* if $(W \cup W', \varphi \cup \varphi')$ is still an admissible atlas.

The key observation which makes the above notion of almost complex embeddings flexible enough under natural transformations is contained in the next lemma.

Lemma 10.4.3   *Let $(U, \mathscr{O}_U/\mathscr{T})$ be the local model of a complex space X, and let $h: U \to U$ be an admissible diffeomorphism of U. Then h^{-1} is also admissible.*

Proof   Suppose that $U \subset \mathbb{C}^n$ and $\mathscr{T} = (z_1, \ldots, z_k)\mathscr{O}_U$ is the ideal of a linear subspace of U. Then the relations (10.4.2), (10.4.3), and (10.4.4) imply that

$$h_i(z) = \sum_{j=1}^{k} z_j a_{ij}(z) \qquad (1 \le i \le k)$$

and

$$h_i(z) = z_i + \sum_{j=1}^{k} z_j b_{ij}(z) \qquad (k < i \le n).$$

Here $a_{ij}(z)$ and $b_{ij}(z)$ are matrices of smooth functions, and $\det(a_{ij}(z)) \ne 0$ on $\mathrm{supp}(\mathscr{O}_U/\mathscr{T})$. By solving this system with respect to z_i, one easily obtains the conclusion that h^{-1} is admissible, too.

In general, let $\mathscr{T} = (f_1, \ldots, f_k)\mathscr{O}_U$ with generators $f_1, \ldots, f_k \in \mathscr{O}(U)$. Then h necessarily has the form

$$h_i(z) - z_i = \sum_{j=1}^{k} f_j(z) a_{ij}(z) \qquad (1 \le i \le n),$$

where $a_{ij} \in \mathscr{E}(U)$ for $1 \le i \le n$ and $1 \le j \le k$.

Next we repeat the familiar desingularization device of any preparation lemma. Namely, let $B \subset \mathbb{C}^k$ be a sufficiently small ball around $0 \in \mathbb{C}^k$, and let us write the elements of $U' = U \times B$ in the form (z, λ) with $z \in U$ and $\lambda \in B$. Then the ideal $\mathcal{F}' = (f_1(z) - \lambda_1, \ldots, f_k(z) - \lambda_k)\mathcal{O}_{U'}$ defines a smooth analytic submanifold of U'. The diffeomorphism $h': U' \to U'$ defined (for B small) by the expressions:

$$h_i'(z, \lambda) = z_i + \sum_{j=1}^{k} \left(f_j(z) - \lambda_j \right) a_{ij}(z) \qquad (1 \le i \le n),$$

$$h_i'(z, \lambda) = \lambda_{i-n} \qquad (n < i \le n + m)$$

is admissible. According to the first part of the proof, h'^{-1} is also admissible. But $h^{-1} = h'^{-1}|_{U \times \{0\}}$, and we obtain that h^{-1} is an admissible transformation. □

A *morphism* between two given almost complex embeddings $\rho: X \hookrightarrow M$ and $\rho': X \hookrightarrow M'$ is a differentiable map $h: M \to M'$ with $\rho' = h\rho$ which is admissible with respect to suitable atlases.

Such a morphism preserves the corresponding sheaves $\mathcal{A}$ and $\mathcal{F}_X$.

It is straightforward to prove the existence of a morphism between two arbitrary embeddings $\rho: X \hookrightarrow M$ and $\rho': X \hookrightarrow M'$ (see Levy 1987a).

A *modification* of an almost complex embedding $\rho: X \hookrightarrow M$ consists of a closed embedding of smooth manifolds $i: M \hookrightarrow M'$ such that there exists an atlas X whose local models corresponding to ρ and $i \circ \rho$ have the property that the commutative diagram

$$
\begin{array}{ccc}
U_j & \overset{\alpha_j}{\hookrightarrow} & U_j' \\
\phi_j \downarrow & & \downarrow \phi_j' \\
M & \overset{i}{\hookrightarrow} & M'
\end{array}
$$

defines an analytic admissible embedding $\alpha_j: U_j \to U_j'$.

Note that the conormal bundle TM'/TM of such a modification defines a complex vector bundle on X. Conversely, every complex vector bundle on X arises as the conormal bundle of a modification of a fixed almost complex embedding $\rho: X \hookrightarrow M$. Moreover, two modifications of ρ are isomorphic if and only if the corresponding conormal bundles are isomorphic. We omit the details of these two facts, which are based on Lemma 10.4.3 and on some standard geometric constructions for vector bundles (see Levy 1987a).

Now we are ready to state the main result of the missing part of the proof of Theorem 10.4.1.

Theorem 10.4.4 (Levy 1987a)   *Let X be a complex space. Then there exists an almost complex embedding of X into a Euclidean space $\mathbb{R}^{2n}$.*

Moreover, any two almost complex embeddings of X are isomorphic modulo suitable modifications.

Proof Let $\rho\colon X \to M$ be an almost complex embedding. Let E be a complement of TM, in the category of complex vector bundles on M, i.e. $TM \oplus E = n \cdot 1$. According to the preceding remarks, there exists a modification $i\colon M \hookrightarrow M'$ of the embedding ρ with E as conormal bundle. By embedding M' into $\mathbb{R}^{2n}$ one obtains in neighbourhoods of $(i \cdot \rho)(X)$ an embedding $X \subset \mathbb{R}^{2n}$.

Before discussing the existence of an almost complex embedding, let us prove the uniqueness assertion. Let $\rho\colon X \to M$ and $\rho'\colon X \to M'$ denote two almost complex embeddings. Let $\mathscr{W} = (W_i)_{i \in I}$ be a covering of X, and let $(U_i, \phi_i)_{i \in I}$, $(U_i', \phi_i')_{i \in I}$ be associated local models on M and M' as in Definition 10.4.2.

Let $h\colon M \to M'$ be a morphism of embeddings which makes the following diagram commutative:

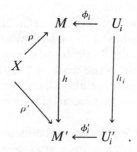

We shall prove that $M \times M'$ is a common modification of M and M'. Let $e\colon M \to M \times M'$ be the embedding $e(x) = (x, h(x))$ ($x \in M$). Analogously one can consider the local embeddings $e_i\colon U_i \to U_i \times U_i'$, $e_i(\lambda) = (\lambda, k_i(\lambda))$, where $k_i\colon U_i \to U_i'$ is a holomorphic map which restricts to the identity on $(W_i, \mathscr{O}_{W_i})$ ($i \in I$). Since h_i is admissible, one observes that $h_i - k_i \in \mathscr{T}_i \mathscr{E}_{V_i}$.

Next one defines local diffeomorphisms $\psi_i\colon U_i \times U_i' \to M \times M'$ by

$$\psi_i(\lambda, \mu) = (\phi_i(\lambda), \phi_i'(\mu + h_i(\lambda) - k_i(\lambda))) \qquad (\lambda \in U_i, \mu \in U_i').$$

Then the diagrams

$$
\begin{array}{ccc}
M & \xleftarrow{\phi_i} & U_i \\
\downarrow{\scriptstyle e} & & \downarrow{\scriptstyle e_i} \\
M \times M & \xleftarrow{\psi_i} & U_i \times U_i'
\end{array}
$$

are commutative. Hence e is a modification of the embedding ρ. Note that the conormal bundle $T(M \times M')/TM$ is isomorphic to TM'.

Similarly one can modify ρ' to $M \times M'$, via the graph $e'\colon M' \to M \times M'$ of a morphism $h': M' \to M$. Because the almost complex embeddings $e' \circ \rho'$ and $e \circ \rho$ have the same topological support and complex tangent bundle on X, they are isomorphic. This completes the proof of the second assertion of Theorem 10.4.4.

In order to prove that there exists an almost complex embedding of X, it suffices to prove, by a familiar Mayer–Vietoris argument, that two local almost complex embeddings $\rho: W \to M$ and $\rho: W' \to M'$ can be patched together to an embedding of $W \cup W' \subset X$.

First, in view of the previous part of the proof, there exists a common modification of the embeddings $W \cap W' \hookrightarrow M$ and $W \cap W' \hookrightarrow M'$. Then one can modify ρ and ρ' in such a way that the new embeddings $\tilde{\rho}: W \to \tilde{M}$ and $\tilde{\rho}': W' \to \tilde{M}'$ are isomorphic on $W \cap W'$. This of course defines in neighbourhoods of $W \cup W'$ an almost complex manifold $\tilde{M} \cup_X \tilde{M}'$, where X is the admissible isomorphism of the two charts of $W \cap W'$. Thus the proof of Theorem 10.4.4 is complete.   □

Step 4 Simplicial systems of sheaves.

This is the last step of the proof of Theorem 10.4.1. Although this step is again rather laborious, it is based on a well-established method in modern analytic geometry. Namely, the basic tool here is the concept of simplicial systems of sheaves originating in the work of Forster and Knorr (1972). For a detailed account of this subject, see also Grauert and Remmert (1984). In the following we only discuss the particular situation imposed by the proof of Theorem 10.4.1.

Let $(X, \mathcal{O}_X)$ be a complex space embedded into the almost complex manifold $M = \mathbb{R}^{2n}$. We consider a finite atlas $\mathcal{W} = (W_i)_{i \in I}$ on X and associated local models $W_i = \mathrm{supp}(\mathcal{O}_{U_i}/\mathcal{T}_i)(U_i \subset \mathbb{C}^n)$ together with the diffeomorphisms $\varphi_i: U_i \to \varphi_i(U_i) \subset M(i \in I)$.

The subsets of I are denoted by $\alpha, \beta, \ldots$. We set $W_\alpha = \bigcap_{i \in \alpha} W_i$ for $\alpha \subset I$. Let us fix coordinates z^α on every local model U_α of W_α, which of course is contained in U_i for each $i \in \alpha$.

Let $\mathcal{F}$ be a coherent $\mathcal{O}_X$-module. On every local model we have a standard resolution

$$K_\bullet\left(z^\alpha - w^\alpha, \mathcal{O}_{U_\alpha} \hat{\otimes} \mathcal{F}(W_\alpha)\right) \to \mathcal{F}|_{W_\alpha} \to 0.$$

By Theorem 7.2.1 (Theorem of Malgrange), we know that the complex

$$K_\bullet\left(z^\alpha - w^\alpha, \mathcal{E}_{U_\alpha} \hat{\otimes} \mathcal{F}(W_\alpha)\right) \to \mathcal{E}_{U_\alpha} \otimes_{\mathcal{O}_{U_\alpha}} \mathcal{F}|_{W_\alpha} \to 0$$

is still exact and that $\mathcal{E} \otimes_{\mathcal{O}} \mathcal{F}$ is a sheaf of Fréchet spaces on U_α.

While the first Koszul complex makes sense only on U_α, the second one can be transferred to the manifold M via the map φ_α. Moreover, we may suppose that the diffeomorphisms φ_i extend to suitable neighbourhoods of U_i, and $\varphi_i(U_i)$, so that there are smooth maps

$$\psi_i: M \to \mathbb{C}^n, \quad \psi_i|_{\varphi_i(U_i)} = \varphi_i^{-1} \qquad (i \in I).$$

Thus the sheaf $\mathcal{E} \otimes_{\mathcal{O}} \mathcal{F}|_{U_i}$ possesses a global resolution

$$K_\bullet\left(z^\alpha \circ \psi_\alpha - w^\alpha, \mathcal{E}_M \,\hat{\otimes}\, \mathcal{F}(W_\alpha)\right) \to (\varphi_\alpha)_* (\mathcal{E} \otimes_{\mathcal{O}} \mathcal{F}|_{W_\alpha}) \to 0.$$

The restriction maps and changes of coordinates from U_α to U_β $(\alpha \supset \beta)$ define a simplicial system of complexes whose cochain complex is denoted in short by $\mathcal{E}\mathbb{K}_\bullet(\mathcal{W}, \mathcal{F})$.

Analogously, the complexes $K_\bullet(z^\alpha - w^\alpha, \mathcal{O}_{U_\alpha} \,\hat{\otimes}\, \mathcal{F}(W_\alpha))$ fit into a simplicial system which produces the complex $\mathbb{K}_\bullet(\mathcal{W}, \mathcal{F})$. Indeed, $\mathcal{F}$ can be regarded as a sheaf of $\mathcal{A}$-modules on M (Definition 10.4.2 and the subsequent remarks) and

$$\mathcal{E}\mathbb{K}_\bullet(\mathcal{W}, \mathcal{F}) \cong \mathcal{E}_M \otimes_{\mathcal{A}} \mathbb{K}_\bullet(\mathcal{W}, \mathcal{F}).$$

According to Theorem 7.2.1, the sheaf $\mathcal{E}_M \otimes_{\mathcal{A}} \mathcal{F}$ is a globally defined Fréchet $\mathcal{E}_M$-module which locally coincides with $(\varphi_\alpha)_* (\mathcal{E}_{U_\alpha} \otimes_{\mathcal{O}_{U_\alpha}} \mathcal{F}|_{W_\alpha})$.

Lemma 10.4.5   (a) *With the above notations, $\mathbb{K}_\bullet(\mathcal{W}, \mathcal{F})$ is a Fredholm complex of $\mathcal{E}_M$-modules on M.*

(b) *The class $[\mathbb{K}_\bullet(\mathcal{W}, \mathcal{F})] \in \mathcal{K}^0(M)$ depends on neither $\mathcal{W}$ nor the local coordinates.*

(c) *For each exact sequence of coherent $\mathcal{O}_X$-modules $0 \to \mathcal{F} \to \mathcal{G} \to \mathcal{H} \to 0$,*

$$[\mathbb{K}_\bullet(\mathcal{W}, \mathcal{G})] = [\mathbb{K}_\bullet(\mathcal{W}, \mathcal{F})] + [\mathbb{K}_\bullet(\mathcal{W}, \mathcal{H})].$$

Proof   (a) We have already remarked that the complex $\mathcal{E}\mathbb{K}_\bullet(\mathcal{W}, \mathcal{F})$ is quasi-isomorphic to $\mathcal{E} \otimes_{\mathcal{A}} \mathcal{F}$. By Corollary 10.1.2 the complex $\mathbb{K}_\bullet(\mathcal{W}, \mathcal{F})$ is $\mathcal{E}$-Fredholm. Moreover, this complex is exact off X.

(b) This assertion follows from the observation that the complexes $\mathbb{K}_\bullet(\mathcal{W}, \mathcal{F})$ and $\mathbb{K}_\bullet(\mathcal{W}', \mathcal{F})$ are quasi-isomorphic for a suitable refinement $\mathcal{W}'$ of the covering $\mathcal{W}$ and after a change of coordinates (see Forster and Knorr 1972).

(c) This assertion is a consequence of Proposition 10.1.8(b). $\square$

According to the last lemma one can unambiguously define the topological class of a coherent sheaf $\mathcal{F}$ by

$$\alpha_X(\mathcal{F}) = [\mathbb{K}_\bullet(\mathcal{W}, \mathcal{F})] \in \mathcal{K}_t^0(M, M \setminus X).$$

Recall that we can choose $M = \mathbb{R}^{2n}$. In this case, we can use Alexander duality to identify the spaces $\mathcal{K}_t^0(M, M \setminus X) \cong \mathcal{K}_0^t(X)$.

For the moment we prefer to work with an arbitrary almost complex embedding $X \subset M$, mainly because of the next lemma.

Lemma 10.4.6   (a) *The class $\alpha_X(\mathcal{F})$ is functorial with respect to modifications of the embedding $X \subset M$.*

(b) *If X is a smooth complex space and $\mathcal{F}$ is a locally free $\mathcal{O}_X$-module, then $\alpha_X(\mathcal{F}) = [\mathcal{F}]$, where the latter is the class of the corresponding vector bundle in $\mathcal{K}^0(X)$.*

Proof   (a) The precise meaning of the above assertion is the following. Let the map $e: M \to M'$ be a modification of the embedding $\rho: X \hookrightarrow M$. Then M'

can be identified with the total space of the conormal vector bundle TM'/TM so that there exists a Thom morphism

$$\mathcal{K}^0(M, M\setminus X) \xrightarrow{\;e_*\;} \mathcal{K}^0(M', M'\setminus X).$$

Then we claim that $e_*\,\alpha_{X,M}(\mathcal{F}) = \alpha_{X,M'}(\mathcal{F})$.

This fact follows from the local description of the Thom morphism e_* as the multiplication with the Koszul complex of the fibre TM'/TM (compare with the proof of Step 1).

(b) For a smooth complex space, one can consider the trivial embedding $X \hookrightarrow X$. In this case it is immediate that the complex $\mathbb{K}_\bullet(\mathcal{W}, \mathcal{F})$ is quasi-isomorphic to $\mathcal{F}$, hence $\alpha_X(\mathcal{F}) = [\mathcal{F}]$.   □

At this point only assertion (b) of Theorem 10.4.1 remains to be proved. For this purpose, let $f\colon X \to Y$ be a proper morphism of complex spaces, and let $\mathcal{F}$ be a coherent $\mathcal{O}_X$-module.

At the beginning we suppose for simplicity that Y is a Stein space, more precisely, a closed analytic subspace of a Euclidean space $\mathbb{C}^m$.

Let $\mathcal{W} = (W_i)_{i \in I}$ denote as above a finite atlas on X with associated local models U_i and diffeomorphisms $\varphi_i\colon U_i \to \mathbb{R}^{2n}$ $(i \in I)$. Note that here $\mathbb{R}^{2n}$ carries an almost complex structure which is different from the canonical one.

According to Lemma 10.4.6, it suffices to prove that $p_*\,\alpha_{X\times Y}(\mathcal{F}) = \alpha_Y(p_*\mathcal{F})$ for the projection $p\colon X\times Y \to Y$. Of course, the sheaf $\mathcal{F}$ is supported by the graph $\{(x, f(x)); x \in X\} \subset X \times Y$, and the restriction of p to $\operatorname{supp}(\mathcal{F})$ is proper.

A local system of charts for $X \times Y$ is given by the diffeomorphisms $\varphi_i \times 1\colon U_i \times \mathbb{C}^m \to \varphi_i(U_i) \times \mathbb{C}^m \subset \mathbb{R}^{2n} \times \mathbb{C}^m$. But, as in the proof of Step 1, it is obvious that the representing complex of $\alpha_{X\times Y}(\mathcal{F})$ is of the form

$$\mathbb{K}_\bullet(\mathcal{W}\times Y, \mathcal{F}) = \left[K_\bullet\!\left(z^\alpha - w^\alpha, \mathcal{O}_{U_\alpha}\,\hat{\otimes}\, K_\bullet\!\left(\lambda - \mu, \mathcal{O}_{\mathbb{C}^m}\,\hat{\otimes}\,\mathcal{F}(W_\alpha)\right)\right)\right]_\alpha.$$

As before, z^α, w^α denote the coordinates on U_α, while λ, μ are coordinates in $\mathbb{C}^m$.

The Gysin morphism $p_*\colon K^0(\mathbb{R}^{2n} \times \mathbb{C}^m) \to K^0(\mathbb{C}^m)$ deletes the Thom class represented locally by the multiplication with the complexes $K_\bullet(z^\alpha - w^\alpha, *)$. So the element $p_*\,\alpha_{X\times Y}(\mathcal{F})$ is represented by the class of the complex attached to the simplicial system of sheaves

$$\left[K_\bullet\!\left(\lambda - \mu, \mathcal{O}_{\mathbb{C}^m}\,\hat{\otimes}\,\mathcal{F}(W_\alpha)\right)\right]_{\alpha \subset I}.$$

On the other hand, the restriction $p_\alpha\colon W_\alpha \times \mathbb{C}^m \mapsto \mathbb{C}^m$ of the projection p is a flat morphism of complex spaces. In virtue of Theorem 4.2.4 the complex $K_\bullet(\lambda - \mu, \mathcal{O}_{\mathbb{C}^m}\,\hat{\otimes}\,\mathcal{F}(W_\alpha))$ is quasi-isomorphic to the sheaf $(p_\alpha)_*(\mathcal{F}|_{W_\alpha})$. But it is clear that the simplicial system $(p_\alpha)_*(\mathcal{F}|_{W_\alpha})$ is equivalent to $p_*\mathcal{F}$.

In conclusion we have proved that $p_*\,\alpha_{X\times Y}(\mathcal{F}) = \alpha_Y(p_*\mathcal{F})$.

For a general target space $(Y, \mathcal{O}_Y)$, one repeats essentially the same argument, using a second decomposition into simplicial systems corresponding to an almost complex embedding of Y. We leave the details to the reader (for a slightly different argument see Levy 1987a).

This finishes a general outline of the proof of Theorem 10.4.1.

10.5 REFERENCES AND COMMENTS

The concept of Fredholm complexes has appeared in several different contexts. Three illustrative sources in this direction are: Segal (1970), Hörmander (1985), and Vasilescu (1982). On the other hand, complexes of Fréchet sheaves with coherent cohomology are familiar objects in the deformation theory of analytic objects (see for instance Bănică and Stănăşilă 1977). When working with Fredholm complexes, the transition from pointwise results to local sheaf-theoretical statements and then to global results is in general difficult. Among the first who observed the significance of these problems for a general spectral theory in several variables was Taylor (1970a). In addition we would like to stress two additional delicate technical points related to Fredholm complexes. Namely, topologically free complexes with Banach or Fréchet fibres are much more difficult to handle than those with Hilbert fibres, and secondly, the singularities in the base space may raise serious new problems. The missing implications in Corollary 10.1.2 are a good example for both types of problems.

Section 10.2 follows with minor modifications the paper Putinar (1982a). In view of the counterexample given in Lemma 10.2.6, it would be interesting to find conditions under which Theorem 10.2.5 remains globally valid.

Compared with the case of a single operator, multivariable index theory has still a series of unexplored areas. One of the basic questions that is not satisfactorily solved is the structure of the analytic set $\sigma(T) \setminus \sigma_e(T)$ attached to a commutative tuple of operators. The local structure of this set determines the index theory of T. The sheaf theoretical methods (see Putinar 1985; Levy 1989) or the approach using geometric measure theory (see Carey and Pincus 1985) offer at present only a partial description of this set and of its relevance for index theory.

A series of subjects related directly to the index theory of systems of operators was not even mentioned in Section 10.3. For instance the C^*-algebraic and K-homological or cyclic cohomological approaches to index theory are very active fields of research which offer precise index formulae for many concrete pseudodifferential operators. A good reference for these areas is Kaminker (1990).

Returning to the abstract Fredholm theory we mention a couple of open questions which have been circulating for more than a decade.

(1) Describe the homotopy of the set of Fredholm commuting n-tuples of operators acting on a separable Hilbert space.

It is known that for $n = 1$ this space is a classifying space for topological K-theory. For partial results in the multidimensional case, see Curto (1981).

(2) Does a Fredholm commutative n-tuple with non-zero index have non-trivial invariant subspaces?

For a series of similar problems see Curto (1988).

The literature devoted to the Riemann–Roch theorem is vast. It was this theorem which motivated Grothendieck to develop K-theory (see Borel and Serre 1958). The incipient stages of the Atiyah–Singer index theorem and the theory of pseudodifferential operators are closely related to the Riemann–Roch theorem for vector bundles on compact complex manifolds (see Hirzebruch 1965; Atiyah and Singer 1968). More recently the Riemann–Roch formalism has found applications to complex singular spaces and arithmetic (see Fulton and Lang 1985; Faltings 1992).

The novelty of Levy's proof of the Riemann–Roch theorem for singular spaces consists in the combination of perturbation results for Fredholm complexes with standard methods from complex geometry, and secondly, in demonstrating the usefulness of the concept of almost complex embeddings of complex spaces.

Appendix 1
Locally convex spaces

Throughout this book we use freely the language and results from the theory of locally convex spaces. Most of our concrete applications are based on the observation that the spaces of smooth functions or analytic functions are in a canonical way nuclear Fréchet spaces. In the following we gather some of the results exploited in the book.

(1) By definition a *locally convex space* is a (not necessarily Hausdorff) topological vector space E over the complex numbers such that E possesses a neighbourhood base at zero consisting of convex sets. For a given locally convex space E, we denote by E' its *dual space*, i.e. the vector space consisting of all continuous linear functionals $u: E \to \mathbb{C}$.

Let E, F be locally convex Hausdorff spaces. A linear map $T: E \to F$ is called *nuclear* if there are sequences $(\lambda_n) \in l^1$, (u_n) in E', and (y_n) in F such that (u_n) is equicontinuous, (y_n) is bounded, and

$$Tx = \sum_{n=1}^{\infty} \lambda_n \langle x, u_n \rangle y_n \tag{A1.1}$$

holds for each $x \in E$. Here of course a family of linear operators $T_\alpha: E \to F$ between locally convex spaces is called *equicontinuous* if for each zero neighbourhood V in F there is a zero neighbourhood U in E with $T_\alpha U \subset V$ for all α, and a subset $B \subset E$ is called *bounded* if for each zero neighbourhood U in E there is a number $\rho > 0$ with $B \subset \rho U$.

The set $N(E, F)$ of all nuclear operators between E and F is a vector space with the property that

$$L(F, F_0) N(E, F) L(E_0, E) \subset N(E_0, F_0),$$

whenever E_0, F_0 are locally convex Hausdorff spaces.

Lemma A.1.1 *Suppose that F is sequentially complete, that $T \in N(E, F)$ is a nuclear operator of the form ($A1.1$), and that $1 \le q \le p \le \infty$. Then T admits a continuous linear factorization:*

$$
\begin{array}{ccc}
E & \xrightarrow{\ T\ } & F \\
\alpha \downarrow & & \uparrow \beta \\
l^p & \xrightarrow{\ S\ } & l^q,
\end{array}
$$

where the action of S is given by $S(x_n) = (|\lambda_n|^{(1/q) - (1/p)} x_n)$.

By applying the last result to the case $p = \infty$ and $q = 1$, one obtains that, for a sequentially complete space F, each nuclear operator $T: E \to F$ is compact.

(2) Let E be a locally convex space. If p is a continuous seminorm on E, then we denote by $\tilde{E}_p$ the Banach space arising as the norm completion of the space $E_p = E/\mathrm{Ker}(p)$ with respect to the norm $p([x]) = p(x)$. If U is an absolutely convex zero neighbourhood in E, then we denote by p_U the *Minkowski functional* of U and set $E_{(U)} = E_{p_U}$.

A locally convex Hausdorff space E is called *nuclear* if for each continuous seminorm p on E there is a continuous seminorm $q \geq p$ such that the continuous linear map $\tilde{E}_q \to \tilde{E}_p$ induced by the identity operator I_E is nuclear.

Lemma A1.2 *A locally convex Hausdorff space E is nuclear if and only if each continuous linear operator $T: E \to X$ into a Banach space X is nuclear.*

Proof Suppose that $T: E \to X$ is a continuous linear operator from a nuclear space into a Banach space. Then there are continuous seminorms $q \geq p$ on E such that $\tilde{E}_q \to \tilde{E}_p$ is nuclear and such that $\|Tx\| \leq p(x)$ for all $x \in E$. Since T factorizes through the nuclear operator $E \to \tilde{E}_p, x \mapsto [x]$, the operator T is nuclear. To prove the converse, note that if for some continuous seminorm p on E the canonical map $E \mapsto \tilde{E}_p$ has a representation of the form (A1.1) such that the linear forms $u_n: E \to \mathbb{C}$ are majorized by a continuous seminorm $q \geq p$, then automatically the canonical map $\tilde{E}_q \to \tilde{E}_p$ is nuclear.   □

Let $T: E \to F$ be a nuclear operator into a sequentially complete space F such that $TE \subset F_0$ for some closed subspace $F_0 \subset F$. If E is nuclear, then the induced operator $T_0: E \to F_0, x \mapsto Tx$, remains nuclear. This follows by combining Lemma A1.1 and Lemma A1.2.

For a nuclear space E, each continuous seminorm p on E is majorized by a continuous seminorm q such that $\tilde{E}_q$ is a separable Hilbert space. To see this, choose a continuous seminorm $r \geq p$ such that $\tilde{E}_r \to \tilde{E}_p$ is nuclear and consider a continuous linear factorization

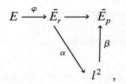

where φ is the canonical map and $\|\beta\| \leq 1$. Let B be the closed unit ball of l^2. The Minkowski functional q of $W = (\alpha \circ \varphi)^{-1}(B)$ majorizes p, and $\alpha \circ \varphi$ induces an isometry $\tilde{E}_q \to l^2$.   □

Using the last observation one can prove the following stability results for nuclear spaces.

Theorem A1.3 *Let E and E_i $(i \in I)$ be nuclear spaces.*

(a) *Each subspace (not necessarily closed) of E is nuclear.*
(b) *Each quotient E/F modulo a closed subspace F of E is nuclear.*
(c) *$\prod_{i \in I} E_i$ is nuclear.*
(d) *If I is countable, then also $\bigoplus_{i \in I} E_i$ is nuclear.*

Proof (a) Let p be a continuous seminorm on E such that $\tilde{E}_p$ is a Hilbert space. Then there is a seminorm $q \geq p$ of the same type such that the induced map $\tilde{E}_q \to \tilde{E}_p$ is nuclear. Since the horizontal maps in the commutative diagram

$$
\begin{array}{ccc}
\tilde{F}_{(p|F)} & \longrightarrow & \tilde{E}_p \\
\uparrow & & \uparrow \\
\tilde{F}_{(q|F)} & \longrightarrow & \tilde{E}_q
\end{array}
$$

are isometries, the left vertical map is nuclear.

(b) Choose seminorms p and q exactly as above. Denote by $\hat{p}$ the induced quotient seminorm on E/F. Then

$$\pi: E_p \to (E/F)_{\hat{p}}, \qquad x + \mathrm{Ker}(p) \mapsto [x] + \mathrm{Ker}(\hat{p})$$

is a continuous, surjective and open map between normed spaces. The same is true for q instead of p. Since the horizontal maps in the commutative diagram

$$
\begin{array}{ccc}
\tilde{E}_p & \xrightarrow{\ \tilde{\pi}\ } & \widetilde{(E/F)}_{\hat{p}} \\
\uparrow & & \uparrow \\
\tilde{E}_q & \xrightarrow{\ \tilde{\pi}\ } & \widetilde{(E/F)}_{\hat{q}}
\end{array}
$$

are continuous surjections between Hilbert spaces, the right vertical map is nuclear.

(c) It suffices to observe that the product topology on $E = \prod_{i \in I} E_i$ is generated by the seminorms $p((x_i)_{i \in I}) = \max_{1 \leq k \leq n} p_k(x_{i_k})$, where n is a natural number and each p_k is a continuous seminorm on E_{i_k}, and that there are canonical isometric isomorphisms

$$\prod_{k=1}^{n} \left(\widetilde{E_{i_k}}\right)_{p_k} \to \tilde{E}_p.$$

(d) Let $T: E = \bigoplus_{i \in \mathbb{N}} E_i \to X$ be a continuous linear map into a Banach space X. Then the induced operators $T_i: E_i \to X$ can be written as

$$T_i x = \sum_{n=0}^{\infty} \lambda_{in} \langle x, u_{in} \rangle y_{in} \qquad (x \in E_i)$$

with $\|y_{in}\| < 1, \sum_{i,n=0}^{\infty} |\lambda_{in}| < \infty$, and such that, for each i, there is a zero neighbourhood U_i in E_i which is mapped into the unit disc in $\mathbb{C}$ by each of the forms u_{in} ($n \in \mathbb{N}$). Let f_{in} be the trivial extension of u_{in} onto E. Choose an enumeration $\pi: \mathbb{N} \to \mathbb{N} \times \mathbb{N}$. Then

$$Tx = \sum_{n=0}^{\infty} \lambda_{\pi(n)} \langle x, f_{\pi(n)} y_{\pi(n)} \rangle \qquad (x \in E)$$

is a representation of the form (A1.1). Note that the absolutely convex hull of $\bigcup_{i \in \mathbb{N}} U_i$ is a zero neighbourhood in E, which is mapped by each of the maps $f_{\pi(n)}$ ($n \in \mathbb{N}$) into the unit disc in $\mathbb{C}$. $\qquad\square$

As a consequence of the above result, each projective limit of an inverse system of nuclear spaces is nuclear, and the inductive limit of a countable direct system of nuclear spaces is nuclear, provided it is Hausdorff.

(3) The above stability result can be used to prove that many of the natural spaces occurring in analysis are nuclear. We consider only the most important examples.

Theorem A1.4   *For each open set U in $\mathbb{R}^n$ (and in $\mathbb{C}^n$), the spaces $\mathscr{E}(U), \mathscr{D}(U)$ (and $\mathscr{O}(U)$, respectively) are nuclear.*   □

Since $\mathscr{O}(U)$ is a closed subspace of $\mathscr{E}(U)$ and since $\mathscr{D}(U)$ is a countable inductive limit of subspaces of $\mathscr{E}(U)$, it suffices to prove the nuclearity of $\mathscr{E}(U)$. We briefly indicate one possible proof of this result. Note that $\mathscr{E}(U)$ is a subspace of a product of spaces of the form

$$C^{\infty}\left(\prod_{j=1}^{n} [a_j, b_j] \right).$$

Hence one only has to prove the nuclearity of spaces of this latter type. An inductive argument allows one to reduce the assertion to the case $n = 1$. But the map

$$C^{\infty}([-1,1]) \to C^{\infty}_{\text{per, even}}(\mathbb{R}), \qquad f \mapsto f \circ \cos$$

is a topological isomorphism onto the subspace of $\mathscr{E}(\mathbb{R})$ consisting of all 2π-periodic, even C^{∞}-functions. Via Fourier transformation

$$C^{\infty}_{\text{per, even}}(\mathbb{R}) \to s, \qquad f \mapsto (\hat{f}(0), \hat{f}(1), \dots),$$

this space is topologically isomorphic to the space s of all rapidly decreasing sequences, i.e of all sequences $x = (x_n)_{n \geq 1}$ in $\mathbb{C}$ with

$$\|x\|_k = \sum_{n=1}^{\infty} n^k |x_n| < \infty \qquad (k \in \mathbb{N}).$$

Using the definition of nuclearity, it is a simple exercise to prove the nuclearity of the locally convex space s.

(4) Let E and F be locally convex spaces, and let $E \otimes F$ be their algebraic tensor product. The sets of the form $\Gamma(U \otimes V)$, where U and V run through all absolutely convex zero neighbourhoods in E and F, and Γ denotes the absolutely convex hull, form a zero neighbourhood basis for the finest locally convex topology on $E \otimes F$ for which the canonical bilinear map

$$E \times F \to E \otimes F, \qquad (x, y) \mapsto x \otimes y \tag{A1.2}$$

is continuous. By definition the π-*tensor product* $E \otimes_{\pi} F$ of E and F is the algebraic tensor product $E \otimes F$ equipped with this locally convex topology. The ε-*tensor product* $E \otimes_{\varepsilon} F$ is the algebraic tensor product $E \otimes F$ equipped with the locally convex topology defined by the seminorms:

$$p_{A,B}(z) = \sup\{|u \otimes v(z)|; \ u \in A, v \in B\}, \tag{A1.3}$$

where A and B run through all equicontinuous subsets of E' and F'. Since the map (A1.2) is continuous with $E \otimes_{\varepsilon} F$ as the space on the right, the π-tensor product topology is, in general, stronger than the ε-tensor product topology. If E and F are Hausdorff, then so are $E \otimes_{\pi} F$ and $E \otimes_{\varepsilon} F$.

Theorem A.1.5   *Suppose that E and F are locally convex Hausdorff spaces and that E is nuclear. Then $E \otimes_{\pi} F = E \otimes_{\varepsilon} F$.*

Proof   The space of continuous bilinear forms on $E \times F$ is canonically isomorphic to the space of continuous linear forms on $E \otimes_{\pi} F$. Let $M \subset (E \otimes_{\pi} F)'$ be equicontinuous. By the bipolar theorem it suffices to show that M is an equicontinuous subset of the dual space of $E \otimes_\varepsilon F$. Each $u \in M$ induces a continuous linear operator $T_u \colon E \to F'_\sigma$ into the weak dual space of F acting as

$$\langle y, T_u x \rangle = u(x, y).$$

By the π-equicontinuity of M, there are absolutely convex zero neighbourhoods U in E and V in F with

$$T_u(U) \subset V° \qquad (u \in M), \tag{A1.4}$$

where $V°$ is the set of all $x' \in F'$ with $|\langle x, x' \rangle| \leq 1$ for each $x \in V$. Since $B = V° \subset F'_\sigma$ is absolutely convex and compact, the space $F'_B = \bigcup_{n \geq 0} nB$ is a continuously embedded Banach subspace of F' with the Minkowski functional of B as norm. Because of (A1.4) the operators T_u ($u \in M$) induce continuous operators $S_u \colon \tilde{E}_{(U)} \to F'_B$, of norm less than or equal to one, such that the diagram

$$
\begin{array}{ccc}
E & \xrightarrow{T_u} & F'_\sigma \\
\varphi_U \downarrow & & \uparrow i \\
\tilde{E}_{(U)} & \xrightarrow{S_u} & F'_B
\end{array}
$$

commutes. The canonical map φ_U has a representation

$$\varphi_U(x) = \sum_{n=1}^{\infty} \lambda_n \langle x, u_n \rangle y_n$$

of the form (A1.1), where in addition we may suppose that the elements y_n belong to the unit ball of $\tilde{E}_{(U)}$. But then, for $u \in M$ and $z \in E \otimes F$, we have

$$|u(z)| = \left| \sum_{n=1}^{\infty} \lambda_n (u_n \otimes v_{n,u})(z) \right| \leq \left(\sum_{n=1}^{\infty} |\lambda_n| \right) p_{A,B}(z),$$

where $v_{n,u} = i \circ S_u(y_n)$ and $A = \{u_n\}$. $\qquad\qquad\square$

Let E and F be locally convex Hausdorff spaces. Then we denote by $E \hat{\otimes}_\varepsilon F$ and $E \hat{\otimes}_\pi F$ the *completed* ε- and π-*tensor products*. Instead of $E \hat{\otimes}_\pi F$ we often write simply $E \hat{\otimes} F$. Suppose that E, F and G, H are locally convex spaces and that $S \in L(E, G), T \in L(F, H)$ are continuous linear operators. Then the tensor product of S and T

$$S \otimes T \colon E \otimes F \to G \otimes H$$

is continuous and linear with respect to the ε-tensor product and π-tensor product topologies, respectively, on both sides. If all spaces are Hausdorff, then $S \otimes T$ extends continuously to the completions. In general, the ε- and π-tensor products of linear maps enjoy quite different (almost complementary) stability properties. Nuclearity ensures that one can take advantage of all these stability properties simultaneously.

A continuous linear operator $T \colon E \to F$ between locally convex spaces is a (*topological*) *homomorphism* if T is open regarded as a map from E onto the range of T equipped with the relative topology of F. A topological homomorphism that is injective (surjective) is called a *topological monomorphism* (*epimorphism*, respectively).

Theorem A1.6   *Let E, F, G, H be locally convex Hausdorff spaces, and let $S \in L(E, G)$, $T \in L(F, H)$ be given operators.*

(a) *If all the spaces are Fréchet spaces and if S, T are surjective, then*

$$S \otimes T: E \hat{\otimes}_\pi F \to G \hat{\otimes}_\pi H$$

is a topological epimorphism between Fréchet spaces, and $\mathrm{Ker}(S \otimes T)$ *is the closure of the space* $\mathrm{Ker}(E \otimes F \xrightarrow{S \otimes T} G \otimes H)$.

(b) *If S and T are topological monomorphisms, then*

$$S \otimes T: E \hat{\otimes}_\varepsilon F \to G \hat{\otimes}_\varepsilon H$$

is a topological monomorphism.

(c) *If all the spaces are complete and if S and T are injective, then so is*

$$S \otimes T: E \hat{\otimes}_\varepsilon F \to G \hat{\otimes}_\varepsilon H.$$

(d) *Let $E \xrightarrow{\alpha} F \xrightarrow{\beta} G$ be an exact sequence between Fréchet spaces, and let H be a Fréchet space. Suppose that F or H is nuclear. Then the following sequence is exact:*

$$E \hat{\otimes}_\pi H \xrightarrow{\alpha \otimes 1} F \hat{\otimes} H \xrightarrow{\beta \otimes 1} G \hat{\otimes}_\varepsilon H.$$

Proof   If E and F are metrizable, then obviously $E \hat{\otimes}_\varepsilon F$ and $E \hat{\otimes}_\pi F$ are Fréchet spaces.

(a) If S, T are surjective maps between Fréchet spaces, then $S \otimes T: E \otimes_\pi F \to G \otimes_\pi H$ is a topological epimorphism and extends, by a standard Fréchet space argument, to a surjective map between the completions. Suppose that $A: X \to Y$ is a continuous linear map between locally convex Hausdorff spaces that induces a topological epimorphism A_0 between dense subspaces $X_0 \subset X$, $Y_0 \subset Y$. Since the map $A_0: X_0 \to Y_0$ is open, we have $\mathrm{Im}(A') = (\mathrm{Ker}(A_0))^\perp$. By the bipolar theorem,

$$\mathrm{Ker}(A) = {}^\perp(\mathrm{Im}(A')) = \overline{\mathrm{Ker}(A_0)}.$$

(b) Using the Hahn–Banach and bipolar theorems, one can show that a continuous linear operator $T: X \to Y$ between locally convex Hausdorff spaces is a topological monomorphism if and only if T' allows the lifting of equicontinuous sets. Hence it suffices to prove that

$$S \otimes T: E \otimes_\varepsilon F \to G \otimes_\varepsilon H$$

is a topologicial monomorphism.

Let us denote by E'_τ the Mackey dual of E, and let $L_e(E'_\tau, F)$ be the space $L(E'_\tau, F)$ equipped with the topology of uniform convergence on all equicontinuous subsets of E'. The topology of this space is generated by the seminorms

$$q_{A,B}(T) = \sup_{\substack{x' \in A \\ y' \in B}} |\langle Tx', y' \rangle| \qquad (A \subset E', B \subset F' \text{ equicontinuous}).$$

The unique linear operator $\Phi: E \otimes_\varepsilon F \to L_e(E'_\tau, F)$ with $\Phi(x \otimes y)(x') = \langle x, x' \rangle y$ is a topological monomorphism, since for $z \in E \otimes F$

$$q_{A,B}(\Phi(z)) = p_{A,B}(z) \qquad (A \subset E', B \subset F' \text{ equicontinuous}).$$

Therefore it suffices to check that the operator

$$L_e(E'_\tau, F) \to L_e(G'_\tau, H), \qquad A \mapsto TAS' \tag{A1.5}$$

is a topological monomorphism, and this is elementary.

(c) Since the injectivity of S and T implies the injectivity of the multiplication operator (A1.5), it suffices to prove that $L_e(E'_\tau, F)$ is complete whenever E and F are complete. To do this, one can first show that the pointwise limit of a Cauchy net (T_α) in $L_e(E'_\tau, F)$ belongs to $L(E'_\sigma, F_\sigma)$, and then show that for arbitrary locally convex Hausdorff spaces E and F this last space coincides with $L(E'_\tau, F)$.

(d) It follows from the previous parts that under the assumptions of part (d) the sequence

$$0 \to (\text{Im } \alpha) \mathbin{\hat\otimes} H \xrightarrow{i \otimes 1} F \mathbin{\hat\otimes} H \xrightarrow{q \otimes 1} (F/\text{Ker } \beta) \mathbin{\hat\otimes} H \to 0$$

is exact. Since the sequence of interest factorizes through this sequence, the proof is complete. $\qquad\square$

Note that for the validity of the kernel formula in part (a) it suffices that S and T are topological epimorphisms between locally convex Hausdorff spaces.

As a typical application of Theorem A1.6, one can show that, for each pseudo-convex open set U in $\mathbb{C}^n$ and each Fréchet space X, the vector-valued $\bar\partial$-sequence

$$0 \to \mathcal{O}(U, X) \to \mathcal{E}^{0,0}(U, X) \xrightarrow{\bar\partial} \mathcal{E}^{0,1}(U, X) \to \cdots \to \mathcal{E}^{0,n}(U, X) \to 0$$

is exact. Here $\mathcal{E}^{0,p}(U, X)$ denotes the space of differential forms of bidegree $(0, p)$ with coefficients in $\mathcal{E}(U, X)$. The above sequence is exact in the case $X = \mathbb{C}$ (see Corollary 4.2.6 in Hörmander 1966). Since there are canonical identifications

$$\mathcal{O}(U, X) = \mathcal{O}(U) \mathbin{\hat\otimes} X \qquad \text{and} \qquad \mathcal{E}(U, X) = \mathcal{E}(U) \mathbin{\hat\otimes} X$$

(see Treves 1967), the exactness in the general case follows from Theorem A1.4 and Theorem A1.6.

(5) A locally convex Hausdorff space E is *semi-reflexive* if $E = (E'_\beta)'$, where E'_β denotes the strong dual of E. If in addition $E = (E'_\beta)_\beta$, then E is *reflexive*. By the theorem of Mackey and Arens, a locally convex Hausdorff space is semi-reflexive if and only if $E'_\beta = E'_\tau$. In view of the definition of the Mackey topology and the bipolar theorem, this in turn is equivalent to the fact that all bounded sets in E are relatively weakly compact. By the bipolar theorem a locally convex Hausdorff space E is *barrelled* if and only if E carries the strong dual topology in the dual system $\langle E, E' \rangle$. Hence being reflexive is equivalent to being semi-reflexive and barrelled.

Lemma A1.7 *Each complete nuclear space E is semi-reflexive.*

Proof Since E is complete, the topological monomorphism

$$\Phi: E \to \prod_p \tilde{E}_p, \qquad x \mapsto (x + \text{Ker}(p))_p,$$

where p runs through all continuous seminorms on E, has closed range. By the nuclearity of E, each bounded set in E is relatively compact in the product space on the right, hence also in E. $\qquad\square$

The proof of Lemma A1.7 shows that in a complete nuclear space each bounded set is relatively compact. A barrelled space with this property is called a *Montel space*. In particular, each nuclear Fréchet space is an (FM)-space, that is, a Fréchet space that is Montel.

In the proof of Theorem A1.6 we saw that, for each pair of locally convex spaces E and F, the canonical map

$$\Phi\colon E \otimes_\varepsilon F \to L_e(E'_\tau, F) \tag{A1.6}$$

is a topological monomorphism. We claim that its range is dense if E is nuclear. Let T be an operator in the space on the right. We fix a closed, absolutely convex and equicontinuous set A in E'_τ and a zero neighbourhood W in F. Then

$$U = {}^\circ A = \{x \in E; |\langle x, x'\rangle| \le 1 \text{ for each } x' \in A\}$$

contains an absolutely convex zero neighbourhood V such that the canonical map $\tilde{E}_{(V)} \to \tilde{E}_{(U)}$ is of the form (A1.1). Its adjoint can be identified with the inclusion $E'_A \to E'_B$, where $B = V^\circ$. This acts as

$$x' \mapsto \sum_{n=1}^\infty \lambda_n \langle y_n, x'\rangle u_n,$$

with suitable bounded sequences (y_n) and (u_n) in $\tilde{E}_{(U)}$ and E'_B, respectively. For a given $\varepsilon > 0$, one can choose finitely many vectors $x_1, \ldots, x_r \in E$ such that

$$x' - \sum_{n=1}^r \lambda_n \langle x_n, x'\rangle u_n \in \varepsilon B$$

for each $x' \in A$. It suffices to choose ε so small that $\varepsilon TB \subset W$, and then to observe that

$$Tx' - \sum_{n=1}^r \lambda_n \langle x_n, x'\rangle Tu_n \in W \qquad (x' \in A).$$

If E and F are complete locally convex Hausdorff spaces and if E is nuclear, then the topological monomorphism (A1.6) extends to a topological isomorphism

$$\Phi\colon E \mathbin{\hat{\otimes}} F \to L_e(E'_\tau, F). \tag{A1.7}$$

Theorem A1.8 *A Fréchet space E is nuclear if and only if its strong dual E' is nuclear.*

Proof Let E be a nuclear Fréchet space. Since $E' = E'_\tau$, to prove the nuclearity of E' it suffices to check that, for each Banach space F, each operator in $L_e(E'_\tau, F) \cong E \mathbin{\hat{\otimes}} F$ is nuclear. But since E and F are Fréchet spaces, each element $z \in E \mathbin{\hat{\otimes}} F$ is of the form

$$z = \sum_{n=1}^\infty \lambda_n x_n \otimes y_n,$$

with $(\lambda_n) \in l^1$ and suitable zero sequences (x_n) in E and (y_n) in F (see Köthe 1979, §41.4.(6)).

A proof of the reverse implication can be found in Treves (1967). $\qquad\qquad\square$

Suppose that E and F are Fréchet spaces, and that E is nuclear. Then there is a canonical topological isomorphism

$$\Phi\colon E' \mathbin{\hat{\otimes}} F \to L_b(E, F),$$

where the subscript b refers to the *topology of uniform convergence on all bounded subsets* of E. To see this, replace E by its strong dual E' in (A1.7), and observe that the equicontinuous sets in E (regarded as the dual space of E') are precisely the bounded subsets of E.

A locally convex Hausdorff space E is a *(DF)-space* if E possesses a fundamental sequence (B_n) of bounded sets (i.e. each bounded set in E is contained in one of the sets B_n) and if each strongly bounded union of equicontinuous subsets of E' is equicontinuous. The strong dual of a (DF)-space is a Fréchet space. Examples of (DF)-spaces are normed spaces and strong duals of Fréchet spaces.

Lemma A1.9 *The strong dual of a complete nuclear (DF)-space E is nuclear.*

Proof Let $B \subset E'_\sigma$ be bounded. Then B is also bounded in $E'_\tau = E'_\beta$. Let (u_n) be a sequence in B. Then $U = {}^\circ\{u_n\}$ is a zero neighbourhood in E. Since E is nuclear, the inclusion

$$E'_{U^\circ} \to E'_B$$

is nuclear (it factors through a nuclear map $E'_{U^\circ} \to E'_{V^\circ}$). Hence $\{u_n\} \subset U^\circ \subset E'_\beta$ is relatively compact. Therefore B is relatively compact, and it is contained in the closure of a bounded countable subset of E'_β. By the definition of (DF)-spaces, the set B is equicontinuous.

Thus we have shown that E is barrelled. But then E carries the strong topology in the dual system $\langle E, E' \rangle$. Therefore the nuclear space E is the strong dual of the Fréchet space E'_β, and the nuclearity of E'_β follows from the previous result. □

Under suitable conditions, the strong dual of a tensor product space is the tensor product of the strong duals. The basic observation is contained in the next result, due to Grothendieck.

Theorem A1.10 *Let E and F be complete locally convex Hausdorff spaces, and let E be nuclear. Suppose that E and F are both (DF)-spaces or both Fréchet spaces. Then, for each bounded absolutely convex set $M \subset E \hat{\otimes} F$, there are sequences $(\lambda_i) \in l^1, (y_i)$ in E, and (φ_i) in $L((E \hat{\otimes} F)_M, F)$ such that (y_i) tends to zero, (φ_i) is equicontinuous, and*

$$u = \sum_{i=1}^{\infty} \lambda_i y_i \otimes \varphi_i(u) \qquad (u \in M). \tag{A1.8}$$

Proof Let $M \subset E \hat{\otimes} F = L_e(E'_\tau, F)$ be as above. Then the system of separately continuous bilinear forms

$$B_T : E' \times F' \to \mathbb{C}, \qquad B_T(x', y') = \langle Tx', y' \rangle \qquad (T \in M)$$

is pointwise bounded. Let D be the closed unit disc in $\mathbb{C}$. Note that, in both cases, E is barrelled. If $B \subset E'$ is bounded, then $V = (MB)^\circ$ is a zero neighbourhood in F' with

$$B_T(B, V) \subset D \qquad (T \in M).$$

If $B \subset F'$ is bounded, then the set

$$U = \bigcap_{y' \in B} \bigcap_{T \in M} B_T(\cdot, y')^{-1}(D)$$

is closed, absolutely convex, and absorbent in E', hence a zero neighbourhood, with

$$B_T(U, B) \subset D \qquad (T \in M).$$

Therefore, in both cases, the family $(B_T)_{T \in M}$ is equicontinuous (see Köthe 1979, §40.2.(1) and §40.2.(10)). Thus we can choose an absolutely convex zero neighbourhood V in E' and an absolutely convex, closed and bounded subset B of F such that

$$T(V) \subset B \qquad (T \in M)$$

and such that $\tilde{E}'_{(V)}$ is a Hilbert space.

Since E' is nuclear, there is a smaller absolutely convex zero neighbourhood $U \subset V$ in E' such that the canonical map $\tilde{E}'_{(U)} \to \tilde{E}'_{(V)}$ is nuclear. Then the adjoint is nuclear and acts (because of the reflexivity of E) as the inclusion map $E_{°V} \to E_{°U}$. In particular, it follows that the inclusion

$$i \colon E_A \to E \qquad (A = {°V})$$

is nuclear. We choose a representation of the form (A1.1),

$$x = \sum_{n=1}^{\infty} \lambda_n \langle x, u_n \rangle y_n, \qquad (x \in E_A) \tag{A1.9}$$

such that (y_n) tends to zero and (u_n) is contained in the unit ball of $(E_A)'$. Note that the norm dual $(E_A)'$ can isometrically be identified with $\tilde{E}'_{(V)}$.

The map Φ induces a continuous linear map

$$\Psi \colon (E \mathbin{\hat{\otimes}} F)_M \to L\left(\tilde{E}'_{(V)}, F_B\right)$$

of norm less than or equal to one between normed spaces. Hence the maps

$$\varphi_n \colon (E \mathbin{\hat{\otimes}} F)_M \to F_B, \qquad u \mapsto \Psi(u)(u_n)$$

are continuous linear with $\|\varphi_n\| \le 1$. To conclude the proof, it suffices to check that

$$\Phi(u) = \Phi\left(\sum_{n=1}^{\infty} \lambda_n y_n \otimes \varphi_n(u) \right) \qquad (u \in M).$$

To verify this identity, it suffices to show that

$$\sum_{n=1}^{\infty} \lambda_n \langle y_n, x' \rangle \langle \varphi_n(u), y' \rangle = y' \circ \Psi(u)[x']$$

holds for all $x' \in E'$ and $y' \in (F_B)'$. But this identity follows from the definition of $\varphi_n(u)$ and (A1.9). □

The following elementary consequence of Theorem A1.10 is the main result in which we are interested.

Corollary A1.11 *Suppose that the conditions of Theorem A1.10 are satisfied. Then for each bounded set $M \subset E \mathbin{\hat{\otimes}} F$, there are absolutely convex, closed and bounded sets $A \subset E$, $B \subset F$ such that M is contained in the image of the closed unit ball under the map*

$$E_A \mathbin{\hat{\otimes}_{\pi}} F_B \to E \mathbin{\hat{\otimes}} F.$$

In particular, it follows that $M \subset \overline{\Gamma(A \otimes B)}$.

Proof We may suppose that the elements of M admit representations of the form (A1.8) such that $\sum_{n=1}^{\infty} |\lambda_n| \le 1$. Then it suffices to define A and B as the closed absolutely convex hulls of the sets $\{y_n\}$ and $\cup(\varphi_n(M); n \geq 1)$, respectively, and to observe that $E_A \hat{\otimes}_\pi F_B$ is the norm completion of $E_A \otimes F_B$ with respect to the norm

$$\|z\| = \inf\left\{ \sum_{i=1}^n p_A(x_i)p_B(y_i); \quad z = \sum_{i=1}^n x_i \otimes y_i \right\},$$

where p_A and p_B are the Minkowski functionals in E_A and F_B.   □

As a consequence, we obtain the following duality result for tensor products of locally convex spaces.

Theorem A1.12 *Let E and F be as in Theorem A1.10. Then the unique continuous linear map $\Phi: (E'_\beta) \hat{\otimes} (F'_\beta) \to (E \hat{\otimes} F)'_\beta$ with*

$$\langle x \otimes y, \Phi(x' \otimes y') \rangle = \langle x, x' \rangle \langle y, y' \rangle \tag{A.1.10}$$

is a topological isomorphism.

Proof Since E and E'_β are nuclear, the ε- and π-tensor product topologies coincide on both sides. By the result stated in Corollary A1.11, the unique linear map $\Phi: (E'_\beta) \otimes_\pi (F'_\beta) \to (E \hat{\otimes} F)'_\beta$ satisfying (A1.10) is continuous. In both cases the space $(E \hat{\otimes} F)'_\beta$ is complete (in the (DF)-space case see Köthe 1979, §41.4.(7)). Hence Φ extends to a continuous linear map $(E'_\beta) \hat{\otimes} (F'_\beta) \to (E \hat{\otimes} F)'_\beta$, again denoted by Φ.

By the remarks preceding Theorem A1.8 (see (A1.7)), there is a canonical identification

$$(E'_\beta) \hat{\otimes} (F'_\beta) \to L_e(E, F'_\beta), \tag{A1.11}$$

where the equicontinuous subsets of E (regarded as the dual space of E'_β) are precisely the bounded sets. Each $u \in (E \hat{\otimes} F)'_\beta$ induces an operator $T_u \in L_e(E, F'_\beta)$ defined by

$$\langle j, T_u(x) \rangle = u(x \otimes y).$$

By Corollary A1.11 the map

$$\Psi: (E \hat{\otimes} F)'_\beta \to L_e(E, F'_\beta), \qquad u \mapsto T_u$$

is a topological monomorphism. This map composed with the inverse of (A1.11) acts as a left inverse for Φ. Hence Ψ, and therefore also Φ, is a topological isomorphism.

□

(6) The proof of the Cartan–Serre theorem given in Chapter 9 makes use of a perturbation result for epimorphisms between Fréchet spaces due to L. Schwartz. We follow the presentation of Grothendieck (1954).

Lemma A1.13 *Let $T: E \to F$ be a topological monomorphism with closed range between locally convex Hausdorff spaces. If $K: E \to F$ is a compact operator, then $S = T + K$ is a topological homomorphism with finite-dimensional kernel and closed range.*

Proof Choose an absolutely convex zero neighbourhood V in E such that $K(V)$ is relatively compact. Then $W = V \cap \text{Ker}(S)$ is a zero neighbourhood in $\text{Ker}(S)$, which is relatively compact because

$$W \subset T^{-1}(\overline{TW}) = T^{-1}(\overline{KW}).$$

Hence $\text{Ker}(S)$ is finite dimensional.

Let M be a topological direct complement of $\mathrm{Ker}(S)$ in E, i.e a closed subspace of E such that $E = \mathrm{Ker}(S) \oplus M$. By passing from E to M, one may suppose from the very beginning that S is injective. To show that S is a topological monomorphism with closed range, it suffices to prove that any net (x_α) in E such that $\lim S(x_\alpha)$ exists in F has itself a convergent subnet.

If there are an index α_0 and a natural number n with $x_\alpha \in nV$ for all $\alpha \geq \alpha_0$, then $K(x_\alpha)$, hence also $T(x_\alpha)$ and (x_α), have a convergent subnet. Let us assume that no such α_0 and n exist, and let us denote by p the Minkowski functional of V. Then there is a subnet $(y_i)_{i \in I}$ of $(x_\alpha)_{\alpha \in A}$ with $\lim p(y_i) = \infty$. Since $\lim S(y_i/p(y_i)) = 0$ and since, for i large enough,

$$y_i/p(y_i) \in (1 + 1/p(y_i))V \subset 2V,$$

by the argument given in the first case, $(y_i/p(y_i))$ would have a convergent subnet. Since S is injective, the limit of this convergent subnet is necessarily zero. But this is clearly not possible. □

Dualizing the last statement, one obtains a perturbation result for epimorphisms. Although more general versions are possible, we confine ourselves to the case of Fréchet spaces. For a locally convex Hausdorff space E, we denote by E_c' the topological dual equipped with the uniform convergence on all compact subsets of E.

Lemma A1.14 (L. Schwartz) *Suppose that $T: E \to F$ is a surjective continuous linear operator between Fréchet spaces and that $K: E \to F$ is a compact operator. Then* $\dim \mathrm{Coker}(T + K) < \infty$.

Proof We claim that $T', K': F_c' \to E_c'$ satisfy the assumptions of the previous lemma.

Since T is surjective, its adjoint T' is injective. If $K \subset E$ is compact and $C = T(K)$, then $C^\circ = (T')^{-1}(K^\circ)$, and hence $T'(C^\circ) = K^\circ \cap \mathrm{Im}(T')$. Thus T allows the lifting of compact sets (see Köthe 1969, §22.2.(7)), and we conclude that $T': F_c' \to E_c'$ is a topological monomorphism. Since F_c' and E_c' are complete, the range of T' is necessarily closed.

If $V \subset E$ is a zero neighbourhood and $K(V)$ is relatively compact in F, then $U = K(V)^\circ$ is a zero neighbourhood in F_c' with $K'(U) \subset V^\circ$. By the Theorem of Arzela and Ascoli, the equicontinuous set $K'(U)$ is relatively compact in E_c'.

As an application of the previous lemma, $(T + K)': F_c' \to E_c'$ has finite-dimensional kernel and closed range. Since the topology of E_c' is coarser than the Mackey topology, the image of $(T + K)'$ is even $\sigma(E', E)$-closed. But then the image of $T + K$ is closed (see Köthe 1979, §33.4.(2)), and the assertion follows from the identification $(F/\mathrm{Im}(T + K))' = \mathrm{Ker}(T + K)'$. □

Appendix 2
Homological algebra

In this appendix, we recall some basic facts from homological algebra, and we fix the conventions and the terminology used throughout this book. As a comprehensive reference, we recommend the classical treatise of Cartan and Eilenberg (1956). Although we do not use the modern language of derived categories, the reader may also profit by consulting the monograph of Gelfand and Manin (1988).

(1) Let A be a commutative ring with identity $1 \in A$. A *complex of A-modules* is a family $(M_n, d_n)_{n \in \mathbb{Z}}$ of left A-modules and A-module morphisms $d_n: M_n \to M_{n-1}$ with the property that $d_n \circ d_{n+1} = 0$ $(n \in \mathbb{Z})$. The complex $(M_n, d_n)_{n \in \mathbb{Z}}$ is denoted in short by $M_\bullet$ or by $(M_\bullet, d_\bullet)$. The maps d_n $(n \in \mathbb{Z})$ are called the *boundary operators* or *differentials* of $(M_\bullet, d_\bullet)$. The complex $M_\bullet$ is *bounded to the right* (*to the left*) if there exists an integer n_0 with $M_n = 0$ for $n > n_0$ (for $n < n_0$, respectively). A complex is *bounded* if it is bounded to the left and to the right.

We also use complexes of A-modules which are indexed by an ascending sequence of integers, such as

$$\cdots \to M^{n-1} \xrightarrow{d^{n-1}} M^n \xrightarrow{d^n} M^{n+1} \to \cdots$$

In this case, the modules and the module morphisms are indexed by upper indices, and the maps $d^n: M^n \to M^{n+1}$ are referred to as the *coboundary operators* (or *differentials*) of the complex $(M^\bullet, d^\bullet)$. Although these lower and upper indices have a geometric explanation (degree and dimension, respectively), we shall only work formally with them. Finally, we call a finite or infinite family of A-module morphisms

$$\cdots \to L \xrightarrow{\alpha} M \xrightarrow{\beta} N \to \cdots$$

a *sequence of A-modules* if the composition of any two successive maps in this family is the zero operator.

The *homology modules* or *homology spaces* of a complex $M_\bullet$ of A-modules are the A-modules

$$H_n(M_\bullet) = \mathrm{Ker}(d_n: M_n \to M_{n-1})/\mathrm{Im}(d_{n+1}: M_{n+1} \to M_n) \qquad (n \in \mathbb{Z}).$$

The complex $M_\bullet$ is *exact* if $H_n(M_\bullet) = 0$ for each $n \in \mathbb{Z}$. Analogously, one defines the *cohomology modules* $H^n(M^\bullet)$ and the notion of exactness for ascending complexes $M^\bullet$.

A *morphism* $f: (M_\bullet, d_\bullet) \to (M'_\bullet, d'_\bullet)$ *of complexes* of A-modules is a family of A-module morphisms $f_n: M_n \mapsto M'_n$ with the property that $d'_n \circ f_n = f_{n-1} \circ d_n$ $(n \in \mathbb{Z})$. A morphism of complexes f as above induces in a natural way the complexes $\mathrm{Ker}\, f$, $\mathrm{Im}\, f$, and $\mathrm{Coker}\, f$. We can iterate the definition of a complex of A-modules and speak of a complex of complexes, and so on. The first important result in this direction is contained in the following lemma.

Lemma A2.1   Let $0 \to M_\bullet \overset{f}{\to} N_\bullet \overset{g}{\to} P_\bullet \to 0$ be a short exact sequence of complexes of A-modules. Then there is a long exact sequence of homology spaces

$$\cdots \to H_n(M_\bullet) \overset{f_*}{\to} H_n(N_\bullet) \overset{g_*}{\to} H_n(P_\bullet) \overset{\delta}{\to} H_{n-1}(M_\bullet) \to \cdots \tag{A2.1}$$

□

For a proof see Cartan and Eilenberg (1956).

If the short exact sequence of complexes in Lemma A2.1 consists of Fréchet A-modules over a Fréchet algebra (see Section 3.2), and if all the boundary operators as well as all the morphisms forming f and g are continuous, then all maps in the sequence (A2.1) are continuous with respect to the canonical (not necessarily Hausdorff) quotient topologies. Moreover, if

$$H_n(M_\bullet) = H_{n-1}(M_\bullet) = 0,$$

then $g_*: H_n(N_\bullet) \to H_n(P_\bullet)$ is a topological isomorphism. The corresponding results also hold for f_* and for δ.

An important particular case of Lemma A2.1 is illustrated by the following diagram:

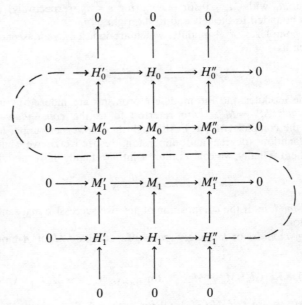

Here all the vertical and horizontal sequences are exact, and the six-term long exact sequence of homology (see the dashed line) induced by the short exact sequence of complexes $0 \to M'_\bullet \to M_\bullet \to M''_\bullet \to 0$ can be used to explain the more general result stated as Lemma A2.1. This particular result is called the *serpent's lemma*.

(2) An A-module P is *projective* if, for any surjective morphism of A-modules $f: M \to N$ and any morphism $g: P \to N$, there exists a lifting $g': P \to M$, that is, a

morphism with $f \circ g' = g$. The situation described in this definition is usually abbreviated by a diagram of the form

$$M \xrightarrow{f} N \longrightarrow 0$$

with dashed arrows g' and g from P below.

The dual object, namely an *injective A-module*, is defined by the property illustrated in the following diagram with exact top row

$$0 \longrightarrow M \xrightarrow{f} N$$

with g down to I and dashed g'.

A free A-module is projective (actually, every projective module is a direct summand of a free module). If the ring A has no zero divisors, then any division A-algebra forms an injective A-module.

Let M be an A-module. Then there is an exact sequence of A-modules

$$\cdots \to P_1 \to P_0 \to M \to 0 \tag{A2.2}$$

with P_n a projective A-module ($n \geq 0$). The complex $P_\bullet$ is called a *projective resolution of the module $M_\bullet$*. Similarly, any A-module admits an *injective resolution*. The sequence (A2.2) may be bounded or not. In the first case, the smallest value of $n \in \mathbb{N}$ for which there exists a projective resolution of the A-module M with $P_{n+1} = 0$ is called the *projective dimension* of M. This numerical invariant, and even the projective resolution of an A-module, is unique (up to a relation called *homotopy equivalence*).

(3) In order to define the *homotopy equivalence*, let $f: M_\bullet \to M'_\bullet$ be a morphism of complexes of A-modules. The morphism f is *homotopically equivalent* to zero if there are A-module morphisms $h_n: M_n \to M'_{n+1}$ with the property that $f_n = h_{n-1} \circ d_n + d'_{n+1} \circ h_n$ ($n \in \mathbb{Z}$):

$$\cdots \longrightarrow M_{n+1} \xrightarrow{d_{n+1}} M_n \xrightarrow{d_n} M_{n-1} \longrightarrow \cdots$$

with maps h_n, f_n, h_{n-1}

$$\cdots \longrightarrow M'_{n+1} \xrightarrow{d'_{n+1}} M'_n \xrightarrow{d'_n} M'_{n-1} \longrightarrow \cdots$$

Two morphisms of complexes $f, g: M_\bullet \to M'_\bullet$ are *homotopically equivalent* if $f - g$ is homotopically equivalent to zero. A complex $M_\bullet$ is *homotopically trivial* or equivalently *split exact* if the identity operator $\mathrm{Id}: M_\bullet \to M_\bullet$ is homotopically equivalent to zero. In other words, a splitting of the complex $(M_\bullet, d_\bullet)$ consists of a family of morphisms of A-modules $h_n: M_n \to M_{n+1}$ ($n \in \mathbb{Z}$) with the property that $x = (h_{n-1} \circ d_n + d_{n+1} \circ h_n)x$ for any $x \in M_n$ and $n \in \mathbb{Z}$.

It is sometimes useful to attach a new complex to a given morphism $f: (M_\bullet, d_\bullet) \to (M'_\bullet, d'_\bullet)$ of complexes of A-modules. This is the *cone of f* consisting of the A-modules

$$C_n(f) = M_{n-1} \oplus M'_n$$

and the differentials $\delta_n: C_n(f) \to C_{n-1}(f)$ acting as the matrix operators

$$\delta_n = \begin{bmatrix} d_{n-1} & 0 \\ (-1)^{n-1} f_{n-1} & d'_n \end{bmatrix} \quad (n \in \mathbb{Z}).$$

Let $M_\bullet[1]$ denote the translation of the complex $M_\bullet$ by one unit: $M_n[1] = M_{n-1}$.

Lemma A2.2 *Let $f: M_\bullet \to M'_\bullet$ be a morphism of complexes of A-modules, and let $C_\bullet(f)$ denote the cone of f. Then there exists a short exact sequence of complexes*

$$0 \to M'_\bullet \to C_\bullet(f) \to M_\bullet[1] \to 0. \tag{A2.3} \ \square$$

For a proof of Lemma A2.2 and for further results illustrating the importance of the concept of the cone of a morphism between complexes of A-modules, see Gelfand and Manin (1988).

A morphism of complexes of A-modules $f: M_\bullet \to M'_\bullet$ is called a *quasi-isomorphism* if its cone $C_\bullet(f)$ is exact, or equivalently, if f induces isomorphisms $f_*: H_n(M_\bullet) \xrightarrow{\sim} H_n(M'_\bullet)$ $(n \in \mathbb{Z})$ at the level of the homology modules. More specifically, the morphism $f: M_\bullet \to M'_\bullet$ is called an *r-quasi-isomorphism*, for a given integer r, if its cone $C_\bullet(f)$ is exact in degree $n \le r$, or equivalently, if $f_*: H_r(M_\bullet) \to H_r(M'_\bullet)$ is onto and $f_*: H_n(M_\bullet) \to H_n(M'_\bullet)$ is bijective for each $n < r$.

Lemma A2.3 *Let $0 \to M'_\bullet \xrightarrow{f} M_\bullet \xrightarrow{g} M''_\bullet \to 0$ be a short exact sequence of complexes of A-modules. Then the maps*

$$C_n(f) \to M''_n, \quad (x, y) \mapsto g_n(y) \qquad (n \in \mathbb{Z})$$

define a quasi-isomorphism $C_\bullet(f) \to M''_\bullet$ of complexes of A-modules. $\square$

Analogous results hold for the cone $C^\bullet(f)$ of a morphism $f: (L^\bullet, u^\bullet) \to (M^\bullet, d^\bullet)$. By definition

$$C^n(f) = M^n \oplus L^{n+1},$$

and the differentials $\delta^n: C^n(f) \to C^{n+1}(f)$ are given by the matrix operators

$$\delta^n = \begin{bmatrix} d^n & (-1)^{n+1} f^{n+1} \\ 0 & u^{n+1} \end{bmatrix}.$$

(4) Let $F: A\text{-Mod} \to \text{Ab}$ be a covariant functor between the category of A-modules (A-Mod) and the category of abelian groups (Ab). The functor F is *additive* if the induced map

$$F_*: \text{Hom}_A(M, M') \to \text{Hom}_{\text{Ab}}(F(M), F(M'))$$

is additive for any pair of A-modules M and M'. The functor F is *exact to the left* if, for any short exact sequence of A-modules

$$0 \to M' \to M \to M'' \to 0, \tag{A2.4}$$

the induced sequence

$$0 \to F(M') \to F(M) \to F(M''),$$

is exact. Similarly, the functor F is said to be *exact to the right* if each exact sequence of the form (A2.4) is transformed by F into an exact sequence

$$F(M') \to F(M) \to F(M'') \to 0.$$

Let M be a fixed A-module. Then the functor $* \otimes_A M$ is exact to the right and the functor $\text{Hom}_A(M, *)$ is exact to the left. A *semi-exact functor* is a functor which is exact to the left or to the right. A functor that is exact to the left and to the right is *exact*.

An A-module M is *flat* if the functor $* \otimes_A M$ is exact. In other words, the A-module M is flat if and only if the induced morphism $N \otimes_A M \to N' \otimes_A M$ is injective for any injective morphism $N \to N'$. Each projective A-module is flat, but the converse implication is not true. For instance, if $\mathcal{O}_0$ and $\mathcal{E}_0$ denote the rings of germs of analytic and C^∞-functions at $0 \in \mathbb{C}$, respectively, then $\mathcal{E}_0$ is a flat $\mathcal{O}_0$-module which is not projective.

(5) One of the central results of homological algebra is the existence of *derived functors* for each additive and semi-exact functor $F: A\text{-Mod} \to \text{Ab}$. To fix our ideas, we shall suppose that F is covariant and exact to the left.

Theorem A2.4 *Let $F: A\text{-Mod} \to \text{Ab}$ be a functor that is covariant, additive, and exact to the left. Then there exist unique 'derived' functors $R^q F: A\text{-Mod} \to \text{Ab}$ $(q \geq 1)$ with the property that, for every exact sequence of A-modules (A.2.4), there is a natural long exact sequence*

$$0 \to F(M') \to F(M) \to F(M'') \to R^1 F(M') \to R^1 F(M) \to \cdots$$

$$\cdots \to R^n F(M') \to R^n F(M) \to R^n F(M'') \to R^{n+1} F(M') \to \cdots \quad \text{(A2.5)}$$

$\square$

The proof of this theorem can be found in Cartan and Eilenberg (1956). A useful criterion for computing the derived functors of a given functor is contained in the next result.

Corollary A2.5 *Under the conditions of Theorem A2.4, let $I^\bullet$ denote an injective resolution of the A-module M. Then $R^q F(M) \cong H^q(F(I^\bullet))$ $(q \in \mathbb{N})$.* $\square$

For functors which are exact to the right, Corollary A2.5 remains true with projective resolutions instead of injective resolutions. The derived functors of $* \otimes_A M$ are denoted by $\text{Tor}_q^A(*, M)$, and the derived functors of $\text{Hom}_A(M, *)$ are denoted by $\text{Ext}_A^q(M, *)$ $(q \in \mathbb{N})$.

The categories encountered in the book are formed by topological modules over some specific topological algebras. In this case the lack of projective or injective objects is compensated for by a strengthening of the notion of an exact sequence. This idea, originally due to Eilenberg and Moore (1965), is developed in the text in the particular cases treated there.

(6) By a *double complex* we shall mean a family $K = (K_{p,q})_{p,q \in \mathbb{Z}}$ of A-modules $K_{p,q}$ together with a system of module homomorphisms

$$\partial': K_{p,q} \to K_{p-1,q}, \quad \partial'': K_{p,q} \to K_{p,q-1} \quad (p, q \in \mathbb{Z})$$

that satisfy $\partial' \partial' = 0$, $\partial'' \partial'' = 0$, and $\partial' \partial'' + \partial'' \partial' = 0$. The A-modules

$$H_{p,q}''(K) = \text{Ker} \, \partial_{p,q}'' / \text{Im} \, \partial_{p,q+1}'',$$

$$H_{p,q}'(K) = \text{Ker} \, \partial_{p,q}' / \text{Im} \, \partial_{p+1,q}'$$

are called the *second* and *first homology groups* of K, respectively. The boundary operators ∂' and ∂'' induce maps

$$\partial': H_{p,q}''(K) \to H_{p-1,q}''(K), \quad [x] \mapsto [\partial' x],$$

$$\partial'': H_{p,q}'(K) \to H_{p,q-1}'(K), \quad [x] \mapsto [\partial'' x].$$

The *iterated homology groups of K* are

$$H'_p H''_q(K) = \mathrm{Ker}(\partial': H''_{p,q}(K) \to H''_{p-1,q}(K))/\partial' H''_{p+1,q}(K),$$

$$H''_q H'_p(K) = \mathrm{Ker}(\partial'': H'_{p,q}(K) \to H'_{p,q-1}(K))/\partial'' H'_{p,q+1}(K).$$

There is a canonical way of associating with each double complex $K = (K_{p,q})_{p,q \in \mathbb{Z}}$ a simple complex, the so-called *total complex* $X = \mathrm{Tot}(K)$ of K. The spaces of the total complex are defined as the 'diagonals'

$$X_n = \prod_{p+q=n} K_{p,q},$$

and the boundary operators $\partial: X_n \to X_{n-1}$ of the total complex act via

$$\partial(x_{p,q})_{p+q=n} = (\partial' x_{p+1,q} + \partial'' x_{p,q+1})_{p+q=n-1}.$$

The double complex is said to possess *bounded diagonals* if for each $n \in \mathbb{Z}$, there are only finitely many integers $p \in \mathbb{Z}$ with $K_{p,n-p} \neq 0$. Using diagram chasing or the theory of spectral sequences (see Mac Lane 1963), one can prove the following central result in the theory of double complexes.

Lemma A2.6 *Let $K = (K_{p,q})_{p,q \in \mathbb{Z}}$ be a double complex with bounded diagonals.*

(a) *Suppose that there is an integer $N \in \mathbb{Z}$ with*

$$H''_{p,q}(K) = 0 \quad for \quad q \neq N.$$

Then for each $n \in \mathbb{Z}$, there is a canonical isomorphism

$$\psi: H_n(X) \to H'_{n-N} H''_N(K).$$

(b) *Suppose that there is an integer $N \in \mathbb{Z}$ with*

$$H'_{p,q}(K) = 0 \quad for \quad p \neq N.$$

Then for each $n \in \mathbb{Z}$, there is a canonical isomorphism

$$\psi: H_n(X) \to H''_{n-N} H'_N(K). \qquad \square$$

Since we need a topological version of this result, we briefly indicate how the isomorphism ψ can be defined. For this purpose, let us suppose that the assumptions of part (a) are satisfied.

We fix $n \in \mathbb{Z}$, and we choose an integer $p > n - N$ such that $K_{i,n-i} = 0$ for $i > p$. Let $a = (a_{i,n-i})_{i \leq p} \in X_n$ be an element with $\partial a = 0$. In other words, we have

$$0 = \partial'' a_{p,n-p}$$

$$0 = \partial' a_{p,n-p} + \partial'' a_{p-1,n-p+1}$$

$$\vdots$$

$$0 = \partial' a_{n-N+1,N-1} + \partial'' a_{n-N,N}$$

$$\vdots$$

Since $n - p < N$, there is a solution of the equation $a_{p,n-p} = \partial'' x_{p,n-p+1}$. Since

$\partial''(a_{p-1,n-p+1} - \partial' x_{p,n-p+1}) = 0$, there is a solution of $a_{p-1,n-p+1} - \partial' x_{p,n-p+1} = \partial'' x_{p-1,n-p+2}$. This procedure can be continued, until one obtains a solution of $a_{n-N+1,N-1} - \partial' x_{n-N+2,N-1} = \partial'' x_{n-N+1,N}$. By assigning to the equivalence class of a in $H_n(X)$ the equivalence class of $a_{n-N,N} - \partial' x_{n-N+1,N}$ in $H'_{n-N}H''_N(K)$, one obtains a map

$$\psi : H_n(X) \to H'_{n-N}H''_N(K).$$

The reader will have no difficulty in checking that ψ is well defined, and in constructing the inverse of ψ.

These explicit descriptions can be used to prove a result about topological double complexes.

Lemma A2.7 *Let* $K = (K_{p,q})_{p,q \in \mathbb{Z}}$ *be a double complex with bounded diagonals consisting of Fréchet A-modules (over a Fréchet algebra A) and continuous A-linear maps.*

(a) *Suppose that all* ∂''*-operators are topological homomorphisms and that all columns are exact except in degree N. Then the maps*

$$\psi : H_n(X) \to H'_{n-N}H''_N(K) \qquad (n \in \mathbb{Z})$$

are topological isomorphisms with respect to the canonical quotient topologies of both sides.

(b) *Suppose that all* ∂'*-operators are topological homomorphisms and that all rows are exact in all degrees except in degree N. Then the maps*

$$\psi : H_n(X) \to H''_{n-N}H'_n(K) \qquad (n \in \mathbb{Z})$$

are topological isomorphisms with respect to the canonical quotient topologies of both sides. $\square$

Appendix 3
K-theory and Riemann–Roch theorems

In this appendix we review some facts and terminology from analytic and topological K-theory, and we discuss various versions of the Riemann–Roch theorem, having different degrees of generality. The appendix serves only as an orientation, and it cannot replace the basic texts in this area, such as Atiyah (1967) or Fulton and Lang (1985). For a concise but relevant survey of K-theory, the reader can also consult Atiyah (1975).

Throughout this appendix we shall consider only complex analytic or algebraic vector bundles. Without explicit mention, all spaces will be supposed Hausdorff, countable at infinity, and (at least) locally compact.

First we recall the definition of the K-theory groups in the category of complex analytic spaces.

Let $(X, \mathscr{O}_X)$ be a complex analytic space, and let $K_a^0(X)$ denote the universal abelian group generated by the isomorphism classes $[E]$ of locally free $\mathscr{O}_X$-modules of finite type E. We set by definition

$$[E] = [E'] + [E'']$$

whenever there exists a short exact sequence of locally free $\mathscr{O}_X$-modules

$$0 \to E' \to E \to E'' \to 0. \tag{A3.1}$$

In fact, $K_a^0(X)$ is a ring with the tensor product as multiplication and with the trivial line bundle $1 = [\mathscr{O}_X]$ as unit.

With each morphism of complex spaces $f: (X, \mathscr{O}_X) \to (Y, \mathscr{O}_Y)$ one associates the natural contravariant (pull-back) map

$$f^*: K_a^0(Y) \to K_a^0(X).$$

Similarly, one defines the universal group $K_0^a(X)$ generated by the classes of coherent $\mathscr{O}_X$-modules, using the same definition of the additive group structure. In general, $K_0^a(X)$ is not a ring, but only a $K_a^0(X)$-module. The reason is that the exactness of short sequences is not preserved when forming the tensor product with arbitrary coherent sheaves.

Both groups, $K_a^0(X)$ and $K_0^a(X)$, are universal objects which represent abelian invariants of vector bundles and coherent sheaves, respectively. Two remarkable examples of such invariants are the following.

The *Chern character* of a vector bundle defines a linear and multiplicative map:

$$\mathrm{ch}: K_a^0(X) \to \bigoplus_{p=0}^{\infty} H^{2p}(X, \mathbb{Q})$$

(see Fulton and Lang 1985, Chapter 1). For a compact complex space $(X, \mathscr{O}_X)$, the *Euler characteristic* $\chi(\mathscr{F})$ of a coherent $\mathscr{O}_X$-module $\mathscr{F}$ is defined by

$$\chi(\mathscr{F}) = \sum_{q=0}^{\infty} (-1)^q \dim_{\mathbb{C}} H^q(X, \mathscr{F}).$$

By the Cartan–Serre theorem, this invariant is well defined (see Theorem 9.1.1). The Euler characteristic is additive with respect to short exact sequences. Hence it defines a group homomorphism

$$\chi \colon K_0^a(X) \to \mathbb{Z}.$$

The relation between these two invariants will be discussed in the second part of this appendix.

In general, there is a natural morphism $K_a^0(X) \to K_0^a(X)$ which associates with a given vector bundle the sheaf of its sections. This morphism is an isomorphism on every smooth projective manifold X, since on such a space every coherent $\mathscr{O}_X$-module admits a finite resolution with vector bundles.

Let $f \colon (X, \mathscr{O}_X) \to (Y, \mathscr{O}_Y)$ be a proper morphism of complex spaces. For every coherent $\mathscr{O}_X$-module $\mathscr{F}$, Grauert's direct-image theorem (see Theorem 9.3.1) shows that the higher direct images $R^q f_* \mathscr{F}$ are coherent $\mathscr{O}_Y$-modules ($q \geq 0$). Hence the map $f_* \colon K_0^a(X) \to K_0^a(Y)$ given by

$$f_*([\mathscr{F}]) = \sum_{q=0}^{\infty} (-1)^q [R^q f_* \mathscr{F}] \qquad (\mathscr{F} \in \mathrm{Coh}(X))$$

is well defined. This direct-image operation turns K_0^a into a covariant functor. In addition, we have the projection formula

$$f_*(\xi f^*(\eta)) = f_*(\xi)\eta, \tag{A3.2}$$

which is valid for each proper holomorphic map $f \colon X \to Y$ and for each pair of elements $\xi \in K_0^a(X)$, $\eta \in K_a^0(Y)$.

The analogous algebraic K-functors are described in their original form (as introduced by Grothendieck), for instance, in Borel and Serre (1958).

Let us turn now to the topological K-theory. Let X be a compact Hausdorff space. The universal abelian group generated by all topological vector bundles of finite rank on X is denoted by $K_t^0(X)$, or simply by $K(X)$ when there is no confusion of notation. The additive operation reduces to

$$[E' \oplus E''] = [E'] + [E''],$$

since every short exact sequence of the form (A3.1) splits at the level of topological vector bundles.

For a locally compact space X, the topological K-group with compact supports is defined by

$$K_t^0(X) = \mathrm{Ker}\Big(K_t^0(\tilde{X}) \xrightarrow{r} K_t^0(\{\infty\}) \Big),$$

where $\tilde{X} = X \cup \{\infty\}$ is the one-point compactification of X and r is the restriction map. The groups K^{-n} are defined by

$$K_t^{-n}(X) = K_t^0(\mathbb{R}^n \times X) \qquad (n \geq 0),$$

and according to the celebrated *Bott Periodicity Theorem* (see Atiyah 1967), there are natural isomorphisms

$$K_t^{-n}(X) \cong K_t^{-n-2}(X) \qquad (n \geq 0).$$

Thus it is natural to define $K_t^n(X) = K_t^{-n}(X)$ for $n > 0$. One can prove that these K-groups form a generalized cohomology theory with products (only the dimension axiom is missing in the system of axioms of Eilenberg and Steenrod). The reduced K-groups (that is, the K-groups modulo a base point) are usually denoted by $\tilde{K}_t^n$ ($n \in \mathbb{Z}$).

For non-compact spaces, there is a second possible definition of a topological K-group, as the inverse limit of the K-groups formed with respect to a compact exhaustion of the given space. In fact, this definition was originally used for non-compact spaces (see Atiyah *et al.* 1964; Atiyah and Hirzebruch 1959, 1962). For spaces having the homotopy type of a finite CW-complex, this K-group (denoted throughout Chapter 10 by $\mathcal{K}$) preserves the basic properties of the K-groups defined before. However, the two spaces are obviously different, as one can see in the case of open balls. Thus, for a locally compact space X that is countable at infinity, we define

$$\mathcal{K}(X) = \varprojlim_{F \in X} K(F).$$

A very useful concept in K-theory, inspired by the theory of elliptic operators, is the so-called *difference construction*. To give the basic example, let X be a locally compact space, and let Y be a closed subset of X. A homomorphism of vector bundles $\alpha: E \to F$ over X, which restricted to Y is an isomorphism, defines the *difference element*

$$d(E, F, \alpha) = [E \cup_\alpha F] - [F \cup_{\mathrm{id}} F] \in \tilde{K}(X/Y) = K(X, Y),$$

where $E \cup_\alpha F$ denotes the vector bundle obtained on X/Y by glueing together $E|_Y$ and $F|_Y$ via α (see Atiyah *et al.* 1964 for details). One proves that the difference elements constructed as above generate the relative group $K(X, Y)$. Similarly, a finite complex of vector bundles on X, which is exact on the subspace Y, determines an element of $K(X, Y)$. For instance, let us consider the Koszul complex of trivial vector bundles on $\mathbb{C}^n$,

$$0 \to \mathbb{C}^n \times \Lambda^0(\mathbb{C}^n) \xrightarrow{d} \mathbb{C}^n \times \Lambda^1(\mathbb{C}^n) \to \cdots \to \mathbb{C}^n \times \Lambda^n(\mathbb{C}^n) \to 0,$$

where the boundary operator d acts as

$$d(v, \xi) = (v, v \wedge \xi) \qquad (v \in \mathbb{C}^n, \xi \in \Lambda^q(\mathbb{C}^n), 0 \le q \le n - 1).$$

This complex is exact on $\{v \ne 0\}$, and thus by the difference construction it defines an element $\tau_n \in K(\mathbb{C}^n, \mathbb{C}^n \setminus \{0\}) = \tilde{K}(S^{2n})$.

This class, called the *Bott element*, generates $\tilde{K}(S^{2n})$ as an abelian group (see Atiyah 1967).

More generally, let $E \xrightarrow{\pi} X$ be a vector bundle over X of rank n, and let us consider, analogously, the twisted Koszul complex on E, regarded as a base space

$$0 \to \Lambda^0(\pi^*E) \xrightarrow{d} \Lambda^1(\pi^*E) \to \cdots \to \Lambda^n(\pi^*E) \to 0.$$

The difference construction yields a class

$$\tau_E \in K(E, E \setminus \{0\}),$$

where 0 denotes the zero section of E. The *Thom isomorphism theorem* in K-theory asserts that the multiplication by τ_E is an isomorphism of $K(X)$-modules

$$K(X) \xrightarrow[\sim]{\tau_E} K(E, E \setminus \{0\}).$$

With the aid of the Thom isomorphism one defines, exactly as in singular cohomology theory, a *Gysin morphism* $f_*: K(X) \to K(Y)$ for every proper differentiable map f between smooth manifolds with the property that the relative cotangent bundle $[T^*X] - [f^*T^*Y]$ carries an almost complex structure (see Atiyah and Hirzebruch 1962).

Compared with the algebraic or analytic theories, the topological K-homology theory was elaborated much later and not without difficulties. Its natural framework appeared only after the new extension theory of (non-commutative) C^*-algebras was well developed (see Brown *et al.* 1973; Kasparov 1988). However, for the purpose of Chapter 10 of this book the analytic formulation of K-homology theory is not needed. It can be avoided by extending the *Alexander duality* from singular cohomology. More precisely, let X be a closed subspace of $\mathbb{C}^n$. Then we set by definition

$$K_q^t(X) = \tilde{K}_t^{q-1}(\mathbb{C}^n \setminus X) \qquad (q \ge 0). \tag{A3.3}$$

Thus K_*^t becomes a K_*^*-module. Furthermore, for a compact complex manifold X, the natural isomorphism $K_0^t(X) \cong K_t^0(X)$ holds. The morphism $a: K_0^t(X) \to K_t^0(X)$ decomposes into factors as follows. Let $X \hookrightarrow \mathbb{C}^n$ be an almost complex embedding of the compact, complex manifold X, and let N denote a tubular neighbourhood of X in $\mathbb{C}^n$ homotopically equivalent to the normal bundle of this embedding. Then the morphism a is obtained as the composition of the following maps:

$$K_0^t(X) = K_t^1(\mathbb{C}^n \setminus X) \overset{\delta}{\longrightarrow} K_t^0(\mathbb{C}^n, \mathbb{C}^n \setminus X) \overset{e}{\underset{\sim}{\longrightarrow}} K_t^0(N, N \setminus X) \overset{\tau_N}{\underset{\sim}{\longleftarrow}} K_t^0(X),$$

where e is the excision isomorphism and τ_N is the Thom isomorphism.

Along the same lines, with every proper map $f: X \to Y$ between topological spaces one associates a functorial morphism $f_*: K_*^t(X) \to K_*^t(Y)$. On compact complex manifolds this push-forward map coincides via the above identification with the Gysin morphism $f_*: K_t^*(X) \to K_t^*(Y)$ (see Fulton and Lang 1985, Chapter 2).

The *Riemann–Roch theorems* were among the first applications of K-theory. In fact, Grothendieck developed algebraic K-theory for this purpose. Next we recall a few statements known as Riemann–Roch theorems. For comprehensive references and details, the reader can consult the monographs of Hirzebruch (1965) and of Fulton and Lang (1985).

Let X be a compact complex manifold, and let E be a holomorphic vector bundle on X. The *Hirzebruch–Riemann–Roch theorem* asserts that the Euler characteristic of the vector bundle E has the following topological expression:

$$\chi(E) = \int_X \mathrm{ch}(E)\mathrm{td}(X). \tag{A3.4}$$

Here $\mathrm{td}(X)$ denotes the Todd characteristic of the complex tangent bundle of X, and the integral means the evaluation of the cohomology class $\mathrm{ch}(E)\mathrm{td}(X)$ on the fundamental cycle of X (see Hirzebruch 1965 for details).

It was Grothendieck who recognized in formula (A3.4) a property of arrows and diagrams, rather than of bundles and cohomology classes. More precisely, let $f: X \to Y$ be a morphism of compact complex manifolds. Then the Chern character and the induced Gysin maps form a diagram

$$
\begin{array}{ccc}
K_t^0(X) & \overset{f_*}{\longrightarrow} & K_t^0(Y) \\
{\scriptstyle ch}\downarrow & & \downarrow{\scriptstyle ch} \\
H^{\mathrm{ev}}(X, \mathbb{Q}) & \overset{f_*}{\longrightarrow} & H^{\mathrm{ev}}(Y, \mathbb{Q})
\end{array}
$$

which in general is non-commutative. A correction of the commutativity of this diagram is expressed in the following *Grothendieck–Riemann–Roch formula:*

$$\mathrm{ch}\, f_*(\xi)\cdot \mathrm{td}(Y) = f_*[\mathrm{ch}(\xi)\cdot \mathrm{td}(X)]. \tag{A3.5}$$

In fact, this formula is purely topological, and it can be proved by elementary computations using only the functorial properties of the Gysin morphisms and the projection formula (A3.2) (see Fulton and Lang 1985, Chapter 2).

Formula (A3.4) can easily be deduced from (A3.5) by choosing Y as a one-point space. Indeed, for a holomorphic vector bundle E on X, one has

$$\mathrm{ch}\, f_*([E]) = \mathrm{ch}\left(\sum_{q=0}^{\infty} (-1)^q R^q f_*(E) \right) = \chi(E),$$

while on the other hand $\mathrm{td}(Y) = 1$ and

$$f_*(\mathrm{ch}(\xi)\mathrm{td}(X)) = \int_X \mathrm{ch}(\xi)\mathrm{td}(X).$$

In the previous computation we implicitly supposed that the diagram

$$
\begin{array}{ccc}
K_a^0(X) & \xrightarrow{f_*} & K_a^0(Y) \\
\varphi_X \downarrow & & \downarrow \varphi_Y \\
K_t^0(X) & \xrightarrow{f_*} & K_t^0(Y)
\end{array}
$$

is commutative. Here φ denotes the functor obtained by forgetting the analytic structure of a given vector bundle. Modulo the purely topological computation contained in formula (A3.5), the commutativity of the diagram (A3.6) represents the core of the *Grothendieck–Riemann–Roch theorem* (see Fulton and Lang 1985, Chapter 5). The next step would be to give up the smoothness of the base spaces X and Y, and therefore to pass from K-cohomology to K-homology. It is the latter one which carries a general push-forward map, and for which the diagram (A3.6) makes sense.

Thus for non-smooth complex spaces or for coherent sheaves on manifolds, it is natural to conjecture a Riemann–Roch theorem in the following form. There exists a functorial transformation $\alpha_X \colon K_0^a(X) \to K_0^t(X)$ in the category of complex spaces such that, for every proper morphism $f \colon X \to Y$, the diagram

$$
\begin{array}{ccc}
K_0^a(X) & \xrightarrow{f_*} & K_0^a(Y) \\
\alpha_X \downarrow & & \downarrow \alpha_Y \\
K_0^t(X) & \xrightarrow{f_*} & K_0^t(Y)
\end{array}
$$

is commutative. Moreover, if X is a projective smooth manifold, then α_X must coincide with φ_X (see Atiyah 1975 for the formulation of this conjecture).

The latter conjecture was proved by Baum, Fulton, and MacPherson for projective (singular) spaces, by Toledo and Tong for compact complex manifolds (and coherent sheaves), and in its full generality by R. Levy (see Chapter 6 in Fulton and Lang 1985 for details and comments).

Appendix 4
Sobolev spaces

In Chapter 6 we need some elementary results for Sobolev-type spaces on open sets in $\mathbb{C}^n$ formed with respect to the $\bar{\partial}$-operator alone. For the convenience of the reader, we gather the basic facts.

Let X be a Banach space. For each open set U in $\mathbb{C}^n$ and each non-negative integer k, the space

$$W^k(U, X) = \{f \in L^2(U, X); \bar{\partial}^j f \in L^2(U, X) \text{ for all } j \in \mathbb{N}^n \text{ with } |j| \le k\}$$

is a Banach space with respect to the norm $\|f\| = (\Sigma_{|j| \le k} \|\bar{\partial}^j f\|_2^2)^{1/2}$. For each $f \in L^2(U, X)$ and each $j \in \mathbb{N}^n$, the derivative $\bar{\partial}^j f = (\partial/\partial \bar{z}_1)^{j_1} \cdots (\partial/\partial \bar{z}_n)^{j_n} f$ can be formed as a distribution. We write $\bar{\partial}^j f \in L^2(U, X)$ if this distribution can be represented by a function in $L^2(U, X)$. Instead of $W^k(U, \mathbb{C})$ we write $W^k(U)$. As usual we denote by $C^k(U)$ the space of all functions f on U such that all partial derivatives of f of order less than or equal to k relative to the $2n$ real coordinates of $\mathbb{C}^n$ exist in the ordinary sense and are continuous.

Completely analogous to the classical *Sobolev embedding lemma* one obtains the following result.

Theorem A4.1 *For each open set U in $\mathbb{C}^n$ and each non-negative integer k, we have the inclusion*

$$W^{k+2n}(U) \subset C^k(U).$$

Proof Let $f \in W^{k+2n}(U)$ be fixed. We briefly indicate how one can show that f has a representative in $C^k(U)$.

(1) Without loss of generality we may suppose that $U = \mathbb{C}^n$ and that f vanishes outside a compact set K in $\mathbb{C}^n$. Otherwise, one can choose an exhaustion $(U_j)_{j \ge 1}$ of U by open sets U_j with $U_j \subset U_{j+1}$ for all j and a sequence (θ_j) in $\mathscr{D}(U)$ such that $\theta_j = 1$ on U_j. Then $\theta_j f$ can be regarded as an element in $W^{k+2n}(\mathbb{C}^n)$, and hence for each j, we could choose a representative $g_j \in C^k(\mathbb{C}^n)$ of $\theta_j f$. Obviously, then there would be a function $g \in C^k(U)$ with $g | U_j = g_j | U_j$ for each j. Consequently, g would be a representative of f.

(2) Suppose that $f \in W^{k+2n}(\mathbb{C}^n)$ vanishes outside a compact set K in $\mathbb{C}^n$. If $\varphi \in \mathscr{D}(\mathbb{C}^n)$ and if $1 \le i \le n$, then $f * \varphi \in \mathscr{D}(\mathbb{C}^n)$ and one can show that $\bar{\partial}_i(f * \varphi) = (\bar{\partial}_i f) * \varphi$ by proving that both sides coincide as distributions. In fact, an inductive argument shows that $\bar{\partial}^j(f * \varphi) = (\bar{\partial}^j f) * \varphi$ for all $j \in \mathbb{N}^n$ with $|j| \le k + 2n$.

(3) Let f be as in the previous step. We claim that there is a constant $c > 0$, only depending on f and the dimension n of the underlying space, such that for each

function $\varphi \in \mathscr{D}(\mathbb{C}^n)$ with $\operatorname{supp}(\varphi) \subset B_1(0)$ (Euclidean ball of radius 1) and each pair of multi-indices $\alpha, \beta \in \mathbb{N}^n$ with $|\alpha| + |\beta| \le k$, the estimate

$$\|\partial^\alpha \bar{\partial}^\beta (f * \varphi)\|_{\infty, \mathbb{C}^n} \le c(\bar{\partial}^{\alpha + \beta + (2, \dots, 2)} f) * \varphi\|_{2, \mathbb{C}^n} \tag{A4.1}$$

holds. First note that because of Stokes' theorem, for $g \in \mathscr{D}(\mathbb{C}^n)$ and $j = 1, \dots, n$,

$$\int |\partial_j g|^2 \, dz = \int (\partial_j g)(\bar{\partial}_j \bar{g}) dz = -\int (\bar{\partial}_j \partial_j g) \bar{g} \, dz$$

$$= -\int (\bar{\partial}_j \partial_j \bar{g}) g \, dz = \int (\partial_j \bar{g})(\bar{\partial}_j g) \, dz = \int |\bar{\partial}_j g|^2 \, dz,$$

where the integration is performed over all of $\mathbb{C}^n$. Hence for g and j as above,

$$\|(\bar{\partial}_j \pm \partial_j) g\|_{2, \mathbb{C}^n} \le 2\|\bar{\partial}_j g\|_{2, \mathbb{C}^n}.$$

If $x_1, \dots, x_n, y_1, \dots, y_n$ denote the $2n$ real variables of $\mathbb{C}^n$, then for $g \in \mathscr{D}(\mathbb{C}^n)$,

$$\|(\partial^{2n} g / \partial x_1 \cdots \partial y_n)\|_{2, \mathbb{C}^n} \le 4^n \|\bar{\partial}^{(2, \dots, 2)} g\|_{2, \mathbb{C}^n}.$$

If $z_j = x_j + iy_j$ for $j = 1, \dots, n$, then

$$g(z) = \int_{-\infty}^{x_1} \cdots \int_{-\infty}^{y_n} (\partial^{2n} g / \partial s_1 \cdots \partial t_n) ds_1 \cdots dt_n.$$

Therefore, using the Cauchy–Schwarz inequality, one obtains the estimate

$$\|g\|_{\infty, \mathbb{C}^n} \le \lambda(\operatorname{supp} g)^{1/2} 4^n \|\bar{\partial}^{(2, \dots, 2)} g\|_{2, \mathbb{C}^n} \tag{A4.2}$$

for each function $g \in \mathscr{D}(\mathbb{C}^n)$. These observations yield the claimed estimate (A4.1).

(4) To conclude the proof, we choose a Dirac sequence (φ_i) in $\mathscr{D}(\mathbb{C}^n)$, and we note that the estimate (A4.1) implies that, for $|\alpha| + |\beta| \le k$, the sequence $\partial^\alpha \bar{\partial}^\beta (f * \varphi_k)$ converges uniformly on all of $\mathbb{C}^n$. □

In the vector-valued case, we prove the following weaker version of the last result.

Corollary A4.2 *For $U \subset \mathbb{C}^n$ open, we have the inclusion*

$$\bigcap_{k \ge 0} W^k(U, X) \subset \mathscr{E}(U, X).$$

Proof Let $f \in L^2(U, X)$ be a function such that $\bar{\partial}^j f \in L^2(U, X)$ for all $j \in \mathbb{N}^n$. We have to show that f has a representative in $\mathscr{E}(U, X)$.

As in the preceding proof, we may suppose that $U = \mathbb{C}^n$ and that f vanishes outside a compact subset of $\mathbb{C}^n$. For every measurable function $g: \mathbb{C}^n \to X$, there is a Lebesgue null set N with $g(\mathbb{C}^n \setminus N)$ separable. Thus we may suppose that X itself is separable.

By Theorem A4.1, for each $u \in X'$, there is a representative $f_u \in \mathscr{D}(\mathbb{C}^n)$ of the L^2-function $u \circ f$. The induced map

$$X' \to \mathscr{D}(\mathbb{C}^n), \qquad u \mapsto f_u$$

is obviously $\mathbb{C}$-linear. If (u_k) is a sequence in X' with w^*-limit 0, then by the dominated convergence theorem and (A4.2), we obtain that

$$\|f_{u_k}\|_{\infty, \mathbb{C}^n} \le M \|u_k \circ (\bar{\partial}^{(2, \dots, 2)} f)\|_{2, \mathbb{C}^n} \xrightarrow{k} 0$$

with a suitable constant $M > 0$. Since X is separable, the closed dual unit ball B_X equipped with the relative w^*-topology is metrizable (§21.3.(4) in Köthe 1969), and it follows from the Krein–Smulian theorem (Corollary IV.6.4 in Schaefer 1966) that there is a function $F: \mathbb{C}^n \to X$ with $f_u(z) = \langle F(z), u \rangle$ for all $z \in \mathbb{C}^n$ and $u \in X'$. As a weak C^∞-function, F belongs to $\mathscr{E}(\mathbb{C}^n, X)$, and $u \circ F = u \circ f$ holds almost everywhere on $\mathbb{C}^n$ for each $u \in X'$.

Since X is separable, there is a sequence (u_k) in X' with the property that each $u \in X'$ is the w^*-limit of a suitable subsequence of (u_k) (see §21.3.(5) in Köthe 1969). Hence there is a null set N in $\mathbb{C}^n$ such that $f = F$ outside N. $\qquad\square$

Let X be a Banach space, and let $U \subset \mathbb{C}^n$ be open. Then for each multi-index $k \in \mathbb{N}^n$, the space

$$W^k(U, X) = \{f \in L^2(U, X); \bar{\partial}^j f \in L^2(U, X) \text{ for all } j \in \mathbb{N}^n \text{ with } 0 \leq j \leq k\}$$

is a Banach space relative to $\|f\| = (\sum_{0 \leq j \leq k} \|\bar{\partial}^j f\|_2^2)^{1/2}$. Exactly as for ordinary Sobolev spaces (see Treves 1967), one obtains the following density results.

Lemma A4.3 *Let $U \subset \mathbb{C}^n$ be open, and let k be either a non-negative integer or a multi-index $k \in \mathbb{N}^n$. Then each function $f \in W^k(U, X)$ vanishing outside a compact subset K of U can be approximated in the norm of $W^k(U, X)$ by functions in $\mathscr{D}(U, X)$.*

Proof If $f \in W^k(U, X)$ is as above, then its trivial extension $\tilde{f}$ belongs to $W^k(\mathbb{C}^n, X)$ and, for $|j| \leq k$ ($0 \leq j \leq k$, respectively) and $\varphi \in \mathscr{D}(\mathbb{C}^n)$, we have

$$\bar{\partial}^j(\tilde{f} * \varphi) = (\bar{\partial}^j \tilde{f}) * \varphi,$$

where $\bar{\partial}^j \tilde{f}$ coincides with the trivial extension of $\bar{\partial}^j f$. If (φ_i) is a Dirac sequence in $\mathscr{D}(\mathbb{C}^n)$, then

$$\bar{\partial}^j(\tilde{f} * \varphi_i) = (\bar{\partial}^j \tilde{f}) * \varphi_i$$

restricted to U converges to $\bar{\partial}^j f$ in $L^2(U, X)$ for $|j| \leq k$ ($0 \leq j \leq k$, respectively). $\qquad\square$

For sufficiently nice domains U in $\mathbb{C}^n$, the last result can be used to prove the density of $\mathscr{E}(\mathbb{C}^n, X)$ in $W^k(U, X)$. We only consider a very particular case.

Lemma A4.4 *Suppose that D is an open polydisc in $\mathbb{C}^n$. Then for each non-negative integer k (for each $k \in \mathbb{N}^n$, respectively), the space $\mathscr{E}(\mathbb{C}^n, X)$ is a dense subspace of $W^k(D, X)$.*

Proof For simplicity, we suppose that D is an open polydisc with centre 0.

For $f \in L^2(D, X)$ and $0 < r < 1$, we denote by f_r the L^2-function on $D_r = (1/r)D$ defined by $f_r(z) = f(rz)$. If $f \in W^k(D, X)$, then an elementary computation shows that $f_r \in W^k(D_r X)$ and that

$$\bar{\partial}^j(f_r) = r^{|j|}(\bar{\partial}^j f)_r$$

for each j with $|j| \leq k$ ($0 \leq j \leq k$, respectively). In particular, for $f \in W^k(D, X)$, we obtain that

$$f = \lim_{r \uparrow 1} (f_r | D)$$

in $W^k(D, X)$. To conclude the proof it suffices to apply the preceding lemma. $\qquad\square$

In Chapter 7 we also consider the standard L^p-*Sobolev spaces* $H^{p,m}(\Omega)$ *of order* $m \in \mathbb{Z}$ on open sets Ω in $\mathbb{R}^n$. To recall the definition, let us fix a real number p with $1 \leq p \leq \infty$ and an integer $m \in \mathbb{Z}$. For $m \geq 0$, the space

$$H^{p,m}(\Omega) = \{u \in \mathscr{D}'(\Omega); D^\alpha u \in L^p(\Omega) \text{ for all } \alpha \in \mathbb{N}^n \text{ with } |\alpha| \leq m\}$$

is a Banach space with respect to the norm

$$\|u\|_{p,m} = \left(\sum_{|\alpha| \leq m} \|D^\alpha u\|_p^p \right)^{1/p}$$

if $p < \infty$ and $\|u\|_{p,m} = \max_{|\alpha| \leq m} \|D^\alpha u\|_\infty$ if $p = \infty$. For $m < 0$, the space $H^{p,m}(\Omega)$ consists of all distributions $u \in \mathscr{D}'(\Omega)$ with the property that there are functions $g_\alpha \in L^p(\Omega)$ ($|\alpha| \leq |m|$) with $u = \sum_{|\alpha| \leq |m|} D^\alpha g_\alpha$. These spaces are Banach spaces with respect to the norm

$$\|u\|_{p,m} = \inf\left(\sum_{|\alpha| \leq |m|} \|g_\alpha\|_p^p \right)^{1/p},$$

where the infimum is formed over all possible representations $u = \sum_{|\alpha| \leq |m|} D^\alpha g_\alpha$ as above. Again, in the case $p = \infty$, the l^p-norm has to be replaced by the maximum norm in the definition of $\| \ \|_{p,m}$.

For more details on these classical Sobolev spaces, we refer the reader to the monograph of Treves (1967).

Bibliography

Agrawal, O. P., Clark, D. N., and Douglas, R. G. (1986). Invariant subspaces in the polydisk. *Pacific J. Math.*, **121**, 1–11.

Albrecht, E. (1972). Funktionalkalküle in mehreren Veränderlichen. Dissertation, University of Mainz.

Albrecht, E. (1974). Funktionalkalküle in mehreren Veränderlichen für stetige lineare Operatoren auf Banachräumen. *Manuscripta Math.*, **14**, 1–40.

Albrecht, E. (1978). On two questions of Colojoară and Foiaş. *Manuscripta Math.*, **25**, 1–15.

Albrecht, E. (1979). On joint spectra. *Studia Math.*, **64**, 263–271.

Albrecht, E. (1981). Spectral decompositions for systems of commuting operators. *Proc. Royal Irish Acad.*, **81A**, 81–98.

Albrecht, E. (1982a). Decomposable systems of operators in harmonic analysis. In *Toeplitz Centennial. Toeplitz Memorial Conference in Operator Theory* (ed. I. Gohberg). Birkhäuser, Basel, pp. 19–35.

Albrecht, E. (1982b). Several variable spectral theory in the non-commutative case. *Banach Center Publications*, **8**, 9–30, Polish Scientific Publ., Warsaw.

Albrecht, E. and Chevreau, B. (1987). Invariant subspaces for l^p-operators having Bishop's property (β) on a large part of their spectrum. *J. Operator Theory*, **18**, 339–372.

Albrecht, E. and Eschmeier, J. (1987). Functional models and local spectral theory. Unpublished manuscript.

Albrecht, E. and Mehta, R. D. (1984). Some remarks on local spectral theory. *J. Operator Theory*, **12**, 285–317.

Albrecht, E. and Vasilescu, F.-H. (1974). Non-analytic local spectral properties in several variables. *Czech. Math. J.*, **24**, 430–443.

Albrecht, E. and Vasilescu, F.-H. (1983). Semi-Fredholm complexes. *Operator Theory: Adv. Appl.*, **11**, 15–39.

Allan, G. R. (1967). On one sided inverses in Banach algebras of holomorphic vector-valued functions. *J. London Math. Soc.* (1), **42**, 463–470.

Ambrozie, C.-G. and Vasilescu, F.-H. (1995). *Banach space complexes*. Kluwer Academic Publishers, Dordrecht.

Apostol, C. (1968). Spectral decompositions and functional calculus. *Rev. Roumaine Math. Pures Appl.*, **13**, 1481–1528.

Apostol, C. (1981). The spectral flavour of Scott Brown's techniques. *J. Operator Theory*, **6**, 3–12.

Apostol, C., Fialkow, L. A., Herrero, D., and Voiculescu, D. (1984). *Approximation of Hilbert space operators*. Res. Notes Math. Vol. 102, Pitman, London.

Arens, R. (1963). The group of invertible elements of a commutative Banach algebra. *Studia Math.*, **11**, 21–23.

Arens, R. and Caldèron, A. P. (1955). Analytic functions of several Banach algebra elements. *Ann. Math.*, **62**, 204–216.

Arveson, W. (1976). *An invitation to C*-algebras*. Springer-Verlag, New York.

Atiyah, M. F. (1967). *K-theory*. Benjamin, New York.

Atiyah, M. F. (1975). *A survey of K-theory*. Lect. Notes in Math. Vol. 575. Springer-Verlag, Berlin, pp. 1–9.

Atiyah, M. F. and Bott, R. (1967–68). A Lefschetz fixed point formula for elliptic complexes. I. *Ann. Math.*, **86**, 347–407; II. *Ann. Math.*, **88**, 451–491.

Atiyah, M. F. and Hirzebruch, F. (1959). Riemann–Roch theorems for differentiable manifolds. *Bull. Amer. Math. Soc.*, **65**, 276–281.

Atiyah, M. F. and Hirzebruch, F. (1962). The Riemann–Roch theorem for analytic embeddings. *Topology*, **1**, 151–166.

Atiyah, M. F. and Singer, I. M. (1968). The index of elliptic operators. III. *Ann. Math.*, **87**, 546–604.

Atiyah, M. F., Bott, R., and Shapiro, A. (1965). Clifford modules. *Topology*, **3**, 3–38.

Axler, S. (1982). Multiplication operators on Bergman spaces. *J. reine angew. Math.*, **336**, 26–44.

Axler, S. (1988). Bergman spaces and their operators. In *Surveys of some recent results in operator theory*, Vol. I (eds. J. B. Conway and B. B. Morrel). Longman Sci. Techn., Harlow, pp. 1–50.

Bănică, C. and Stănăşilă, O. (1977). *Méthodes algébriques dans la théorie globale des espaces complexes*. Gauthier-Villars, Paris.

Bartle, R. G. and Graves, L. M. (1952). Mappings between function spaces. *Trans. Amer. Math. Soc.*, **72**, 400–413.

Baum, P., Fulton, W., and MacPherson, R. (1979). Riemann–Roch and topological K-theory for singular varieties. *Acta. Math.*, **143**, 155–192.

Bercovici, H., Foiaş, C., and Pearcy, C. (1985). *Dual algebras with applications to invariant subspaces and dilation theory*. CBMS Regional Conference Series in Math., Vol. 56, Amer. Math. Soc., Providence, Rhode Island.

Bercovici, H., Foiaş, C., and Pearcy, C. (1988). Two Banach space methods and dual operator algebras. *J. Funct. Analysis*, **78**, 306–345.

Berezanskii, Yu. M. (1968). Expansion in eigenfunctions of self-adjoint operators. *Amer. Math. Soc.*, Providence, Rhode Island.

Bingener, J. and Kosarew, S. (1987). *Lokale Modulräume in der analytischen Geometrie*. Vieweg Verlag, Braunschweig.

Bishop, E. (1957). Spectral theory for operators on a Banach space. *Trans. Amer. Math. Soc.*, **86**, 414–445.

Bishop, E. (1959). A duality theorem for an arbitrary operator. *Pacific J. Math.*, **9**, 379–394.

Borel, A. and Serre, J. P. (1958). Le théorème de Riemann–Roch (d'après Grothendieck). *Bull. Soc. Math. France*, **86**, 97–136.

Bourbaki, N. (1967). *Théories spectrales*. Hermann, Paris.

Boutet de Monvel, L. and Guillemin, V. (1981). *The spectral theory of Toeplitz operators*. Ann. Math. Studies Vol. 99, Princeton, New Jersey.

Bredon, G. E. (1967). *Sheaf theory*. McGraw-Hill, New York.

Brown, L. G., Douglas, R. G., and Fillmore, P. A. (1973). *Unitary equivalence modulo the compact operators and extensions of C*-algebras*. Lect. Notes Math. Vol. 345. Springer-Verlag, Berlin.

Brown. S. (1978). Some invariant subspaces for subnormal operators. *Integral Equations Operator Theory*, **1**, 310–333.

Brown, S. (1987). Hyponormal operators with thick spectra have invariant subspaces. *Ann. Math.*, **125**, 93–103.

Brown, S. and Chevreau, B. (1988). Toute contraction à calcul fonctionel isometrique est reflexive. *C.R. Acad. Sci. Paris*, **307**, 185–188.

Bungart, L. (1964). Holomorphic functions with values in locally convex spaces and applications to integral formulas. *Trans. Amer. Math. Soc.*, **111**, 317–344.

Carey, R. W. and Pincus, J. D. (1977). Mosaics, principal functions, and mean motion in von Neumann algebras. *Acta Math.*, **138**, 153–218.

Carey, R. W. and Pincus, J. D. (1985). Principal currents. *Integral Equations Operator Theory*, **8**, 614–640.

Cartan, H. (1940). Sur les matrices holomorphes de *n* variables complexes. *J. Math. Pures Appl.*, **19**, 1–26.

Cartan, H. (1961). Familles d'espaces complexes et fondements de la géometrie analytique. *Séminaire E.N.S.* 1960–61.

Cartan, H. and Eilenberg, S. (1956). *Homological algebra*. Princeton University Press, Princeton, New Jersey.

Cartan, H. and Serre, J. P. (1953). Un théorème de finitude concernant les variétés analytiques compactes. *C.R. Acad. Sci. Paris*, **237**, 128–130.

Chirka, E. M. (1982). Regularity of the boundaries of analytic sets (in Russian). *Mat. Sbornik*, **117**, 291–336; English translation: *Math. USSR, Sb.* **45**, 291–335 (1983).

Chirka, E. M. (1985). *Complex analytic sets* (in Russian). Nauka, Moscow.

Colojoară, I. and Foiaş, C. (1968). *Theory of generalized spectral operators*. Gordon and Breach, New York.

Connes, A. (1985). Non commutative differential geometry. *Publ. Math. IHES*, **62**, 257–360.

Connes, A. (1994). *Noncommutative geometry*. Academic Press, San Diego.

Connes, A. and Moscovici, H. (1990). Cyclic cohomology, the Novikov conjecture and hyperbolic groups. *Topology*, **29**, 345–388.

Conway, J. B. (1991). *The theory of subnormal operators*. Math. Surveys and Monographs Vol. 38. Amer. Math. Soc., Providence, Rhode Island.

Conway, J. B. and Dudziak, J. J. (1990). Von Neumann operators are reflexive. *J. reine angew. Math.*, **408**, 34–56.

Curto, R. E. (1981). Fredholm and invertible tuples of operators. The deformation problem. *Trans. Amer. Math. Soc.*, **226**, 129–159.

Curto, R. E. (1988). Applications of several complex variables to multiparameter spectral theory. In *Surveys of some recent results in operator theory*. Vol. II (J. B. Conway and B. B. Morrel (eds.)). Longman Sci. Techn., Harlow, pp. 25–90.

Curto, R. E. and Muhly, P. S. (1985). C^*-algebras of multiplication operators on Bergman spaces. *J. Funct. Analysis.*, **64**, 315–329.

Curto, R. E. and Salinas, N. (1985). Spectral properties of cyclic subnormal *m*-tuples. *Amer. J. Math.*, **107**, 113–138.

Dales, H. G. (1978). Automatic continuity: a survey. *Bull. London Math. Soc.*, **10**, 129–183.

Deimling, K. (1974). *Nichtlineare Gleichungen und Abbildungsgrade*. Springer-Verlag, Berlin.

Dirac, P. A. M. (1930). *The principles of quantum mechanics*. Clarendon Press, Oxford.

Douady, A. (1966). Le problème des modules pour les sous-espaces analytiques compacts d'un espace analytique donné. *Ann. Inst. Fourier*, **16**, 1–95.

Douady, A., Frish, J., and Hirschowitz, A. (1972). Recouvrements privilégiés, *Ann. Inst. Fourier*, **22**, 59–96.

Douglas, R. G. (1980). *C^*-algebra extensions and K-homology*. Ann. Math. Studies Vol. 95. Princeton University Press, Princeton, New Jersey.

Douglas, R. G. and Paulsen, V. (1989). *Hilbert modules over function algebras*. Pitman Res. Notes Math. Vol. 219, Harlow.

Douglas, R. G. and Yan, K. (1990). On the rigidity of Hardy submodules. *Integral Equations Operator Theory*, **13**, 350–363.

Douglas, R. G. and Yan, K. (1992). A multidimensional Berger–Shaw inequality. *J. Operator Theory*, **27**, 205–217.

Douglas, R. G. and Voiculescu, D. (1981). On the smoothness of sphere extensions. *J. Operator Theory*, **6**, 103–111.

Douglas, R. G., Paulsen, V. I., Sah, C. H., and Yan. K. (1995). Algebraic reduction and rigidity for Hilbert modules. *Amer. J. Math.*, **117**, 75–92.

Dunford, N. (1954). Spectral operators. *Pacific J. Math.*, **4**, 321–354.

Dunford, N. and Schwartz, J. T. (1958). *Linear operators*. Part I. Wiley-Interscience, New York.

Dunford, N. and Schwartz, J. T. (1963). *Linear operators*. Part II. Wiley-Interscience, New York.

Dunford, N. and Schwartz, J. T. (1971). *Linear operators*. Part III. Wiley-Interscience, New York.

Duren, P., Khavinson, D., Shapiro, H. S., and Sundberg, C. (1992). Contractive zero-divisors in Bergman spaces. *Contemp. Math.*, **137**, 217–220.

Eilenberg, S. and Moore, J. C. (1965). *Foundation of relative homological algebra*. Mem. Amer. Math. Soc. Vol. 55, Providence, Rhode Island.

Eschmeier, J. (1982a). Operator decomposability and weakly continuous representations of locally compact abelian groups. *J. Operator Theory*, **7**, 201–208.

Eschmeier, J. (1982b). Local properties of Taylor's analytic functional calculus. *Invent. Math.*, **68**, 103–116.

Eschmeier, J. (1983). Equivalence of decomposability and 2-decomposability for several commuting operators. *Math. Ann.*, **262**, 305–312.

Eschmeier, J. (1984a). Some remarks concerning the duality problem for decomposable systems of commuting operators. *Operator Theory: Advances and Appl.*, **14**, 115–123, Birkhäuser, Basel.

Eschmeier, J. (1984b). Are commuting systems of decomposable operators decomposable? *J. Operator Theory*, **12**, 213–219.

Eschmeier, J. (1985). Spectral decompositions and decomposable multipliers. *Manuscripta math.*, **51**, 201–224.

Eschmeier, J. (1987a). Analytic spectral mapping theorems for joint spectra. *Operator Theory: Advances and Appl.*, **24**, 167–181, Birkhäuser, Basel.

Eschmeier, J. (1987b). *Analytische Dualität und Tensorprodukte in der mehrdimensionalen Spektraltheorie*. Schriftenreihe des Math. Inst. der Universität Münster, Heft 42, 2. Serie, 132 pp., Münster.

Eschmeier, J. (1988a). A decomposable Hilbert space operator which is not strongly decomposable. *Integral Eq. Operator Theory*, **11**, 162–172.

Eschmeier, J. (1988b). Tensor products and elementary operators. *J. reine angew. Math.*, **390**, 47–66.

Eschmeier, J. (1989). Operators with rich invariant subspace lattices. *J. reine angew. Math.*, **396**, 41–69.

Eschmeier, J. (1992a). Analytically parametrized complexes of Banach spaces. In *Proc. Conf. on Elementary Operators and Applications* (ed. M. Mathieu), pp. 163–177, World Sci., Singapore.

Eschmeier, J. (1992b). Bishop's condition (β) and joint invariant subspaces. *J. reine angew. Math.*, **426**, 1–22.

Eschmeier, J. (1993). Representations of AlgLat(T). *Proc. Amer. Math. Soc.*, **117**, 1013–1021.

Eschmeier, J. and Prunaru, B. (1990). Invariant subspaces for operators with Bishop's property (β) and thick spectrum. *J. Funct. Analysis*, **94**, 196–222.

Eschmeier, J. and Putinar, M. (1984). Spectral theory and sheaf theory. III. *J. reine angew. Math.*, **354**, 150–163.

Eschmeier, J. and Putinar, M. (1987). The sheaf $\mathscr{E}$ is not topologically flat over $\mathscr{O}$. *Ann. University Craiova*, **XV**, 1–4.

Eschmeier, J. and Putinar, M. (1988). Bishop's condition (β) and rich extensions of linear operators. *Indiana Univ. Math. J.*, **37**, 325–348.

Eschmeier, J. and Putinar, M. (1989). On quotients and restrictions of generalized scalar operators. *J. Funct. Analysis*, **84**, 115–134.

Eschmeier, J. and Putinar, M. (1993). Spectra of analytic Toeplitz tuples on Bergman spaces. *Acta Math. Sci.* (Szeged), **57**, 85–101.

Eschmeier, J. and Putinar, M. (1995). The finite fibre problem and an index formula for elementary operators. *Proc. Amer. Math. Soc.*, **123**, 743–746.

Fainshtein, A. S. (1980). On the joint essential spectrum of a family of linear operators (in Russian). *Funct. Anal. Appl.*, **14**, 83–84.

Fainshtein, A. S. and Shulman, V. S. (1980). On Fredholm complexes of Banach spaces (in Russian). *Funct. Anal. Appl.*, **14**, 87–88.

Faltings, G. (1992). *Lectures on the arithmetic Riemann–Roch theorem*. Ann. Math. Studies Vol. 127. Princeton University Press, Princeton, New Jersey.

Fischer, G. (1976). *Complex analytic geometry*. Lect. Notes Math. Vol. 538. Springer-Verlag, Berlin.

Floret, K. and Wloka, J. (1968). *Einführung in die Theorie der lokalkonvexen Räume*. Lect. Notes Math. Vol. 56. Springer-Verlag, Berlin.

Foiaş, C. (1963). Spectral maximal spaces and decomposable operators in Banach spaces. *Arch. Math.*, **14**, 341–349.

Foiaş, C. (1968). Spectral capacities and decomposable operators. *Rev. Roum. Math. Pures Appl.*, **13**, 1539–1545.

Foiaş, C. (1970). On the maximal spectral spaces of a decomposable operator. *Rev. Roum. Math. Pures. Appl.*, **15**, 1599–1606.

Foiaş, C. (1972). On the scalar part of a decomposable operator. *Rev. Roum. Math. Pures Appl.*, **17**, 1181–1198.

Foiaş, C. and Vasilescu, F.-H. (1970). On the spectral theory of commutators. *J. Math. Analysis Appl.*, **31**, 473–486.

Foiaş, C. and Vasilescu, F.-H. (1974). Non-analytic local functional calculus. *Czech. Math. J.*, **24**, 270–283.

Forster, O. (1967). Zur Theorie des Steinschen Algebren und Moduln. *Math. Z.*, **97**, 376–405.

Forster, O. (1974). Funktionentheoretische Hilfsmittel in der Theorie der kommutativen Banachalgebren. *Jber. Deutsch. Math.-Verein.*, **76**, 1–17.

Forster, O. and Knorr, K. (1971). Ein Beweis des Grauertschen Bildgarbensatzes nach Ideen von B. Malgrange. *Manuscripta math.*, **5**, 19–44.

Forster, O. and Knorr, K. (1972). Relativ-analytische Räume und die Kohärenz von Bildgarben. *Invent. Math.*, **16**, 113–160.

Freed, D. (1988). An index theorem for families of Fredholm operators parametrized by a group. *Topology*, **27**, 279–300.

Frisch, J. (1967). Points de platitude d'un morphisme d'espaces analytiques complexes. *Invent. Math.*, **4**, 118–138.

Frunză, Şt. (1975a). The Taylor spectrum and spectral decompositions. *J. Funct. Analysis*, **19**, 390–421.

Frunză, Şt. (1975b). An axiomatic theory of spectral decompositions of operators. I (in Romanian). *St. Cerc. Math.*, **27**, 655–711.

Frunză, Şt. (1976). Spectral decompositions and duality. *Illinois J. Math.*, **20**, 314–321.

Frunză, Şt. (1977). A characterization for the spectral capacity of a finite system of operators. *Czech. Math. J.*, **27**, 356–362.

Fulton, W. and Lang, S. (1985). *Riemann–Roch algebra*. Springer-Verlag, New York.

Gamelin, T. W. (1970). Localization of the corona problem. *Pacific J. Math.*, **34**, 73–81.

Gamelin, T. W. (1980). Wolff's proof of the corona theorem. *Israel J. Math.*, **37**, 113–119.

Garnett, J. B. (1981). *Bounded analytic functions*. Academic Press, New York.

Gelfand, S. I. and Manin, Yu. I. (1988). *Methods of homological algebra*. Vol. I., Nauka, Moscow.

Gleason, A. M. (1963). The Cauchy–Weil theorem. *J. Math. Mech.*, **12**, 429–444.

Gleason, A. M. (1964). Finitely generated ideals in Banach algebras. *J. Math. Mech.*, **13**, 125–132.

Godement, R. (1958). *Théorie des faisceaux*. Hermann, Paris.

Gohberg, I. and Krein, M. G. (1970). *Theory and applications of Volterra operators in Hilbert space*. Transl. Amer. Math. Soc. Vol. 24. Providence, Rhode Island.

Golovin, V. D. (1986). *Homology of analytic sheaves and duality theorems*. Nauka, Moscow.

Gramsch, B. (1967). Funktionalkalkül mehrerer Veränderlicher in lokalbeschränkten Algebren. *Math. Ann.*, **174**, 311–344.

Grauert, H. (1960). *Ein Theorem der Analytischen Garbentheorie und die Modulräume Komplexer Strukturen*. Publ. Math. IHES Vol. 5.

Grauert, H. and Remmert, R. (1979). *Theory of Stein spaces*. Springer-Verlag, Berlin.

Grauert, H. and Remmert, R. (1984). *Coherent analytic sheaves*. Springer-Verlag, Berlin.

Griffiths, P. A. and Adams, J. (1974). *Topics in algebraic and analytic geometry*. Princeton University Press, Princeton, New Jersey.

Grothendieck, A. (1954). *Espaces vectoriels topologiques*. Departamento de Matemática da Universidade de São Paulo.

Grothendieck, A. (1955). *Produits tensoriels topologiques et espaces nucléaires*. Mem. Amer. Math. Soc. Vol. 16. Providence, Rhode Island.

Guichardet, A. (1966). Sur l'homologie et la cohomologie des algèbres de Banach. *C.R. Acad. Sci. Paris*, Ser. A., **262**, 38–41.

Guichardet, A. (1980). *Cohomologie des groupes topologiques et des algèbres de Lie*. Cedic/Fernand Nathan, Paris.

Hakim, M. and Sibony, N. (1980). Spectre de $A(\Omega)$ pour des domaines bornés faiblement pseudoconvexes réguliers. *J. Funct. Analysis*, **37**, 127–135.

Harte, R. (1972). Spectral mapping theorems. *Proc. Royal Irish Acad.*, **72A**, 89–107.

Harte, R. (1988). *Invertibility and singularity for bounded linear operators*. Marcel Dekker, New York.

Harvey, R. and Lawson, B. (1975). On boundaries of complex analytic varieties. I, *Ann. Math.*, **102**, 233–290.

Helemskii, A. Ya. (1964). Description of the annihilator extensions of commutative Banach algebras (in Russian). *Dokl. Acad. Nauk SSSR*, **157**, 60–62.

Helemskii, A. Ya. (1970). On the homological dimension of normed modules over Banach algebras (in Russian). *Matem. Sbornik*, **81**, 430–444.

Helemskii, A. Ya. (1986). *The homology of Banach and topological algebras* (in Russian). Moscow Univ. Press, Moscow; English translation, Kluwer Academic Publishers, Dordrecht, 1989.

Helemskii, A. Ya. (1989). *Banach and locally convex algebras*: *general theory, representations, homology* (in Russian). Nauka, Moscow; English translation, Clarendon Press, Oxford, 1993.

Helton, J. W. and Howe, R. (1975). Trace of commutators of integral operators. *Acta Math.*, **135**, 271–305.

Henkin, G. M. and Leiterer, J. (1983). *Theory of functions on complex manifolds*. Akademie Verlag, Berlin.

Herrero, D. (1982). *Approximation of Hilbert space operators*. Res. Notes Math. Vol. 72. Pitman, London.

Hirzebruch, F. (1965). *Topological methods in algebraic geometry*. Springer-Verlag, Berlin.

Hochschild, G. (1956). Relative homological algebra. *Trans. Amer. Math. Soc.*, **82**, 246–249.

Hoffman, K. (1962). *Banach spaces of analytic functions*. Prentice-Hall, Englewood Cliffs, New Jersey.

Hörmander, L. (1965). L^2-estimates and existence theorems for the $\bar{\partial}$-operator. *Acta Math.*, **113**, 89–152.

Hörmander, L. (1966). *An introduction to complex analysis in several variables*. Van Nostrand, Princeton, New Jersey.

Hörmander, L. (1967). Generators for some rings of analytic functions. *Bull. Amer. Math. Soc.*, **73**, 943–949.

Hörmander, L. (1983). *The analysis of linear partial differential operators*. II— *Differential operators with constant coefficients*. Springer-Verlag, Berlin.

Hörmander, L. (1985). *The analysis of linear partial differential operators*. III— *Pseudodifferential operators*. Springer-Verlag, Berlin.

Johnson, B. E. (1969). Continuity of derivations on commutative Banach algebras. *Amer. J. Math.*, **91**, 1–10.

Johnson, B. E. (1972). *Cohomology in Banach algebras*. Mem. Amer. Math. Soc. Vol. 127, Providence, Rhode Island.

Kadison, R. V. and Ringrose, J. R. (1971). Cohomology of operator algebras, I.— Type I von Neumann algebras. *Acta Math.*, **126**, 227–243.

Kaminker, J. (ed.) (1990). *Geometric and topological invariants of elliptic operators*. Contemp. Math. Vol. 105, Amer. Math. Soc., Providence, Rhode Island.

Kamowitz, H. (1962). Cohomology groups of commutative Banach algebras. *Trans. Amer. Math. Soc.*, **102**, 352–372.

Karoubi, M. (1987). *Homologie cyclique et K-théorie*. Astérisque, Vol. 149, Soc. Math. France, Paris.

Kashiwara, M. and Schapira, P. (1990). *Sheaves on manifolds*. Springer-Verlag, Berlin.

Kasparov, G. G. (1988). Equivariant KK-theory and the Novikov conjecture. *Invent. Math.*, **91**, 147–201.

Kato, T. (1966). *Perturbation theory for linear operators*. Springer-Verlag, Berlin.

Kaup, B. and Kaup, L. (1983). *Holomorphic functions of several variables*: *an introduction to the fundamental theory*. de Gruyter, Berlin–New York.

Kaup, L. (1967). Eine Künnethformel für Fréchetgarben. *Math. Z.*, **97**, 158–168.

Kaup, L. (1968). Das topologische Tensorprodukt kohärenter analytischer Garben. *Math. Z.*, **106**, 273–292.

Khenkin, G. M. (1990). The method of integral representations in complex analysis. In vol. *Several complex variables*. I (ed. A. G. Vitushkin). Springer-Verlag, Berlin.

Kiehl, R. and Verdier, J. L. (1971). Ein einfacher Beweis des Kohärenzsatzes von Grauert. *Math. Ann.*, **195**, 24–50.

Klimek, M. (1991). *Pluripotential theory*. London Mathematical Society Monographs, Oxford University Press, Oxford.

Knapp, A. (1986). *Representation theory of semisimple groups. An overview based on examples*. Princeton University Press, Princeton, New Jersey.

Knorr, K. (1971). Der Grauertsche Kohärenzsatz. *Invent. Math.*, **12**, 118–172.

Kohn, J. J. (1973). Global regularity for $\bar{\partial}$ on weakly pseudoconvex manifolds. *Trans. Amer. Math. Soc.*, **181**, 273–292.

Koranyi, A. (1993). A survey of spaces of holomorphic functions on bounded symmetric domains. *Contemp. Math.*, to appear.

Köthe, G. (1969). *Topological vector spaces* I. Springer-Verlag, Berlin.

Köthe, G. (1979). *Topological vector spaces* II. Springer-Verlag, New York.

Kramm, B. (1980). *Komplexe Funktionen-Algebren*. Bayreuther Math. Schriften Heft 5, Universität Bayreuth.

Kramm, B. (1984). Nuclearity (resp. Schwarzity) helps to embed holomorphic structure into spectra. *Contemp. Math.*, **32**, 143–162.

Lang, S. (1984). *Algebra*, Second Edition. Addison-Wesley, Reading, Massachusetts.

Lange, R. (1980). A purely analytic criterion for a decomposable operator. *Glasgow Math. J.*, **21**, 69–70.

Lange, R. and Wang, S. (1992). *New approaches in spectral decompositions*. Contemp. Math. Vol. 128, Amer. Math. Soc., Providence, Rhode Island.

Laursen, K. B. and Neumann, M. M. (1986). Decomposable operators and automatic continuity. *J. Operator Theory*, **15**, 33–51.

Laursen, K. B. and Neumann, M. M. (1992). Decomposable multipliers and applications to harmonic analysis. *Studia Math.*, **101**, 193–214.

Leiterer, J. (1978). Banach coherent analytic Fréchet sheaves. *Math. Nachr.*, **85**, 91–109.

Levy, R. (1983). Cohomological invariants for essentially commuting systems of operators (in Russian). *Funct. Anal. Appl.*, **17**, 79–80.

Levy, R. (1987a). The Riemann–Roch theorem for complex spaces. *Acta Math.*, **158**, 149–188.

Levy, R. (1987b). A new proof of the Grauert direct image theorem. *Proc. Amer. Math. Soc.*, **99**, 535–542.

Levy, R. (1989). Algebraic and topological K-functors of commuting n-tuple of operators. *J. Operator Theory*, **21**, 219–254.

Li, Y-S. (1992). Corona problem of several complex variables. *Contemp. Math.*, **137**, 307–328.

Łojasiewicz, S. (1964). Triangulation of semi-analytic sets. *Ann. Scuola Norm. Sup. Pisa (3)*, **18**, 449–474.

Mac Lane, S. (1963). *Homology*. Springer-Verlag, Berlin.

Malgrange, B. (1955-56). Existence et approximation des solutions des equations aux derivées partielles et des equations de convolution. *Ann. Inst. Fourier*, **6**, 271–355.

Malgrange, B. (1958). *Lectures on the theory of functions of several complex variables*. Tata Inst. Fund. Research, Bombay.

Malgrange, B. (1966). *Ideals of differentiable functions*. Oxford University Press, Oxford.

Mallios, A. (1986). *Topological algebras: selected topics*. North Holland, Amsterdam, New York.

Mantlik, F. (1988). Parameterabhängige lineare Gleichungen in Banach und Frécheträumen. Dissertation. Universität Dortmund.

Mantlik, F. (1990). Linear equations depending differentiably on a parameter. *Integral Eq. Operator Theory*, **13**, 231–250.

Markoe, A. (1972). Analytic families of differential complexes. *J. Funct. Analysis*, **9**, 181–188.

Mather, J. N. (1968). Stability of C^∞-mappings. I: The division theorem. *Ann. Math.*, **87**, 89–104.

Mityagin, V. S. and Henkin, G. M. (1971). Linear problems of complex analysis (in Russian). *Uspehi Mat. Nauk*, **26**, 94–152. English transl.: *Russian Math. Surv.*, **26**, 99–164.

Nagel, A. (1974). On algebras of holomorphic functions with boundary values. *Duke Math. J.*, **41**, 527–535.

Narasimhan, R. (1984). *Complex analysis in one variable*. Birkhäuser Verlag, Boston, Massachusetts.

Nikolskii, N. K. (1986). *Lectures on the shift operator*. Springer-Verlag, Berlin.

Nirenberg, L. (1971). *A proof of Malgrange preparation theorem*. Lect. Notes Math. Vol. 209, Springer-Verlag, Berlin, pp. 97–105.

Nöbeling, G. (1961). Über die Derivierten des Inversen und des Direkten Limes einer Modulfamilie. *Topology*, **1**, 47–61.

O'Brian, N. R., Toledo, D., and Tong, Y. L. L. (1981). Hirzebruch–Riemann–Roch for coherent sheaves. *Amer. J. Math.*, **103**, 253–271.

Ohsawa, T. (1988). On the extension of L^2-holomorphic functions. II. *Publ. RIMS*, Kyoto University, **24**, 265–275.

Ohsawa, T. (1989). Complete Kähler manifolds and function theory of several complex variables. Preprint.

Oka, K. (1984). *Mathematical papers* (R. Remmert, ed.). Springer-Verlag, Berlin.

Olin, R. and Thomson, J. (1980). Algebras of subnormal operators. *J. Funct. Analysis*, **37**, 271–301.

Onishchik, A. L. (1990). *Methods in the theory of sheaves and Stein spaces*. Encl. Math. Sci. Vol. 10 (S. G. Gindikin, G. M. Khenkin, eds.), Several complex variables IV, Springer-Verlag, Berlin.

Øvrelid, N. (1971). Generators of the maximal ideals of $A(D)$. *Pacific J. Math.*, **39**, 219–223.

Palamodov, V. P. (1961). The structure of polynomial ideals and their quotients in spaces of infinitely differentiable functions (in Russian). *Dokl. Akad. Nauk SSSR*, **141**, 1302–1305.

Palamodov, V. P. (1968). Differential operators on analytic coherent sheaves (in Russian). *Mat. Sbornik*, **77**, 390–422.

Palamodov, V. P. (1971). Homological methods in the theory of locally convex spaces (in Russian). *Uspehi Mat. Nauk*, **26**, 3–65. English translation: *Russian Math. Surveys* **26**, 1–64 (1971).

Palamodov, V. P. (1990). *Deformations of complex spaces*. Encl. Math. Sci. Vol. 10. (S. G. Gindikin, G. M. Khenkin, eds.), Several complex variables. IV, Springer-Verlag, Berlin.

Pourcin, G. (1975). Sous-espaces privilegiés d'un polycilindre. *Ann. Inst. Fourier*, **25**, 151–193.

Putinar, M. (1979). Three papers on several variable spectral theory. Preprint Series in Math., INCREST.

Putinar, M. (1980). Functional calculus with sections of an analytic space. *J. Operator Theory*, **4**, 297–306.

Putinar, M. (1982a). The superposition property for Taylor's functional calculus. *J. Operator Theory*, **7**, 149–155.

Putinar, M. (1982b). Some invariants for semi-Fredholm systems of essentially commuting operators. *J. Operator Theory*, **8**, 65–90.

Putinar, M. (1983a). Spectral theory and sheaf theory. I. In vol.: *Dilation theory, Toeplitz operators and other topics*. Birkhäuser, Basel, Boston, Stuttgart, pp. 283–298.

Putinar, M. (1983b). Uniqueness of Taylor's functional calculus. *Proc. Amer. Math. Soc.*, **89**, 647–650.

Putinar, M. (1984a). Elements of spectral theory of Stein algebra representations (in Romanian). *Studii Cerc. Math.*; part I, **36** (3), 193–220; part II, **36**, 293–310; part III, **36**, 387–408.

Putinar, M. (1984b). Spectral inclusion for subnormal n-tuples. *Proc. Amer. Math. Soc.*, **90**, 405–406.

Putinar, M. (1984c). Hyponormal operators are subscalar. *J. Operator Theory*, **12**, 385–395.

Putinar, M. (1984d). Functional calculus and the Gelfand transformation. *Studia Math.*, **79**, 83–86.

Putinar, M. (1985). Base change and the Fredholm index. *Integral Eq. Operator Theory*, **8**, 674–692.

Putinar, M. (1986). Spectral theory and sheaf theory. II. *Math. Z.*, **192**, 473–490.

Putinar, M. (1990a). Spectral theory and sheaf theory. IV. *Proc. Symp. Pure. Math.*, **51**, 273–293.

Putinar, M. (1990b). Invariant subspaces of several variable Bergman spaces. *Pacific J. Math.*, **147**, 355–364.

Putinar, M. (1991). Une démonstration du théorème de platitude de Malgrange dans le cas analytique complexe. *Rev. Roum. Math. Pures Appl.*, **36**, 261–270.

Putinar, M. (1992). Quasi-similarity of tuples with Bishop's property (β). *Integral Eq. Operator Theory*, **15**, 1047–1052.

Putinar, M. (1994). On the rigidity of Bergman submodules. *Amer. J. Math.*, **116**, 1421–1432.

Putinar, M. and Salinas, N. (1992). Analytic transversality and Nullstellensatz in Bergman spaces. *Contemp. Math.*, **137**, 367–381.

Radjabalipour, M. (1978). Decomposable operators. *Bull. Iran. Math. Soc.*, **9**, 1–49.

Radulescu, F. (1988). On smooth extensions of odd dimensional spheres and multi-dimensional Helton and Howe formula. *J. reine angew. Math.*, **386**, 145–171.

Ramirez de Arellano, E. (1970). Ein Divisionsproblem und Randintegraldarstellungen in der komplexen Analysis. *Math. Ann.*, **184**, 172–187.

Ramis, J.-P. and Ruget, G. (1974). Résidus et dualité. *Invent. Math.*, **26**, 89–131.

Range, R. M. (1986). *Holomorphic functions and integral representations in several complex variables*. Springer-Verlag, Berlin.

Remmert, R. (1956). Projektionen analytischer Mengen. *Math. Ann.*, **130**, 410–441.

Remmert, R. (1960). Analytic and algebraic dependence of meromorphic functions. *Amer. J. Math.*, **82**, 891–899.

Robertson, A. P. and Robertson, W. (1964). *Topological vector spaces*. Cambridge University Press, Cambridge.

Rossi, H. (1963). On envelopes of holomorphy. *Comm. Pure Appl. Math.*, **16**, 9–17.

Royden, H. L. (1963). Function algebras. *Bull. Amer. Math. Soc.*, **69**, 281–298.

Rubel, L. A. and Shields, A. L. (1966). The space of bounded analytic functions on a region. *Ann. Inst. Fourier* (Grenoble), **16**, 235–277.

Rudin, W. (1969). *Function theory in polydiscs*. Benjamin, New York.

Rudin, W. (1980). *Function theory in the unit ball of $\mathbb{C}^n$*. Springer-Verlag, New York.

Salinas, N. (1989). The $\bar{\partial}$-formalism and the C^*-algebra of the Bergman n-tuple. *J. Operator Theory*, **22**, 325–343.

Salinas, N., Sheu, A., and Upmeier, H. (1989). Toeplitz operators on pseudoconvex domains and foliation C^*-algebras. *Ann. Math.*, **130**, 531–565.

Sarason, D. (1966). Invariant subspaces and unstarred algebras. *Pacific J. Math.*, **17**, 511–517.

Schaefer, H. (1966). *Topological vector spaces*. Macmillan, New York.

Schwartz, L. (1955). Division par une fonction holomorphe sur une variété analytique complexe. *Summa Brasiliensis Math.*, **3**, 181–209.

Segal, G. (1970). Fredholm complexes. *Quart. J. Math. Oxford Ser.*, **21**, 385–402.

Séminaire Banach (1972). Lecture Notes Math., Vol. 277. Springer-Verlag, Berlin.

Séminaire de géométrie analytique (1974). (eds. A. Douady and J. L. Verdier). Astérisque Vol. 16. Soc. Math. France, Paris.

Serre, J.-P. (1955). Faisceaux algebriques coherents. *Ann. Math.*, **61**, 197–278.

Serre, J.-P. (1965). *Algèbre locale. Multiplicités*. Lect. Notes Math., Vol. 11. Springer-Verlag, Berlin.

Shabat, B. V. (1985). *Introduction to complex analysis*, Vol. II (in Russian). Nauka, Moscow.

Shilov, G. E. (1953). On the decomposition of a commutative normed ring into a direct sum of ideals. *Math. USSR Sb.*, **32**, 353–364; *Amer. Math. Soc. Transl.*, **1** (1955), 37–48.

Sibony, N. (1987). Problème de la couronne pour les domaines pseudoconvexes à bord lisse. *Ann. Math.*, **126**, 675–682.

Singer, I. (1981). *Bases in Banach spaces*, II. Springer-Verlag, Berlin.

Skoda, H. (1972). Application des techniques L^2 à la théorie des idéaux d'une algèbre de fonctions holomorphes avec poids. *Ann. Scient. Ec. Norm. Sup.*, **5**, 545–579.

Słodkowski, Z. (1977). An infinite family of joint spectra. *Studia Math.*, **61**, 239–255.

Słodkowski, Z. and Żelazko, W. (1974). On joint spectra of commuting families of operators. *Studia Math.*, **50**, 127–148.

Taylor, J. L. (1970a). A joint spectrum for several commuting operators. *J. Funct. Analysis*, **6**, 172–191.

Taylor, J. L. (1970b). The analytic functional calculus for several commuting operators. *Acta Math.*, **125**, 1–48.

Taylor, J. L. (1972a). Homology and cohomology for topological algebras. *Adv. Math.*, **9**, 147–182.

Taylor, J. L. (1972b). A general framework for a multi-operator functional calculus. *Adv. Math.*, **9**, 184–252.

Taylor, J. L. (1976). Topological invariants of the maximal ideal space of a Banach algebra. *Adv. Math.*, **19**, 149–206.

Tillmann, H. G. (1963). Eine Erweiterung des Funktionalkalküls für lineare Operatoren. *Math. Ann.*, **151**, 424–430.

Tougeron, J. C. (1972). *Idéaux de fonctions différentiables*. Springer-Verlag, Berlin.

Treves, F. (1967). *Topological vector spaces, distributions and kernels*. Academic Press, New York.

Upmeier, H. (1987). Index theory for Toeplitz operators on bounded symmetric domains. *Bull. Amer. Math. Soc.*, **16**, 109–112.

Vasilescu, F.-H. (1977). A characterization of the joint spectrum in Hilbert spaces. *Rev. Roum. Math. Pures Appl.*, **22**, 1003–1009.

Vasilescu, F.-H. (1978a). A Martinelli type formula for the analytic functional calculus. *Rev. Roum. Math. Pures Appl.*, **23**, 1587–1605.

Vasilescu, F.-H. (1978b). On pairs of commuting operators. *Studia Math.*, **62**, 203–207.

Vasilescu, F.-H. (1979). Stability of the index of a complex of Banach spaces. *J. Operator Theory*, **2**, 247–275.

Vasilescu, F.-H. (1980). The stability of the Euler characteristic for Hilbert complexes. *Math. Ann.*, **248**, 109–116.

Vasilescu, F.-H. (1982). *Analytic functional calculus and spectral decompositions*. Reidel, Dordrecht.

Vasilescu, F.-H. (1992). An operator valued Poisson kernel. *J. Funct. Analysis*, **110**, 47–72.

Vogt, D. (1987). On the functors $\text{Ext}^1(E, F)$ for Fréchet spaces. *Studia Math.*, **85**, 163–197.

Vogt, D. (1989). Topics on projective spectra of (LB)-spaces. In vol. *Advances in the theory of Fréchet spaces* (ed. T. Terziouglu), Kluwer, Dordrecht.

Waelbroeck, L. (1954). Le calcul symbolique dans les algèbres commutatives. *J. Math. Pures Appl.*, **33**, 147–186.

Wang, S. and Liu, G. (1984). On the duality theorem of bounded S-decomposable operators. *J. Math. Anal. Appl.*, **99**, 150–163.

Wells, R. O. Jr. (1973). *Differential analysis on complex manifolds*. Prentice-Hall, Englewood Cliffs.

Whitney, H. (1972). *Complex analytic varieties*. Addison-Wesley, Reading, Massachusetts.

Wrobel, V. (1988). Tensor products of linear operators in Banach spaces and Taylor's joint spectrum. I, *J. Operator Theory*, **16**, 273–283; II, *J. Operator Theory*, **19**, 3–24.

Xia, D. (1983). *Spectral theory of hyponormal operators*. Birkhäuser, Basel.

Zenger, Ch. (1968). On convexity properties of the Bauer field of values of a matrix. *Numerische Math.*, **12**, 96–105.

Zhu, K. (1990). *Operator theory in function spaces*. Marcel Dekker, New York.

Index